Open Channel Hydraulics

Terry W. Sturm
Georgia Institute of Technology

Boston   Burr Ridge, IL   Dubuque, IA   New York   San Francisco   St. Louis
Bangkok   Bogotá   Caracas   Kuala Lumpur   Lisbon   London   Madrid   Mexico City
Milan   Montreal   New Delhi   Santiago   Seoul   Singapore   Sydney   Taipei   Toronto

OPEN CHANNEL HYDRAULICS, SECOND EDITION

Published by McGraw-Hill, a business unit of The McGraw-Hill Companies, Inc., 1221 Avenue of the Americas, New York, NY 10020. Copyright © 2010 by The McGraw-Hill Companies, Inc. All rights reserved. Previous edition © 2001. No part of this publication may be reproduced or distributed in any form or by any means, or stored in a database or retrieval system, without the prior written consent of The McGraw-Hill Companies, Inc., including, but not limited to, in any network or other electronic storage or transmission, or broadcast for distance learning.

Some ancillaries, including electronic and print components, may not be available to customers outside the United States.

This book is printed on acid-free paper.

1 2 3 4 5 6 7 8 9 0 DOC/DOC 0 9

ISBN 978-0-07-339787-0
MHID 0-07-339787-3

Global Publisher: *Raghothaman Srinivasan*
Sponsoring Editor: *Debra B. Hash*
Director of Development: *Kristine Tibbetts*
Developmental Editor: *Lorraine K. Buczek*
Senior Marketing Manager: *Curt Reynolds*
Senior Project Manager: *Kay J. Brimeyer*
Senior Production Supervisor: *Laura Fuller*
Associate Design Coordinator: *Brenda A. Rolwes*
Cover Designer: *Studio Montage, St. Louis, Missouri*
(USE) Cover Image: *Rio Pastaza, Ecuador:* © *Geoff Sturm (http://geoffsturm.com)*
Lead Photo Research Coordinator: *Carrie K. Burger*
Compositor: *Laserwords Private Limited*
Typeface: *10/12 Times Roman*
Printer: *R. R. Donnelley Crawfordsville, IN*

All credits appearing on page or at the end of the book are considered to be an extension of the copyright page.

Quote on page v: Maclean, Norman. *A River Runs Through It.* © 1976, University of Chicago Press. Reprinted with permission of University of Chicago Press.

Library of Congress Cataloging-in-Publication Data

Sturm, Terry W.
  Open channel hydraulics / Terry W. Sturm.—2nd ed.
    p. cm.
  Includes index.
  ISBN 978-0-07-339787-0—ISBN 0-07-339787-3 (hard copy : alk. paper) 1. Channels (Hydraulic engineering) 2. Hydraulics. I. Title.
  TC175.S774 2010
  627'.23—dc22

                                          2008054594

www.mhhe.com

To my family, my inspiration at the river source,

and to the memory of
my brother Tim (1949–1998) and my father Everett (1923–1999),
traveling companions along the forever winding river of life.

Eventually, all things merge into one, and a river runs through it. The river was cut by the world's great flood and runs over rocks from the basement of time. On some of the rocks are timeless raindrops. Under the rocks are the words, and some of the words are theirs.

I am haunted by waters.

—Norman Maclean, *A River Runs Through It*

BRIEF TABLE OF CONTENTS

CONTENTS

PREFACE

WHAT'S NEW IN THE SECOND EDITION

In this second edition, material has been added on long-throated flumes as flow measuring devices, flood control channel design including the Corps of Engineers riprap revetment design procedure, computation of canal delivery curves, reservoir outflow control using gated spillways, HEC-RAS implementation of WSPRO and other bridge backwater computation methods, HEC-RAS unsteady flow module for flood routing, and an introduction to pier and abutment scour countermeasures. Existing material on turbulent flow resistance, grass lining of channels for stability, and hydraulic jump in rough channels has been expanded and updated.

Another important change in the second edition is the addition of Chapter 11 on Computational Fluid Dynamics (CFD) contributed by my colleague at Georgia Tech, Dr. Thorsten Stoesser. Three-dimensional CFD techniques are becoming more and more important in hydraulic engineering and more accessible to the practicing engineer as well. In this chapter, Dr. Stoesser introduces CFD as applied to open channel flow and demonstrates its tremendous potential to add to our knowledge of three-dimensional turbulent flow in open channels and improve our design of stream restoration measures, floodplain management schemes and hydraulic structures. As hydraulic engineers, we now have CFD as a third tool, in addition to experimental and theoretical techniques, to attack the challenging problems of hydraulic engineering in the 21st century. It is hoped that this chapter will encourage practicing engineers to further explore the possibilities of CFD.

Finally, I have increased both the number of illustrative examples and end-of-chapter exercises by 25 percent in this revision of the textbook.

PURPOSE OF TEXT

The study of open channel hydraulics is a challenging and exciting endeavor because of the influence of gravity on free surface flows. The position of the free surface is not known *a priori,* and counterintuitive phenomena can occur from the viewpoint of the first-time student of open channel flow. This book offers a study of gravity flows starting from a firm foundation in modern fluid mechanics that includes both experimental results and numerical computation techniques. The development of the subject matter proceeds from basic fundamentals to selected applications with numerous worked-out examples. Experimental results and their comparison with theory are used throughout the book to develop an understanding of free-surface flow phenomena. Computational tools range from spreadsheets to computer programs to solve more difficult problems. Some computer programs are provided in Visual BASIC, both as learning tools and as examples to encourage the use of computational methods regardless of the platform available in a very dynamic environment. In addition, several well-known computer packages available in the public domain are demonstrated and discussed to inform users with respect to the methodologies employed and their limitations.

This book has grown out of instructional and research materials developed over several years and used in a graduate course sequence in open channel flow and sediment transport, as well as in a continuing education course that I have taught at the Georgia Institute of Technology. Because of its unique focus on fundamentals as well as applications, and experimental results as well as numerical analysis, this book should fill a niche between exhaustive handbooks and purely academic treatises on the subject of open channel hydraulics.

ORGANIZATION OF CHAPTERS

The basic equations of continuity, energy, and momentum are derived for open channel flow in the first chapter, from the viewpoint of both a finite control volume and an infinitesimal control volume, although the complete derivation of the general unsteady form of the differential momentum equation is saved for Chapter 7. Dimensional analysis is introduced in some detail in the first chapter because of its use throughout the book. This is followed by Chapters 2 and 3 on the specific energy concept and the momentum function, respectively, and their applications to open channel flow problems. Design of open channels for uniform flow is examined in Chapter 4 with a detailed consideration of the estimation of flow resistance. Applications include the design of channels with vegetative and rock riprap linings, and the design of storm and sanitary sewers. Chapter 5, on gradually varied flow, emphasizes modern numerical solution techniques. The methodology for water-surface profile computation used in current computer programs promulgated by federal agencies is discussed, and example problems are given. The design of hydraulic structures, including spillways, culverts, and bridges, is the subject of Chapter 6. Accepted computer programs used in such design are introduced and their methodologies thoroughly explored. Chapters 7, 8, and 9 develop current techniques for the solution of the one-dimensional Saint-Venant equations of unsteady flow and their simplifications. In Chapter 7, the Saint-Venant equations are derived, and the method of characteristics is introduced for the simple wave problem as a means of understanding the mathematical transformation of the governing equations into characteristic form. The numerical techniques of explicit and implicit finite differences and the numerical method of characteristics are given in Chapter 8, with applications to hydroelectric transients in headraces and tailraces, the dam-break problem, and flood routing in rivers. Chapter 9 covers simplified methods of flow routing including the kinematic wave method, diffusion method, and the Muskingum-Cunge method. The complex subject of alluvial channel flows that have a movable bed as well as an adjustable free surface is explored in Chapter 10. This chapter emphasizes the important links among sediment discharge, bed forms, and flow resistance that are essential to an understanding of open channel flow in rivers. Also covered in Chapter 10 are alluvial channel adjustments in slope, form, and shape; and bed scour in response to the flow blockage caused by bridge foundations. Finally, the book concludes with a chapter introducing the methods of CFD applied to open channel flow that includes a case study.

The book includes an appendix to supplement the text material. It is a general discussion of some selected numerical techniques that can be used throughout the book. The book website contains some example computer programs for the computation of normal and critical depth in prismatic channels, including compound channels, and computation of water surface profiles. These programs are written in Visual BASIC as learning aids for more extensive programming exercises at the end of several chapters. Additional programs for solution of the more advanced exercises on unsteady flow computations can be found on this website.

TEACHING WITH THIS TEXTBOOK

Open Channel Hydraulics is intended for advanced undergraduates and first-year graduate students in the general fields of water resources and environmental engineering. Chapters 1 through 5 and Chapters 7 through 9 provide sufficient material for a semester course in open channel hydraulics covering both steady and unsteady flow. The book also can be used for a first-year graduate course or a senior elective course on hydraulic structures and river hydraulics, utilizing Chapters 4, 5, 6, 9, 10, and 11. This material, which includes several applications and example problems, should be useful to the practitioner charged with the responsibility for such tasks as floodplain management, spillway design for small reservoirs, culvert and sewer design for drainage, investigation of stability and flow resistance of alluvial streams, and estimation of bridge backwater and scour. Because of this applied focus of the book, it should be a useful addition to a consulting engineer's library as well as a practical textbook on the fundamentals of open channel flow.

Each chapter contains worked-out example problems to aid in the understanding of the text material. Where possible, solutions are given in dimensionless form in graphs to provide an intuitive understanding of the physics of the problem and the behavior of its solution over a wide range of variables. At the end of each chapter, exercises are presented that involve application of the material in the chapter as well as student exploration of further ramifications of the text material. In some chapters, actual laboratory results are given for data reduction and presentation by students to experimentally verify text material.

ONLINE RESOURCES

Visit the text website at www.mhhe.com/sturm <http://www.mhhe.com/sturm> for various resources available to instructors and students.

ACKNOWLEDGMENTS

I am indebted to more people than I can enumerate here for the completion of this project. My initial motivation for preparing for an academic career in hydraulics dates back to a keynote address that I heard delivered by Hunter Rouse, who was an accomplished orator as well as writer, at a conference held at the University of

Iowa. The subject was the careers of famous hydraulicians including their foibles as well as achievements. I later graduated from the University of Iowa under the late Jack Kennedy, who was a continuing inspiration to a struggling Ph.D. student. I am much indebted to the continuing encouragement given by the late Ben C. Yen at the University of Illinois, where I received my B.S. and M.S. degrees in Civil Engineering, and Edward R. Holley at the University of Texas at Austin over the course of my career. C. Samuel Martin served as mentor and colleague for many years at Georgia Tech. The encouragement and research collaboration of my late colleague Amit Amirtharajah has been invaluable. I owe much to the previous treatises on open channel hydraulics by Ven Te Chow and F. M. Henderson, as do many other authors, as well as practitioners. Review comments on the 1st edition by Johnny Morris, Larry Mays, and the late Ben C. Yen, and suggestions by Edward R. Holley led to an improved manuscript, although I bear the responsibility for any errors or shortcomings that remain. I express my gratitude to Mark Landers of the USGS for locating and providing copies of the river slides by Barnes.

I want to thank the following reviewers who provided many excellent suggestions for this revision, although I was unable to implement all of them due to space limitations.

Brian Barkdoll, *Michigan Technological University*
Yee-Meng Chiew, *Nanyang Technological University*
Keith E. Dennett, *University of Nevada*
Rollin H. Hotchkiss, *Brigham Young University*
G. Padmanabhan, *North Dakota State University*
Tim J. Ward, *University of New Mexico*

My students have been a continuing source of motivation for me to try to explain complex aspects of open channel hydraulics with clarity. I have learned much from their curiosity and probing questions about the details of open channel flow phenomena.

Finally, I am forever indebted to my wife, Candy, whose patience, love, and support brought me through this project and its revision, and to my grown children, Geoffrey, Sarah, and Christy, who continually inspire and renew me.

Basic Principles

1.1 INTRODUCTION

Open channel hydraulics is the study of the physics of fluid flow in conveyances in which the flowing fluid forms a free surface and is driven by gravity. The primary case of interest in this book is water as the flowing fluid having an interface or free surface formed with the ambient atmosphere, but the basic principles also apply to other cases, such as density-stratified flows. Natural open channels include brooks, streams, rivers, and estuaries. Artificial open channels are exemplified by storm sewers, sanitary sewers, and culverts flowing partly full, as well as drainage ditches, irrigation canals, aqueducts, and flood diversion channels. Applications of open channel hydraulics range from the design of artificial channels for beneficial purposes such as irrigation, drainage, water supply, and wastewater conveyance to the analysis of flooding in natural waterways to delineate floodplains and assess flood damages for a flood of specified frequency. Principles of open channel hydraulics also are utilized to describe the transport and fate of environmental contaminants, including those carried by sediments in motion, as well as to predict flood surges caused by dam breaks or hurricanes.

1.2 CHARACTERISTICS OF OPEN CHANNEL FLOW

Although the basic principles of fluid mechanics are applicable to open channel flow, such flow is considerably more complex than closed conduit flow due to the free surface. The relevant forces causing and resisting motion and the inertia must form a balance such that the free surface is a streamline along which the pressure is constant and equal to atmospheric pressure. This extra degree of freedom in open

1

channel flow means that the flow boundaries no longer are fixed by the conduit geometry, as in closed conduit flow, but rather the free surface adjusts itself to accommodate the given flow conditions.

Another important characteristic of open channel flow is the extreme variability encountered in cross-sectional shape and roughness. Conditions range from a circular gravity sewer flowing partly full to a natural river channel with a floodplain subject to overbank flow. Roughness heights in the gravity sewer correspond to those encountered in closed conduit flow, while roughness elements such as brush, vegetation, and deadfalls in natural open channels make the roughness extremely difficult to quantify. Even in the case of the circular gravity sewer, resistance to flow is complicated by the change in cross-sectional shape as the depth changes. In alluvial channels, the boundary itself is movable, giving rise to bed forms that provide a further contribution to flow resistance.

Because of the free surface, gravity is the driving force in open channel flow. The ratio of inertial to gravity forces in open channel flow is the most important governing dimensionless parameter. It is called the *Froude number*, defined by

$$\mathbf{F} = \frac{V}{(gD)^{1/2}} \tag{1.1}$$

in which V is the mean velocity, D is a length scale related to depth, and g is gravitational acceleration. In some instances the Reynolds number also is important, as in closed conduit flow, but one of the few simplifications in natural open channels is the existence of a large Reynolds number so that viscous effects assume less importance. Flow resistance in this case can be dominated by form resistance, which is associated with asymmetric pressure distributions resulting from flow separation. The success of Manning's equation in characterizing open channel flow resistance in fact depends on the existence of a Reynolds number large enough that the Manning's resistance factor is invariant with Reynolds number.

1.3 SOLUTION OF OPEN CHANNEL FLOW PROBLEMS

The complexities offered by open channel flow often can be dealt with through a combination of theory and experiment, as in other branches of fluid mechanics. The basic principles of continuity, energy conservation, and force-momentum flux balance must be satisfied, but we often must resort to experiments to complete the solution of the problem. The resulting relationships can be quite complicated, especially when the variability of the cross-sectional geometry is considered.

In the not-too-distant past, the design of open channels was achieved with the aid of numerous nomographs and graphical relationships because of the nonlinearity of the governing equations combined with complex geometry. More extensive analysis of unsteady flow problems or gradually varied flow problems associated with river floodplains required mainframe computers. Presently, the proliferation of personal computers and engineering workstations has provided much greater accessibility and flexibility for simple as well as complex problems in open channel hydraulics. Programs that are truly interactive with immediate feedback of results in the form of screen graphics can be written with ease. The hydraulic engineer can investigate a wide array of design solutions and their implications in a completely interactive mode in the modern engineering workstation. On the other hand, such ease of use sometimes leads to misinformed applications of accepted computer programs in open channel hydraulics.

1.4   PURPOSE

The theme of this book is to present modern numerical techniques for the solution of open channel flow problems as well as to emphasize experimental results and their application in free surface flows. The occurrence of a mobile bed surface caused by sediment transport in alluvial channels is treated as well because it is intertwined with the flow resistance problem. In addition, focus is placed on the application of basic principles of fluid mechanics to the formulation of open channel flow problems, so that the assumptions and limitations of the numerical models now widely available are made clear. The combination of theoretical, experimental, and numerical techniques applied to open channel flow provides a synthesis that has become the hallmark of modern fluid mechanics.

1.5   HISTORICAL BACKGROUND

The following discussion relies on the excellent historical treatment of hydraulics by Rouse and Ince (1957), to which the reader is referred for further details.

From the advent of civilization, the conveyance of water in open channels has been used to meet basic needs, such as irrigation for the Egyptians and Mesopotamians, water supply for the Romans, and waste disposal for Europeans in the Middle Ages, with the disastrous results of waterborne disease transmission. In some cases, artificial open channels were constructed, while in others natural river channels were utilized to convey water and wastes.

The Egyptians used a dam for water diversion and gravity flow through canals to distribute water from the Nile River, and the Mesopotamians developed canals to transfer water from the Euphrates to the Tigris rivers, but there is no recorded evidence of any understanding of the theoretical flow principles involved. The Chinese are known to have devised a system of dikes for protection from flooding several thousand years ago. Evidence of water supply pipes and brick conduits for drainage dated to 3000 years B.C. has been found in the Indus River valley. The success of these early, extensive hydraulic works was likely the result of experience only.

Roman aqueducts were used to transport water from springs to distribution reservoirs. The aqueducts were rectangular, masonry canals supported by masonry arches, and they conformed to the natural topography in longitudinal slope. The water discharge in the aqueducts was measured as the cross-sectional area of flow with no regard for the velocity or slope producing the velocity. Although the existence of a conservation principle was recognized, the conserved quantity of volume flux was misunderstood. Yet, these aqueducts served their engineering purpose, albeit inefficiently and uneconomically in modern terms.

The philosophical approach of the Greeks toward physical phenomena was revived by the Scholasticism of the Middle Ages, and it remained for Leonardo da Vinci to introduce the experimental method in open channel flow during the Renaissance. Leonardo's prolific writings included observations of the velocity distribution in rivers and a correct understanding of the continuity principle in streams with narrowing width. Some early experimental results on pipe and channel flow were reported by Du Buat in 1816, but the experimental work on canals begun by Darcy and completed by Bazin in the late 19th century, and Bazin's experiments on weirs, were unsurpassed at the time and remain an enduring legacy to the experimental approach.

The problem of open channel flow resistance was recognized as important by many engineers in the 18th and 19th centuries. The work of Chezy on flow resistance began in 1768, originating from an engineering problem of sizing a canal to deliver water from the Yvette River to Paris. The resistance coefficient attributed to him, however, was introduced much later because his work dealt only with ratios of the independent variables of slope and hydraulic radius to the 1/2 power in a relationship for velocity ratios in different streams. His work was not published until the 19th century. The Manning equation for open channel flow resistance, about which much will be said in this book, has a complex historical development but was based on field observations. The Irish engineer Robert Manning actually discarded the formula because of its nonhomogeneity in favor of a more complex one in 1889, and Gauckler in 1868 preceded Manning in introducing a formula of the type that now bears the name of Manning.

The theoretical approach to open channel flow rests on the firm foundation built by Newton, Leibniz, Bernoulli, and Euler, as in other branches of fluid mechanics; but one of its early fruits was the analytical solution of the equation of gradually varied flow by Bresse in 1860 and the correct formulation of the momentum equation for the hydraulic jump, which he attributed to the 1838 lecture notes of Belanger. In addition, Julius Weisbach extended the sharp-crested weir equation in 1841 to a form similar to that used today. By the end of the 19th century, many of the elements of the modern approach to open channel flow, which includes both theory and experiment, had been established.

The work of Bakhmeteff, a Russian emigre to the United States, had perhaps the most important influence on the development of open channel hydraulics in the early 20th century. Of course, the foundations of modern fluid mechanics (boundary layer theory, turbulent velocity and resistance laws) were being laid by Prandtl and his students, including Blasius and von Kármán, but Bakhmeteff's contributions dealt specifically with open channel flow. In 1932, his book on the subject was published, based on his earlier 1912 notes developed in Russia (Bakhmeteff 1932). His book concentrated on "varied flow" and introduced the notion of specific energy, still an important tool for the analysis of open channel flow problems. In Germany at this time, the contributions of Rehbock to weir flow also were proceeding, providing the basis for many further weir experiments and weir formulas.

By the mid-20th century, many of the gains in knowledge in open channel flow had been consolidated and extended in the books by Rouse (1950), Chow (1959), and Henderson (1966), in which extensive references can be found. These books set the stage for applications of modern numerical analysis techniques and experimental instrumentation to problems of open channel flow.

1.6 DEFINITIONS

In a *steady* open channel flow, the depth and velocity at a point do not change as a function of time. In the more general case of *unsteady* flow, both velocity and depth vary with time, as in the case of the passage of a flood wave in a river as shown in Figure 1.1a relative to a fixed observer on the riverbank. The change in velocity and

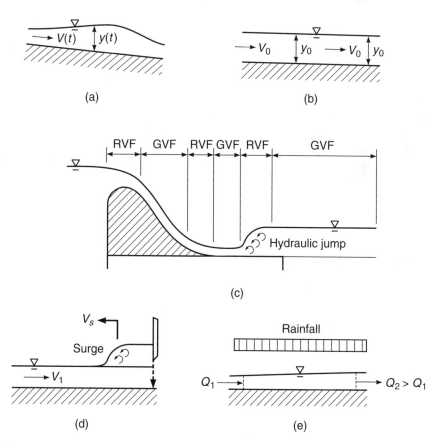

FIGURE 1.1
Types of open channel flow: **(a)** unsteady; **(b)** steady, uniform; **(c)** steady, gradually varied (GVF) and steady, rapidly varied (RVF); **(d)** unsteady, rapidly varied; **(e)** spatially varied.

depth in a large river may occur so gradually and over such long distances that the observer can see only a gradual rise and fall of river stage. If the flood wave results from a dam break, on the other hand, an abrupt change in depth and velocity and a distinct wave front or surge may be observed. In the former case, only near the peak of the flood wave could the flow be considered approximately steady, or quasi-steady, allowing steady flow analyses.

Spatial variations in velocity and depth in the flow direction are distinguished by the terms *uniform* and *nonuniform*. In a uniform flow, the mean cross-sectional velocity and depth are constant in the flow direction, as shown in Figure 1.1b. This flow condition is difficult to create in the laboratory and rarely occurs in the field, but often is used as the basis for open channel design. It requires the existence of a channel of uniform geometry and slope in the flow direction; that is, a prismatic channel. The nonuniform flow condition can be divided into two types: *gradually varied* and *rapidly varied*. Gradually varied flow is nonuniform flow, but the curvature of the free surface and of the accompanying streamlines is so slight that the transverse pressure distribution at any station along the flow can be approximated as hydrostatic. This assumption allows the flow to be treated with one-dimensional forms of the governing differential equations in which we are concerned with variation of the flow variables in the flow direction only. Fortunately, most river flows can be treated in this manner. Rapidly varied flow, on the other hand, is not amenable to this approach and often requires application of the momentum equation in control volume form as in the hydraulic jump or a two-dimensional formulation of the governing differential equations as in the highly curvilinear flow over a spillway crest. Examples of gradually varied and rapidly varied flow are shown in Figures 1.1c and 1.1d.

Spatially varied flow really is a class of nonuniform flow but owes its nonuniformity to variation in the flow discharge in the direction of motion as well as to an imbalance of gravity and resisting forces. Examples of spatially varied flow include side channel spillways and continuous rainfall additions to gutter flow, as shown in Figure 1.1e.

1.7   BASIC EQUATIONS

The basic equations of fluid mechanics are applied to open channel flow with some modifications due to the free surface. These equations are the continuity, momentum, and energy equations, which can be derived directly from the Reynolds transport theorem applied to a fixed control volume as shown in Figure 1.2a. The Reynolds transport theorem is derived in many elementary fluid mechanics textbooks (Crowe et al. 2009; White 2008) and is given by

$$\frac{dB}{dt} = \frac{d}{dt}\int_{cv} b\rho\,d\forall + \int_{cs} b\rho(\mathbf{V}\cdot\mathbf{n})\,dA \qquad (1.2)$$

in which B = system property; t = time; b = the intensive value of B per unit mass m, dB/dm; ρ = fluid density; $\forall$ = volume of the control volume (cv); $\mathbf{V}$ = velocity

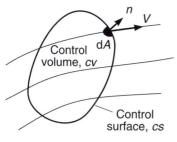

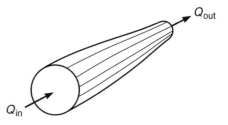

(a)

(b)

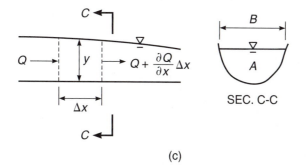

(c)

(d)

FIGURE 1.2
Control volumes **(a)** arbitrary control volume; **(b)** streamtube; **(c)** river reach; **(d)** streamline.

vector; $\mathbf{n}$ = outward normal unit vector; and A = area of the control surface (cs). The volume integral on the right hand side of Equation 1.2 sums up the values of the property per unit mass b over each mass element given by ρdV. In the surface integral in Equation 1.2, $(\rho \mathbf{V} \cdot \mathbf{n})\, dA$ represents the mass flux through an elemental area dA on the control surface. The dot product of the velocity vector with the unit outward normal $(\mathbf{V} \cdot \mathbf{n})$ determines the component of the velocity perpendicular to the surface since only that component can carry the property through the surface. Furthermore, the dot product is positive for outward fluxes and negative for inward fluxes into the control volume. Thus, the surface integral sums up the products of the property per unit mass b and the mass flux over the control surface to give the net outward flux of the property. In summary, Equation 1.2 states that the time rate of change of the system property is the sum of the time rate of change of the property inside the control volume and the net outward flux of the property through the control surface.

The Reynolds transport theorem can be applied to the properties of mass, momentum, and energy to obtain the control volume form of the corresponding governing conservation equations. The control volume forms of the equations can be simplified for the case of steady, one-dimensional flow and used in the analysis of many open channel flow problems.

In the case of mass m, the property $B = m$ and it follows that $dB/dt = 0$ and $b = dB/dm = 1$, so that

$$0 = \frac{d}{dt} \int_{cv} \rho \, d\forall + \int_{cs} \rho (\mathbf{V} \cdot \mathbf{n}) \, dA \tag{1.3}$$

which means simply that the time rate of change of mass inside the control volume in the first term must be balanced by the net outward mass flux through the control surface expressed by the second term. Now, in the case of steady flow of an incompressible fluid for the one-dimensional streamtube shown in Figure 1.2b, we have the familiar form of the continuity equation:

$$\int_{cs} (\mathbf{V} \cdot \mathbf{n}) \, dA = 0 = \Sigma Q_{out} - \Sigma Q_{in} \tag{1.4}$$

in which ΣQ = summation of the volume fluxes in or out of the control volume. The mean cross-sectional velocity, V_s, is defined as the volume flux divided by the cross-sectional area of flow perpendicular to the streamlines such that the volume flux can be written as

$$Q = \int_{cs} v_s \, dA = V_s A \tag{1.5}$$

in which v_s is the point velocity in the streamline direction; V_s is the mean cross-sectional velocity; and A is the cross-sectional area of flow.

Equation 1.3 also can be written in differential form for the general case of unsteady open channel flow of an incompressible fluid. If the control volume is considered to have a differential length Δx, as shown in Figure 1.2c, then as Δx approaches zero, Equation 1.3 becomes

$$\frac{\partial A}{\partial t} + \frac{\partial Q}{\partial x} = 0 \tag{1.6}$$

At any cross section, the time rate of change of flow area due to unsteadiness as the free surface rises or falls must be balanced by a spatial gradient in the volume flux Q in the flow direction. For steady flow, $\partial A/\partial t$ is zero by definition and $\partial Q/\partial x$ then also must become zero, which implies that the volume flux Q is constant along the channel, in agreement with Equation 1.4. The differential form of the continuity equation as given by Equation 1.6 will be applied in the numerical analysis of unsteady open channel flow in Chapter 8.

If we turn now to the property of momentum, the fundamental property B in the Reynolds transport theorem becomes a vector quantity defined by the linear momentum $\mathbf{B} = m\mathbf{V}$, in which m = mass and $\mathbf{V}$ = velocity vector. The total derivative $d\mathbf{B}/dt$ is exactly the vector sum of forces $\Sigma\mathbf{F}$ acting on the control volume according to Newton's second law. In this case, $d\mathbf{B}/dm = \mathbf{V}$ and the Reynolds transport theorem for a fixed control volume becomes a vector equation, which can be written as

$$\Sigma\mathbf{F} = \frac{d}{dt}\int_{cv} \mathbf{V}\rho\, d\forall + \int_{cs} \mathbf{V}\rho(\mathbf{V}\cdot\mathbf{n})\, dA \qquad (1.7)$$

Equation 1.7 states that the vector sum of forces acting on the control volume is equal to the time rate of change of linear momentum inside the control volume plus the net momentum flux out of the control volume through the control surface. In fact, this equation can be thought of simply as Newton's second law applied to a fluid. It is crucial to note that Equation 1.7 is a vector equation that represents three separate equations, written in each coordinate direction with the appropriate components of each vector quantity.

For the special case of the streamtube control volume in Figure 1.2b, the steady, one-dimensional form of the momentum equation in the stream direction, s, is given by

$$\Sigma F_s = \int_{cs} \rho v_s(\mathbf{V}\cdot\mathbf{n})\, dA = \Sigma\,(\beta\rho Q V_s)_{\text{out}} - \Sigma\,(\beta\rho Q V_s)_{\text{in}} \qquad (1.8)$$

in which v_s is the point velocity in the streamtube direction; V_s is the mean velocity; and β is the momentum flux correction coefficient to account for a nonuniform velocity distribution. The momentum equation as given by Equation 1.8 states that the vector sum of external forces in the streamtube direction is equal to the momentum flux out of the control volume in the s direction minus the momentum flux into the control volume in the s direction.

The momentum flux correction coefficient β in Equation 1.8 is defined by

$$\beta = \frac{\displaystyle\int_A v_s^2\, dA}{V_s^2 A} \qquad (1.9)$$

to correct for the substitution of the mean velocity squared for the point velocity squared and bringing it outside the integral in Equation 1.8. In turbulent flow in prismatic channels, the value of β is not significantly greater than the value of unity, which is the value for a uniform velocity distribution. In other open channel flow situations such as immediately downstream of a bridge pier, or in a river channel with floodplain flow, the value of β cannot be taken as unity because of the highly nonuniform velocity distributions in these situations.

It is important to note that the volume flux, Q, has been substituted for AV_s in Equation 1.8 and that the remaining V_s in the momentum flux term is the component of mean velocity in the direction in which the forces are summed. The outward volume flux takes a positive sign from $(\mathbf{V} \cdot \mathbf{n})$ because of the positive outward unit vector, and a negative sign goes with the inward volume flux for the same reason. The sign of V_s depends on the chosen positive direction for the force summation. If the forces are being summed in a direction x that is different from the streamtube direction, the volume flux remains unchanged but the component velocity is taken in the x direction with the appropriate sign. In the x direction, Equation 1.8 becomes

$$\Sigma F_x = \Sigma (\beta \rho Q V_x)_{\text{out}} - \Sigma (\beta \rho Q V_x)_{\text{in}} \tag{1.10}$$

If the momentum equation is applied to a differential control volume along a streamline, as in Figure 1.2d, and only pressure and gravity forces are considered, the result is Euler's equation for an incompressible, frictionless fluid:

$$-\frac{\partial p}{\partial s} - \rho g \frac{\partial z}{\partial s} = \rho \frac{\partial v_s}{\partial t} + \rho v_s \frac{\partial v_s}{\partial s} \tag{1.11}$$

in which p = pressure; z = elevation; v_s = streamline velocity; t = time; and s = coordinate in the streamline direction. If only steady flow is considered and Euler's equation is integrated along a streamline, the result is the familiar Bernoulli equation written here in terms of head between any two points along the streamline:

$$\frac{p_1}{\gamma} + z_1 + \frac{v_1^2}{2g} = \frac{p_2}{\gamma} + z_2 + \frac{v_2^2}{2g} \tag{1.12}$$

in which γ is the specific weight of water = ρg. In this form, the Bernoulli equation terms have dimensions of energy or work per unit weight of fluid, and so it is truly a work-energy equation derived from, but independent of, the momentum equation. The terms are scalars and represent pressure work, potential energy, and kinetic energy in that order. For applications to open channel flow, we need to expand the equation from a streamline to a streamtube and include the energy head loss term due to friction, h_f, for a real fluid, which results in

$$\frac{p_1}{\gamma} + z_1 + \alpha_1 \frac{V_1^2}{2g} = \frac{p_2}{\gamma} + z_2 + \alpha_2 \frac{V_2^2}{2g} + h_f \tag{1.13}$$

This expansion of the Bernoulli equation to a streamtube with head loss included is called the *extended Bernoulli equation* or the *energy equation*. It requires the assumption of a hydrostatic pressure distribution at points 1 and 2, because this means that the piezometric head ($p/\gamma + z$) is a constant across the cross section. In open channel flows, z is often the elevation of the channel bottom so that p/γ becomes the flow depth, y. The use of the mean velocity in the velocity head term necessitates a kinetic energy flux correction coefficient defined by

$$\alpha = \frac{\int_A v_s^3 \, dA}{V_s^3 A} \tag{1.14}$$

to account for a nonuniform velocity distribution. As we shall see in succeeding chapters, the value of α can be significantly larger than unity in rivers with overbank flow and therefore cannot be neglected.

To emphasize the independence of the extended Bernoulli or energy equation from the momentum equation, it should be pointed out that the energy equation can be derived in a more general way from the Reynolds transport theorem and the first law of thermodynamics:

$$\frac{dE}{dt} = \frac{dQ_h}{dt} - \frac{dW_s}{dt} - \frac{dW_p}{dt} = \frac{d}{dt}\int_{cv} e\rho\, d\forall + \int_{cs} e\rho(\mathbf{V}\cdot\mathbf{n})\, dA \qquad (1.15)$$

in which B has been replaced by the total energy E; Q_h = the heat transfer to the fluid; W_s = the shaft work done by the fluid on hydraulic machines; W_p = the flow work; and e is dE/dm = the internal energy ($\hat{u}$) plus kinetic energy ($v^2/2$) plus potential energy (gz) per unit mass. The rate of flow work is the rate at which the fluid pressure force does work, and it is given by

$$\frac{dW_p}{dt} = \int_{cs} p(\mathbf{V}\cdot\mathbf{n})\, dA \qquad (1.16)$$

If the flow is steady, and the pressure work term is moved to the right-hand side of Equation 1.15 where it can be incorporated into the surface integral, the result is

$$\frac{dQ_h}{dt} - \frac{dW_s}{dt} = \int_{cs}\left(\hat{u} + \frac{v^2}{2} + gz + \frac{p}{\rho}\right)\rho\,(\mathbf{V}\cdot\mathbf{n})\, dA \qquad (1.17)$$

which states that the rate of heat transfer into the fluid minus the rate of shaft work done by the fluid is equal to the net flux of energy, including the net rate of doing pressure work, directed out of the control volume. Integrating Equation 1.17 over the streamtube control volume of Figure 1.2b, dividing through by the weight flux (γQ), and rearranging the terms, we have

$$\frac{p_1}{\gamma} + z_1 + \alpha_1\frac{V_1^2}{2g} + h_s = \frac{p_2}{\gamma} + z_2 + \alpha_2\frac{V_2^2}{2g} + \left[\frac{\hat{u}_2 - \hat{u}_1}{g} - \frac{q_h}{g}\right] \qquad (1.18)$$

in which h_s = the shaft work term represented as the energy head added to the fluid by a pump ($+$) or subtracted from the fluid by a turbine ($-$); α = the kinetic energy flux correction coefficient resulting from the nonuniformity of the velocity distribution across the inlet and outlet of the streamtube identified by points 1 and 2, respectively; and q_h = the heat transferred to the fluid per unit mass between points 1 and 2. Ignoring the shaft work term for comparison, it is obvious that the term in square brackets in Equation 1.18, derived from the first law of thermodynamics and the Reynolds transport theorem, represents the difference in mechanical energy between points 1 and 2 and is identical to the head loss due to friction in Equation 1.13, h_f. Therefore, for steady, one-dimensional flow of an incompressible fluid, the head loss term represents the irreversible change in internal energy and the energy converted into heat due to viscous dissipation. Under these conditions, the energy equation is identical whether it is derived by integrating the momentum equation along a flow path, which results in a work-energy equation, or obtained directly from the conservation of energy principle stated by the first law of thermodynamics.

The continuity equation is a statement of the conservation of mass. Likewise, the energy equation expresses conservation of energy. It is a scalar equation and in the form of work/energy because of the spatial integration of $\Sigma\mathbf{F} = m\mathbf{a}$. The momentum equation also comes from Newton's second law applied to a fluid but is a vector equation that states that the sum of forces in any coordinate direction is equal to the change in momentum flux in that direction. In the control volume form, the momentum equation can be applied to quite complicated flow situations, as long as the external forces on the control volume can be quantified. The energy equation, on the other hand, requires the capability of quantifying energy dissipation inside the control volume.

Often, all three fundamental equations are applied simultaneously to solve what otherwise would be intractable problems. The hydraulic jump is an example in which the momentum and continuity equations are applied first to obtain the sequent depth (depth after the jump), and then the energy equation is employed to solve for the unknown energy loss.

Even experienced hydraulicians sometimes misapply the momentum and energy equations. The cardinal rule is that the energy equation must include all significant energy losses and the momentum equation must include all significant forces. Breaking this rule sometimes leads to conflicting results from the momentum and energy equations because of misapplication rather than a breakdown of the fundamental physical laws.

1.8   A NOTE ON TURBULENCE

Turbulence is ubiquitous in the natural environment, particularly in open channel flows in rivers and streams. Turbulence manifests itself as small fluctuations with time and space in flow variables such as velocity. It originates in an open channel flow due to random disturbances that are unstable and energetic enough to resist viscous damping. It was the insight of Osborne Reynolds to form the ratio of the inertia forces to the viscous forces, now known as the Reynolds number, the magnitude of which determines the critical point beyond which turbulence persists in the flow. For a Reynolds number larger than about 2100 in pipe flow or open channel flow, the disturbances grow to form fully developed turbulence, which can be described as a collection of eddies or swirls with sizes varying continuously from large to small scales that are responsible for the random fluctuations in flow variables just discussed. What is essential for the characterization of turbulence, however, is the existence of fluctuating vorticity, or rotation of fluid elements in the flow at many scales, and the enhanced mixing and increased energy dissipation that accompanies it. For the lack of these characteristics, ocean waves, which generate an unsteady time record of periodically fluctuating velocity, cannot be considered to be turbulent.

In this book, our interest in turbulence is with respect to its efficient mixing properties in open channel flow. Turbulent mixing effected by eddies determines the open channel velocity distribution and flow resistance as discussed in Chapter 4. Because the scale of turbulent eddies is so much larger than the molecular scale, which is responsible for momentum exchange and shear stress in a laminar flow, turbulent flow resistance is larger and increases more rapidly with mean flow velocity than for the laminar case. In addition, streamwise velocities near a solid boundary are larger than corresponding laminar velocities because of the greater efficiency of the turbulent mixing process. The vertical distribution of suspended sediment in an open channel flow transporting sediment is determined also by the properties of turbulence as discussed in Chapter 10. Methods for numerical modeling of turbulence will be discussed in detail in Chapter 11.

1.9   SURFACE VS. FORM RESISTANCE

Flow resistance in fluid flow can result from two fundamentally different physical processes, which take on special meaning when we discuss open channel flow resistance coefficients. Surface resistance is the traditional form of resistance resulting from surface friction or shear stress at a solid boundary. Integration of the shear stress over the surface area of the circular cylinder in Figure 1.3, for example, would result in surface drag.

Surface resistance alone cannot account for the measured flow resistance of a blunt object, such as a circular cylinder. Because of the phenomenon of flow separation of a real fluid as the boundary layer encounters an adverse pressure gradient, an asymmetric pressure distribution occurs around the circular cylinder, leading to form drag as shown in Figure 1.3 with higher pressure on the upstream face of the cylinder than on the downstream face in the zone of separation. In contrast, inviscid flow theory predicts a symmetric pressure distribution and no form drag (as well as no surface drag) on the cylinder, as shown in Figure 1.3. If the component of the pressure force in the flow direction is obtained by integrating the real fluid pressure distribution around the sphere, the result is a form drag or form resistance that is completely separate from surface drag. The total drag then is the sum of the surface drag and form drag. The magnitude of the form drag depends highly on the point of separation, which is different in the laminar and turbulent cases, as shown by Figure 1.3. The turbulent boundary layer persists farther around the cylinder before separating because of the higher velocities near the boundary. In open channel flow, the resistance offered by large roughness elements or alluvial bed forms may be due largely to form resistance. This point will be discussed in more detail in Chapters 4 and 10.

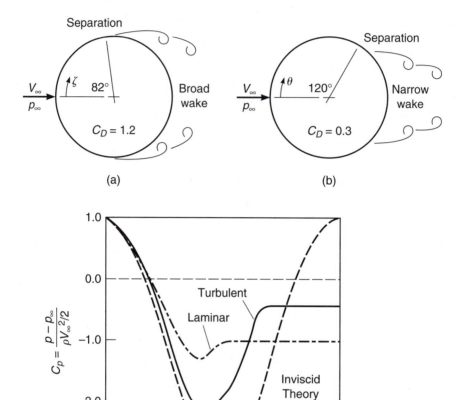

FIGURE 1.3
Separation and form resistance in real fluid flow around a circular cylinder: **(a)** laminar separation; **(b)** turbulent separation; **(c)** real and ideal fluid pressure distributions (White 2008).

1.10 DIMENSIONAL ANALYSIS

The purpose of dimensional analysis, which will be used extensively in this book, is to reduce the number of independent variables in an open channel flow problem or any other fluid mechanics problem by transforming the dependent variable and several independent variables that form a functional relationship into a smaller number of dimensionless ratios. This reduces the number of experiments involved in developing an experimental relationship, since only the independent dimensionless parameters need to be varied rather than each individual independent variable. Rather than varying the velocity, depth, and gravitational acceleration independently in a hydraulic jump experiment, for example, it is necessary to vary only the Froude number, which is a dimensionless combination of these variables, and present the results for the ratio of depths before and after the jump in terms of the Froude number. In addition, the dimensionless variables often represent ratios of forces, such as inertia and gravity, so that the magnitude of a particular dimensionless variable and its variation in a given experiment relate to an understanding of the physics of the flow situation. Furthermore, presentation of experimental results in terms of dimensionless variables generalizes the results to a wider range of applications and confirms the validity of the dimensionless ratios chosen to model a particular flow phenomenon.

If the governing equations can be completely formulated for a given problem, the equations can be nondimensionalized to deduce the embedded dimensionless parameters of importance. For example, application of the momentum equation to a hydraulic jump and nondimensionalization of the resulting equation for the depth after the jump results directly in the appearance of the Froude number as the only independent dimensionless parameter for this problem. The necessary condition for nondimensionalization of an equation is dimensional homogeneity, which simply requires every term to have the same dimensions in any properly posed equation describing a physical phenomenon. Once the governing equations are transformed into dimensionless form, the solution can be obtained in terms of the resulting dimensionless variables, either analytically or numerically, for a completely general solution. This solution can be applied to similar flow situations under conditions different from those for which the results were obtained, so long as the ranges of the dimensionless variables are the same.

In some cases, equations of open channel flow such as the Manning's equation or the head-discharge equation for flow over a weir at first may not appear to be dimensionally homogeneous. In these cases, some "constant" must have dimensions for the equation to be dimensionally homogeneous. If the equation for discharge Q over a sharp-crested weir, for example, is written as a constant C_1 times $LH^{3/2}$, where L is the crest length and H is the head on the crest, it is clear that the equation is not dimensionally homogeneous unless C_1 has dimensions of length to the 1/2 power divided by time. These in fact are the dimensions of the square root of the gravitational acceleration, g, which has been incorporated implicitly into the value of C_1. This practice requires that the coefficient C_1 take on a different numerical value for different systems of units, which is less desirable than leaving the original equation in terms of the gravitational acceleration.

As an example of nondimensionalization of the governing equations, the inviscid flow solution shown in Figure 1.3 can be obtained from an application of Bernoulli's equation between the approach flow (variables with a subscript of ∞) and any point on the circumference of the cylinder:

$$p_\infty + \rho \frac{V_\infty^2}{2} = p + \rho \frac{v^2}{2} \tag{1.19}$$

If the equation is nondimensionalized, there results

$$\frac{p - p_\infty}{\rho \dfrac{V_\infty^2}{2}} = C_p = 1 - \left(\frac{v}{V_\infty}\right)^2 \tag{1.20}$$

in which C_p is defined as a dimensionless pressure coefficient. The solution for the pressure coefficient is obtained by substituting the inviscid flow solution for the circumferential velocity $v = 2V_\infty \sin\theta$ into Equation 1.20 with the result

$$C_p = 1 - 4\sin^2\theta \tag{1.21}$$

Equation 1.21 gives the theoretical distribution of the dimensionless pressure coefficient C_p shown in Figure 1.3. Thus, if the governing equation of a fluid mechanics problem is known, then the equation itself can be made dimensionless, as in Equation 1.20, and the resulting solution also will be dimensionless.

In many problems of open channel flow, the theoretical solution is not directly applicable without the addition of experimental results to evaluate unknown parameters, or it may not be possible to formulate and solve the governing equations in very complicated flows. This requires a different approach for obtaining the important dimensionless parameters of the problem. In the case of drag on a circular bridge pier, for example, specification of the experimental drag coefficient is necessary to calculate the drag force, which includes both surface and form drag, the latter of which is not easily calculated from the governing equations. Presentation of the experimental results for the drag force in dimensionless form requires a general technique such as that afforded by the Buckingham Π theorem (see, for example, White 2008). The Buckingham Π theorem can be stated as follows:

> If a physical process involves a functional relationship among n variables, which can be expressed in terms of m basic dimensions, it can be reduced to a relation between $(n - m)$ dimensionless variables, or Π terms, by choosing m repeating variables, each of which is combined in turn with the remaining variables to form the Π terms as products of the variables taken to the appropriate powers. The m repeating variables must contain among them all basic dimensions found in all the variables but cannot themselves form a Π term.

In mathematical terms, if a dependent variable A_1 can be expressed in terms of $(n - 1)$ independent variables as

$$A_1 = f(A_2, A_3, \ldots, A_n) \tag{1.22}$$

then the Buckingham Π theorem allows the n variables to be expressed as a functional relation among $(n - m)$ Π groups:

$$\phi\left(\Pi_1, \Pi_2, \ldots, \Pi_{n-m}\right) = 0 \tag{1.23}$$

The basic dimensions usually are taken as mass (M), length (L), and time (T), although force (F), length, and time are an equally valid choice. The force dimension is uniquely related to the remaining dimensions by Newton's second law; that is, $F = MLT^{-2}$. In certain instances, the fundamental dimensions may be fewer than three; for example, only length and time may be involved. When choosing repeating variables, it is important to recognize that it is better not to choose the dependent variable as a repeating variable, so that it will appear in only one Π term.

If, for example, $n = 5$ and $m = 3$ with M, L, and T as the basic dimensions, the two Π terms can be found from

$$[\Pi_1] = M^0 L^0 T^0 = [A_2]^{x_1}[A_3]^{y_1}[A_4]^{z_1}[A_1] \tag{1.24}$$

$$[\Pi_2] = M^0 L^0 T^0 = [A_2]^{x_2}[A_3]^{y_2}[A_4]^{z_2}[A_5] \tag{1.25}$$

in which the square brackets denote "dimensions of" the enclosed variables; and A_2, A_3, and A_4 have been chosen as repeating variables. By substituting the dimensions of the variables into the right-hand sides of Equations 1.24 and 1.25 and equating the exponents on M, L, and T on both sides of the equations, the resulting algebraic equations can be solved for the unknown exponents and the resulting Π terms.

Now consider the drag problem for a completely immersed cylinder in which the drag force, D, can be expressed in terms of the cylinder diameter, d; the cylinder length, l_c; the approach velocity, V_∞; the fluid density, ρ; and the fluid viscosity, μ:

$$D = f_1\,(d, l_c, V_\infty, \rho, \mu) \tag{1.26}$$

A total of six variables with all three basic dimensions (M, L, T) are represented, so there will be three Π terms. The repeating variables are chosen to be the density, velocity, and cylinder diameter, which contain among them M, L, and T as basic dimensions but do not themselves form a dimensionless group. The cylinder diameter and length could not be chosen together as repeating variables because they would form a Π group. First, the drag force is combined with powers of the repeating variables, either algebraically or by inspection, to yield the first Π term; then the same process is repeated for the cylinder length and the fluid viscosity. The result is given by

$$\frac{D}{\rho d^2 V_\infty^2} = f_2\left(\frac{l_c}{d}, \mathbf{Re}\right) \tag{1.27}$$

which gives the dimensionless drag ratio in terms of the Reynolds number, $\mathbf{Re} = \rho V_\infty d/\mu$ and the ratio of cylinder length to diameter, l_c/d. Traditionally, the drag ratio is redefined as a more general drag coefficient, applicable to other shapes of immersed objects as $D/(\rho A V_\infty^2/2)$, with A in the coefficient of drag defined as the frontal area of the immersed object projected onto a plane perpendicular to the oncoming flow ($l_c \times d$). Also, a factor of 2 is added to the definition of the drag coefficient as a matter of tradition. For an infinitely long cylinder, the ratio l_c/d no longer has an influence because there are no end effects, so the experimental coefficient of drag is determined from the Reynolds number alone and used to calculate the drag force.

The choice of the repeating variables is not unique, so there are equally valid alternative forms of the Π groups. If, for example, the repeating variables were chosen to be μ, V_∞, and d in the cylinder drag problem, the result would be

$$\frac{D}{\mu V_\infty d} = f_3\left(\mathbf{Re}, \frac{l_c}{d}\right) \tag{1.28}$$

However, the alternate dependent Π group in (1.28) could be deduced from taking the product of the drag ratio and Reynolds number in (1.27). In the same manner, the justification for replacing d^2 in the denominator of the drag ratio in (1.27) with the frontal area is that the drag ratio in (1.27) can be divided by l_c/d and replaced by the result. In general, it is possible to state that a new Π group can be formed as

$$\Pi_1' = \Pi_1^a \Pi_2^b \Pi_3^c \tag{1.29}$$

and used to replace one of the original Π groups.

In the more general case of several bridge piers, each with diameter d and spacing s between piers and in open channel flow with a finite depth of water y_0, the formation of gravity surface waves around the piers may give rise to additional flow resistance so that the drag force can be written as

$$D = f_4(d, s, y_0, V_\infty, \rho, \mu, g) \tag{1.30}$$

in which the gravitational acceleration has been added to the list of variables. Alternatively, the specific weight γ could be added to the list instead of g, but the ratio γ/ρ, which is equal to g, then would appear in the dimensionless group related to the gravity force. Now, there are eight variables and still three basic dimensions resulting in five Π groups that can be expressed as

$$\frac{D}{\rho \, d y_0 V_\infty^2} = f_5\left(\frac{d}{s}, \frac{d}{y_0}, \mathbf{Re}, \mathbf{F}\right) \tag{1.31}$$

The additional geometric variable results in an additional geometric ratio, and the introduction of the gravitational force necessarily brings into play the Froude number, $\mathbf{F}$. The relative importance of the Π groups on the right-hand side of (1.31) would be determined by experiments.

The existence of the free surface in open channel flow inevitably involves the gravity force, either through the formation of surface waves, the existence of a component of the body force in the flow direction, or a differential pressure force due to changes in depth. Therefore, a dimensional analysis of an open channel flow problem includes the gravitational acceleration in the list of variables, and the Froude number necessarily emerges as an important dimensionless parameter, as discussed previously.

The choice of independent and dependent variables is crucial to the success of dimensional analysis. There can be only one dependent variable, and the independent variables must not be redundant; that is, one of the independent variables cannot be obtained from some combination of the others. The inclusion of extra independent variables that are truly independent is not fatal because the experimental results will show which of the resulting dimensionless groups is unimportant, but failing to include a significant independent variable can give an incomplete experimental relationship. Ultimately, such decisions are made in the course of research on a particular problem and may involve trial and error to arrive at the final set of important dimensionless ratios.

1.11 COMPUTER PROGRAMS

Some computer programs are given on the book website in Visual BASIC code, which is applicable to the Microsoft Windows environment. Visual BASIC is an event-driven language composed of both form modules, which contain the graphical user interface, and standard modules, which contain the computational code. The programs on the website include standard modules that consist of numerical procedures or subprograms. They can be converted easily to other languages such as Fortran or C, combined with form modules in Visual BASIC for input and output, or incorporated into Excel spreadsheets using Visual BASIC for Applications. The purpose here is to develop the core methodology for the use of numerical analysis to solve open channel flow problems. To this end, Appendix A contains some basic material on numerical methods that will be used throughout the text. The example programs found on the website are intended to serve as learning tools to explore the application of numerical techniques to open channel flow problems.

REFERENCES

Bakhmeteff, B. A. *Hydraulics of Open Channel Flow.* New York: McGraw-Hill, 1932.

Chow, V. T. *Open Channel Hydraulics.* New York: McGraw-Hill, 1959.

Crowe, C. T., D. F. Elger, B. C. Williams, and J. A. Roberson, *Engineering Fluid Mechanics,* 9th ed. New York: John Wiley & Sons, Inc., 2009.

Henderson, F. M. *Open Channel Flow.* New York: Macmillan, 1966.

Rouse, Hunter, ed. *Engineering Hydraulics.* Iowa City: Iowa Institute of Hydraulic Research, 1950.

Rouse, Hunter, and Simon Ince. *History of Hydraulics.* Iowa City: Iowa Institute of Hydraulic Research, 1957.

White, F. M. *Fluid Mechanics,* 6th ed. New York: McGraw-Hill, 2008.

EXERCISES

1.1. Classify each of the following flows as steady or unsteady from the viewpoint of the observer:

Flow	Observer
(*a*) Flow of river around bridge piers.	(1) Standing on bridge.
	(2) In boat, drifting.
(*b*) Movement of flood surge downstream.	(1) Standing on bank.
	(2) Moving with surge.

1.2. At the crest of an ogee spillway, as shown in Figure 1.1c, would you expect the pressure on the face of the spillway to be greater than, less than, or equal to the hydrostatic value? Explain your answer.

1.3. On the Internet, find a photograph of the Hoover Dam overflow spillway near Las Vegas, Nevada. The flow coming over the spillway is collected in a channel that runs perpendicular to the incoming flow. How would you classify the flow in the collection channel during a flood overflow? Explain your answer.

1.4. The river flow at an upstream gauging station is measured to be 1500 m³/s, and at another gauging station 3 km downstream, the discharge is measured to be 750 m³/s at the same instant of time. If the river channel is uniform, with a width of 300 m, estimate the rate of change in the water surface elevation in meters per hour. Is it rising or falling?

1.5. A paved parking lot section has a uniform slope over a length of 100 m (in the flow direction) from the point of a drainage area divide to the inlet grate, which extends across the lot width of 30 m. Rainfall is occurring at a uniform intensity of 10 cm/hr. If the detention storage on the paved section is increasing at the rate of 60 m³/hr, what is the runoff rate into the inlet grate?

1.6. If the lake level upstream of the spillway in Figure 1.1c is 55 m above the channel floor at the base of the spillway just upstream of the hydraulic jump, estimate the depth and velocity there for a flow rate of 1000 m³/s and a spillway width of 30 m. What is the value of the Froude number? Neglect the approach velocity in the lake and the head losses on the spillway.

1.7. A rectangular channel 6 m wide with a depth of flow of 3 m has a mean velocity of 1.5 m/s. The channel undergoes a smooth, gradual contraction to a width of 4.5 m.
(*a*) Calculate the depth and velocity in the contracted section.
(*b*) Calculate the net fluid force on the walls and floor of the contraction in the flow direction.
In each case, identify any assumptions that you make.

1.8. A bridge has cylindrical piers 1 m in diameter and spaced 15 m apart. Downstream of the bridge where the flow disturbance from the piers is no longer present, the flow depth is 2.9 m and the mean velocity is 2.5 m/s.
(*a*) Calculate the depth of flow upstream of the bridge assuming that the pier coefficient of drag is 1.2.
(*b*) Determine the head loss caused by the piers.

1.9. A symmetric compound channel in overbank flow has a main channel with a bottom width of 30 m, side slopes of 1:1, and a flow depth of 3 m. The floodplains on either side of the main channel are 300 m wide and flowing at a depth of 0.5 m. The mean velocity in the main channel is 1.5 m/s, while the floodplain flow has a mean velocity of 0.3 m/s. Assuming that the velocity variation within the main channel and the floodplain subsections is much smaller than the change in mean velocities between subsections, find the value of the kinetic energy correction coefficient α.

1.10. The power law velocity distribution for fully rough, turbulent flow in an open channel is given by

$$\frac{u}{u_*} = a\left(\frac{z}{k_s}\right)^{1/6}$$

in which u = point velocity at a distance z from the bed; u_* = shear velocity = $(\tau_0/\rho)^{1/2}$; τ_0 = bed shear stress; ρ = fluid density; k_s = equivalent sand grain roughness height; and a = constant.

(a) Find the ratio of the maximum velocity, which occurs at the free surface where z = the depth, y_0, to the mean velocity for a very wide channel.
(b) Calculate the values of the kinetic energy correction coefficient α and the momentum flux correction coefficient β for a very wide channel.

1.11. The velocity distribution for laminar flow in an open channel is given by

$$\frac{u}{u_*} = \frac{u_*}{\nu}\left(z - \frac{z^2}{2y_0}\right)$$

in which ν = kinematic viscosity; y_0 = depth of flow; and the other variables are as defined in Exercise 1.10. Answer questions (a) and (b) of Exercise 1.10 for this laminar velocity distribution.

1.12. An alternative expression for the velocity distribution in fully rough, turbulent flow is given by the logarithmic distribution

$$\frac{u}{u_*} = \frac{1}{\kappa}\ln\left(\frac{z}{z_0}\right)$$

in which κ = the von Karman constant = 0.40; $z_0 = k_s/30$; and the other variables are the same as defined in Exercise 1.10. Show that α and β for this distribution in a very wide channel are given by

$$\alpha = 1 + 3\varepsilon^2 - 2\varepsilon^3$$

$$\beta = 1 + \varepsilon^2$$

in which $\varepsilon = (u_{max}/V) - 1$; u_{max} = maximum velocity; and V = mean velocity.

1.13. In a hydraulic jump in a rectangular channel of width b, the depth after the jump y_2 is known to depend on the following variables:

$$y_2 = f[y_1, q, g]$$

in which y_1 = depth before the jump; q = discharge per unit width = Q/b; and g = gravitational acceleration. Complete the dimensional analysis of the problem.

1.14. The backwater Δy caused by bridge piers in a bridge opening is thought to depend on the pier diameter and spacing, d and s, respectively; downstream depth, y_0; downstream velocity, V; fluid density, ρ; fluid viscosity, μ; and gravitational acceleration, g. Complete the dimensional analysis of the problem.

1.15. The longitudinal velocity, u, near the fixed bed of an open channel depends on the distance from the bed, z; the kinematic viscosity, ν; and the shear velocity, $u_* = (\tau_0/\rho)^{0.5}$ in which τ_0 is the wall shear stress. Develop the dimensional analysis for the point velocity, u.

1.16. In the very slow motion of a fluid around a sphere, the drag force on the sphere, D, depends on the sphere diameter, d; the velocity of the approach flow, V; and the fluid viscosity, μ. Complete the dimensional analysis. How many dimensionless groups are there and what are the implications for the corresponding values of the group(s)? Why was the fluid density not included in the list of variables?

1.17. The discharge over a sharp-crested weir, Q, is a function of the head on the weir crest, H; the crest length, L; the height of the crest, P; density, ρ; viscosity, μ; surface tension, σ; and gravitational acceleration, g. Carry out the dimensional analysis using ρ, g, and H as repeating variables. If it is known that Q is directly proportional to crest length, L, how would you alter the dependent Π group?

1.18. The terminal fall velocity, w_f, of a sphere in a stationary fluid of infinite extent is a function of the fluid density, ρ; the reduced gravitational acceleration, $(\rho_s/\rho - 1)g$, in which $\rho_s = $ density of the sphere and $g = $ gravitational acceleration; the dynamic fluid viscosity, μ; and the sphere diameter, d. Complete the dimensional analysis for the fall velocity as the dependent variable with ρ, d, and μ as repeating variables. Repeat the dimensional analysis with $(\rho_s/\rho - 1)$ and g taken as separate independent variables.

2

Specific Energy

2.1 DEFINITION OF SPECIFIC ENERGY

The concept of specific energy as introduced by Bakhmeteff (1932) has proven to be very useful in the analysis of open channel flow. It arises quite naturally from a consideration of steady flow through a transition defined by a gradual rise in the channel bottom elevation, as shown in Figure 2.1. For given approach flow conditions of velocity and depth, the unknown depth, y_2, after a channel bottom rise of height Δz is of interest. If for the moment we neglect the energy loss, the energy equation combined with continuity can be written as

$$y_1 + \frac{Q^2}{2gA_1^2} = y_2 + \frac{Q^2}{2gA_2^2} + \Delta z \tag{2.1}$$

in which y = depth; Q = discharge; A = cross-sectional area of flow; and $\Delta z = z_2 - z_1$ = change in bottom elevation from cross-section 1 to 2. Now, it is apparent that the sum of depth and velocity head must change by the amount Δz and that the change must result in an interchange between depth and velocity head such that the energy equation is satisfied. If *specific energy* is defined as the sum of depth and velocity head, it follows that the possible solutions of the problem for the depth depend on the variation of specific energy with depth. In fact, there are two real solutions for the depth in this problem, and the plot of depth as a function of specific energy clarifies which solution will prevail. Such a plot for constant discharge Q is called the *specific energy diagram.*

A more formal definition of *specific energy* is the height of the energy grade line above the channel bottom. In uniform flow, for example, the energy grade line, by definition, is parallel to the channel bottom, so that the specific energy is constant in the flow direction. In this case, the decrease in potential energy head as the channel

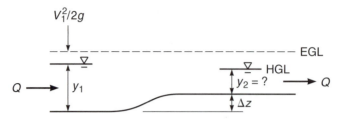

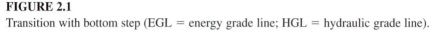

FIGURE 2.1

Transition with bottom step (EGL = energy grade line; HGL = hydraulic grade line).

bottom elevation decreases in the flow direction is the same as the drop in the energy grade line due to energy head loss caused by boundary friction. In Figure 2.1, the specific energy decreases in the flow direction, but it would be equally possible for the specific energy to increase in the flow direction by dropping rather than raising the channel bottom. The total energy always must remain constant or decrease, but the specific energy can increase as well. In gradually varied flow, a continuous change in specific energy with flow direction leads to a classification of gradually varied flow profiles, in Chapter 5, according to the interchange between depth and velocity head. We show that the rate at which specific energy changes in the flow direction in gradually varied flow is determined by the excess or deficit of the work done by gravity in comparison to the energy loss due to boundary resistance.

Because the specific energy arises in connection with the determination of depth changes in one-dimensional flow, certain restrictions are inherent in its definition. First, the specific energy is defined at cross sections where the flow is gradually varied, so that the depth is identical to the pressure head at the channel bottom; that is, the free surface represents the hydraulic grade line. What happens between two points at which specific energy is defined is not restricted by this assumption, however, as evidenced by the situation in Figure 2.1. Second, the water surface and energy grade line are assumed to be horizontal across the cross section, so that a single value of velocity head corrected by the kinetic energy flux correction coefficient α suffices for the entire cross section. With these two restrictions in mind, the definition of specific energy, E, becomes

$$E = y + \frac{\alpha V^2}{2g} \tag{2.2}$$

in which y = flow depth; α = kinetic energy flux correction; and V = mean cross-sectional velocity.

A third restriction on the definition in Equation 2.2 occurs in the case of a channel with a large slope angle θ, as shown in Figure 2.2. In this case it no longer is obvious how the depth should be measured (vertically as y or perpendicular to the channel bottom as d) nor, in fact, whether either of these definitions of depth is the correct representation of the pressure head, p/γ. This can be clarified by considering the force balance between the gravity and pressure force perpendicular to the channel bottom in Figure 2.2, whereby $p/\gamma = d \cos\theta$, in which γ is the specific weight

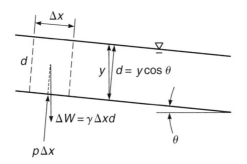

FIGURE 2.2
Depth and pressure head on a steep slope.

of the fluid. Furthermore, it should be noted from the geometry in Figure 2.2 that $d = y \cos\theta$. The correct expression for specific energy must be written as

$$E = d \cos\theta + \frac{\alpha V^2}{2g} = y \cos^2\theta + \frac{\alpha V^2}{2g} \qquad (2.3)$$

As a practical matter, $\cos^2\theta$ does not vary from unity by more than 1 percent if $\theta < 6°$, so that the approximate form shown in Equation 2.2 is valid for all except the steepest channels, such as a spillway chute.

2.2 SPECIFIC ENERGY DIAGRAM

Now we are ready to consider the actual functional variation of depth y with specific energy, E, in the graphical form called the *specific energy diagram*. At first, it will be convenient to consider the case of a rectangular channel of width b. The flow rate per unit of width q can be defined for the rectangular channel as Q/b, where Q = total channel discharge. Continuity then allows us to write the velocity, V, as q/y, and so the specific energy for a rectangular channel with $\alpha = 1$ is

$$E = y + \frac{q^2}{2gy^2} \qquad (2.4)$$

It is apparent from Equation 2.4 that there indeed is a unique functional variation between y and E for a constant value of q, and it is sketched as the specific energy diagram in Figure 2.3. Note from Equation 2.4 that, as y becomes very large, E approaches y, so that the straight line $y = E$ is an asymptote of the upper limb of the specific energy curves shown in Figure 2.3. In addition, it can be shown that, as y approaches zero, E becomes infinitely large, implying that the E axis is an asymptote of the lower limb of the specific energy curve. Between these two limits, the specific energy must have a minimum value for a given value of flow rate per unit of width q. In other words, flows with a specific energy less than the minimum

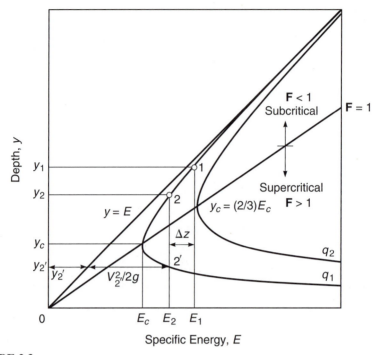

FIGURE 2.3
Specific energy diagram for a transition in a rectangular channel with a smooth, upward bottom step, Δz.

value for a given q are physically impossible. The critical depth, y_c, corresponding to the condition of minimum specific energy, E_c, can be found by differentiating the expression for specific energy in Equation 2.4 with respect to y and setting the result to zero:

$$\frac{dE}{dy} = 0 = 1 - \frac{q^2}{gy_c^3} \tag{2.5}$$

Now for the critical depth, y_c, we have

$$y_c = \left[\frac{q^2}{g}\right]^{1/3} \quad \text{(rectangular channel)} \tag{2.6}$$

which indicates that critical depth is a function of only the flow rate per unit width q for a rectangular channel. Furthermore, with the help of Equation 2.6, the value of minimum specific energy, E_c, is given by

$$E_c = y_c + \frac{q^2}{2gy_c^2} = \frac{3}{2} y_c \quad \text{(rectangular channel)} \tag{2.7}$$

and shown in Figure 2.3 as the locus of values of critical depth and minimum specific energy for each specific energy curve defined by its own unique value of q.

Because both y_c and E_c increase as q increases, the specific energy curves move upward and to the right in Figure 2.3 as q increases and $q_2 > q_1$.

The physical meaning of the specific energy diagram is not nearly so clear as its mathematical interpretation. It is obvious from Figure 2.3 that, mathematically, there are two possible values of depth for a given value of specific energy. The physical meaning of these two depths becomes clear from a rearrangement of Equation 2.6 as

$$\frac{q^2}{gy_c^3} = \frac{V_c^2}{gy_c} = \mathbf{F}^2 = 1 \tag{2.8}$$

from which we can conclude that the critical depth condition is specified by the value of the Froude number, $\mathbf{F}$, becoming unity.

If we further observe that the celerity, c, of a very small amplitude disturbance at the water surface is $(gy)^{1/2}$, a physical interpretation of the meaning of the two limbs of the specific energy curves in Figure 2.3 is possible. First, we assume in Figure 2.4 that a small amplitude disturbance in shallow water of depth y is propagated at a celerity c relative to still water. If we superimpose a velocity c in the opposite direction, this becomes a steady flow problem with constant energy, so that $(y + c^2/2g) = $ constant and therefore

$$dy + \frac{c}{g} \, dc = 0 \tag{2.9}$$

Then, with the aid of continuity for steady flow, $cy = $ constant; and we have that $c \, dy + y \, dc = 0$, which can be combined with Equation 2.9 to prove that $c = (gy)^{1/2}$ with respect to the still water as a reference frame.

Now, for depth $y < y_c$, the Froude number, V/c, must be greater than one and the velocity $V > c$. In other words, the flow velocity is greater than the celerity of a small surface disturbance and so sweeps any disturbance downstream. This flow regime is called *supercritical*, or *rapid flow* and is characterized by relatively small depths and large velocities, as can be seen in Figure 2.3. The upper regime of flow, on the other hand, has $y > y_c$ and the Froude number less than unity. Therefore, the flow velocity, $V < c$, and wave disturbances can travel both upstream and downstream in this regime, which is called *subcritical flow*. Subcritical flow has relatively large depths and small velocities; for this reason, it also sometimes is called

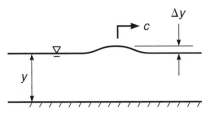

FIGURE 2.4
Water surface disturbance of small amplitude with celerity c.

tranquil flow. We can conclude that subcritical flow is a flow regime in which the depth control, or boundary condition, can exert its influence in the upstream direction, while in supercritical flow, a control can influence the flow profile only in the downstream direction. These observations become important later when we consider the computation of flow profiles.

Finally, we can return to the original problem posed by the transition in Figure 2.1. If the upstream flow is subcritical, as indicated by point 1 in Figure 2.3, which depth is the proper solution for point 2, for which $E_2 = E_1 - \Delta z$? The lower depth $y_{2'}$ can be reached only by a decrease in specific energy to its minimum value, followed by an increase in specific energy. Because this is physically impossible, the correct solution for the unknown depth is the subcritical one, y_2. As the flow passes over the rise in the channel bottom, the depth will decrease and the water surface elevation will dip to accommodate the increase in velocity head due to acceleration.

2.3   CHOKE

A limiting condition for the transition shown in Figure 2.1 occurs if $\Delta z > \Delta z_c$, where Δz_c is the difference between the approach specific energy and the minimum specific energy. If this difference is exceeded, it would appear that the specific energy must become less than the minimum value, a condition already shown to be impossible. In response to this dilemma, the flow responds with a rise in the water surface and the available specific energy upstream of the transition, as shown in Figure 2.5. In fact, the specific energy rise in Figure 2.5 is just sufficient to force flow through the transition at the critical depth. Any further increases in Δz will cause a corresponding increase in the upstream specific energy, while the depth in the transition will remain critical. This condition, referred to as a *choke,* illustrates quite dramatically the extra degree of freedom afforded by the adjustment of the free surface in open channel flow.

The step height required to just cause choking can be developed from the energy equation applied from the approach section to the critical section over the step:

$$\Delta z_c = E_1 - E_c = E_1 - 1.5y_c \tag{2.10}$$

If Equation 2.10 is divided by the approach depth, y_1, the result for the dimensionless critical step height depends on the approach Froude number, $\mathbf{F}_1$, alone:

$$\frac{\Delta z_c}{y_1} = 1 + \frac{\mathbf{F}_1^2}{2} - 1.5\mathbf{F}_1^{2/3} \text{ (rectangular channel)} \tag{2.11}$$

Equation 2.11 is plotted in Figure 2.6. For an approach Froude number of 0.1, for example, the critical step height for choking is 68 percent of the approach depth but rapidly becomes a smaller fraction of the approach depth as the approach Froude number increases.

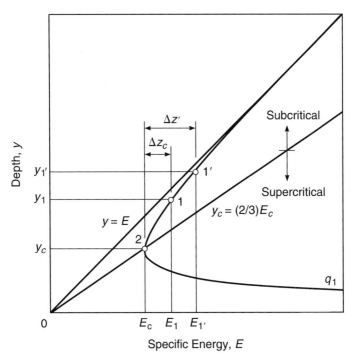

FIGURE 2.5
Choking in transition with a smooth, upward bottom step.

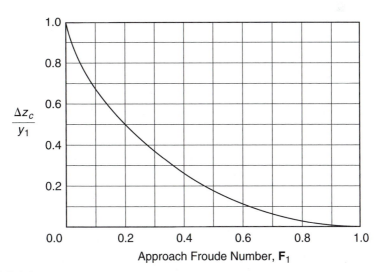

FIGURE 2.6
Critical step height for choking in a transition with a bottom step in a rectangular channel.

29

Example 2.1

For an approach flow in a rectangular channel with depth of 2.0 m (6.6 ft) and velocity of 2.2 m/s (7.2 ft/s), determine the depth of flow over a gradual rise in the channel bottom of $\Delta z = 0.25$ m (0.82 ft). Repeat the solution for $\Delta z = 0.50$ m (1.64 ft).

Solution.   First, it is necessary to know whether the approach flow is supercritical or subcritical, which is ascertained most easily by simply calculating the critical depth for a flow rate per unit width of $q = 2 \times 2.2 = 4.4$ m²/s (47.4 ft²/s):

$$y_c = (4.4^2/9.81)^{1/3} = 1.25 \text{ m } (4.10 \text{ ft})$$

from which it is obvious that the approach flow is subcritical, because $y_1 > y_c$. The approach flow Froude number also could be calculated:

$$F_1 = 2.2/(9.81 \times 2)^{1/2} = 0.5$$

Now the energy equation written between the approach flow and the section of maximum step height (0.25 m) is

$$E_1 = 2 + 2.2^2/19.62 = 2.25 = 0.25 + y_2 + 4.4^2/(19.62 \times y_2^2)$$

which can be solved by trial. Only roots larger than the critical depth of 1.25 m (4.10 ft) are sought. The result is $y_2 = 1.62$ m (5.32 ft). Note that the absolute elevation of the water surface drops by the amount $(2 - 0.25 - 1.62) = 0.13$ m (0.43 ft). If the step height increases to 0.5 m (1.64 ft), the available specific energy is the approach specific energy (2.25 m) less the step height of 0.5 m, or 1.75 m (5.74 ft), which is less than the minimum specific energy of $(1.5 \times 1.25) = 1.88$ m (6.17 ft). This means that a choke occurs in which the depth over the step becomes critical (1.25 m) and the upstream depth increases as given by the solution of

$$y_1 + 4.4^2/(19.62 \times y_1^2) = 0.5 + 1.88 = 2.38 \text{ m } (7.81 \text{ ft})$$

The result is $y_1 = 2.17$ m (7.12 ft), which results in an upstream increase in depth of 0.17 m (0.56 ft). The critical step height, which will just cause choking, can be obtained from Figure 2.6 or Equation 2.11 for a Froude number of 0.5 as $\Delta z_c/y_1 = 0.18$, from which $\Delta z_c = 0.36$ m (1.18 ft).

2.4   DISCHARGE DIAGRAM

Transitions in channel width also can be analyzed by the specific energy concept. For the rectangular channel, however, it is no longer true that the flow rate per unit width q remains constant. Suppose the channel width changes from b_1 in the approach subcritical flow to b_2 in the contracted section, as shown in Figure 2.7a. With negligible energy loss, the energy equation simply states that $E_1 = E_2$, but this requires that the flow regime move from one specific energy curve downward to another that is appropriate for the new value of q, as shown in Figure 2.7b by the points 1 and 2.

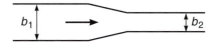

(a) Plan View of Width Contraction

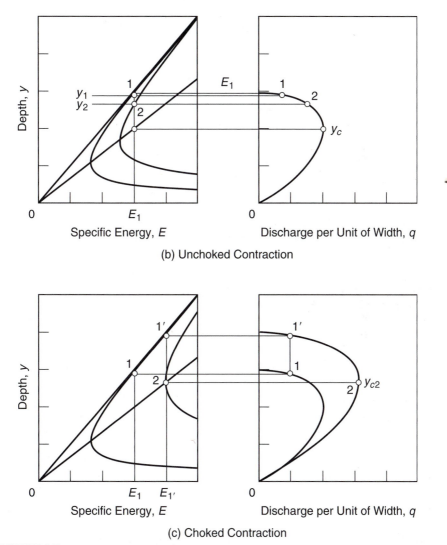

(b) Unchoked Contraction

(c) Choked Contraction

FIGURE 2.7

Specific energy and discharge diagrams for contraction in width in a rectangular channel.

An alternative way of viewing the change in flow regime in a width contraction can be gained by writing the energy equation and noting that the quantity that remains constant in this instance is not q but rather the specific energy, E (neglecting energy losses and assuming a horizontal channel bottom). Therefore, if a discharge function for a given specific energy, E_1, is defined by

$$q = y [2g (E_1 - y)]^{1/2} \tag{2.12}$$

then it is obvious that there is a unique functional relation between the discharge per unit width q and depth y for the rectangular channel for a constant value of specific energy. The function is shown in Figure 2.7b alongside the specific energy diagram with the approach and contracted sections identified as points 1 and 2, respectively. Two specific energy curves are shown: one corresponding to the upstream width b_1 and flow rate per unit width q_1; the other for the contracted section with width b_2 and flow rate per unit width q_2. The decrease in depth from point 1 to point 2 occurs at constant specific energy, as shown in the specific energy diagram, and corresponds to an increase in discharge per unit width in the discharge diagram. The discharge function given by (2.12) has a maximum that can be found by setting $dq/dy = 0$ and solving for y to obtain $(2/3)E_1$. This is precisely the relation for critical depth derived previously for a rectangular channel, which means that critical depth not only is the depth of minimum specific energy for constant q but also can be interpreted as the depth of maximum discharge for a given specific energy. In Figure 2.7b, the critical depth associated with the given specific energy in the specific energy diagram has been transferred across horizontally to the maximum q in the discharge diagram. Figure 2.7b also shows that the position of point 1 on the approach specific energy curve determines the available specific energy and establishes a single discharge diagram for that value of specific energy because $y = E$ when $q = 0$ in the discharge diagram.

Choking can be caused in a contraction by decreasing the width to a value such that the available specific energy no longer is sufficient to pass the flow through the contraction without an increase in the upstream depth. This is illustrated in Figure 2.7c by the points 1, 1′, and 2. Point 1 must move up the specific energy curve to the point 1′ upstream of the contraction with an increase in specific energy in Figure 2.7c. This establishes a new discharge curve in Figure 2.7c with a new value of maximum discharge and a new critical depth, shown by point 2. The flow regime passes from the new upstream depth $y_{1'}$ to y_{c2} in both the specific energy and discharge diagrams but in different ways, as shown in Figure 2.7c. Also apparent from Figure 2.7c is that, once the choking criterion is exceeded, further decreases in the downstream width b_2 cause the depth at point 1′ to continue increasing asymptotically to the straight line $y = E$ as the approach velocity head becomes nearly negligible. In this instance, the critical depth in the contracted section approaches two-thirds of the approach depth for a rectangular channel.

Another interpretation of the discharge diagram is shown very clearly by the example in Figure 2.8, in which flow from a reservoir into a short horizontal channel or over a broad-crested weir is controlled by a sluice gate. The reservoir level establishes the fixed value of specific energy, and raising the sluice gate in the channel causes an increase in discharge as the depth of flow upstream of the gate decreases. Simultaneously, the depth downstream of the gate increases to maintain the same discharge. The discharge reaches its maximum value when the upstream depth becomes critical. Beyond this value, the gate no longer has any

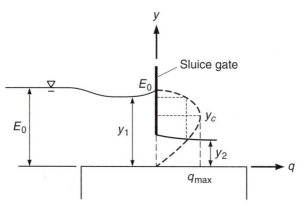

FIGURE 2.8
Discharge diagram for flow under a sluice gate on a broad-crested weir.

influence and the discharge cannot be increased further without raising the reservoir level. At the maximum discharge, the depth in the rectangular, horizontal channel becomes two-thirds of the head in the reservoir if the approach velocity head is negligible.

If the approach flow to a contraction is supercritical, specific energy analysis still applies in the general case without choking, but oblique standing waves can complicate the analysis. If choking occurs due to a contraction, two limiting cases are possible for a supercritical approach flow, as shown in Figure 2.9. Choking condition A is caused by the occurrence of a hydraulic jump upstream of the contraction followed by passage through the critical depth in the contracted opening. Choking condition B, on the other hand, is the result of going directly from the supercritical state to the critical depth for the contraction. Between conditions A and B, choking may or may not occur (point 3 or 2', for example). These two conditions are analyzed in more detail in the following chapter.

2.5   CONTRACTIONS AND EXPANSIONS WITH HEAD LOSS

The general equation governing contractions and expansions with a subcritical approach flow at cross-section 1 is the energy equation with head losses included, as given by

$$y_1 + \frac{Q^2}{2gA_1^2} = \Delta z + y_2 + \frac{Q^2}{2gA_2^2} + K_L \left| \frac{1}{A_2^2} - \frac{1}{A_1^2} \right| \frac{Q^2}{2g} \qquad (2.13)$$

in which Δz is positive for an upward step. Energy losses are considered and expressed as a minor loss coefficient, K_L, times the difference in velocity heads between the two cross sections. The abrupt expansion has the highest energy loss

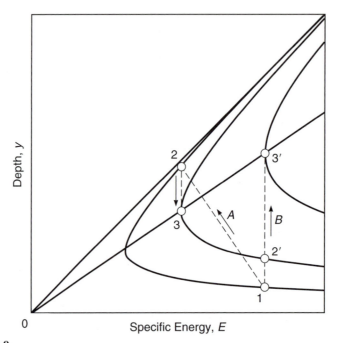

FIGURE 2.9
Choking modes for contraction with supercritical approach flow.

because of flow separation and viscous dissipation of mean flow energy in the separated zone. Henderson (1966) has shown from a combined energy and momentum analysis that the expression for the head loss in an abrupt open channel expansion is given by

$$h_L = \frac{V_1^2}{2g}\left[\left(1 - \frac{b_1}{b_2}\right)^2 + \frac{2\mathbf{F}_1^2 b_1^3(b_2 - b_1)}{b_2^4}\right] \tag{2.14}$$

in which the subscripts 1 and 2 represent the approach and expanded sections, respectively, as shown in Figure 2.10. Equation 2.14 assumes that the depth at cross-section 1 equals the depth at cross-section 2 and that the pressure distribution at cross-section 2 is hydrostatic across the full width b_2, including the separation zone. The momentum equation then is written between cross-sections 2 and 3, and the energy equation from 1 to 3 gives the head loss. The first term on the right hand side of (2.14) is identical to the expression for head loss in an abrupt pipe expansion, while the second term is the open channel flow term with its dependence on Froude number $\mathbf{F}_1$. For $\mathbf{F}_1 < 0.5$, the second term is small compared to the first, so that it can be neglected under this condition and $y_1 = y_2 = y_3$. Then, for an expression for head loss like that given in Equation 2.13 to be consistent with Equation 2.14 with the second term neglected, K_L must be given by

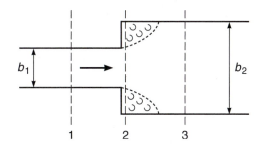

FIGURE 2.10
Plan view of abrupt open channel expansion.

$$K_L = \frac{1 - \dfrac{b_1}{b_2}}{1 + \dfrac{b_1}{b_2}} \qquad (\mathbf{F}_1 < 0.5) \tag{2.15}$$

in which K_L varies from approximately 0.8 to 0.05 as b_1/b_2 increases from 0.1 to 0.9. Gradual tapering of the expansion at a rate of 1:4 (lateral:longitudinal) results in a head loss coefficient that is only about 30 percent of the value given by Equation 2.15. Energy losses are smaller in the case of contractions than expansions. Henderson (1966) reports values of either 0.11 or 0.23 times the downstream velocity head, depending on whether the contractions are rounded or square edged, respectively. For rivers, the HEC-RAS manual (2008) suggests a value for K_L of 0.3 for gradual expansions and a value of 0.1 for gradual contractions. The default values for WSPRO (Shearman et al. 1986; Shearman 1990) are 0.5 for expansions and 0.0 for contractions.

The actual effect of head losses in the specific energy analysis of contractions and expansions depends on their relative magnitude in comparison with the approach specific energy. In a contraction followed by an expansion, as in the open channel venturi meter shown in Figure 2.11, or in a bridge contraction, the contraction energy loss may be considerably smaller than the expansion loss, as shown in the specific energy diagram. The overall effect of the total head loss is an upstream approach depth at point 1 that is larger than the downstream tailwater depth at point 3, even though choking is not occurring. As the tailwater is lowered from point 3 to 3′ for the same total discharge, choking occurs at some point, as shown in Figure 2.11 at point 2′. Choking also can occur as the contracted section width gets smaller for the same total discharge and the same tailwater. Further decreases in contracted width cause the depth to remain critical in the contracted section, although critical depth itself also is increasing as the width b_2 decreases and q_2 increases. This causes backwater, a rise in upstream depth. In this case, the flow regime passes to supercritical downstream of the contracted section followed by a hydraulic jump to the fixed tailwater.

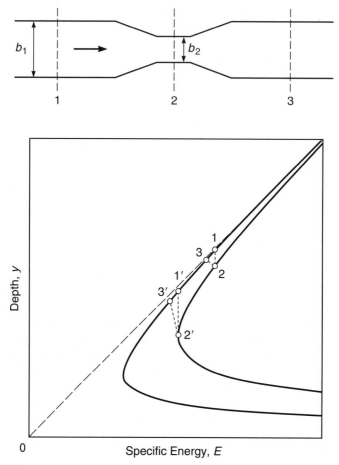

FIGURE 2.11
Open channel contraction followed by an expansion with head loss.

2.6   CRITICAL DEPTH IN NONRECTANGULAR SECTIONS

Specific energy for nonrectangular sections must be formulated before deriving the critical condition as the point of minimum specific energy. Specific energy in any nonrectangular section of area A and depth y, as shown in Figure 2.12, can be expressed as

$$E = y + \alpha \frac{Q^2}{2gA^2} \qquad (2.16)$$

Differentiating with respect to y and setting $dE/dy = 0$ results in

$$\frac{dE}{dy} = 0 = 1 - \frac{\alpha Q^2}{gA^3} \frac{dA}{dy} \qquad (2.17)$$

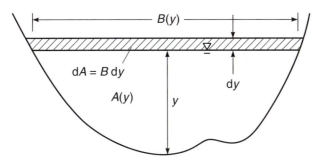

FIGURE 2.12
Geometric properties of general nonrectangular section.

From Figure 2.12, we see that $dA/dy = B$, in which $B(y)$ is the top width at the water surface and a function of y. The condition for minimum specific energy and critical depth then is

$$\frac{\alpha Q^2 B_c}{g A_c^3} = 1 \tag{2.18}$$

in which the subscript c indicates that A and B are functions of the critical depth, y_c. If we define the hydraulic depth $D = A/B$ and substitute $V = Q/A$, the Froude number for a nonrectangular channel is defined and has the value of unity at the critical condition:

$$\mathbf{F} = \frac{V}{(gD/\alpha)^{1/2}} \tag{2.19}$$

The value of the minimum specific energy can be obtained from Equations 2.16 and 2.18 and is given by

$$E_c = y_c + \frac{D_c}{2} \tag{2.20}$$

in which α does not appear explicitly but nevertheless is involved in the determination of y_c and therefore E_c.

The computation of critical depth for the nonrectangular channel is a matter of solving Equation 2.18 for the geometry of a particular cross-sectional shape. The appropriate geometric elements needed for the trapezoidal, triangular, circular, and parabolic cross sections are listed in Table 2.1. An exact solution is available for both the triangular and parabolic cases, but the trapezoidal and circular sections require the solution of a nonlinear algebraic equation to obtain the critical depth.

A graphical solution in nondimensional terms is possible for both the trapezoidal and circular cases (Henderson 1966). The trapezoidal section, for example, requires the solution of Equation 2.18 after substitution of the appropriate geometric expressions:

$$\frac{\alpha Q^2}{g} = \frac{A_c^3}{B_c} = \frac{[y_c(b + my_c)]^3}{(b + 2my_c)} \tag{2.21}$$

TABLE 2.1   Geometric elements for channels of different shape (y = flow depth)

Section	Area, A	Wetted Perimeter, P	Top Width, B
Rectangular	by	$b + 2y$	b
Trapezoidal	$y(b + my)$	$b + 2y(1 + m^2)^{1/2}$	$b + 2my$
Triangular	my^2	$2y(1 + m^2)^{1/2}$	$2my$
Circular†	$(\theta - \sin\theta)\, d^2/8$	$\theta d/2$	$d\sin(\theta/2)$
Parabolic‡	$(2/3)\, By$	$(B/2)[(1 + x^2)^{1/2} + (1/x)\ln(x + (1 + x^2)^{1/2})]$	$B_1(y/y_1)^{1/2}$

†$\theta = 2\cos^{-1}[1 - 2\,(y/d)]$
‡$x = 4y/B$

in which b is the bottom width of the trapezoidal section with side slopes of $m{:}1$ (horizontal:vertical). To present the solution of Equation 2.21 graphically, the following dimensionless variables are defined for the trapezoidal channel:

$$Z_{\text{trap}} = \frac{Qm^{3/2}}{(g/\alpha)^{1/2}b^{5/2}}; \qquad y' = \frac{my_c}{b} \qquad (2.22)$$

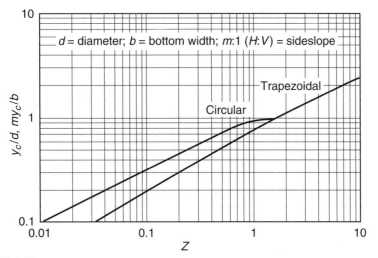

FIGURE 2.13
Critical depth for trapezoidal and circular channels; $Z_{circ} = Q/[g^{1/2}d^{5/2}]$; $Z_{trap} = Qm^{3/2}/[g^{1/2}b^{5/2}]$ (Henderson 1966). (*Source: OPEN CHANNEL FLOW by Henderson, © 1966. Reprinted by permission of Prentice-Hall, Inc., Upper Saddle River, NJ.*)

Equation 2.21 can be made dimensionless with these variables to produce

$$Z_{trap} = \frac{[y'(1 + y')]^{3/2}}{(1 + 2y')^{1/2}} \tag{2.23}$$

This relation, plotted in Figure 2.13, can be used to find critical depth directly for a trapezoidal channel. A similar relation has been developed for the circular case, also plotted in Figure 2.13 (Henderson 1966). For the circular section the dimensionless variables are redefined as

$$Z_{circ} = \frac{Q}{(g/\alpha)^{1/2}d^{5/2}}; \qquad y' = \frac{y_c}{d} \tag{2.24}$$

in which d = conduit diameter. The value of α has been shown as unity in the definition of Z in Figure 2.13, which is a reasonable assumption for a prismatic channel.

The minimum specific energy can be determined and plotted for the trapezoidal and circular sections as well (Henderson 1966). For the trapezoidal section with the dimensionless variables as defined in Equation 2.22 and with $E' = mE_c/b$, Equation 2.20 in dimensionless form is given by

$$E' = y' + \frac{y'(1 + y')}{2(1 + 2y')} \tag{2.25}$$

Now, because both E' and Z are unique functions of y', E' can be given as a function of Z, as in Figure 2.14. A similar relation can be developed for the circular section, also shown in Figure 2.14, with $E' = E_c/d$.

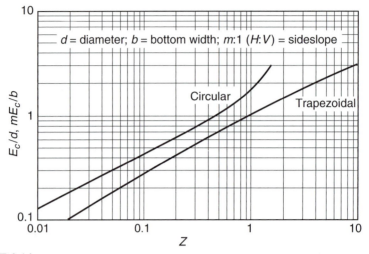

FIGURE 2.14
Minimum specific energy for trapezoidal and circular channels $Z_{circ} = Q/[g^{1/2}d^{5/2}]$; $Z_{trap} = Qm^{3/2}/[g^{1/2}b^{5/2}]$ (Henderson 1966). (*Source: OPEN CHANNEL FLOW by Henderson, © 1966. Reprinted by permission of Prentice-Hall, Inc., Upper Saddle River, NJ.*)

Example 2.2

Find the critical depth in a trapezoidal channel with a 20 ft (6.1 m) bottom width and 2:1 side slopes if $Q = 1000$ cfs (28.3 m³/s). Use the bisection technique and compare the solution with that from Figure 2.13.

Solution. The bisection procedure developed in Appendix A can be used to find critical depth if the function $F(y)$ is properly defined. The equation to be satisfied is Equation 2.18, so take $F(y)$ to be

$$F(y) = Q - \frac{g^{1/2}A^{3/2}}{B^{1/2}} \tag{2.26}$$

The Visual BASIC program Y0YC that solves (2.26) for critical depth is given on the book website. The data input consists of the discharge, and the channel bottom width and side slope. The main procedure establishes the initial interval for the root search (Y1 and Y2) and the specified relative error criterion ER. It then calls the BISECTION subprocedure, which in turn calls the function subprocedure F which evaluates (2.26) for each iteration. The iteration continues until the error criterion is satisfied. Note that the critical depth evaluation requires only the channel geometric parameters (b and m) and the discharge, Q, while the normal depth computation (to be discussed in Chapter 4) also requires the channel slope, S, and roughness coefficient, n. The result for this example is 3.740 ft (1.14 m), which can be checked with the graphical technique of Figure 2.13 by calculating Z_{trap}:

$$Z_{trap} = 1000 \times 2^{3/2}/(32.2^{1/2} \times 20^{5/2}) = 0.28$$

Then, from Figure 2.13, $my_c/b = 0.37$ and $y_c = 3.7$ ft (1.1 m).

Example 2.3

The trapezoidal channel of Example 2.2 ($b = 6.1$ m [20.0 ft], side slopes $= 2{:}1$, and $Q = 28.3$ m³/s [1000 cfs]) gradually contracts in width to a trapezoidal channel with the same side slopes but having a bottom width, $b = 4.0$ m (13.1 ft). There is no change in channel bottom elevation in the contraction. If the approach flow depth is 1.50 m (4.92 ft), find the flow depth in the contracted section. Neglect the head loss.

Solution. The approach flow depth of $y_1 = 1.50$ m is subcritical because it is greater than the critical depth in the approach channel section of $y_{c1} = 1.14$ m determined in Example 2.2. The approach flow specific energy is

$$E_1 = y_1 + \frac{Q^2}{2gA_1^2} = 1.5 + \frac{(28.3)^2}{19.62[1.50(6.10 + 2 \times 1.50)]^2} = 1.72 \text{ m (5.64 ft)}$$

Next find the critical depth and minimum specific energy in the contracted section in order to check for the possibility of choking. The critical depth is determined by setting the Froude number squared to unity, $Q^2 B_c/(gA_c^3) = 1.0$, which is rearranged as

$$\frac{A_c^{3/2}}{B_c^{1/2}} = \frac{Q}{\sqrt{g}} = \frac{28.3}{\sqrt{9.81}} = 9.035$$

$$\frac{[y_{c2}(b_2 + my_{c2})]^{3/2}}{(b_2 + 2my_{c2})^{1/2}} = \frac{[y_{c2}(4.0 + 2y_{c2})]^{3/2}}{(4.0 + 4y_{c2})^{1/2}} = 9.035$$

and solved by trial and error, or alternatively by Figure 2.13, to give $y_{c2} = 1.36$ m (4.46 ft). Then the minimum specific energy at the contracted section is calculated from

$$E_{c2} = y_{c2} + \frac{A_c}{2B_c} = 1.36 + \frac{1.36(4.0 + 2 \times 1.36)}{2(4.0 + 4 \times 1.36)} = 1.84 \text{ m (6.04 ft)}$$

Now to check for a choking condition, it can be observed that $E_1 < E_{c2}$, which means that indeed the contraction will choke with $y_2 = y_{c2} = 1.36$ m (4.46 ft). The new upstream depth, y_1', can be determined from $E_1' = E_{c2}$, which upon substitution becomes

$$y_1' + \frac{Q^2}{2gA_1'^2} = y_1' + \frac{(28.3)^2}{19.62[y_1'(6.10 + 2y_1')]^2} = 1.84$$

Solving, the result for the new upstream depth is $y_1' = 1.68$ m (5.51 ft), which is an increase in upstream water surface elevation of 0.18 m (0.59 ft).

2.7   OVERBANK FLOW

In some situations, the foregoing elementary relationships for the occurrence of critical flow no longer apply in the form given. One example of interest is river overbank flow, a shallow flow over wide floodplains combined with a main channel flow that is out of bank. In this case, it no longer is permissible to neglect α, because of large nonuniformities between the velocities in the overbank and main channel. When Equation 2.16 is differentiated with respect to depth y to obtain an expression for critical depth, the variation of α with y must be considered:

$$\frac{dE}{dy} = 1 - \frac{\alpha Q^2 B}{gA^3} + \frac{Q^2}{2gA^2}\frac{d\alpha}{dy} \tag{2.27}$$

Now if dE/dy is set to zero, a compound channel Froude number can be defined (Blalock and Sturm, 1981):

$$\mathbf{F}_c = \left(\frac{\alpha Q^2 B}{gA^3} - \frac{Q^2}{2gA^2}\frac{d\alpha}{dy}\right)^{1/2} \tag{2.28}$$

The first term on the right hand side of Equation 2.28 leads to the conventional definition of the Froude number, while the second term represents the contribution of a nonconstant value of the kinetic energy correction coefficient, α. The cross-section is divided into a main channel and floodplain subsections for the computation of α, which depends on the assumption that the energy grade line is horizontal across the cross section so that the energy grade line slope is the same in each subsection. It further is assumed that the slope of the energy grade line, S_e, can be formulated as $S_e = Q^2/K^2$, in which K is the total channel conveyance as defined by a uniform flow formula such as Manning's equation, which is discussed in more detail in Chapter 4. The conveyance depends only on the geometric and roughness properties of the cross section. Under these assumptions, we have $Q^2/K^2 = Q_i^2/k_i^2$, or

$$\frac{Q_i}{Q} = \frac{k_i}{K} \tag{2.29}$$

in which Q_i = subsection flow rate; k_i = subsection conveyance; Q = total discharge; and K = total conveyance = Σk_i. The definition of α can be expressed:

$$\alpha = \frac{\sum_i (Q_i/a_i)^3\, a_i}{(Q/A)^3\, A} \tag{2.30}$$

in which a_i = subsection area and A = total cross-sectional area. In Equation 2.30, it has been assumed that the primary contribution to α is the difference in velocity between subsections. Substitution of Equation 2.29 into Equation 2.30 then leads to the definition

$$\alpha = \frac{\sum_i (k_i^3/a_i^2)}{K^3/A^2} \tag{2.31}$$

in which k_i = the conveyance of the ith subsection; a_i = the area of the ith subsection; and $K = \Sigma k_i$ = the conveyance of the total cross section. The conveyance of the ith subsection is calculated from a uniform flow formula such as Manning's equation.

Differentiating the kinetic energy correction coefficient as defined by Equation 2.31 and substituting into Equation 2.28 leads to a working definition of the compound channel Froude number:

$$\mathbf{F}_c = \left[\frac{Q^2}{2gK^3} \left(\frac{\sigma_2 \sigma_3}{K} - \sigma_1 \right) \right]^{1/2} \tag{2.32}$$

in which

$$\sigma_1 = \sum_i \left[\left(\frac{k_i}{a_i} \right)^3 \left(3t_i - 2r_i \frac{dp_i}{dy} - 3 \frac{a_i}{n_i} \frac{dn_i}{dy} \right) \right] \tag{2.33a}$$

$$\sigma_2 = \sum_i \left(\frac{k_i^3}{a_i^2} \right) \tag{2.33b}$$

$$\sigma_3 = \sum_i \left[\left(\frac{k_i}{a_i} \right) \left(5t_i - 2r_i \frac{dp_i}{dy} - 3 \frac{a_i}{n_i} \frac{dn_i}{dy} \right) \right] \tag{2.33c}$$

in which a_i, p_i, r_i, t_i, n_i, and k_i represent the flow area, wetted perimeter, hydraulic radius, top width, roughness coefficient, and conveyance of the ith subsection, respectively, and K = total conveyance. All the terms on the right hand side are evaluated in the course of water surface profile computations except dp_i/dy, which can be evaluated as shown in Figure 2.15 because the cross section is composed of a series of ground points connected by straight lines. At any given water surface elevation, only those portions of the boundary that intersect the free surface are considered to contribute to dp_i/dy. At the point of minimum specific energy, $\mathbf{F}_c$ can be expected to have a value of unity so that Equations 2.32 and 2.33 can be used to solve for critical depth in a compound channel.

For a specific range of discharge in some compound channel cross sections, multiple values of critical depth can exist with one minimum in the specific energy occurring in the overbank flow case and the other occurring in the case of main channel flow alone. Blalock and Sturm (1981) demonstrated the validity of the

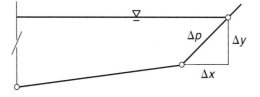

FIGURE 2.15
Evaluation of dp/dy at the water surface intersection with the channel bank (Blalock and Sturm 1981). (*Source: M. E. Blalock and T. W. Sturm, "Minimum Specific Energy in Compound Open Channel," J. Hyd. Div., © 1981, ASCE. Reproduced by permission of ASCE.*)

compound channel Froude number in correctly predicting multiple points of minimum specific energy by investigating the hypothetical cross section A, as shown in Figure 2.16, for a fixed discharge of 5000 cfs (142 m³/s). For this discharge the cross section has two points of minimum specific energy (C1 and C2), as can be seen in Figure 2.17. The compound channel Froude number is equal to unity at the critical depths, corresponding to points of minimum specific energy, as shown in Figure 2.18. In addition, Figure 2.18 shows that more conventional definitions of Froude number give incorrect values of the critical depth. The Froude number, $\mathbf{F}_\alpha$, is defined by Equation 2.19, as is $\mathbf{F}$ with $\alpha = 1.0$.

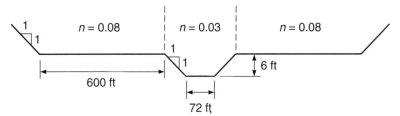

FIGURE 2.16

Hypothetical compound channel cross-section A.

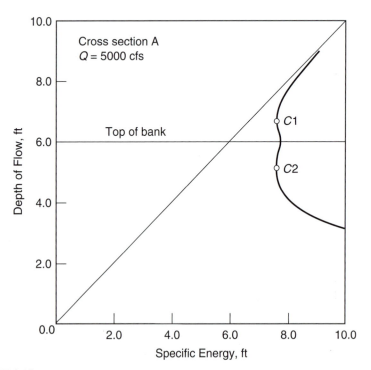

FIGURE 2.17

Specific energy diagram for cross-section A (Blalock and Sturm 1981).
(*Source: M. E. Blalock and T. W. Sturm, "Minimum Specific Energy in Compound Open Channel," J. Hyd. Div.,* © *1981, ASCE. Reproduced by permission of ASCE.*)

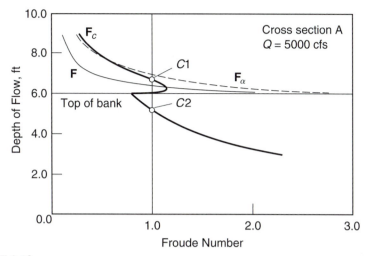

FIGURE 2.18
Froude numbers for cross-section A (Blalock and Sturm 1981). (*Source: M. E. Blalock and T. W. Sturm, "Minimum Specific Energy in Compound Open Channel," J. Hyd. Div., © 1981, ASCE. Reproduced by permission of ASCE.*)

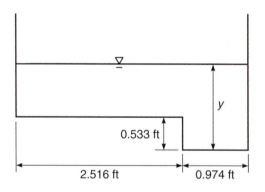

FIGURE 2.19
Experimental compound channel cross-section (Blalock and Sturm 1981). (*Source: M. E. Blalock and T. W. Sturm, "Minimum Specific Energy in Compound Open Channel," J. Hyd. Div., © 1981, ASCE. Reproduced by permission of ASCE.*)

The concept of two points of minimum specific energy, as illustrated by cross-section A in Figure 2.17, was investigated experimentally by Blalock and Sturm (1981) in a tilting flume with the cross section shown in Figure 2.19. Uniform flow was established in the flume for various slopes at an average constant discharge of 1.69 cfs (ft^3/s). Detailed velocity distributions were measured to compute α and the specific energy at each measured depth of flow. The experimental results are shown in Table 2.2, in which two points of minimum specific energy (Runs 2 and 8) are predicted by a value of unity for the compound channel Froude number within the experimental uncertainty. The two values of critical depth also correspond to minimum values of the momentum function (Blalock and Sturm 1983), as explained in Chapter 3.

TABLE 2.2   Experimental values of compound channel Froude number for various depths of flow in the cross-section of Figure 2.19 with an average discharge of 1.692 cfs (0.0479 m³/s)

Run	y, ft	α	E, ft	F_c
1	0.650	1.192	0.718	0.70
4	0.625	1.198	0.702	0.82
2	**0.600**	**1.224**	**0.700**	**0.97**
3	0.567	1.238	0.701	1.25
10	0.533	1.093	0.704	0.82
7	0.500	1.087	0.700	0.90
8	**0.467**	**1.096**	**0.690**	**1.00**
9	0.433	1.100	0.701	1.13

Source: Data from Blalock and Sturm 1981.

The compound channel Froude number also can be derived by setting $V = c$, where c is the wave celerity in a compound section, in the equations of the characteristics of the general unsteady form of either the energy or momentum equation (Blalock and Sturm 1983; Chaudhry and Bhallamudi 1988). Once an expression for the wave celerity c is developed from the characteristics of the unsteady energy or momentum equation (see Chapter 7), the compound channel Froude number can be defined as V/c with a result identical to that of minimizing the specific energy or momentum functions. Könemann (1982) also suggests an expression for the compound channel Froude number by minimizing the expression for specific energy, except that the terms involving the rate of change of wetted perimeter with respect to depth of flow, dp_i/dy, are neglected. Interpretation of the flow regime of the separate floodplain and main channel subsections has been proposed by Schoellhamer, Peters, and Larock (1985) using a subdivision Froude number; however, the compound channel Froude number given herein applies to the entire cross section for the purpose of water surface profile computation, as discussed in Chapter 5.

For a particular compound channel geometry and roughness, it is possible to establish a range of values of the discharge (if any) over which multiple critical depths can be expected (Sturm and Sadiq 1996). The key to such a determination is to recognize that curves of depth versus compound channel Froude number can be made dimensionless and independent of discharge Q. The bank-full Froude number for the main channel is defined by

$$F_1 = \frac{QB_1^{1/2}}{g^{1/2}A_1^{3/2}} \tag{2.34}$$

in which the subscript 1 refers to bank-full values of the geometric parameters. Dividing either (2.28) or (2.32) by F_1 effectively removes the influence of discharge, so that the curve for F_c/F_1 can be plotted as a function of y/y_1 alone, as shown in Figure 2.20 for cross-section A. To find critical depth y_c, F_c is set to a value of unity, so that it is obvious from Figure 2.20 that there is a range of values of $1/F_1$ and, therefore, a range

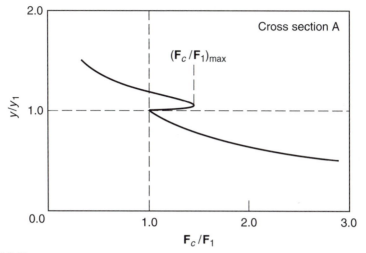

FIGURE 2.20
Dimensionless compound channel Froude number for cross-section A.

of discharges, over which two values of critical depth exist, one in overbank flow and the other in main channel flow alone. (The intermediate depth is a local maximum in specific energy rather than a point of minimum specific energy.) Because $1/\mathbf{F}_1$ decreases with increasing discharge, we can see from Figure 2.20 that an upper limit is placed on the discharge Q_U beyond which only one critical depth exists for the case of overbank flow. The limit Q_U occurs when $\mathbf{F}_c = \mathbf{F}_1$ and for $\mathbf{F}_c = 1$; hence, Q_U can be calculated from the condition $\mathbf{F}_1 = 1$ as

$$Q_U = \frac{\sqrt{g}\, A_1^{3/2}}{\sqrt{B_1}} \tag{2.35}$$

The lower limiting discharge Q_L for the discharge range of multiple critical depths occurs when $\mathbf{F}_c/\mathbf{F}_1$ takes on a maximum value as shown in Figure 2.20. In this case, $\mathbf{F}_1$ for $Q = Q_L$ can be expressed as Q_L/Q_U from (2.35) and combined with the condition $\mathbf{F}_c = 1$. We have

$$\left(\frac{\mathbf{F}_c}{\mathbf{F}_1}\right)_{max} = \frac{Q_U}{Q_L} \tag{2.36}$$

The value of $(\mathbf{F}_c/\mathbf{F}_1)_{max}$ can be generated from a series of values of depth y for any discharge, although it is convenient to use a discharge corresponding to $Q = Q_U$. Equations 2.35 and 2.36 provide the means for isolating the root search for critical depth when multiple critical depths exist. A nonlinear algebraic equation solver, such as the interval-halving technique, can be applied to solve $\mathbf{F}_c = 1$ when the bounds of the root search are properly defined. Alternatively, Chaudhry and Bhallamudi (1988) propose an iterative numerical procedure to solve the equation given by $\mathbf{F}_c = 1$, in which $\mathbf{F}_c$ is defined from the momentum equation, and provide a detailed procedure for a symmetrical, rectangular compound channel.

The computation of critical depth with the compound channel Froude number defined by Equation 2.32 requires the determination of the geometric properties of the natural cross section. An algorithm to accomplish this task is shown in the Visual BASIC procedure Ycomp given on the book website. The algorithm requires an input data file of distance-elevation pairs between which a straight line variation is assumed. In addition, the distances at which subsection boundaries are located and the values of Manning's n in each subsection must be specified. The various quantities necessary for the evaluation of the compound channel Froude number by Equations 2.32 and 2.33 are computed. The procedure can be used to evaluate the critical depth in a compound channel or a simple natural channel cross section, as illustrated by the following example.

Example 2.4

For cross-section A, previously defined in Figure 2.16, find the discharge range of multiple critical depths, if any, and determine the critical depth for discharges of 4000, 5000, and 6500 cfs (113, 142, and 184 m³/s).

Solution.   First, the values of the cross-sectional area and top width for bank-full flow are determined to be $A_1 = 468$ ft² (43.5 m²) and $B_1 = 84.0$ ft (25.6 m). Then, the upper limiting discharge, Q_U, is calculated as

$$Q_U = \frac{\sqrt{32.2}\,(468)^{3/2}}{\sqrt{84}} = 6268 \text{ cfs } (177.6 \text{ m}^3/\text{s})$$

The value of $(F_c/F_1)_{max} = 1.446$ is calculated from a series of increasing values of y/y_1, as shown previously in Figure 2.20. The lower limiting discharge is given by

$$Q_L = \frac{6268}{1.446} = 4335 \text{ cfs } (122.8 \text{ m}^3/\text{s})$$

Therefore, two values of critical depth should be expected in the range from 4335 to 6268 cfs (122.8 to 177.6 m³/s) for cross-section A.

The equation to be solved for critical depth is given by setting the compound channel Froude number, F_c, in Equation 2.32 equal to unity and defining a new function given by $F(y) = F_c - 1 = 0$. The only difficulty is in computing the geometric properties required for the evaluation of F_c. This can be accomplished by assuming straight lines between surveyed ground points and computing the geometric properties as a summation of those for regular geometric figures from one ground point to the next. This has been done in the function subprocedure FC shown in the program Ycomp on the book website. Otherwise, the evaluation of critical depth proceeds as in Example 2.2 using the bisection subprocedure. Note that the unknown variable sought in the bisection subprocedure is the critical water surface elevation rather than the critical depth. The code module requires a data file of the cross-section ground points and the subsection breakpoints and roughness coefficients. The program output for $Q = 5000$ cfs (142 m³/s) gives critical depths of 5.182 ft (1.579 m) and 6.740 ft (2.054 m). For $Q = 4000$ cfs (113 m³/s), there is only the main channel value of critical depth equal to 4.480 ft (1.365 m), while only the overbank critical depth of 7.194 ft (2.193 m) exists for $Q = 6500$ cfs (184 m³/s).

   The procedure logic for calculating multiple critical depths in the program
Ycomp on the book website is illustrated by the flowchart in Figure 2.21. The value
of the upper limiting discharge, *QU,* relative to the given *Q* determines the existence
of a lower or upper critical depth, which is then calculated. The upper critical depth
is designated as YC2, while the lower critical depth is YC1. Their values are set
equal to (-1) to indicate that they do not exist. The value of $\mathbf{F^+}$ is the compound
channel Froude number evaluated at a depth of 1.02 times the bankfull depth to
determine if the Froude number is increasing, as in the case of multiple critical

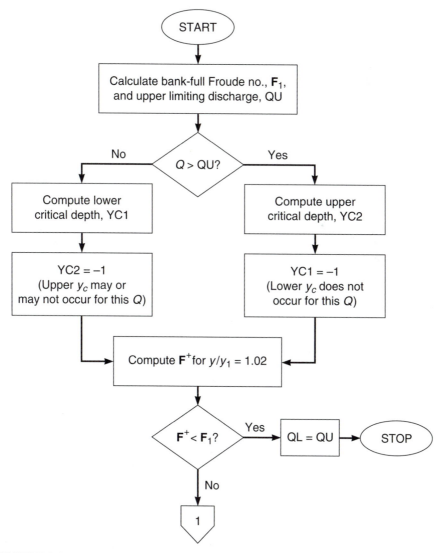

FIGURE 2.21
Flowchart for finding multiple critical depths.

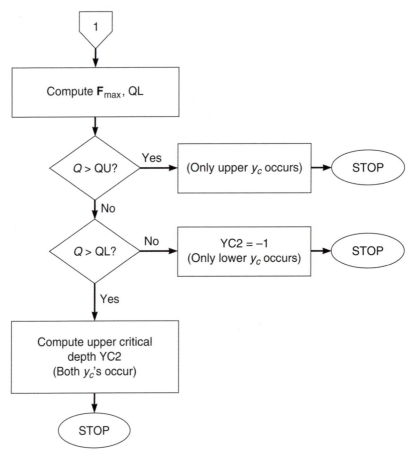

FIGURE 2.21 (*continued*)

depths, or not. If it is not increasing above bankfull depth, then only one critical depth exists and $QL = QU$. If it is increasing, then the maximum value of the compound channel Froude number $\mathbf{F}_{max}$ is needed to calculate the lower limiting discharge, QL, for the possible case of multiple critical depths for the given Q. Once both QU and QL are known, then decisions are made about the existence of only a lower critical depth, of only an upper critical depth, or of both.

2.8   WEIRS

The occurrence of critical depth is put to good use in the design of open channel flow measuring devices. By creating an obstruction, critical depth is forced to occur and a unique relationship between depth and discharge results. This is the principle upon which weir design is based. The very extensive set of experimental results developed for weirs accounts for their continuing popularity as flow measuring devices.

Sharp-Crested Rectangular Notch Weir

The sharp-crested weir equation can be derived with respect to Figure 2.22 by first assuming (1) no head losses, (2) atmospheric pressure across section *AB,* and (3) no vertical contraction of the nappe. Under these idealized assumptions, the velocity along any streamline at section *AB* is given by $v = (2gh)^{1/2}$, where h = vertical distance below the energy grade line. This velocity distribution can be integrated over the cross-section *AB* to obtain a theoretical value of discharge per unit of width, q_t:

$$q_t = \int_{\frac{V_0^2}{2g}}^{\frac{V_0^2}{2g}+H} \sqrt{2gh}\, dh \tag{2.37}$$

in which V_0 = approach flow velocity; and H = approach flow head on the crest of the weir. Carrying out the integration, the result is

$$q_t = \frac{2}{3}\sqrt{2g}\left[\left(1 + \frac{V_0^2}{2gH}\right)^{3/2} - \left(\frac{V_0^2}{2gH}\right)^{3/2}\right]H^{3/2} \tag{2.38}$$

If the term in square brackets, which expresses the effect of the approach velocity head, is combined with contraction and head loss effects into a discharge coefficient, C_d, the actual total discharge is given by

$$Q = \frac{2}{3}\sqrt{2g}\,C_d L H^{3/2} \tag{2.39}$$

in which L is the length of the notch crest perpendicular to the flow; and H is the head measured above the crest. The discharge coefficient can be determined only by experiments. In the United States, it is customary to simplify Equation 2.39 as

$$Q = C_1 L H^{3/2} \tag{2.40}$$

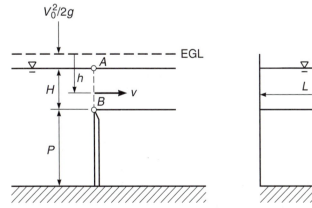
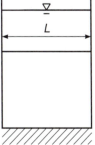

FIGURE 2.22
Idealized flow over a rectangular, sharp-crested weir.

in which Q is in cubic feet per second and H is in feet, but the dimensionless form of C_d in (2.39) is preferred.

With the geometric variables defined as in Figure 2.23, a dimensional analysis for the coefficient C_d yields

$$C_d = f\left(\frac{L}{b}, \frac{H}{L}, \frac{H}{P}, \textbf{Re}, \textbf{We}\right) \tag{2.41}$$

in which L is the crest length perpendicular to the flow; b is the approach chan-nel width; H is the head above the notch; P is the height of the notch crest above the channel bottom; **Re** is the Reynolds number; and **We** is the Weber number. One of the earliest experimental relations for C_d was given for the suppressed weir ($L/b = 1$) by Rehbock (Henderson 1966), in which he neglected viscous and surface tension effects so that C_d was given as a function of H/P alone:

$$C_d = 0.611 + 0.08\frac{H}{P} \tag{2.42}$$

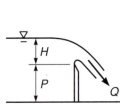
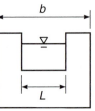

(a) Definition Sketch

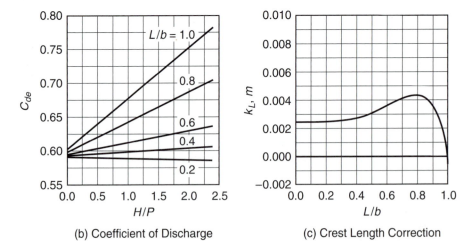

(b) Coefficient of Discharge   (c) Crest Length Correction

FIGURE 2.23
Head-discharge relationship for a sharp-crested weir (Kindsvater and Carter 1957). (*Source: C. E. Kindsvater and R. W. C. Carter, "Discharge Characteristics of Rectangular Thin-Plate Weirs," J. Hyd. Div., © 1957, ASCE. Reproduced by permission of ASCE.*)

In the suppressed weir, there are no lateral contraction effects on the weir nappe, so that the coefficient of discharge does not depend on L/b. Furthermore, Rehbock's formula reflects no influence of H/L on the discharge coefficient.

Based on experimental results obtained at Georgia Tech, Kindsvater and Carter (1957) proposed that the Reynolds number and Weber number effects can be included in the head-discharge relationship by making small corrections to the head, H, and crest length, L. By doing so, they derived from their experimental results an effective coefficient of discharge, C_{de}, that depended only on H/P and L/b, as shown in Figure 2.23. Their relationship is given in the form of Equation 2.39 as

$$Q = \frac{2}{3}\sqrt{2g}\,C_{de}L_eH_e^{3/2} \tag{2.43}$$

in which

$$L_e = L + k_L \tag{2.44a}$$

$$H_e = H + k_H \tag{2.44b}$$

where C_{de} and k_L are given in Figures 2.23b and 2.23c, respectively, and k_H was found to be nearly constant with a value of 0.001 m (0.003 ft). The crest-length correction, k_L, is maximum at $L/b = 0.8$ with a value of 0.0043 m (0.014 ft), as shown in Figure 2.23c. Equations for C_{de} based on the Kindsvater-Carter data are given as a function of the lateral contraction ratio, L/b, and the vertical contraction ratio, H/P, in Table 2.3. Kindsvater and Carter (1957) found that there was a negligible influence of H/L on the discharge coefficient.

Kindsvater and Carter (1957) constructed their sharp-crested weir notches, not with a knife edge but with an upstream square edge having a top width of 1.6 mm (1/16 in.) and a downstream bevel. The head for the sharp-crested weir should be measured at a distance of three to four times the maximum head measured upstream of the weir plate (Bos 1989).

The suppressed weir, or full-width weir, with $L/b = 1.0$ must have provisions for aeration of the underside of the nappe, because some air is entrained by the nappe, which affects the discharge coefficient due to subatmospheric pressure in the pocket underneath the nappe. Undesirable oscillations of the nappe also can result from irregular air supply rates to the pocket. To ensure full aeration, Bos (1989) suggests that the tailwater remain at least 0.05 m (0.16 ft) below the weir crest.

For precise measurements, Kindsvater and Carter (1957) recommended a limitation of $H/P < 2$, with P no less than 9 cm (0.3 ft). If H/P exceeds 5, then the weir itself no longer is the control section, and so large values of H/P should definitely be avoided.

Sharp-Crested Triangular Notch Weir

The triangular or V-notch sharp-crested weir defined in Figure 2.24a provides a precise measurement of discharge over a wide range of discharges. Utilizing the same approach as for the derivation of the head-discharge relationship for a rectangular

TABLE 2.3   Coefficients of discharge for the Kindsvater-Carter formula

L/b	C_{de}
1	0.602 + 0.075 H/P
0.9	0.599 + 0.064 H/P
0.8	0.597 + 0.045 H/P
0.7	0.595 + 0.030 H/P
0.6	0.593 + 0.018 H/P
0.5	0.592 + 0.011 H/P
0.4	0.591 + 0.0058 H/P
0.3	0.590 + 0.0020 H/P
0.2	0.589 − 0.0018 H/P
0.1	0.588 − 0.0021 H/P
0	0.587 − 0.0023 H/P

Source: Data from Kindsvater and Carter 1957; Bos 1989.

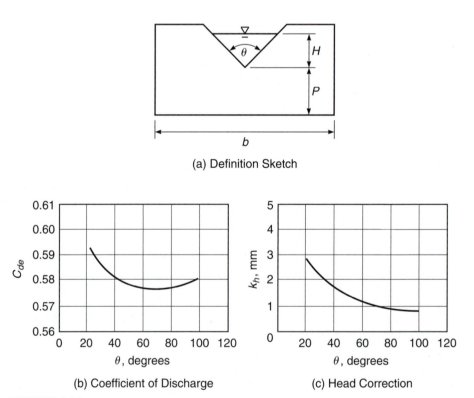

(a) Definition Sketch

(b) Coefficient of Discharge

(c) Head Correction

FIGURE 2.24

Triangular, sharp-crested weir (Bos 1989). (*Source: Bos, M. G. 1989 "Discharge Measurements Structures," ILRI Publication 20, 3rd Revised Edition, Wageningen, The Netherlands, 320 p.*)

sharp-crested weir, it can be shown that the head-discharge relationship for a V-notch weir is given by

$$Q = C_d \frac{8}{15} \sqrt{2g} \, \tan\frac{\theta}{2} H^{5/2} \tag{2.45}$$

in which θ is the notch angle. The weir can either be fully contracted or partially contracted for narrow approach channels. Equation 2.45 can be presented in the Kindsvater-Carter form in which the head, H, is replaced by an effective head, $H_e = H + k_h$, and the coefficient of discharge C_{de} becomes independent of Reynolds and Weber number effects (Bos 1989). The value of C_{de}, shown in Figure 2.24b, varies between approximately 0.58 and 0.59 as a function of θ only, provided that $H/P \leq 0.4$ and $P/b \leq 0.2$ to ensure the fully contracted case. The value of k_h varies from 1 to 3 mm (0.0033 to 0.01 ft) as a function of the notch angle, θ, as shown in Figure 2.24c. For a partially contracted V-notch weir, sufficient data for the discharge coefficient is available only for the case of $\theta = 90°$, and the discharge coefficient varies with H/P and P/b for this case, as given by Bos (1989).

Broad-Crested Weir

The broad-crested weir has a finite crest length parallel to the flow. In addition, the crest is long enough that parallel flow and critical depth occur at some point along the crest, as shown in Figure 2.25 for a rectangular, broad-crested weir. If the energy equation is applied from the approach flow to the critical section on the crest and energy losses are neglected, we have

$$H_e = y_c + \frac{Q^2}{2gA_c^2} = \frac{3}{2}\left[\frac{(Q/L)^2}{g}\right]^{1/3} \tag{2.46}$$

in which H_e is the energy head on the crest as shown in Figure 2.25a; that is, $H_e = H + V_0^2/2g$, in which V_0 is the approach velocity. If the energy losses are absorbed in a discharge coefficient, C_d, and we solve for Q in terms of the head, H, the result is

$$Q = C_v C_d \frac{2}{3}\left[\frac{2}{3}g\right]^{1/2} LH^{3/2} \tag{2.47}$$

in which the approach velocity coefficient $C_v = (H_e/H)^{3/2}$. Equation 2.47 can be solved for the discharge, assuming that $C_v = 1$, and then the approach velocity head can be calculated to update the value of C_v for a second calculation of Q. Alternatively, C_v can be related to the variable $C_d A^*/A_1$, in which C_d = weir discharge coefficient; $A^* = LH$ = flow area in the control section of the weir with a water surface height corresponding to the upstream head, H; and A_1 = flow cross-sectional area in the approach section where H is measured = $L(H + P)$ for a suppressed weir (Bos 1989). The resulting relationship between C_v and $C_d A^*/A_1$ is

$$\frac{C_d A^*}{A_1} = \frac{[C_v^{2/3} - 1]^{1/2}}{0.385 C_v} \tag{2.48}$$

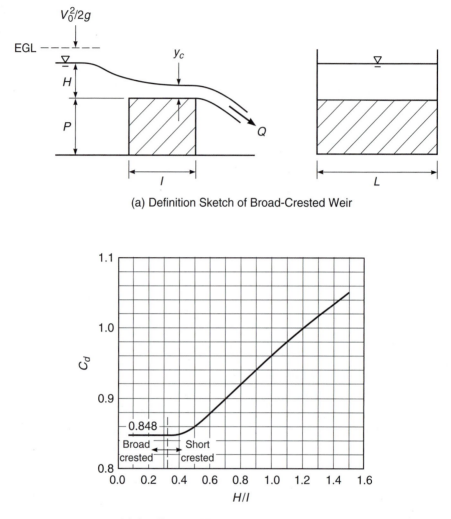

(a) Definition Sketch of Broad-Crested Weir

(b) Coefficient of Discharge for $H/(H + P) \leq 0.35$

FIGURE 2.25
Coefficient of discharge for rectangular broad-crested and short-crested weirs (Bos 1989).
(*Source: Bos, M. G. 1989 "Discharge Measurements Structures," ILRI Publication 20, 3rd Revised Edition, Wageningen, The Netherlands, 320 p.*)

Equation 2.48 is plotted in Figure 2.26 so that C_v can be estimated directly.

For the broad-crested weir, an additional geometric variable, l, the length of the crest parallel to the flow, is introduced into the dimensional analysis for C_d, and it can be shown that

$$C_d = f\left(\frac{H}{P}, \frac{H}{l}\right)$$

(2.49)

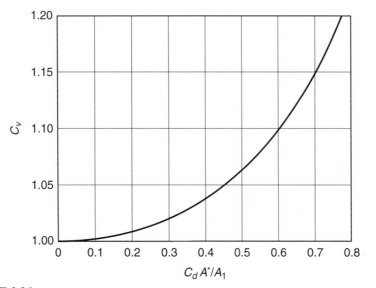

FIGURE 2.26
Approach velocity correction coefficient for a broad-crested weir (Bos 1989). (*Source: Bos, M. G. 1989 "Discharge Measurements Structures," ILRI Publication 20, 3rd Revised Edition, Wageningen, The Netherlands, 320 p.*)

as given in Figure 2.25b (Bos, 1989) for $H/(H + P) \geq 0.35$. In fact, whether the broad-crested weir behaves as expected depends on the value of H/l. The following ranges of behavior can be delineated:

1. $0.08 < H/l < 0.33$, broad crested.
2. $0.33 < H/l < 1.5$, short crested.
3. $H/l > 1.5$, sharp crested.

In the range of broad-crested behavior, the crest is long enough in the flow direction to obtain parallel flow at the critical section and a theoretical discharge coefficient, $C_d = 1.0$, but friction losses reduce the experimental value of the discharge coefficient, C_d, to 0.848 as long as the weir remains broad crested and $H/(H + P) \leq 0.35$. In the short-crested range of behavior, the flow is curvilinear along the entire crest of the weir and the coefficient of discharge actually increases, as shown in Figure 2.25b.

The advantage of a broad-crested weir is that the tailwater can be above the crest of the weir without affecting the head-discharge relationship as long as the control section is unaffected. The limit of tailwater height, H_t, above the crest of the weir so that the discharge does not decrease by more than 1 percent is called the *modular limit*. The modular limit usually is expressed in terms of the ratio H_t/H_e where H_e is the upstream head on the crest of the weir, and it has a value of $H_t/H_e < 0.66$ for a rectangular broad-crested weir (Bos 1989).

Long-Throated Flumes

Flow measuring flumes are critical depth devices that usually include a lateral contraction with or without a bed contraction, that is, a rise in the channel bottom. The long-throated flume is similar to a broad-crested weir in that the crest is long enough in the flow direction to create a critical section at which the streamlines are nearly straight and parallel. The advantage of flumes is that debris and sediment are more easily passed downstream than for weir obstructions. More specialized flumes such as the Parshall flume have been in use for some time; however, the Parshall flume design depends on an experimentally determined head-discharge relationship with a critical section created by an abrupt drop in the channel bottom in addition to a lateral contraction. Standard designs based solely on experimental data have been developed for a number of specified crest widths depending on the range of flow rates to be measured (Bos 1989). By contrast, the long-throated flume has the advantage of having an analytical head-discharge relationship, which can be combined with a relatively universal coefficient of discharge obtained from experimental results for a wide variety of crest shapes (Clemmens et al. 2001).

The long-throated flume is illustrated in Figure 2.27 for a rectangular throat at the sill. Several standard dimensions are shown in the figure. The entrance transition

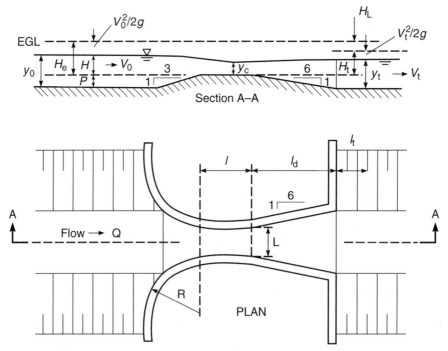

FIGURE 2.27
Long-throated flume definition sketch (Bos 1989).

is recommended to be either plane surfaces with a 3:1 rate of contraction or cylindrical surfaces with a radius of $2H_{e \ max}$ as shown in Figure 2.27. The downstream expansion is recommended to be plane surfaces having a 6:1 rate of expansion. The gauging station for measuring the upstream head H should be located at a distance of 2 to 3 $H_{e \ max}$ upstream of the flume. The maximum energy head is based on the maximum discharge expected to be measured in the channel or canal. The length of the sill in the flow direction should be greater than 1.5 $H_{e \ max}$ to allow the use of the parallel flow assumption over the sill.

The head-discharge relationship for the long-throated flume is derived in a manner analogous to the development of Equation 2.47 for a rectangular, broad-crested weir. By rearranging the energy equation written between the approach flow section and the throat of the flume, and including a discharge coefficient C_d to account for energy losses, an expression for the discharge can be written as (see Figure 2.27)

$$Q = C_d A_c \sqrt{2g(H_e - y_c)} \tag{2.50}$$

in which H_e is the total approach energy head including the approach flow velocity head as for the broad-crested weir. Now all that is needed to generate the head-discharge relationship for the flume is an analytical equation for the value of the critical depth y_c in terms of H_e for a particular throat geometry and an experimental expression for the coefficient of discharge. In the case of a rectangular throat, for example, the required relationship is $y_c = (2/3) H_e$ as given by Equation 2.7, where H_e is equal to the minimum specific energy E_c, which occurs on the sill in the throat of the flume. Upon substitution of this relationship into Equation 2.50, we recover Equation 2.47 for a rectangular broad-crested weir in terms of H_e. Similarly, the head-discharge relationship for a flume with a throat having a triangular cross section can be derived from Equation 2.50 and the relationship $y_c = 0.8 H_e$. In the case of trapezoidal and circular cross sections at the throat of the flume, the value of y_c/H_e is not a constant but dependent on the geometry as shown in Figure 2.28. The scaling of the curves in Figure 2.28 is similar to that found in Figure 2.14, which allows all trapezoidal shapes to be represented by a single curve. The ratio of y_c to H_e for a trapezoidal flume throat varies from 0.67 to 0.8, while for the circular case, it ranges from 0.75 down to 0.6 as the circular section begins to approach full conditions and submerges at $H_e/d \approx 1.5$. Figure 2.14 can also be used to estimate the flume discharge for trapezoidal and circular throats by multiplying the value of Q obtained from that figure by the coefficient of discharge.

The coefficient of discharge for long-throated flumes having a variety of common shapes and sizes is given by (Bos 1989)

$$C_d = \left(\frac{H_e}{l} - 0.07 \right)^{0.018} \tag{2.51}$$

which is valid over the interval of $0.1 \le H_e/l \le 0.7$ with C_d varying from 0.94 to 0.99. As a result of the interval of application extending into the short-crested range, and the attempt to develop a universal curve, the uncertainty in C_d is approximately $\pm 4\%$ at the 95% confidence limits.

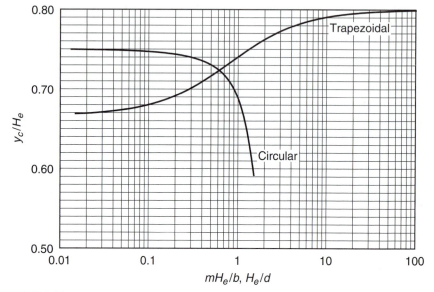

FIGURE 2.28
Ratio of critical depth to energy head for trapezoidal and circular channels.

There are three main design criteria for the long-throated flume as outlined by Clemmens et al. (2001): (1) the modular limit should not be exceeded to ensure that the critical section at the sill remains unsubmerged; (2) a minimum freeboard of 0.2 H_{max} between the upstream water surface and the banks of the canal should be satisfied; and (3) the approach flow Froude number should be less than 0.5 to avoid surface waves and instability. The primary design trade-off is between the submergence and freeboard criteria. The contraction should be large enough to avoid submergence of the critical section on the sill but not so large as to cause the upstream freeboard criterion to be violated.

Evaluating the modular limit so as to avoid submergence of the critical section is based on estimating head loss in the flume. The head loss can be written as (Clemmens et al. 2001):

$$H_L = \Delta H_a + \Delta H_f + \Delta H_{ex} \qquad (2.52)$$

in which ΔH_a = the friction loss from the approach channel to the sill; ΔH_f = the friction loss in the diverging channel from the sill to the tailwater; and ΔH_{ex} = the expansion loss at the exit from the sill. The first component of head loss can be obtained from the discharge coefficient by noting that $E_c/H_e = (C_d)^{1/u}$, where u is the exponent on the head for a particular shape of the control section at the throat (1.5 for rectangular), so that

$$\frac{\Delta H_a}{H_e} = 1 - (C_d)^{1/u} \qquad (2.53)$$

The friction loss in the transition to the exit section can be estimated from Manning's equation or the Darcy-Weisbach equation (see Chapter 4). It is generally considerably less than the expansion loss. Clemmens et al. (2001) recommend an equation for friction loss based on flow over flat plates, which can be modified to:

$$\Delta H_f = f\frac{l_d}{2g} \times \frac{1}{2}\left(\frac{V_c^2}{4R_c} + \frac{V_t^2}{4R_t}\right) + f\left(\frac{l_t}{4R_t}\right)\frac{V_t^2}{2g} \tag{2.54}$$

in which l_d = length of the diverging transition; l_t = length of the tailwater section required for full velocity head recovery; R_c and R_t = hydraulic radii (area divided by wetted perimeter) of the critical section and the tailwater section, respectively; and f = Darcy-Weisbach friction factor (see Chapter 4). The flow lengths as shown in Figure 2.27 are estimated from

$$l_d = 6P \tag{2.55}$$

$$l_t = 10\left(P + \frac{l}{2}\right) - l_d \tag{2.56}$$

in which the expansion ratio has been assumed to be 6 and P is the downstream sill height. Finally, the expansion loss is estimated from an equation of the form of the first term of Equation 2.14 for an abrupt expansion and is given by

$$\Delta H_{ex} = K_{ex}\frac{(V_c - V_t)^2}{2g} \tag{2.57}$$

in which K_{ex} = 0.66 for an expansion ratio of 6:1 (Bos 1989).

If the critical section at the throat of the flume is to be the control for determining the discharge, then the tailwater cannot be allowed to submerge the critical section. Flow in a flume for which the tailwater does not cause submergence is called modular, and the ratio of the maximum allowable tailwater energy head to the upstream energy head is called the modular limit. At the modular limit, the tailwater energy head is exactly equal to the upstream energy head, which is determined by the critical flow section, minus the total head loss. In order to prevent submergence and the modular limit from being exceeded, we must require that

$$H_{t\,actual} \leq H_e - [1 - (C_d)^{1/u}]H_e - \Delta H_f - \Delta H_{ex} \tag{2.58}$$

Because the tailwater is needed to calculate the head losses, the exact value of the modular limit, ML = $H_{t\,max}/H_e$, is obtained by iterating on the maximum tailwater head until Equation 2.58 is an equality that then establishes the maximum allowable height of tailwater.

Example 2.5

A long-throated flume with a rectangular sill width L of 0.30 m (0.98 ft) and a sill height P of 0.15 m (0.49 ft) is placed in a trapezoidal channel having a bottom width of 0.75 m (2.46 ft) and side slopes of 1:1. The tailwater depth is 0.31 m (1.0 ft) just downsteam of the flume at a maximum expected discharge of 0.05 m³/s (1.76 cfs). Determine the upstream head and the sill length in the direction of flow, and then check the modular limit.

Solution. Assume a value of $H_e/l = 0.40$, which determines the sill length in the flow direction once the approach flow energy head is calculated. This value is well above the allowable minimum of 0.1 and depends in part on the magnitude of the minimum discharge and the corresponding minimum head that can be measured with an acceptable degree of uncertainty. From Equation 2.51, the value of the coefficient of discharge C_d is 0.98. Then from Equation 2.50 for a rectangular sill, we have

$$H_e = \left[\frac{Q}{C_d \frac{2}{3}\sqrt{\frac{2}{3}gL}} \right]^{2/3} = \left[\frac{0.05}{(0.98) \times \frac{2}{3}\sqrt{\frac{2}{3}g} \times (0.3)} \right]^{2/3} = 0.215 \text{ m}$$

and the length of the sill is $l = 0.215/0.40 = 0.54$ m (1.77 ft). The value of the approach flow head H follows from calculation of the approach flow depth, y_0. By definition, $E_0 = H_e + P = 0.215 + 0.15 = 0.365$ m so that

$$y_0 + \frac{Q^2}{2gA_0^2} = y_0 + \frac{0.05^2}{19.62 \times [y_0(0.75 + y_0)]^2} = 0.365$$

which can be solved to give $y_0 = 0.364$ m (1.19 ft). Then it follows that $A_0 = 0.406$ m²; $V_0 = Q/A_0 = 0.123$ m/s; and $H = y_0 - P = 0.214$ m (0.702 ft). For this example, the approach velocity head is practically negligible, and the approach Froude number is

$$F_0 = \frac{V_0}{\sqrt{gD_0}} = \frac{0.123}{\sqrt{9.81 \times 0.406/(0.75 + 2 \times 0.364)}} = 0.08$$

which is well below the limit of 0.5. The required freeboard is $0.2 H = 0.043$ m (0.14 ft).

Checking the modular limit requires calculation of the head losses through the flume. The flow lengths of the diverging section and tailwater section are

$$l_d = 6P = 6 \times 0.15 = 0.90 \text{ m}$$

$$l_t = 10\left(P + \frac{l}{2}\right) - l_d = 10\left(0.15 + \frac{0.54}{2}\right) - 0.90 = 3.3 \text{ m}$$

The velocity in the critical section on the sill is $V_c = Q/A_c = 0.05/(0.3 \, y_c) = 1.17$ m/s in which $y_c = (2/3) H_e = 0.143$ m. For the downstream tailwater section, assume that the maximum tailwater is equal to the actual tailwater for the first iteration. Then the velocity for a depth of 0.31 m is $V_t = Q/A_t = 0.05/[0.31(0.75 + 0.31)] = 0.15$ m/s. The hydraulic radius in the critical section is

$$R_c = \frac{A_c}{P_c} = \frac{0.3(0.143)}{[0.3 + 2(0.143)]} = 0.073 \text{ m}$$

while the hydraulic radius of the tailwater section is given by

$$R_t = \frac{A_t}{P_t} = \frac{0.31(0.75 + 0.31)}{0.75 + 2\sqrt{2}(0.31)} = 0.202 \text{ m}$$

Now for the friction loss, we have by substituting into Equation 2.54 with $f \cong 0.016$ for a smooth surface over this Reynolds number range:

$$\Delta H_f = 0.016 \times \frac{0.90}{19.62} \times \frac{1}{2}\left(\frac{1.17^2}{4 \times 0.073} + \frac{0.15^2}{4 \times 0.202}\right)$$

$$+ 0.016 \times \frac{3.3}{4 \times 0.202} \times \frac{0.15^2}{19.62} = 0.0018 \text{ m}$$

The expansion loss comes from substituting into Equation 2.57 to produce

$$\Delta H_{ex} = 0.66 \times \frac{(1.17 - 0.15)^2}{19.62} = 0.035 \text{ m}$$

Finally, we check the possibility of submergence from Equation 2.58 to give

$$H_e - \Delta H_a - \Delta H_f - \Delta H_{ex} = 0.215 - (1 - 0.98^{1/1.5})(0.215)$$

$$- 0.0018 - 0.035 = 0.175$$

The energy loss through the structure gives a tailwater energy head of 0.175 m, which is acceptable because it is greater than the actual tailwater head of 0.16 m relative to the sill (not including the velocity head, which is very small for this example). However, to find the modular limit, additional iterations on the maximum tailwater head are needed. The result of iterating on y_t until Equation 2.58 is an equality is that $y_{t\,max} = 0.325$ m (1.07 ft) and $H_{t\,max}$ is practically unchanged at 0.175 m (0.574 ft) so that the modular limit, ML = $H_{t\,max}/H_e$ = 0.175/0.215 = 0.814. A downstream tailwater depth exceeding 0.325 m for the maximum discharge would cause submergence of the critical section.

2.9 ENERGY EQUATION IN A STRATIFIED FLOW

Let us suppose now that the flows over obstacles that we have been considering in this chapter occur at the bottom of a deep reservoir of depth D as a result of a plunging gravity current of higher density, as shown in Figure 2.29. The ambient density in the reservoir is ρ_a and that of the gravity current is ρ_b. Such flows occur naturally as the result of a density stratifying agent such as temperature or salt. If a small obstacle of height Δz is on the bottom of the reservoir, then we can write the energy equation for the lower flow layer as before, taking into account the

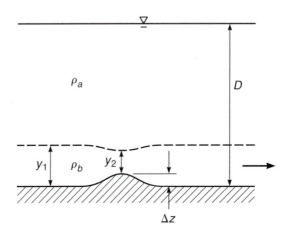

FIGURE 2.29
Two-layer density stratified flow over a step in the channel bottom, $D >> y_1$ and $\rho_b > \rho_a$.

additional pressure of the overlying stagnant layer. The energy equation from the approach flow to a point over the obstacle is

$$\rho_a g(D - y_1) + \rho_b g y_1 + \rho_b \frac{V_1^2}{2} = \rho_a g(D - y_2 - \Delta z)$$

$$+ \rho_b g(y_2 + \Delta z) + \rho_b \frac{V_2^2}{2} \quad (2.59)$$

Collecting terms and dividing by ρ_b results in

$$\frac{\Delta \rho}{\rho} g y_1 + \frac{V_1^2}{2} = \frac{\Delta \rho}{\rho} g(y_2 + \Delta z) + \frac{V_2^2}{2} \quad (2.60)$$

in which $\Delta \rho / \rho = (\rho_b - \rho_a)/\rho_b$. This equation is identical to the previous results for single-layer flow if the gravitational acceleration is replaced by the reduced gravitational acceleration $(\Delta \rho / \rho)g = g'$. The specific energy then is written as $E' = y + V^2/2g'$ and the Froude number from taking $dE'/dy = 0$ is

$$\mathbf{F}_D = \frac{V}{(g'y)^{1/2}} \quad (2.61)$$

in which $\mathbf{F}_D$ is called the *densimetric Froude number*. Note that the Froude number previously defined for a single-layer flow of water really is just a special case of the two-layer flow of water under air, in which $\Delta \rho / \rho \approx 1$.

The densimetric Froude number represents the ratio of inertial force to buoyancy force, which is just another manifestation of the influence of gravity. In movable-bed channels, which are treated in Chapter 10, yet another form of the densimetric Froude number, called the *sediment number,* is encountered. It uses the sediment grain diameter as the length scale and symbolizes the ratio of inertial force to the submerged weight of a sediment grain. We encounter the Froude number throughout the remainder of the text; for example, in hydraulic jumps, uniform flow, gradually varied flow, and unsteady flow.

REFERENCES

Bakhmeteff, B. A. *Hydraulics of Open Channel Flow.* New York: McGraw-Hill, 1932.

Blalock, M. E., and T. W. Sturm. "Minimum Specific Energy in Compound Open Channel." *J. Hyd. Div.,* ASCE, 107, no. 6 (1981), pp. 699–717.

Blalock , M. E., and T. W. Sturm. Closure to "Minimum Specific Energy in Compound Open Channel." *J. Hyd. Div.,* ASCE, 109, no. 3 (1983), pp. 483–87.

Bos, M. G. *Discharge Measurement Structures.* ILRI Publication 20, 3rd Revised Edition. Wageningen, the Netherlands (1989).

Chaudhry, M. H., and S. M. Bhallamudi. "Computation of Critical Depth in Symmetrical Compound Channels." *J. Hydr. Res.,* 26, no. 4 (1988), pp. 377–96.

Clemmens, A. J., T. L. Wahl, M. G. Bos, and J. A. Replogle. *Water Measurement with Flumes and Weirs,* ILRI Publication 58. Wageningen, the Netherlands (2001).

HEC-RAS Hydraulic Reference Manual, version 4.0. Davis, CA: U.S. Army Corps of Engineers, Hydrologic Engineering Center, 2008.

Henderson, F. M. *Open Channel Flow.* New York: Macmillan, 1966.

Kindsvater, C. E., and R. W. C. Carter. "Discharge Characteristics of Rectangular Thin-Plate Weirs." *J. Hyd. Div.,* ASCE, 83, no. HY6 (1957), pp. 1453-1 to 35.

Könemann, N. Discussion of "Minimum Specific Energy in Compound Open Channel." *J. Hyd. Div.,* ASCE, 108, no. HY3 (1982), pp. 462–64.

Schoellhamer, D. H., J. C. Peters, and B. E. Larock. "Subdivision Froude Number." *J. Hydr. Engrg.,* ASCE, 111, no. 7 (1985), pp. 1099–1104.

Shearman, J. O., W. H. Kirby, V. R. Schneider, and H. N. Flippo. "Bridge Waterways Analysis Model: Research Report." Federal Highway Administration, *Report No. FHWA/RD-86/108.* U.S. Dept. of Transportation, Washington, D.C., 1986.

Shearman, J. O. "User's Manual for WSPRO—A Computer Model for Water Surface Profile Computations." Federal Highway Administration, *Report FHWA-IP-89-027.* U.S. Dept. of Transportation, Washington, D.C., 1990.

Sturm, T. W., and Aftab Sadiq. "Water Surface Profiles in Compound Channel with Multiple Critical Depths." *J. Hydr. Engrg.,* ASCE, 122, no. 12 (1996), pp. 703–10.

EXERCISES

2.1. Water is flowing at a depth of 10 ft with a velocity of 10 ft/s in a channel of rectangular section. Find the depth and change in water surface elevation caused by a smooth upward step in the channel bottom of 1 ft. What is the maximum allowable step size so that choking is prevented? (Use a head loss coefficient = 0.)

2.2. The upstream conditions are the same as in Exercise 2.1 with a smooth contraction in width from 10 ft to 9 ft and a horizontal bottom. Find the depth of flow and change in water surface elevation in the contracted section. What is the greatest allowable contraction in width so that choking is prevented? (Head loss coefficient = 0.)

2.3. The upstream conditions in a rectangular channel are the same as in Exercise 2.1 with a smooth contraction in width from 10 ft to 8 ft. How much should the channel bottom drop to maintain a constant water surface elevation through the transition? (Head loss coefficient = 0)

2.4. Determine the downstream depth in the transition and the change in water surface elevation if the channel bottom rises 0.15 m and the upstream conditions are a velocity of 4.5 m/s and a depth of 0.6 m.

2.5. Determine the downstream depth in a subcritical transition if $Q = 262$ cfs and the channel bottom rises 3.279 ft in going from an upstream circular channel to a downstream rectangular channel. The upstream circular channel has a diameter of 9.18 ft and a depth of flow of 7.34 ft. The downstream rectangular channel has a width of 6.56 ft. Neglect the head loss.

2.6. Determine the upstream depth of flow in a subcritical transition from an upstream rectangular flume that is 49 ft wide to a downstream trapezoidal channel with a width of 75 ft and side slopes of 2:1. The transition bottom drops 1 ft from the upstream flume to the downstream trapezoidal channel. The flow rate is 12,600 cfs, and the depth in the downstream trapezoidal channel is 22 ft. Use a head loss coefficient of 0.5.

2.7. In a horizontal rectangular flume, suppose that a smooth "bump" with a height of 0.33 ft has been placed on the channel bottom. The discharge per unit width in the flume is 0.4 cfs/ft. Determine the depth at the obstruction for a tailwater depth of 1.0 ft and negligible head losses. Sketch the results on a specific energy diagram.

2.8. A rectangular channel 3.6 m wide contracts to a 1.8-m wide rectangular channel and then expands back to the 3.6 m width. The contraction is gradual enough that head losses can be neglected, but the expansion loss coefficient is 0.5. The discharge through the transition is 10 m³/s. If the downstream depth at the reexpanded section is 2.4 m, calculate the depths at the approach section and the contracted section. Show the positions of the depth and specific energy for all three sections on a specific energy diagram.

2.9. The head upstream of a circular culvert having a diameter of 6.0 ft is 5.0 ft above the culvert invert. If critical depth occurs at the culvert entrance, what is the discharge if the approach velocity head is negligible? Suppose that an impervious plug of mud and debris blocks the lower 2.0 ft of the culvert entrance above the invert in the form of a horizontal sill. What will the discharge be for the same head of 5.0 ft above the invert? Neglect entrance energy losses.

2.10. Determine the discharge in a circular culvert on a steep slope if the diameter is 1.0 m and the upstream head is 1.3 m with an unsubmerged entrance. Also calculate the critical depth. Neglect entrance losses. Repeat for a box culvert that is 1.0 m square.

2.11. An open channel has a semicircular bottom and vertical, parallel walls. If the diameter, d, is 3 ft, calculate the critical depth and the minimum specific energy for two discharges, 10 cfs and 30 cfs.

2.12. Derive an exact solution for critical depth in a parabolic channel and place it in dimensionless form. Repeat the procedure for a triangular channel.

2.13. Show that the ratio of critical depth to minimum specific energy, y_c/E_c, is 0.80 for a triangular channel and 0.75 for a parabolic channel.

2.14. A parabolic-shaped irrigation canal has a top width of 10 m at a bank-full depth of 2 m. Calculate the critical discharge, Q_c (i.e., the discharge for which the depth of uniform flow is equal to critical depth) for a uniform flow depth of 1.0 m. If $Q < Q_c$ for the uniform flow depth of 1.0 m, will the uniform flow be supercritical or subcritical?

2.15. A USGS study of natural channel shapes in the western United States reports an average ratio of maximum depth to hydraulic depth in the main channel (with no overflow) of $y/D = 1.55$ for 761 measurements.
   (a) Calculate the ratio of maximum depth to hydraulic depth for a (1) triangular channel, (2) parabolic channel, (3) rectangular channel. What do you conclude?
   (b) Calculate the discharge for a bank-full Froude number of $\mathbf{F}_1 = 1.0$ if $y/D = 1.55$ and $B_1 = 100$ ft for $y_1 = 10$ ft. What is the significance of this discharge?

2.16. A natural channel cross section has a bank-full cross-sectional area of 45 m² and a top width of 37.5 m. The maximum value of $\mathbf{F}_c/\mathbf{F}_1$ has been calculated to be 1.236. Find the discharge range, if any, within which multiple critical depths could be expected.

2.17. The main channel of North Fork, Peachtree Creek in Atlanta can be approximated as a parabolic channel with a bank-full depth of 8.0 ft and a bank-full top width of 50 ft. There are symmetric floodplains on either side of the main channel that are perfectly flat, each with a width of 150 ft. If the flow rate is 3500 cfs, is it possible for there to be multiple critical depths for this cross section? Use the computer program **Ycomp** on the book website to calculate the critical depth(s) for $Q = 3500$ cfs and $Q = 3000$ cfs.

2.18. Design a broad-crested weir for a laboratory flume with a width of 15 in. The discharge range is 0.1 to 1.0 cfs. The maximum approach flow depth is 18 in. Determine the height of the weir and the weir length in the flow direction. Plot the expected head-discharge relationship.

2.19. Plot and compare the head-discharge relationships for a rectangular, sharp-crested weir having a crest length of 1.0 ft in a 5-ft wide channel with that for a 90° V-notch, sharp-crested weir if both weir crests are 1 ft above the channel bottom. Consider a head range of 0–0.5 ft.

2.20. Derive the head-discharge relationship for a triangular, broad-crested weir and a corresponding relationship for C_v analogous to Equation 2.48.

2.21. Derive the head-discharge relationship for a truncated, triangular, sharp-crested weir with notch angle θ and vertical walls that begin at a height of h_1 above the triangular crest. Assume that $H > h_1$.

2.22. A trapezoidal flume has a bottom width of 1.0 m and side slopes of 1:1. A sill with a height of 0.5 m is placed in the flume forming a trapezoidal critical control section. The length of the sill is 1.5 m in the flow direction. Calculate the discharge if the approach flow head is measured to be 0.60 m above the sill.

2.23. Find the upstream head in the long-throated rectangular flume of Example 2.5 for the minimum discharge of 0.02 m³/s, which has a tailwater depth of 0.225 m. The sill length in the flow direction is 0.54 m and the height is 0.15 m as in Example 2.5. Also check if submergence will occur.

2.24. A rectangular canal has a bottom width of 6.0 ft. A circular broad-crested weir is placed in the canal by constructing a headwall across the canal through which a 3.0 ft diameter circular pipe is placed. The pipe is horizontal with the invert located 0.5 ft above the bottom of the canal, and it has a length of 7.5 ft. If the upstream head on the weir is measured to be 1.5 ft relative to the invert of the pipe, calculate the discharge in the canal.

2.25. Modify the computer program Y0YC on the book website to calculate the critical depth in a circular channel.

2.26. Write a computer program that computes the depth in a width contraction and the upstream depth given a subcritical tailwater depth as in Figure 2.11. Assume that the channel is rectangular at all three sections and make provision for a head-loss coefficient that is nonzero; include a check for possible choking.

2.27. A laboratory experiment has been conducted in a horizontal flume in which a sharp-crested weir plate has been installed to determine the head-discharge relationship for a rectangular, sharp-crested weir. With reference to Figure 2.23, $P = 0.506$ ft, $L = 0.25$ ft, and $b = 1.25$ ft. The discharge was measured by a bend meter for which the calibration is given by $Q = 0.075 \, \Delta h^{0.523}$, in which Q = discharge in cubic feet per second (cfs); Δh = manometer deflection in inches of water; and the uncertainty in the calibration is ± 0.003 cfs. The head on the crest of the weir was measured by a point gauge and is given in the data table that follows. An upstream view of the weir nappe can be seen in Figure 2.30.

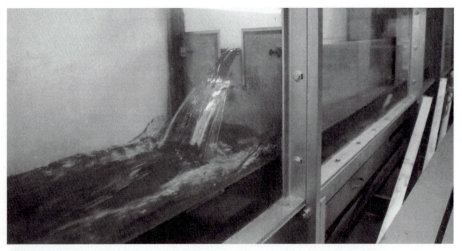

FIGURE 2.30
Upstream view of the flow over a rectangular, sharp-crested weir (photograph by G. Sturm).

Δ h, in.	H, ft
13.2	0.498
11.5	0.476
11.2	0.474
8.3	0.425
8.0	0.421
6.2	0.384
6.1	0.386
4.3	0.333
4.2	0.334
2.4	0.272
2.0	0.257

(a) Plot the head on the vertical scale and the discharge on the horizontal scale of log-log axes and obtain a least-squares regression fit forcing the inverse slope to be the theoretical value of 3/2. What are the single best-fit value of C_d and the standard error in C_d? Compare the standard error of the "Q estimate" with the uncertainty in the bend-meter calibration.

(b) Calculate the discharge first using the Kindsvater-Carter relationship and then using the single best-fit value of C_d. Compare both sets of results with the measured discharges by calculating the percent differences and also plotting the measured vs. calculated discharges.

Momentum

3.1 INTRODUCTION

The momentum equation in control-volume form is a valuable tool in open channel flow analysis. It often is applied in situations involving complex internal flow patterns with energy losses that initially are unknown. The advantage of the momentum equation is that the details of the internal flow patterns in a control volume are immaterial. It is necessary only to be able to quantify the forces and momentum fluxes at the control surfaces that form the boundaries of the control volume. This property of the momentum equation allows it to be used in a complementary fashion with the energy equation to solve for unknown energy losses in otherwise intractable problems.

3.2 HYDRAULIC JUMP

The most common application of the momentum equation in open channel flow is the analysis of the hydraulic jump. The hydraulic jump, an abrupt change in depth from supercritical to subcritical flow, always is accompanied by a significant energy loss. A counterclockwise roller rides continuously up the surface of the jump, entraining air and contributing to the general complexity of the internal flow patterns illustrated in Figure 3.1. Turbulence is produced at the boundary between the incoming jet and the roller. The turbulent eddies dissipate energy from the mean flow, although there is a lag distance in the downstream direction between the point of maximum production of turbulence and maximum dissipation of energy (Rouse, Siao, and Nagaratnam 1958). Furthermore, the kinetic energy of the turbulence is rapidly dissipated along with the mean flow energy in the downstream direction, so

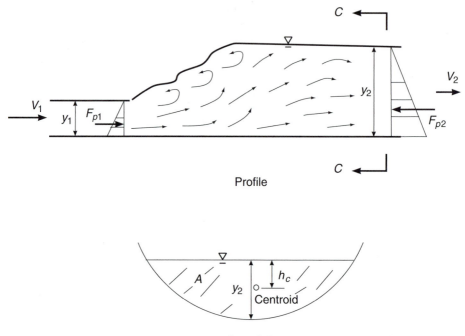

FIGURE 3.1
Application of the momentum equation to a hydraulic jump in a nonrectangular channel.

that the turbulent kinetic energy is small at the end of the jump. Turbulence measurements in free hydraulic jumps of low Froude number using an acoustic Doppler velocimeter have confirmed a rapid decline in turbulence intensities and kinetic energy of the turbulence in the streamwise direction followed by a leveling off at a distance of approximately 10–12 times the sequent depth, y_2 (Liu et al. 2004). Although the rate of decline is not as rapid as indicated by the earlier experiments of Rouse (1958) in air, the complexities of the internal flow pattern of the hydraulic jump are dissipated in a relatively short zone, which makes the momentum equation applied to a finite control volume ideal for analyzing the jump.

If any general nonrectangular cross section is considered as shown in Figure 3.1, a control volume is chosen such that the hydraulic jump is enclosed at the upstream and downstream boundaries, where the flow is nearly parallel. This choice of control volume boundaries allows the assumption of a hydrostatic pressure force at the entrance and exit of the control volume. Also assumed is that the velocity profiles are nearly uniform at the upstream and downstream cross sections, with the result that the momentum correction coefficient $\beta = 1$. The boundary shear over the relatively short length of the jump is neglected in comparison to the change in pressure force. Finally, the jump is assumed to occur in a horizontal channel. Under these assumptions, the momentum equation in the flow direction becomes

$$F_{p1} - F_{p2} = \rho Q(V_2 - V_1) \tag{3.1}$$

in which F_p = hydrostatic force; ρQV = momentum flux; and the subscripts 1 and 2 refer to the upstream and downstream cross sections, respectively. The hydrostatic force is expressed as $\gamma h_c A$, in which h_c is the distance below the free surface to the centroid of the area on which the force acts, as shown in Figure 3.1, and the mean velocity, $V = Q/A$, from the continuity equation. With these substitutions and dividing Equation 3.1 by the specific weight, γ, there results

$$A_1 h_{c1} + \frac{Q^2}{gA_1} = A_2 h_{c2} + \frac{Q^2}{gA_2} \tag{3.2}$$

We see from this rearrangement of the equation that, if we define a function M, which we will call the *momentum function*, as

$$M = A h_c + \frac{Q^2}{gA} \tag{3.3}$$

then its equality upstream and downstream of the hydraulic jump can be used to determine the sequent depth, which is the depth after the jump, if the upstream conditions are given, or vice versa. More precisely, the momentum function is force plus momentum flux divided by the specific weight of the fluid, and this quantity is conserved across the hydraulic jump.

The distance from the free surface to the centroid of the flow section, h_c, is a unique function of the depth, y, and the geometry of the cross section. For example, the momentum function for the trapezoidal section is given by

$$M = \frac{by^2}{2} + \frac{my^3}{3} + \frac{Q^2}{gy(b + my)} \tag{3.4}$$

in which b = bottom width; m = sideslope ratio; and y = flow depth as defined in Table 3.1. The trapezoidal section has been divided into a rectangle and two triangles, and the additive property of the first moment of the area about the free surface has been used to obtain the expression for Ah_c. The momentum function definitions for several other prismatic cross sections also are given in Table 3.1.

The momentum equation can be placed in dimensionless form and solved numerically for the sequent depth. If M_1 is known for the trapezoidal section from incoming flow conditions, for example, then setting $M_1 = M_2$ and nondimensionalizing results in

$$\frac{1.5\Lambda^2}{y_1'} + \Lambda^3 + \frac{3Z^2}{\Lambda y_1'^4(1 + \Lambda y_1')} = \frac{1.5}{y_1'} + 1 + \frac{3Z^2}{y_1'^4(1 + y_1')} \tag{3.5}$$

in which $\Lambda = y_2/y_1$; $y_1' = my_1/b$; and $Z^2 = Q^2 m^3/gb^5$. Equation 3.5 can be solved directly for Z and then plotted as $y_2/y_1 = f(y_1', Z)$ as shown in Figure 3.2 where $Z = Z_{\text{trap}}$. Similarly, the solution for the sequent depth ratio for the circular case can be given as shown in Figure 3.3 with $Z_{\text{circ}}^2 = Q^2/gd^5$. Implicit equations for y_2/y_1 and their graphical solutions in a form similar to that of Figures 3.2 and 3.3 for trapezoidal and circular channels were proposed by Massey (1961) and Thiruvengadam (1961), respectively.

TABLE 3.1   Momentum function for channels of different shapes (y = flow depth)

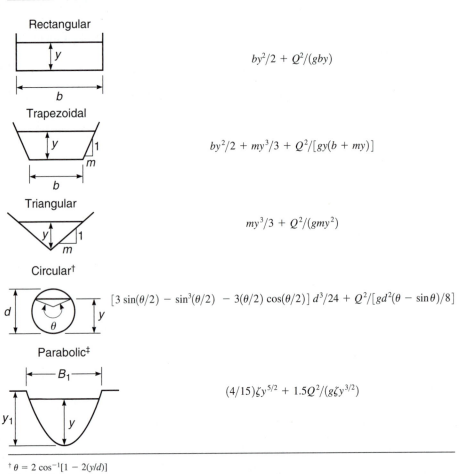

Rectangular

$$by^2/2 + Q^2/(gby)$$

Trapezoidal

$$by^2/2 + my^3/3 + Q^2/[gy(b + my)]$$

Triangular

$$my^3/3 + Q^2/(gmy^2)$$

Circular[†]

$$[3\sin(\theta/2) - \sin^3(\theta/2) - 3(\theta/2)\cos(\theta/2)]\,d^3/24 + Q^2/[gd^2(\theta - \sin\theta)/8]$$

Parabolic[‡]

$$(4/15)\zeta y^{5/2} + 1.5Q^2/(g\zeta y^{3/2})$$

[†] $\theta = 2\cos^{-1}[1 - 2(y/d)]$
[‡] $\zeta = B_1/y_1^{1/2}$

   To solve the nonlinear algebraic equations for the sequent depth ratio numerically, a function $F(y) = M_1 - M_2$ is defined and solved by interval halving or some other nonlinear algebraic equation solver. The critical depth must be found first, however, to limit the root search to the appropriate subcritical or supercritical solution.

   For the rectangular cross section, there is an exact solution for the sequent depth ratio that depends only on the upstream Froude number. Setting the values of the momentum function per unit width upstream and downstream of the jump equal and rearranging, we have

$$\frac{y_1^2}{2} - \frac{y_2^2}{2} = \frac{q^2}{g}\left[\frac{1}{y_2} - \frac{1}{y_1}\right] \tag{3.6}$$

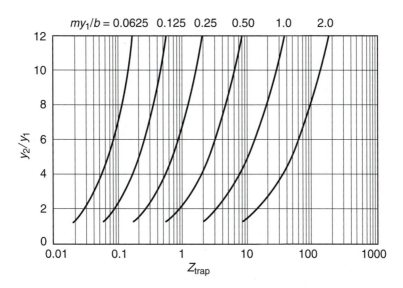

FIGURE 3.2

Sequent depth ratio for a hydraulic jump in a trapezoidal channel ($Z_{trap} = Qm^{3/2}/[g^{1/2}b^{5/2}]$).

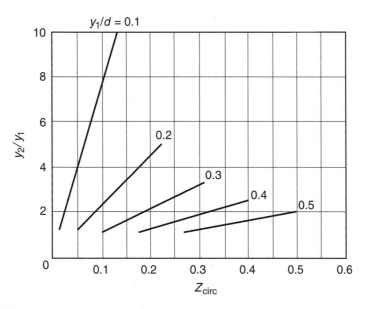

FIGURE 3.3

Sequent depth ratio for a hydraulic jump in a circular channel ($Z_{circ} = Q/[g^{1/2}d^{5/2}]$).

With some algebraic manipulation and nondimensionalization, Equation 3.6 becomes a quadratic equation:

$$\Lambda^2 + \Lambda - 2\mathbf{F}_1^2 = 0 \tag{3.7}$$

in which $\Lambda = y_2/y_1$ and $\mathbf{F}_1 =$ the approach flow Froude number $= (q^2/gy_1^3)^{1/2}$. The solution to Equation 3.7 is given by the quadratic formula

$$\Lambda = \tfrac{1}{2}\left[-1 + \sqrt{1 + 8\mathbf{F}_1^2}\,\right] \tag{3.8}$$

The unknown energy loss can be obtained as the difference between the upstream and downstream values of the specific energy, $E_L = E_1 - E_2$. In dimensionless form, this is

$$\frac{E_L}{E_1} = 1 - \frac{\Lambda + \mathbf{F}_1^2/2\Lambda^2}{1 + \mathbf{F}_1^2/2} \tag{3.9}$$

Equations 3.8 and 3.9 are shown in Figure 3.4 and compared with experimental data obtained by Bradley and Peterka (1957) of the U.S. Bureau of Reclamation in a comprehensive study of stilling basins. The data were obtained in five flumes with

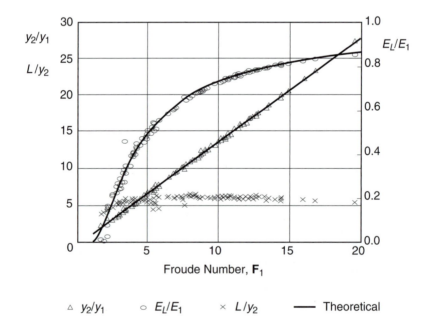

FIGURE 3.4

Comparison of theory and experiment for a hydraulic jump in a rectangular channel: $y_2/y_1 =$ sequent depth ratio; $E_L/E_1 =$ energy loss ratio; $L/y_2 =$ jump length ratio. (*Data from Bradley and Peterka 1957.*)

the upstream Froude numbers having values between approximately 2 and 20. The agreement between the experimental data and the theoretical Equations 3.8 and 3.9 is quite good, confirming the initial assumptions made in the momentum analysis. The length of the jump L, which can be determined only experimentally, also is shown in Figure 3.4. The jump length was defined in the experiments somewhat qualitatively, as the distance from the front of the jump to either the point where the jet left the floor or a point on the water surface immediately downstream of the roller, whichever was larger. Based on this data, the jump length often is defined as six times the depth after the jump.

Graphical solutions for the hydraulic jump in triangular and parabolic channels can be obtained in the same manner as for the trapezoidal channel. In both cases, the Froude number is the only independent dimensionless parameter. Silvester (1964) summarized some experimental data for the sequent depth ratio and the energy loss in triangular and parabolic channels, and reasonable agreement with the momentum solutions was demonstrated. The sequent depth for the triangular, parabolic, and rectangular channels can be compared directly on the basis of the actual approach flow Froude number for a nonrectangular channel, as in Figure 3.5. We can see that the magnitude of the sequent-depth ratio for the same Froude number increases as the channel cross section becomes "fuller" from triangular to parabolic to rectangular. The ratio of the energy loss to the available upstream energy E_L/E_1 is compared for the triangular, parabolic, and rectangular channels in Figure 3.6, and they are remarkably close to each other.

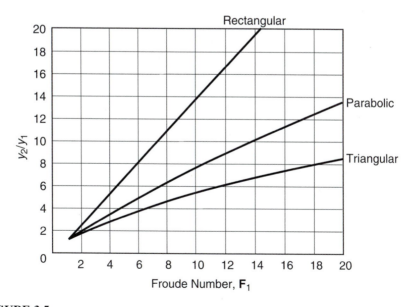

FIGURE 3.5
Comparison of sequent depth ratios in rectangular, parabolic, and triangular channels.

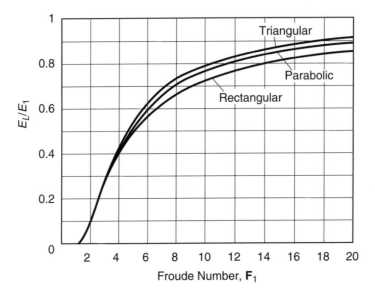

FIGURE 3.6
Comparison of energy losses in a hydraulic jump in rectangular, parabolic, and triangular channels.

The momentum equation also has been applied to the circular, or radial, hydraulic jump (Koloseus and Ahmad 1969). The major difference between the jump in a prismatic, rectangular channel and the radial jump is that the hydrostatic forces on the walls of the radially expanding channel have a component in the radial direction. This, in turn, requires that the surface profile of the jump be known. The simplest assumption, which is adopted for Figure 3.7, is to take the effective jump profile to be linear. Arbhabhirama and Abella (1971) assumed an elliptic water surface profile, but Khalifa and McCorquodale (1979) showed that air entrainment shifts the effective profile as determined by the hydraulic grade line toward the linear shape. The sequent depth ratio for a radial jump ($r_0 = r_2/r_1 = 2$) is compared with the rectangular channel jump ($r_2/r_1 = 1$) in Figure 3.7. We see that the radial jump has a smaller sequent depth ratio for the same approach Froude number but a larger energy loss. Lawson and Phillips (1983) as well as Khalifa and McCorquodale (1979) have demonstrated reasonably good experimental agreement with the theoretical sequent depth ratio and relative energy loss when assuming the linear jump profile.

The appearance of the hydraulic jump, as well as the sequent depth ratio and the dimensionless energy loss, is a function of the approach Froude number. At Froude numbers between 1.0 and 1.7, the depth is close to critical and a train of standing waves called an undular jump appears downstream, but it is more aptly described as a free surface instability with almost no energy dissipation. As shown in Figure 3.8, a prejump stage occurs for Froude numbers between 1.7 and 2.5,

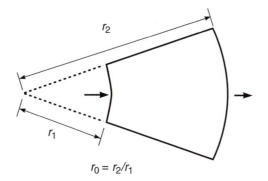

(a) Definition Sketch

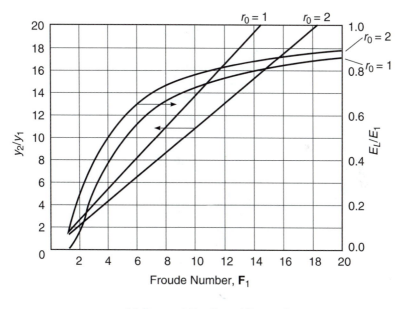

(b) Sequent Depth and Energy Loss

FIGURE 3.7
Sequent depth and energy loss ratios for a radial hydraulic jump.

which is a precursor to the transition stage having Froude numbers between 2.5 and 4.5. In this transition stage, the entering jet oscillates from the channel bottom to the free surface, creating surface waves for long distances downstream. Jumps with Froude numbers between 4.5 and 9 are well balanced and stable, because the jet

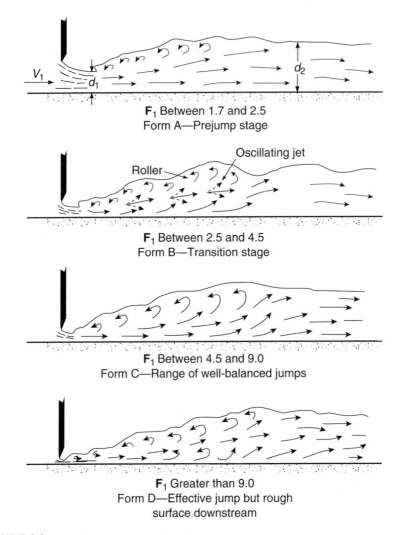

F₁ Between 1.7 and 2.5
Form A—Prejump stage

F₁ Between 2.5 and 4.5
Form B—Transition stage

F₁ Between 4.5 and 9.0
Form C—Range of well-balanced jumps

F₁ Greater than 9.0
Form D—Effective jump but rough
surface downstream

FIGURE 3.8
Appearance of a hydraulic jump for different Froude number ranges (U.S. Bureau of Reclamation 1987).

leaves the channel bottom at approximately the same point as the end of the surface roller. For an approach Froude number in excess of 9, the downstream water surface can be rough, but large energy losses can be expected.

It is instructive to consider the shape of the momentum function, since it obviously is a function of depth y alone for a given Q and geometry in much the same fashion as the specific energy function. If we consider the rectangular channel, for example, the momentum function per unit of channel width is given by

$$\frac{M}{b} = \frac{y^2}{2} + \frac{q^2}{gy} \tag{3.10}$$

which has two branches and a minimum. As y approaches zero, the momentum function per unit of width approaches infinity, while it approaches the parabola $y^2/2$ as y becomes very large. The minimum value of the momentum function is obtained by differentiating with respect to y and setting the result to zero:

$$y - \frac{q^2}{gy^2} = 0 \tag{3.11}$$

If we solve for y, we obtain the expression for critical depth for a rectangular channel derived from a consideration of minimum specific energy. Therefore, critical depth occurs not only at the minimum value of specific energy for a given discharge, Q, but also at the minimum value of the momentum function.

The correspondence between the specific energy and momentum functions is illustrated in Figure 3.9 for a hydraulic jump in a rectangular channel with the functions given in dimensionless form. Clearly, conservation of the momentum function as required by the hydraulic jump analysis requires an energy loss. Hence, the jump can

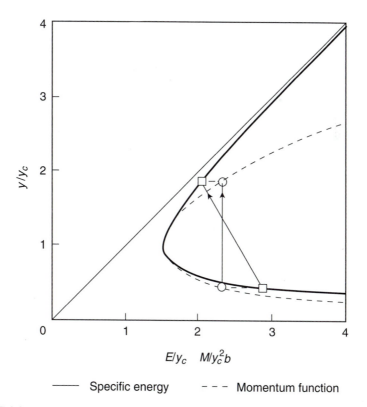

FIGURE 3.9
Hydraulic jump sequent depths on specific energy and momentum diagrams nondimensionalized by critical depth.

only occur from the supercritical to the subcritical state as an abrupt transition between the two flow regimes. Also note that the sequent depth ratio, y_2/y_1, and the energy loss increase for smaller values of the approach depth. As the approach depth decreases, the velocity head increases; and so the Froude number must increase. In other words, the dimensionless specific energy and momentum diagrams confirm the increase in sequent depth ratio and energy loss with Froude number found previously from the solutions of the energy and momentum equations and shown in Figure 3.4.

The general case for the minimum value of the momentum function can be derived for any nonrectangular section for which $\beta = 1 = $ constant. Setting the derivative of the momentum function with respect to y to zero yields

$$\frac{dM}{dy} = \frac{d}{dy}(Ah_c) - \frac{Q^2B}{gA^2} = 0 \tag{3.12}$$

in which dA/dy has been replaced by the top width, B. Using the definition of the first moment of the area and the Leibniz rule, it can be shown that the first term of the derivative is equal to the flow area, A, from which it is obvious that the minimum value of the momentum function occurs when the Froude number squared for the nonrectangular channel is equal to unity; that is, $Q^2B/gA^3 = 1.0$.

Although general agreement between experimental results and the momentum theory for the hydraulic jump has been demonstrated, it is useful to consider the effects of the assumptions made in the analysis. Harleman (1959) concluded from the data of Rouse, Siao, and Nagaratnam (1958) that the effect of assuming a uniform velocity distribution and neglecting the turbulence at the two end sections of the hydraulic jump indeed is small. Rajaratnam (1965), however, showed from his analysis of the jump as a wall jet that the integrated boundary shear stress can affect the sequent depth ratio. Leutheusser and Kartha (1972) generalized and extended this conclusion by conducting experiments on jumps with fully developed inflows and undeveloped inflows. From the two-dimensional Reynolds equations, they derived an integrated form of the hydraulic jump equation that eliminates the conventional assumptions by lumping them into a single factor, ε:

$$\frac{\Lambda(\Lambda + 1)}{2\mathbf{F}_1^2} = 1 + \varepsilon \tag{3.13}$$

in which $\Lambda = y_2/y_1$ and $\mathbf{F}_1 = $ approach Froude number. For $\varepsilon = 0$, we recover the result given in Equation 3.7. On the other hand, if we consider the influence of the mean shear stress over the length, L, of the jump, ε is given by

$$\varepsilon = -\frac{C_f}{2}\frac{L}{y_2}\frac{\Lambda^2}{\Lambda - 1} \tag{3.14}$$

in which $C_f = $ overall skin friction coefficient. Leutheusser and Kartha (1972) showed from their experimental results that ε has essentially no influence on the sequent depth ratio for approach Froude numbers less than 10. For greater values of the Froude number, however, the developed-inflow jump had a smaller sequent depth ratio than predicted by Equation 3.8 due to the influence of the boundary shear force. Furthermore, the developed-inflow jump was longer and lower than in the undeveloped-inflow case, which Leutheusser and Kartha suggest is due to the tendency for the undeveloped inflow to separate, thus reducing the boundary shear. It must also be pointed out that the jump length in Equation 3.14 is defined as the point at which no further changes are observed in the centerline velocity distribution in the downstream direction. The dimensionless length, L/y_2, has a value of approximately 16 for the fully developed inflow and a typical value of C_f is 1×10^{-3}. These experimental values result in a value of ε of approximately -0.1 and a relative error in the sequent depth ratio of less than 10 percent at a Froude number of 10. Note that $(1 + \varepsilon)$ appears in the solution for Λ as a factor multiplying the Froude number in Equation 3.8; that is, $8(1 + \varepsilon) \mathbf{F}_1^2$.

If the effect of boundary shear is relatively small for hydraulic jumps in smooth channels, it may not necessarily be negligible in the case of a channel with significant boundary roughness. Experiments by Hughes and Flack (1984) confirm this to be the case for both strip roughness and gravel beds. Their laboratory results showed that both the length and sequent depth of a hydraulic jump are reduced by large roughness elements. A bed of $\frac{1}{2}$ to $\frac{3}{8}$ in. gravel, for example, resulted in a 15 percent reduction in the sequent depth ratio predicted for a smooth channel at a Froude number of 7. Carollo et al. (2007) obtained an experimental expression for ε in Equation 3.14 for hydraulic jumps over rough beds formed by gravel, corrugated aluminum sheets with the corrugations perpendicular to the flow direction (Ead and Rajaratnam 2002), and the rough beds of Hughes and Flack (1984). The equation for ε is a best-fit curve given by

$$\varepsilon = -\frac{2}{\pi} \arctan \left[0.8 \left(\frac{k_s}{y_1} \right)^{0.75} \right] \tag{3.15}$$

in which k_s = roughness height. The uncertainty in values of $\Lambda = y_2/y_1$ computed using Equation 3.15 are $\pm 15\%$. The length of the jump roller was given by $L/y_2 = 4.6 [1.0 - 1.0/\Lambda]$ so that the jump length varied from about $3.6y_2$ to $4.1y_2$ for well-balanced jumps ($4.5 < \mathbf{F}_1 < 9.0$) on rough beds in comparison to $6y_2$ for smooth beds.

The effect of boundary shear on the hydraulic jump is similar to the effect of form roughness provided by baffle blocks on the floor of a stilling basin. The obstruction causes a lower sequent depth ratio at the same Froude number and makes the jump position more stable. The effect of the obstruction on the momentum balance is illustrated in Figure 3.10, in which clearly the decrease in the value of the momentum function from the supercritical to subcritical state in the hydraulic jump must be exactly equal to the drag force of the obstruction, p_f, divided by the fluid specific weight, γ.

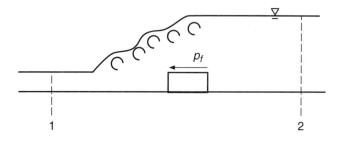

(a) Hydraulic Jump with Baffle Blocks

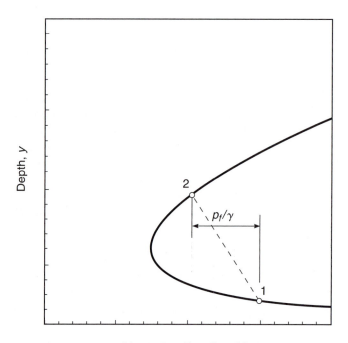

Momentum Function, M

(b) Reduction in Momentum Function

FIGURE 3.10
Decrease in momentum function for a hydraulic jump due to the external force of blocks on the control volume.

3.3   STILLING BASINS

Hydraulic jumps are used extensively as energy dissipation devices for spillways because of the large percentage of incoming energy of the supercritical flow that is lost (see Figure 3.4). The stilling basin, located at the downstream end of the spillway or the spillway chute, usually is constructed of concrete. It is intended to hold the jump within the basin, stabilize it, and reduce the length required for the jump to occur. The resulting low-velocity subcritical flow released downstream prevents erosion and undermining of dam and spillway structures.

Generalized designs of stilling basins have been developed by the U.S. Bureau of Reclamation and others, based on experience, field observations, and laboratory model studies. Special appurtenances are placed within the stilling basin to help achieve its purpose. Chute blocks placed at the entrance to the stilling basin tend to split the incoming jet and block a portion of it to reduce the basin length and stabilize the jump. The end sill is a gradual rise at the end of the basin to further shorten the jump and prevent scour downstream, which may result from the high velocities that develop near the floor of the basin. The sill can be solid or dentated. Dentation diffuses the jet at the end of the basin. Baffle blocks are placed across the floor of the basin at specified spacings to further dissipate energy by the impact of the high velocity jet. However, the blocks can be used for only relatively low velocities of incoming flow; otherwise, cavitation damage may result.

With reference to the types of jumps that can form as a function of the Froude number of the incoming flow (see Figure 3.8), the Bureau of Reclamation has developed several standard stilling basin designs (U.S. Bureau of Reclamation, 1987), three of which are shown in Figures 3.11, 3.12, and 3.13. For incoming Froude numbers from 1.7 to 2.5, the jump is weak and no special appurtenances are required. This is called the *Type I basin*. In the Froude number range from 2.5 to 4.5, a transition jump forms with considerable wave action. The *Type IV basin* is recommended for this jump, as shown in Figure 3.11. It has chute blocks and a solid end sill but no baffle blocks. The recommended tailwater depth is 10 percent greater than the sequent depth to help prevent sweepout of the jump. Because considerable wave action can remain downstream of the basin, this jump and basin are sometimes avoided altogether by widening the basin to increase the Froude number. For Froude numbers greater than 4.5, either *Type III* or *Type II basins,* as shown in Figures 3.12 and 3.13, are recommended. The Type III basin shown in Figure 3.12 includes baffle blocks, and so it is limited to applications where the incoming velocity does not exceed 60 ft/s. For velocities exceeding 60 ft/s, the Type II basin shown in Figure 3.13, which has no baffle blocks and a dentated end sill, is suggested. It is slightly longer than the Type III basin, and the tailwater is recommended to be 5 percent greater than the sequent depth to help prevent sweepout.

Matching the tailwater and sequent depth curves over a range of operating discharges is one of the most important aspects of stilling basin design. If the tailwater is lower than the sequent depth of the jump, the jump may be swept out of the basin, which then no longer serves its purpose because dangerous erosion is likely to occur downstream of the basin. On the other hand, a tailwater elevation that is higher than the sequent depth causes the jump to back up against the spillway chute and "drown out" or be submerged, so that it no longer dissipates as much energy. The ideal situation is one in which the sequent depths perfectly match the tailwater over the full range of operating discharges, but this is unlikely to occur. Instead, the basin floor elevation is set to match sequent

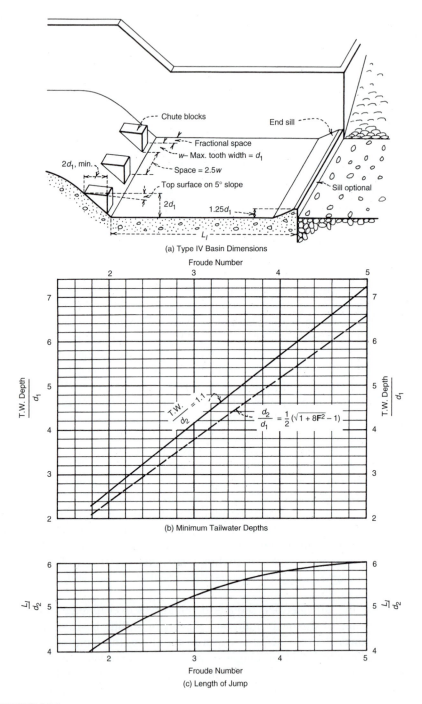

(a) Type IV Basin Dimensions

(b) Minimum Tailwater Depths

(c) Length of Jump

FIGURE 3.11

Type IV stilling basin characteristics for Froude numbers between 2.5 and 4.5; d_1, d_2 = sequent depths (U.S. Bureau of Reclamation 1987).

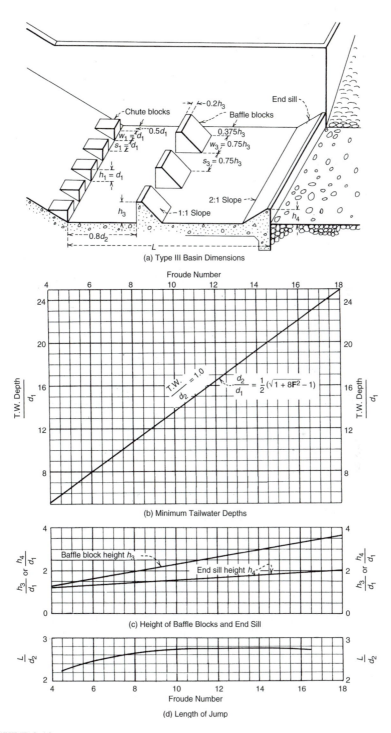

(a) Type III Basin Dimensions

(b) Minimum Tailwater Depths

(c) Height of Baffle Blocks and End Sill

(d) Length of Jump

FIGURE 3.12

Type III stilling basin characteristics for Froude numbers above 4.5 where incoming velocity is $V_1 \leq 60$ ft/s; d_1, d_2 = sequent depths (U.S. Bureau of Reclamation 1987).

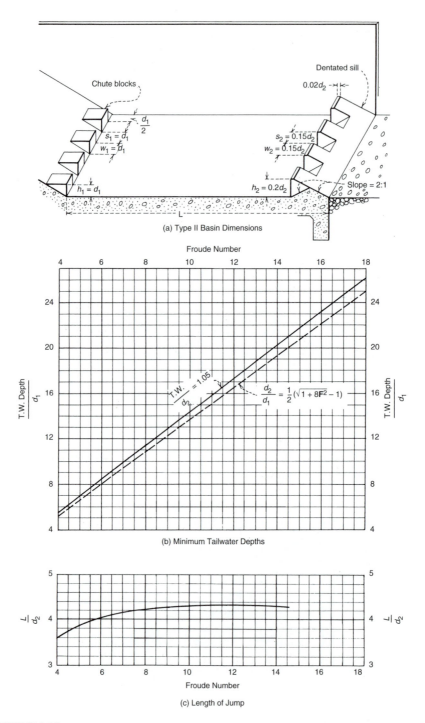

(a) Type II Basin Dimensions

(b) Minimum Tailwater Depths

(c) Length of Jump

FIGURE 3.13

Type II stilling basin characteristics for Froude numbers above 4.5; d_1, d_2 = sequent depths (U.S. Bureau of Reclamation 1987).

depth and tailwater elevations at the maximum design discharge at point A, as shown in Figure 3.14a, and the basin can be widened as shown in the figure to help improve the match at lower discharges while erring on the submerged side rather than the sweep-out side. If the sequent depth curve is shaped as shown in Figure 3.14b, the tailwater and sequent depth elevations would have to be matched for a lower discharge than the maximum, such as point B in the figure, to ensure sufficient tailwater for all discharges.

Setting the floor elevation of the stilling basin and selection of the type of basin to use depends on predicting the flow and velocity at the toe of the spillway and hence the energy loss over the spillway. Some general design guidance is provided in the *Design of Small Dams* (U.S. Bureau of Reclamation 1987). If the stilling basin is located immediately downstream of the crest of an overflow spillway or if the spillway chute is no longer than the hydraulic head, no loss at all is recommended. Here, the hydraulic head is defined as the difference in elevation between the reservoir water surface and the downstream water surface at the entrance to the stilling basin. If the spillway chute length is between one and five times the hydraulic head, an energy loss of 10 percent of the hydraulic head is suggested. For spillway chute lengths in excess of five times the hydraulic head, a 20 percent loss of hydraulic head should be considered. For more accurate estimates of head loss, the equation of gradually varied flow can be solved along a spillway chute of constant slope, as described in Chapter 5, except in the vicinity of the crest where the flow is not gradually varied and the boundary layer is not fully developed. For this region, the two-dimensional Navier-Stokes equations in boundary layer form must be solved numerically (Keller and Rastogi 1977).

Example 3.1

A stilling basin at the toe of a spillway has a width of 180 ft and carries a design discharge of 24,500 cfs when the upstream lake elevation is 32.0 ft. At this discharge, the tailwater elevation in the channel downstream of the spillway is 25.6 ft. Determine the elevation of the floor and choose the type and dimensions of the stilling basin. Neglect spillway head loss.

Solution.   The discharge per unit width received by the stilling basin is $q = 24,500/180 = 136$ cfs/ft (12.6 m²/s) and the corresponding critical depth is $y_c = (q^2/g)^{1/3} = (136^2/32.2)^{1/3} = 8.31$ ft (2.53 m). Start by assuming the stilling basin floor elevation $Z = 10.0$ ft. Then writing the energy equation from the upstream reservoir to the toe of the spillway, we have

$$32.0 = y_1 + \frac{q^2}{2gy_1^2} + Z = y_1 + \frac{287.2}{y_1^2} + 10.0$$

from which a trial and error solution yields the depth of flow at the toe of the spillway, $y_1 = 3.99$ ft (1.22 m) and from continuity the velocity is $V_1 = q/y_1 = 136/3.99 = 34.1$ ft/s (10.4 m/s). The resulting value of the supercritical Froude number is $V_1/(gy_1)^{1/2} = 3.01$. For this Froude number, a Type IV stilling basin is indicated in which the tailwater depth should be 1.1 times the value from Equation 3.8 (see Figure 3.11). Solving for the sequent depth of the hydraulic jump from Equation 3.8, the result is

$$y_2 = \frac{y_1}{2}\left[-1 + \sqrt{1 + 8F_1^2}\right] = \frac{3.99}{2}\left[-1 + \sqrt{1 + 8 \times 3.01^2}\right] = 15.1 \text{ ft}$$

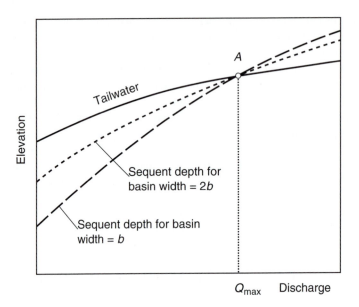

(a) Case A

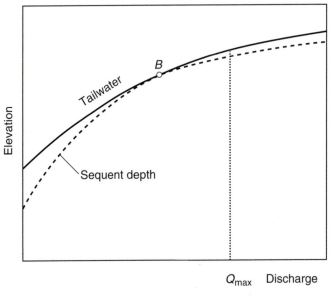

(b) Case B

FIGURE 3.14

Matching tailwater elevation to water surface elevation corresponding to sequent depth of a hydraulic jump (U.S. Bureau of Reclamation 1987).

from which we determine that the tailwater depth must be $1.1 \times 15.1 = 16.6$ ft. The corresponding water surface elevation is $Z + 1.1y_2 = 10.0 + 16.6 = 26.6$ ft, which is higher than the downstream tailwater elevation of 25.6 ft. The jump will sweep out of the basin and likely cause unacceptable downstream erosion. Repeat with a lower basin elevation of $Z = 8.5$ ft to obtain $y_1 = 3.82$ ft (1.16 m), $V_1 = 35.6$ ft/s (10.8 m/s), and $\mathbf{F}_1 = 3.21$. Then the sequent depth is $y_2 = 15.54$ ft, and the required tailwater elevation is $Z + 1.1y_2 = 8.5 + 1.1 \times 15.54$ ft $= 25.6$ ft, which matches the given tailwater elevation. Thus, the required floor elevation of the basin is 8.5 ft (2.59 m). From Figure 3.11, the length of the stilling basin for a Froude number of 3.2 is $5.4\, y_2 = 84$ ft (25.6 m); the sill height is $1.25\, y_1 = 4.8$ ft (1.5 m); and the chute blocks have a width, $w = 3.5$ ft or 1.1 m ($<y_1$) with a spacing of $2.5w = 8.8$ ft (2.7 m).

3.4 SURGES

Although a consideration of surges rightfully belongs in a discussion of unsteady flow, surges can be analyzed by the methods of this chapter by transforming them from an unsteady flow problem to a steady one. This transformation, as shown in Figure 3.15, is accomplished by superimposing a surge velocity, V_s, to the right so that the surge becomes stationary. From this viewpoint, which is that of an observer

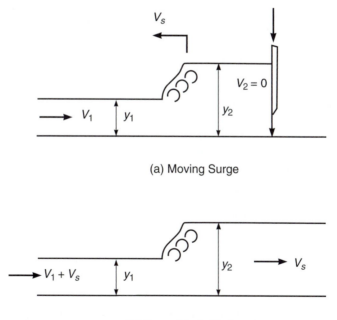

(a) Moving Surge

(b) Surge Made Stationary

FIGURE 3.15
The moving surge in (a) is reduced to the stationary jump in (b).

moving at the speed of the surge, the problem is nothing more than the steady-flow formation of a hydraulic jump.

Surges occur in many open channel flow situations. The abrupt closing of a sluice gate at the downstream end of the channel, shown in Figure 3.15, would create a surge as shown. Other examples include the shutdown of a hydroelectric turbine and the resulting surge in the headrace, a tidal bore, and the surge created in the downstream river channel by an abrupt dam break.

By making the surge stationary, the steady-flow form of the continuity and momentum equations can be applied to Figure 3.15b. The continuity equation for a rectangular channel of unit width is

$$(V_1 + V_s)y_1 = (V_2 + V_s)y_2 \tag{3.16}$$

which can be rewritten in the form

$$V_s = \frac{V_1 y_1 - V_2 y_2}{y_2 - y_1} \tag{3.17}$$

In this form, the continuity equation states that the net flow rate through the surge is given by the rate of volume increase effected by the surge movement.

The momentum equation written for the stationary surge in Figure 3.15b becomes

$$\frac{(V_1 + V_s)^2}{gy_1} = \frac{1}{2}\frac{y_2}{y_1}\left(1 + \frac{y_2}{y_1}\right) \tag{3.18}$$

This is of the same form as the hydraulic jump equation except that the velocity of flow, V_1, has been replaced by $(V_1 + V_s)$. In effect, the left hand side of the equation represents the Froude number as seen by the moving observer. Because $y_2/y_1 > 1$, the Froude number as seen by an observer moving with the surge is supercritical even though the flow in front of the surge could be supercritical or subcritical as seen by a stationary observer. It also can be concluded that the flow behind the surge is subcritical from the viewpoint of the moving observer. A further conclusion that can be drawn from Equation 3.18 is that, as y_2/y_1 approaches one for an infinitesimal surface disturbance, the celerity of that disturbance in still water ($V_1 = 0$) is given by $(gy_1)^{1/2}$ as derived previously in Chapter 2 from an energy argument.

Equations 3.17 and 3.18 provide only two equations in the three unknowns: y_2, V_2, and V_s. The third equation required for solution often comes from a specified boundary condition. In the case of a gate slamming shut at the downstream end of the channel in Figure 3.15, for example, the necessary condition is $V_2 = 0$.

Example 3.2

A steady flow occurs in a rectangular channel upstream of a sluice gate. The velocity is 1.0 m/s (3.3 ft/s) and the depth of flow is 3.0 m (9.8 ft) just upstream of the gate. If the sluice gate suddenly is slammed shut, what are the height and speed of the upstream surge?

Solution.   From continuity, Equation 3.17, and after substitution of $y_1 = 3.0$, $V_1 = 1.0$, and $V_2 = 0$, we obtain:

$$V_s = \frac{3.0}{y_2 - 3.0}$$

Equation 3.18, the momentum equation, then gives

$$(1.0 + V_s)^2 = 14.72 \frac{y_2}{3.0} \left(1 + \frac{y_2}{3.0}\right)$$

These two equations can be solved by trial by first substituting a value of y_2 in the second equation that is greater than y_1 and solving for V_s, which then can be compared with the value of V_s from the first equation. Iteration is continued until the values of V_s are equal. Alternatively, a function could be formed by substituting V_s from (3.17) into (3.18) and rearranging so that the right hand side is zero. The zero of the function then could be determined from a nonlinear algebraic equation solver. In either case, the result is $y_2 = 3.58$ m (11.8 ft) and $V_s = 5.20$ m/s (17.1 ft/s). This speed of the surge is what would be seen by a stationary observer, while an observer moving with the speed of the surge would see a Froude number of $(V_1 + V_s)/(gy_1)^{0.5} = 1.14$ in front and a Froude number of $V_s/(gy_2)^{0.5} = 0.88$ behind the surge.

3.5   BRIDGE PIERS

Momentum analysis can be useful when applied to the obstruction caused by bridge piers in river flow. The resulting obstruction leads to backwater effects upstream in subcritical flow and can even cause choking.

Two types of flow are shown in Figure 3.16. Type I is a subcritical approach flow with a decrease in depth when passing through the constriction with the flow remaining subcritical. In Type II flow, choking occurs with critical depth existing in the constriction. For Type I flow, the momentum equation can be written between sections 1 and 4, in Figure 3.16, to give

$$M_1 = M_4 + D/\gamma \tag{3.19}$$

in which M_1 and M_4 are the momentum function values at sections 1 and 4, and D is the drag force exerted by the piers. For known conditions at the downstream section 4 and with the drag force given as $D = C_D \rho A_p V_1^2/2$, the change in depth or backwater $h_1^* = (y_1 - y_4)$ can be determined. In the expression for drag, A_p is the frontal area of the pier and C_D is the drag coefficient with a value between 1.5 and 2.0 for a blunt shape. Substituting into Equation 3.19 for a rectangular channel, we have

$$\frac{y_1^2}{2} + \frac{q^2}{gy_1} = \frac{y_4^2}{2} + \frac{q^2}{gy_4} + \frac{C_D a y_1 V_1^2}{2gs} \tag{3.20}$$

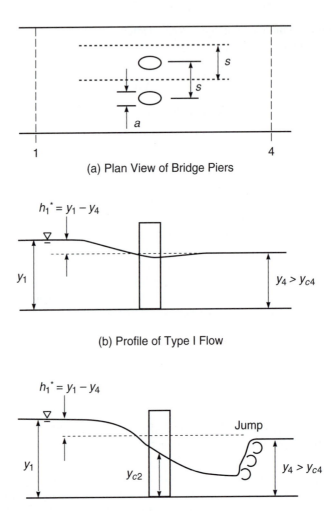

(a) Plan View of Bridge Piers

(b) Profile of Type I Flow

(c) Profile of Type II Flow

FIGURE 3.16
Flow between bridge piers.

in which a = pier width, and s = pier spacing. Equation 3.20 can be nondimensionalized in terms of the downstream Froude number, $\mathbf{F}_4$, to produce

$$\mathbf{F}_4^2 = \frac{\lambda(\lambda + 1)(\lambda + 2)}{C_D a/s + 2\lambda} \tag{3.21}$$

in which $\lambda = h_1^*/y_4$, which is the ratio of the backwater to the downstream depth. Equation 3.21 is plotted in Figure 3.17, from which the backwater caused by piers can be estimated, provided their coefficient of drag is known.

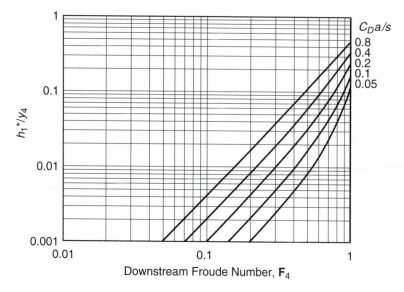

$C_D a/s$
0.8
0.4
0.2
0.1
0.05

h_1^*/y_4

Downstream Froude Number, $\mathbf{F}_4$

FIGURE 3.17
Solution for backwater caused by bridge piers in Type I flow.

Example 3.3

A bridge is supported by elliptical piers having a width of 1.5 m (4.9 ft) and a spacing of 15.0 m (49.2 ft). The piers have a drag coefficient of 2.0. If the downstream depth and velocity are 1.90 m (6.23 ft) and 2.40 m/s (7.87 ft/s), respectively, what is the backwater caused by the piers?

Solution.   First, the value of $C_D a/s = 2.0(1.5)/15 = 0.2$. The downstream Froude number is $2.4/(9.81 \times 1.9)^{0.5} = 0.56$. From Figure 3.17, h_1^*/y_4 is approximately 0.04, from which the backwater $h_1^* = 0.04(1.9) = 0.076$ m (0.25 ft). If Equation 3.21 is solved numerically, the value of $h_1^*/y_4 = 0.041$, which confirms the graphical solution. While this backwater value may seem small, it could represent a significant increase in the area flooded upstream of the bridge in very flat, wide floodplains. Also clear from Figure 3.17 is that the smaller are the flow blockage (a/s) and the downstream Froude number, the less backwater that will develop.

3.6   SUPERCRITICAL TRANSITIONS

Transitions for which the approach flow is supercritical offer a design challenge because of the existence and propagation of standing wave fronts. The reason for the occurrence of standing wave fronts in supercritical flow can be visualized from the viewpoint of an observer riding on a small particle or disturbance moving at a constant speed, V, in still water, as shown in Figure 3.18. At each instant of time, the disturbance, P, sends a circular wave front outward that moves at a speed equal

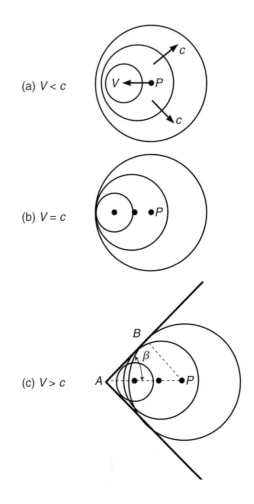

FIGURE 3.18
Movement of a point disturbance, P, at speed V in still water (c = wave celerity).

to the wave celerity, c. If $V < c$, as in Figure 3.18a, the wave fronts outdistance the moving disturbance, so that no pileup or addition of wave fronts occurs. On the other hand, for cases of $V > c$, as in Figure 3.18c, the disturbance, P, moves faster than the wave fronts. The result is an accumulation of wave fronts, the outer locus of which forms a straight line at an angle β to the path of the disturbance that can be defined as a standing wave front. In between these two extremes, $V = c$, and the standing wave front is perpendicular to the path of the disturbance, P, as shown in Figure 3.18b. If the disturbance is infinitesimal, so that $c = (gy)^{0.5}$, where y = flow depth, then obviously V/c is the Froude number and case (a) in Figure 3.18 is for subcritical flow, while case (c) represents supercritical flow.

The supercritical case in Figure 3.18c can be analyzed in more detail to determine the angle β. In the time t_1, the distance moved by point P to point A is Vt_1 while, at the same time, the initial wave front grows from P to point B over a radial distance given by ct_1. Then, $\sin \beta = 1/\mathbf{F}$, where $\mathbf{F}$ is the Froude number. If the viewpoint of the observer is changed, the fluid moves at a speed V in the supercritical case and any boundary irregularity, such as that caused by a change in wall direction in the contraction shown in Figure 3.19, gives rise to a standing wave front at an angle, β_1, relative to the original flow direction. This analysis indicates that larger Froude numbers result in smaller angles of deflection of the standing wave front, but the possibility of a finite height of the standing wave front is not considered. In this circumstance, the analysis must be modified.

Design of Supercritical Contraction

Consider a straight-walled contraction, as shown in Figure 3.19, with a wall angle of θ, an approach supercritical Froude number $\mathbf{F}_1$, and contraction ratio $r\,(= b_3/b_1)$. Standing waves of finite height are formed at the initial change in wall direction having an angle of β_1. They meet at the centerline of the contraction and are reflected back to the wall with an angle of $(\beta_2 - \theta)$. The goal of good contraction design, as outlined by Ippen and Dawson (1951), is to choose the value of θ for given values of Froude number and contraction ratio that minimize the transmission of the standing waves downstream. This can be accomplished if the combined length of the first two sets of standing waves terminates precisely at the physical end of the transition, so that subsequent reflections downstream are cancelled out by the negative disturbances emanating from the end of the contraction.

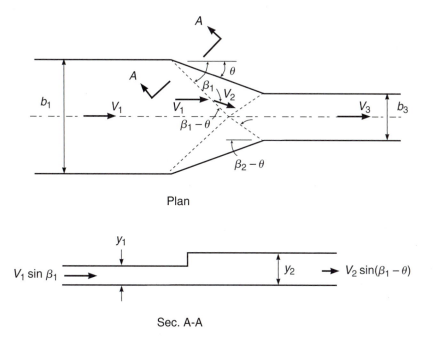

Plan

Sec. A-A

FIGURE 3.19

Straight-wall contraction in a supercritical flow with standing oblique waves.

This design problem can be solved by applying the momentum and continuity equations across the wave fronts in much the same way as for the hydraulic jump. With reference to section A-A in Figure 3.19, the continuity equation across the wave front is given by

$$V_1 y_1 \sin \beta_1 = V_2 y_2 \sin(\beta_1 - \theta) \tag{3.22}$$

and the momentum equation components parallel and perpendicular to the wave front are given by, respectively,

$$V_1 \cos \beta_1 = V_2 \cos(\beta_1 - \theta) \tag{3.23}$$

$$\sin \beta_1 = \frac{1}{\mathbf{F}_1} \left[\frac{y_2}{2y_1} \left(\frac{y_2}{y_1} + 1 \right) \right]^{1/2} \tag{3.24}$$

Now Equations 3.22, 3.23, and 3.24 can be solved for β_1, V_2, and y_2, given the value of the contraction angle θ and the approach Froude number $\mathbf{F}_1$. With these results, the solution can be repeated across the second set of standing wave fronts in Figure 3.19 to obtain β_2, V_3, and y_3. First Equation 3.22 is divided by Equation 3.23 to yield

$$\frac{y_2}{y_1} = \frac{\tan \beta_1}{\tan(\beta_1 - \theta)} \tag{3.25}$$

which is substituted into Equation 3.24 to obtain

$$\sin \beta_1 = \frac{1}{\mathbf{F}_1} \left[\frac{1}{2} \frac{\tan \beta_1}{\tan(\beta_1 - \theta)} \left(\frac{\tan \beta_1}{\tan(\beta_1 - \theta)} + 1 \right) \right]^{1/2} \tag{3.26}$$

Finally, Equation 3.22 is written in terms of the upstream and downstream Froude numbers relative to the first wave front, $\mathbf{F}_1$ and $\mathbf{F}_2$, respectively, to give

$$\frac{\mathbf{F}_2}{\mathbf{F}_1} = \frac{\sin \beta_1}{\sin(\beta_1 - \theta)} \frac{1}{\left(\dfrac{y_2}{y_1} \right)^{3/2}} \tag{3.27}$$

Equation 3.26 is solved for β_1 for given values of θ and $\mathbf{F}_1$. Then the values of y_2 and $\mathbf{F}_2$ can be obtained from (3.25) and (3.27), respectively. If this solution procedure is repeated across the second wave front, the values of β_2, V_3, and y_3 follow.

The solution just obtained does not necessarily minimize transmission of waves downstream. An additional condition required is for the total length of the transition, L, to be exactly equal to the sum of the lengths of the two sets of standing wave fronts, L_1 and L_2:

$$L = \frac{b_1 - b_3}{2 \tan \theta} = \frac{b_1}{2 \tan \beta_1} + \frac{b_3}{2 \tan(\beta_2 - \theta)} \tag{3.28}$$

However, as shown by Sturm (1985), the condition given by (3.28) is entirely equivalent to satisfying continuity through the transition, as given by

$$\frac{1}{r} = \frac{b_1}{b_3} = \frac{\mathbf{F}_3}{\mathbf{F}_1}\left(\frac{y_3}{y_1}\right)^{3/2} \tag{3.29}$$

With Equation 3.29, the solution procedure determines a unique value of r for minimization of wave transmission, as well as β_2, V_3, and y_3, when values of θ and $\mathbf{F}_1$ are given. The solution curves are shown in Figure 3.20. Solution curves of $\theta = f(r, \mathbf{F}_1)$ also are given by Harrison (1966) and Subramanya (1982). For a given $\mathbf{F}_1$, either r or θ can be given but not both (Harrison 1966; Sturm 1985). For example, from given values of r and $\mathbf{F}_1$, Figure 3.20 determines the unique values of θ and y_3, while Equation 3.29 can be solved for the corresponding $\mathbf{F}_3$. The result is minimization of wave transmission.

In the lower half of Figure 3.20, the choking conditions A and B described in Chapter 2 are shown. Choking condition A is based on the occurrence of a hydraulic jump upstream of the contraction followed by passage through critical depth in the contraction. Curve A is obtained by conserving the momentum function for the hydraulic jump upstream of the contraction and specific energy through the contraction itself. The second choking criterion, given by curve B in Figure 3.20, is for the case of $\mathbf{F}_3$ becoming equal to 1, so that the flow goes directly from the approach supercritical flow to critical depth in the contraction but with energy loss included. Curve B is derived from the solution procedure just described for $\mathbf{F}_3$ approaching a value of unity so that energy loss is inherently included, as shown by Sturm (1985). If θ lies between curves A and B, choking may or may not occur, depending on the existence of a hydraulic jump, but if it is to the right of curve B, choking definitely will occur.

The energy loss associated with choking condition B can be derived by writing the energy equation between sections 1 and 3 in Figure 3.19, including the unknown energy loss ΔE. Then, the specific energy at section 3 is set equal to its minimum value for which $\mathbf{F}_3 = 1$, and the equation is solved for ΔE in dimensionless form to produce (Sturm 1985)

$$\frac{\Delta E}{E_1} = \frac{1}{\left(1 + \dfrac{\mathbf{F}_1^2}{2}\right)}\left[1 + \frac{\mathbf{F}_1^2}{2} - \frac{3}{2}\left(\frac{\mathbf{F}_1}{r_c}\right)^{2/3}\right] \tag{3.30}$$

in which r_c = critical contraction ratio given by curve B for which $\mathbf{F}_3 = 1$. Values of $\Delta E/E_1$ along curve B exceed 0.1, or 10 percent, only for values of Froude number $\mathbf{F}_1$ in excess of approximately 4.

Example 3.4

A straight-walled rectangular contraction has an approach channel width of 3.0 m (9.8 ft) and a contracted width of 1.5 m (4.9 ft). The approach flow has a depth of 0.10 m (0.33 ft) and a velocity of 3.0 m/s (9.8 ft/s). What are the values of downstream depth and velocity and what should the contraction angle and length be to minimize transmission of standing waves? Will choking occur?

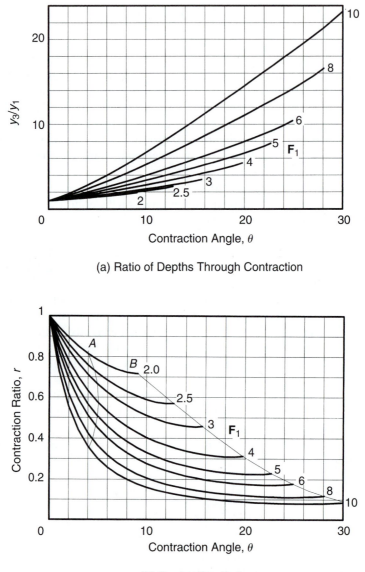

(a) Ratio of Depths Through Contraction

(b) Contraction Ratio, r

FIGURE 3.20
Supercritical contraction with the minimization of standing waves: A, B = choking criteria (Sturm 1985). (*Source: T. W. Sturm, "Simplified Design of Contractions in Supercritical Flow," J. Hydr. Engrg., © 1985, ASCE. Reproduced by permission of ASCE.*)

Solution.   The approach Froude number is $V_1/(gy_1)^{0.5} = 3.0/(9.81 \times 0.1)^{0.5} = 3.0$, while $r = b_3/b_1 = 1.5/3 = 0.5$. Then, from Figure 3.20, $\theta = 11°$ and $y_3/y_1 = 2.5$ approximately, so that $y_3 = 2.5 \times 0.1 = 0.25$ m (0.82 ft). From Equation 3.29, $\mathbf{F}_3 = (1/r)(\mathbf{F}_1)/(y_3/y_1)^{1.5} = (1/0.5) \times 3.0/2.5^{1.5} = 1.52$ and so $V_3 = 1.52 \times (9.81 \times 0.25)^{0.5} = 2.38$ m/s (7.81 ft/s).

The length of the transition from (3.28) with $\theta = 11°$ is $(b_1 - b_3)/(2 \tan \theta) = (3 - 1.5)/(2 \tan 11°) = 3.86$ m (12.7 ft). The solution lies in the region between curves A and B so that choking is possible. Note that $\theta < 5°$ is required for choking not to occur under any circumstance. In this example, $\theta = 5°$ would necessitate limiting the contraction ratio, r, to approximately 0.67 for an approach Froude number of 3 and the contraction length would increase to 8.57 m (28.1 ft).

Design of Supercritical Expansion

In some instances, it may be desirable to design an expansion for supercritical flow at points where high velocity, supercritical flow issues from sluice gates, spillways, or steep chutes. As described by Chow (1959), the flow will separate if the expansion is too abrupt; and the transition may be too long if the flow is forced to expand too gradually. In addition, local standing waves may emanate from the walls of the transition and combine at the centerline with further propagation downstream. Rouse, Bhootha, and Hsu (1951) studied this problem both experimentally and analytically. They suggest a two-part wall curvature, as shown in Figure 3.21. The most efficient shape for the divergent portion of the expansion is given by

$$\frac{z}{b_1} = \frac{1}{2}\left[\frac{1}{4}\left(\frac{x}{b_1\mathbf{F}_1}\right)^{3/2} + 1\right] \tag{3.31}$$

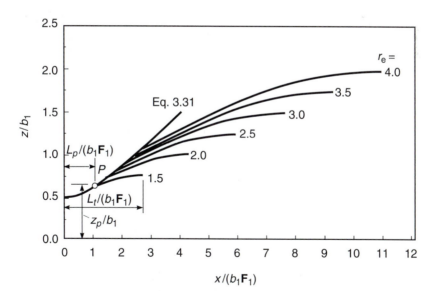

FIGURE 3.21
Generalized boundary curves for expansion (best fit of Rouse et al. (1951) curves by Mazumdar and Hager (1993). (*Source: H. Rouse, B. V. Bhootha, and E. Y. Hsu, "Design of Channel Expansions," © 1951, ASCE. Reproduced by permission of ASCE.*)

in which z = lateral position of the wall from the centerline of the expansion; b_1 = approach channel width; x = longitudinal coordinate measured from the beginning of the expansion; and $\mathbf{F}_1$ = Froude number of the approach flow. This curve continues diverging in the downstream direction, which requires the second portion of the transition wall geometry downstream of point P in Figure 3.21, for example. It consists of a reverse curvature obtained from an analysis of the positive and negative disturbances from the wall to promote cancellation of their effects and elimination of the propagation of excessive standing waves.

The length of the first portion of the transition, L_p, and the total length of the transition, L_t, are given by Mazumder and Hager (1993) to be

$$\frac{L_p}{b_1\mathbf{F}_1} = 0.7r_e \tag{3.32}$$

$$\frac{L_t}{b_1\mathbf{F}_1} = 1 + 3.25(r_e - 1) \tag{3.33}$$

in which r_e = expansion ratio = b_2/b_1, and the lengths L_t and L_p are as defined in Figure 3.21. The geometry of the reverse curvature downstream of point P is given approximately by a best fit of the original boundary curves of Rouse, Bhootha, and Hsu (1951) obtained by Mazumder and Hager (1993):

$$\frac{z - z_p}{b_2/2 - z_p} = \sin\left[90° \frac{x - L_p}{L_t - L_p}\right] \tag{3.34}$$

in which z_p is determined from Equation 3.31 for $x = L_p$ given by Equation 3.32. Mazumder and Hager (1993) experimentally studied expansions designed according to the generalized Rouse et al. boundary curves and concluded that the maximum Froude number can be as much as 2.5 times the design Froude number without significantly changing the wave heights or the pattern of standing waves. As a practical matter, this means that the expansion can be shorter, because it can be designed for a smaller Froude number.

REFERENCES

Arbhabhirama, A., and A. Abella. "Hydraulic Jump Within Gradually Expanding Channel." *J. Hyd. Div.*, ASCE 97, no. HY1 (1971), pp. 31–41.

Bradley, J. N., and A. J. Peterka. "The Hydraulic Design of Stilling Basins: Hydraulic Jumps on a Horizontal Apron (Basin I)." *J. Hyd. Div.*, ASCE 83, no. HY5 (1957), pp. 1–24.

Carollo, F. G., V. Ferro, and V. Pampalone. "Hydraulic Jumps on Rough Beds." *J. Hydr. Engrg.*, ASCE 133, no. 9 (2007), pp. 989–99.

Chow, Ven Te. *Open-Channel Hydraulics*. New York: McGraw-Hill, 1959.

Ead, S. A., and N. Rajaratnam. "Hydraulic Jumps on Corrugated Beds." *J. Hydr. Engrg.*, ASCE 128, no. 7 (2002), pp. 656–63.

Harleman, D. R. F. Discussion of "Turbulence Characteristics of the Hydraulic Jump," by H. Rouse, T. T. Siao, and S. Nagaratnam. *Transactions of the ASCE*, vol. 124. 1959, pp. 959–62.

Harrison, A. J. M. "Design of Channels for Supercritical Flow." *Proc. Inst. of Civil Engrs.* 35 (1966), pp. 475–90.

Hughes, W. J., and J. E. Flack. "Hydraulic Jump Properties over a Rough Bed." *J. Hydr. Engrg.*, ASCE 110, no. 12 (1984), pp. 1755–71.

Ippen, A. T., and J. H. Dawson. "Design of Channel Contractions." In *High Velocity Flow in Open Channels: A Symposium, Transactions of the ASCE*, vol. 116. 1951, pp. 326–46.

Keller, R. J., and A. K. Rastogi. "Design Chart for Predicting Critical Point on Spillways." *J. Hyd. Div.*, ASCE 103, no. HY12 (1977), pp. 1417–29.

Khalifa, A. M., and J. A. McCorquodale. "Radial Hydraulic Jump." *J. Hyd. Div.*, ASCE 105, no. HY9 (1979), pp. 1065–78.

Koloseus, H. J., and D. Ahmad. "Circular Hydraulic Jump." *J. Hyd. Div.*, ASCE 95, no. HY1 (1969), pp. 409–22.

Lawson, J. D., and B. C. Phillips. "Circular Hydraulic Jump." *J. Hydr. Engrg.*, ASCE 109, no. 4 (1983), pp. 505–18.

Leutheusser, H. J., and V. Kartha. "Effects of Inflow Condition on Hydraulic Jump." *J. Hyd. Div.*, ASCE 98, no. HY8 (1972), pp. 1367–85.

Liu, M., N. Rajaratnam, and D. Z. Zhu. "Turbulence Structure of Hydraulic Jumps of Low Froude Numbers." *J. Hydr. Engrg.*, ASCE 130, no. 6 (2004), pp. 511–20.

Massey, B. S. "Hydraulic Jump in Trapezoidal Channels, An Improved Method." *Water Power* (June 1961), pp. 232–37.

Mazumder, S. K., and W. H. Hager. "Supercritical Expansion Flow in Rouse Modified and Reverse Transitions." *J. Hydr. Engrg.*, ASCE 119, no. 2 (1993), pp. 201–19.

Rajaratnam, N. "The Hydraulic Jump as a Wall Jet." *J. Hyd. Div.*, ASCE 91, no. HY5 (1965), pp. 107–32.

Rouse, H., B. V. Bhootha, and E. Y. Hsu. "Design of Channel Expansions." *Transactions of the ASCE*, vol. 116. 1951, pp. 1369–85.

Rouse, H., T. T. Siao, and S. Nagaratnam. "Turbulence Characteristics of the Hydraulic Jump." *J. Hyd. Div.*, ASCE 84, no. HY1 (1958), pp. 1528-1 to 1528-30.

Silvester, R. "Hydraulic Jump in All Shapes of Horizontal Channels." *J. Hyd. Div.*, ASCE 90, no. HY1 (1964), pp. 23–55.

Sturm, T. W. "Simplified Design of Contractions in Supercritical Flow." *J. Hydr. Engrg.*, ASCE 111, no. 5 (1985), pp. 871–75.

Subramanya, K. *Flow in Open Channels*, vol. 2. New Delhi, India: Tata McGraw-Hill, 1982.

Thiruvengadam, A. "Hydraulic Jump in Circular Channels." *Water Power* (December 1961), pp. 496–97.

U.S. Bureau of Reclamation, Department of the Interior. *Design of Small Dams*, 3rd ed., a Water Resources Technical Publication. Washington, DC: U.S. Government Printing Office, 1987.

EXERCISES

3.1. A hydraulic jump is to be formed in a trapezoidal channel with a base width of 20 ft and side slopes of 2:1. The upstream depth is 1.25 ft and $Q = 1000$ cfs. Find the downstream depth and the head loss in the jump. Solve by Figure 3.2 and verify by manual calculations. Compare the results for the sequent depth ratio and relative head loss with those in a rectangular channel of the same bottom width and approach Froude number.

3.2. Determine the sequent depth for a hydraulic jump in a 3 ft diameter storm sewer with a flow depth of 0.6 ft at a discharge of 5 cfs. Solve by Figure 3.3 and verify by manual calculations.

3.3. Derive the relationship between the sequent depth ratio and approach Froude number for a triangular channel and verify with Figure 3.5. Repeat the derivation for a parabolic channel.

3.4. A flume with a triangular cross section contains water flowing at a depth of 0.15 m and at a discharge of 0.30 m³/s. The side slopes of the flume are 2:1. Determine the sequent depth for a hydraulic jump.

3.5. A parabolic channel has a bank-full depth of 2.0 m and a bank-full width of 10.0 m. If the downstream sequent depth of a hydraulic jump in the channel is 1.5 m for a flow rate of 8.5 m³/s, what is the upstream sequent depth?

3.6. A hydraulic jump occurs on a sloping rectangular channel that has an angle of inclination, θ. The sequent depths are d_1 and d_2 measured perpendicular to the channel bottom. Assume that the jump has a length, L_j, and a linear profile. Derive the solution for the sequent depth ratio, and show that it is identical to the solution for a horizontal slope if the upstream Froude number, $\mathbf{F}_1$, is replaced by the dimensionless number, $\mathbf{G}_1$, given by

$$\mathbf{G}_1 = \frac{\mathbf{F}_1}{\sqrt{\cos\theta - \dfrac{L_j \sin\theta}{d_2 - d_1}}}$$

3.7. The supercritical flow at the toe of a spillway has a depth of 1.12 ft and velocity of 60 ft/s.
(a) Calculate the required tailwater depths and the required lengths of a Type III and a Type II stilling basin and compare with those for a free jump without blocks or bottom roughness.
(b) If a stilling basin without blocks is constructed of grouted rock riprap with a roughness height of 0.25 ft, what would be the sequent depth and length of a hydraulic jump in the basin?

3.8. The discharge of water over a spillway 40 ft wide is 10,000 cfs into a stilling basin of the same width. The lake level behind the spillway has an elevation of 200 ft, and the river water surface elevation downstream of the stilling basin is 100 ft. Assuming a 10 percent energy loss in the flow down the spillway, find the invert elevation of the floor of the stilling basin so that the hydraulic jump forms in the basin. Select the appropriate U.S. Bureau of Reclamation (USBR) stilling basin and sketch it showing all dimensions.

3.9. Recalculate the elevation Z of the floor of the stilling basin in Exercise 3.8 for $Q = 10,000$ cfs if the width of the basin is 80 ft instead of 40 ft. For discharges less than the design discharge, will the elevation of the water surface after the jump be greater or less for a width of 80 ft in comparison to 40 ft?

3.10. A spillway chute and the hydraulic jump stilling basin at the end of the chute are rectangular in shape with a width of 80 ft. The floor of the stilling basin is at an elevation 787.6 ft above the datum. The incoming flow has a depth of 2.60 ft at a design discharge of 9500 cfs. Within the basin are 15 baffle blocks, each 2.5 ft high and 2.75 ft wide.
(a) Assuming an effective coefficient of drag of 0.5 for the baffle blocks, based on the upstream velocity and combined frontal area of the blocks, calculate the sequent depth and compare with the sequent depth without baffle blocks.
(b) What is the energy loss in the basin with and without the blocks?
(c) If the tailwater elevation for $Q = 9500$ cfs is 797.6, will the stilling basin perform as designed? Explain your answer.

3.11. A horizontal laboratory flume has a width of 1.25 ft and a maximum flow depth of 2.5 ft upstream of a vertical sluice gate. If a free hydraulic jump is positioned just downstream of the gate, calculate its supercritical Froude number and the sequent depth of the jump for $Q = 0.5$ cfs and 2.0 cfs if the gate opening is adjusted so that the depth upstream of the gate is the same with a value of 2.5 ft for both discharges.

3.12. In a short, horizontal, rectangular flume in the laboratory, the depth just downstream of a sluice gate at the upstream end of the flume is 1.0 cm and the depth just upstream of the sluice gate is 60 cm. The width of the flume is 38 cm. If the tailgate height is 15 cm, over which there is a free overfall, will a hydraulic jump occur or will it be submerged?

3.13. A circular culvert has a diameter of 1.8 m. The discharge is 2.8 m³/s. Near the downstream end of the culvert, the flow is uniform with a depth of 0.70 m. The tailwater is 1.2 m above the outlet invert. Will a hydraulic jump occur in the culvert, or will it be swept out?

3.14. A steady flow is occurring in a rectangular channel, and it is controlled by a sluice gate. The upstream depth is 1.0 m, and the upstream velocity is 3.0 m/sec. If the gate is slammed shut abruptly, determine the depth and speed of the resulting surge.

3.15. The depths upstream and downstream of a sluice gate in a rectangular channel are 8 ft and 2 ft, respectively, for a steady flow.
(a) What is the value of the flow rate per unit of width q?
(b) If q in part (a) is reduced by 50 percent by an abrupt partial closure of the gate, what will be the height and speed of the surge upstream of the gate?

3.16. Write both the momentum and energy equations for the subcritical case of flow through bridge piers of diameter a and spacing s. If the head loss in the energy equation is written as $K_L V_1^2/2g$, in which K_L is the head loss coefficient and V_1 is the approach velocity, show that $K_L = C_D a/s$ for the special case that $(y_1 - y_4)$ is very small (Figure 3.16).

3.17. For a river flow between bridge piers 3 m in diameter with a spacing of 20 m, determine the backwater using the momentum method if the downstream depth is 4.0 m and the downstream velocity is 1.9 m/s. Assume a coefficient of drag of 2.0 for the bridge piers.

3.18. The main river channel downstream of a bridge can be approximated as a rectangular cross section in which the flow depth is 2.0 m and the velocity is 3.0 m/s. The bridge has circular piers with a diameter of 1.0 m in the main channel. What pier spacing s would cause the Type II choking case shown in Figure 3.16 and what would be the value of the backwater?

3.19. A straight-walled contraction connects two rectangular channels 12 ft and 6 ft wide. The discharge through the contraction is 200 cfs, and the depth of the approach flow is 0.7 ft. Calculate the downstream depth, Froude number, and the length of the contraction that will minimize standing waves. Will choking be a problem?

3.20. For Exercise 3.19, calculate the wave-front angles β_1 and β_2. What variables do they depend on? Produce a generalized plot for β in terms of the dimensionless variables on which it depends.

3.21. The approach flow in a supercritical contraction has a discharge per unit width of 6.0 m²/s and a depth of 0.6 m. Calculate the limiting discharges per unit width in the contracted section and the values of r for Type A and Type B choking assuming no head loss except that in the hydraulic jump for Type A. Compare the limiting value of r for Type B choking with Figure 3.20 and explain any difference.

3.22. A supercritical transition expands from a width of 1.0 m to 3.0 m, and the approach flow depth and velocity are 0.64 m and 10 m/s for maximum design conditions, respectively. Calculate the downstream depth, and design and plot the transition geometry. What would be the length of the expansion if it were designed for a Froude number that is 40 percent of the design value?

3.23. Write a computer program that finds the sequent depth for a hydraulic jump in a trapezoidal channel. First, compute critical depth and determine if the given depth is subcritical or supercritical to limit the root search. Then, use the bisection method to find the sequent depth.

3.24. The following data for a hydraulic jump have been measured in a laboratory flume by two different lab teams. The flume has a width of 38 cm. The upstream sluice gate was set to produce a supercritical flow for a given measured discharge, and the tailgate was adjusted until the hydraulic jump was positioned at the desired location in the flume. The depths measured by a point gauge upstream and downstream of the jump, y_1 and y_2, respectively, are given in the following table as is the discharge measured by a calibrated bend meter with an uncertainty of ± 0.0001 m³/s. The estimated uncertainty in the upstream depths is ± 0.02 cm, while the downstream depths have a larger estimated uncertainty of ± 0.30 cm due to surface waves. Photographs of the flume and hydraulic jumps at selected Froude numbers are shown in Figure 3.22.

	Team A			Team B	
y_1, cm	y_2, cm	Q, m³/s	y_1, cm	y_2, cm	Q, m³/s
1.41	15.8	0.0165	1.47	15.4	0.0166
1.62	14.2	0.0165	1.76	14.5	0.0166
2.22	12.3	0.0165	2.20	12.8	0.0166
3.10	10.2	0.0165	2.63	11.0	0.0166
1.30	13.3	0.0131	1.21	13.0	0.0125
1.38	13.0	0.0131	1.45	11.7	0.0125
1.59	11.9	0.0131	1.70	10.9	0.0125
2.10	9.9	0.0131	1.87	10.0	0.0125

(a) Calculate and plot the sequent depth ratios as a function of the Froude numbers of the experimental data and compare them with the theoretical relationship for a hydraulic jump in a rectangular channel.

(b) Calculate and plot the dimensionless energy loss E_L/E_1 as function of Froude number for the experimental data and compare it with the theoretical relationship.

(a) Froude number = 1.9

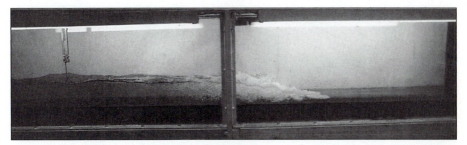

(b) Froude number = 4.4

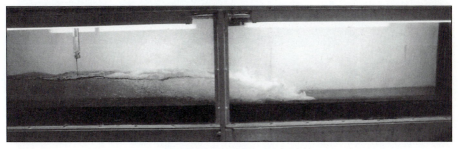

(c) Froude number = 7.2

FIGURE 3.22
Hydraulic jump with different upstream Froude numbers (photographs by G. Sturm).

(*c*) Estimate the experimental uncertainty in y_2/y_1 and the Froude number and plot error bars on your graphs. Does the estimated uncertainty account for the differences between measured and theoretical values? Are the results for Team A and Team B consistent?

(*d*) Based on the photos in Figure 3.22, describe the appearance of the jump as a function of Froude number and indicate the relative energy loss E_L/E_1 for each photo.

4

Uniform Flow

4.1 INTRODUCTION

Uniform flow in open channels often is used as a design condition to determine the dimensions of artificial channels. The design discharge is set by considerations of acceptable risk and frequency analysis, and the channel slope and cross-sectional shape are determined by topography, soil conditions, and availability of land. Specification of the resistance coefficient results then in a unique value of the depth of uniform flow, known as the *normal depth*. The determination of normal depth establishes the position of the free surface and the required channel depth necessary to complete the design of the channel dimensions. The resistance coefficient is a vital link in this design process, and its estimation has commanded the attention of hydraulic engineers since the 19th century. An understanding of its variation with the surface roughness of the channel developed slowly, and only in recent times have other factors that influence its value been studied.

The hydraulic resistance of conduits flowing full is one of the most extensively studied areas in hydraulic engineering, but many difficulties remain for the case of flow resistance in open channels. In the case of full pipe flow, Nikuradse's experiments on sand-grain roughened pipes, and the subsequent work by Colebrook (1939) and Moody (1944), led to the development of the friction factor-Reynolds number plot, now known as the *Moody diagram,* in which relative roughness is a parameter. The Moody diagram has been applied with considerable success by practicing engineers to the problem of determining pipe flow resistance. Flow resistance in open channels, on the other hand, has been more difficult to quantify. A much wider range of roughness is encountered in open channel flow, and the extra degree of freedom offered by the free surface in open channel flow gives rise to the complex effects of nonuniformity, cross-sectional shape, and surface waves.

The importance of the resistance coefficient goes beyond its use in channel design for uniform flow. The computation of flood stages in gradually varied flow and of the movement of translatory waves in unsteady flow depend on an accurate estimate of the resistance coefficient. Much of our present understanding of the resistance coefficient is due to a combination of theory and experiment applied to uniform flow, but much remains to be learned about flow resistance in gradually varied and unsteady flow.

Determining flow resistance in movable-bed channels is especially challenging because of bed forms, such as dunes and ripples, that create form resistance that varies with the flow conditions. Further discussion of this case can be found in Chapter 10.

4.2   DIMENSIONAL ANALYSIS

Because of the significant role played by experimental work in establishing values of flow resistance, it is useful to begin with a dimensional analysis of the problem. For a channel of any general shape, we write the functional dependence of the mean boundary shear stress τ_0 as (Rouse 1965)

$$\tau_0 = f_1(\rho, \mu, g, V, R, k, \mathbf{C}, \mathbf{N}, \mathbf{U}) \tag{4.1}$$

in which ρ = fluid density; μ = fluid viscosity; g = gravitational acceleration; V = mean cross-sectional flow velocity; R = hydraulic radius, which is a characteristic length scale of the flow, defined as flow area divided by wetted boundary perimeter; and k = measure of roughness element height. The last three parameters on the right of Equation 4.1 already are dimensionless. The parameter $\mathbf{C}$ reflects the effect of cross-sectional shape; $\mathbf{N}$ indicates the degree of nonuniformity of flow; and $\mathbf{U}$ represents unsteadiness effects. Dimensional analysis of the functional relation given as Equation 4.1 yields

$$\frac{\tau_0}{\rho V^2} = f_2\left(\mathbf{Re} = \frac{\rho V(4R)}{\mu}, \mathbf{Rr} = \frac{k_s}{4R}, \mathbf{F}, \mathbf{C}, \mathbf{N}, \mathbf{U} \right) \tag{4.2}$$

in which $\mathbf{Re}$ = Reynolds number; $\mathbf{Rr}$ = relative roughness; $\mathbf{F}$ = Froude number; and $\mathbf{C}$, $\mathbf{N}$, and $\mathbf{U}$ already have been defined. The length scale used in the Reynolds number and relative roughness is four times the hydraulic radius, and this will be justified subsequently. From the control-volume form of the momentum equation applied to a steady, uniform pipe flow and the Darcy-Weisbach equation, which defines the friction factor, f, in terms of pipe diameter as the length scale, the dependent dimensionless parameter on the left of Equation 4.2 becomes

$$\frac{\tau_0}{\rho V^2} = \frac{f}{8} \tag{4.3}$$

in which V is the cross-sectional mean velocity. Equation 4.3 can be taken as the definition of the Darcy-Weisbach friction factor f, and we want the definition of f to remain the same for open channel flow. The functional relation of Equation 4.2 is the basis for the Moody diagram, which gives values of the friction factor, f, in pipe flow as a function of Reynolds number and relative roughness with the influences represented by the remaining dimensionless parameters in Equation 4.2 neglected. In open channel flow, the Reynolds number often is large, so that the flow is in the fully rough turbulent regime and the primary independent parameter is the relative roughness.

4.3   MOMENTUM ANALYSIS

Consider a control volume of length ΔL in steady, uniform flow, as shown in Figure 4.1. By definition, the hydrostatic forces, F_{p1} and F_{p2}, are equal and opposite. In addition, the mean velocity is invariant in the flow direction, so that the change in momentum flux is zero. Thus, the momentum equation reduces to a balance between the gravity force component in the flow direction and the resisting shear force:

$$\gamma A \Delta L \sin \theta = \tau_0 P \Delta L \tag{4.4}$$

in which γ = specific weight of the fluid; A = cross-sectional area of flow; τ_0 = mean boundary shear stress; and P = wetted perimeter of the boundary on which the shear stress acts. If Equation 4.4 is divided through by $P\Delta L$, the hydraulic radius $R = A/P$ appears as an intrinsic variable from the momentum analysis. Physically, it represents the ratio of flow volume to boundary surface area, or shear stress to unit weight, in the flow direction. Equation 4.4 can be written as

$$\tau_0 = \gamma R \sin \theta = \gamma R S \tag{4.5}$$

if we replace $\sin\theta$ with $S = \tan\theta$ for small values of θ. Furthermore, if we solve Equation 4.5 for the bed slope, which equals the energy grade line slope, h_f/L, in steady uniform flow, and express the shear stress in terms of the pipe flow definition of friction factor f from Equation 4.3, we have

$$\frac{h_f}{L} = \frac{\tau_0}{\gamma R} = \frac{f\rho V^2/8}{\gamma R} = \frac{f}{4R}\frac{V^2}{2g} \tag{4.6}$$

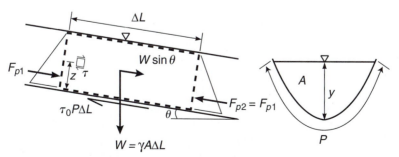

FIGURE 4.1
Force balance in uniform flow.

from which it is evident that the appropriate length scale in the Darcy-Weisbach equation for open channel flow is $4R$ to replace the pipe diameter. It would seem reasonable to use $4R$ as the length scale in the Reynolds number and relative roughness as well. The unexpected benefit of this approach is that friction factors in turbulent open channel flow are similar (but not exactly the same) to those obtained from the Moody diagram developed from pipe flow results. In other words, the hydraulic radius embodies much of the effect of channel shape on friction factor but not all of it. The effects of nonuniform shear stress distribution and secondary currents also are related to shape and must be accounted for separately.

Before applying uniform flow formulas to the design of open channels, the background of their development is considered. First, Chezy's and Manning's formulas for steady, uniform flow in open channels are presented. Then, the equations for the friction factor as a function of Reynolds number and relative roughness in pipe flows are reviewed and extended to open channel flow. Finally, the effects of the Froude number, nonuniformity, and cross-sectional shape on open channel flow resistance are explored.

4.4 BACKGROUND OF THE CHEZY AND MANNING FORMULAS

While Equation 4.5 gives a formula for the calculation of mean shear stress in uniform flow, the problem of determining the depth of uniform flow for a given discharge requires an additional uniform flow formula. Historically, such formulas have been presented for velocity of flow as a function of hydraulic radius and slope. If Equation 4.6 is solved for velocity, we have

$$V = \left[\frac{8g}{f}\right]^{1/2} \sqrt{RS} = C\sqrt{RS} \tag{4.7}$$

in which C is called the Chezy C in honor of Antoine Chezy, who first proposed this formula. Chezy, a French engineer, was charged with the task of designing a water supply canal from the Yvette River to the city of Paris in 1768. His final recommendations in 1775 contained the Chezy formula written in terms of ratios of velocities of two rivers; and in a later memorandum in 1776, he gave the formula for velocity as we now know it. He presented a constant value for C, but he realized that it varied from one river to another. Unfortunately, Chezy's work was not published and so did not become widely known until after 1897, when it was published by Herschel in the United States (Biswas 1970). Du Buat, a contemporary of Chezy, arrived at the same uniform flow formula some four years later than Chezy, although he concluded that the effects of boundary roughness could be neglected. His work was published in an early book on hydraulics. Many other uniform flow formulas of the "universal type" with no variation of the coefficients with roughness were proposed in the early 19th century such as those of Eytelwein and Prony (Dooge 1992).

Within this context of extensive work on uniform flow in the first half of the 19th century, Robert Manning began his career as a drainage engineer in 1846. He was self-taught and greatly admired the French writings on hydraulics. The uniform flow formula bearing Manning's name was not proposed until the end of his career, when in 1889 at the age of 73 he presented it in a paper while he was still chief engineer of the Board of Works of Ireland. His formula was based primarily on the pioneering work of Darcy and Bazin on outdoor experimental canals from 1855 to 1860. This work was published by Bazin in 1865 after the death of Darcy and it showed conclusively that the Chezy C depended on the nature of the surface roughness of the canal boundaries (Dooge 1992).

In his 1889 paper, Manning selected seven well-known uniform flow formulas for velocity in an open channel expressed as a function of hydraulic radius and slope. He calculated the velocities over a range of hydraulic radii from 0.25 to 30 m from each formula for a given slope and analyzed the mean of the results. From these preliminary results, he concluded that the velocity was proportional to the hydraulic radius to the $\frac{4}{7}$ power and to the slope to the $\frac{1}{2}$ power, but he realized there might be a more generally applicable value of the exponent on hydraulic radius. He then took the crucial step of analyzing the results of some selected experiments of Bazin on semicircular canals lined with cement and with a sand-cement mixture. Manning concluded that the exponent in both cases was very close to the fraction $\frac{2}{3}$. The resulting uniform flow formula was given as formula V in the 1889 paper:

$$V = C_1 R^{2/3} S^{1/2} \qquad (4.8)$$

in which the subscript on C has been added to distinguish the coefficient from the Chezy C. Manning proceeded to compare the results of this formula with 170 observations, of which 104 were those of Bazin. He concluded that this formula performed better than those of Bazin and Kutter, the latter of which was very popular at the time.

Manning was dissatisfied with formula V however because of its lack of dimensional homogeneity and the necessity of taking a cube root in the evaluation of the velocity. He therefore proposed a second formula, formula I, which overcame these objections, although it used the barometric pressure head to achieve an artificial nondimensionality. It performed nearly as well as formula V and seemed to be Manning's formula of choice. Ironically, this formula has been discarded and formula V bears Manning's name. Manning concluded his paper with the following statement: "Although the author makes no pretension to mathematical skill, he may, in conclusion, be allowed to express the hope that the results of his labors, such as they are, may advance, even in a small degree, a science, in the study and practice of which he has spent a long professional life."

The dissemination of the uniform flow formula that now bears Manning's name was greatly enhanced by Flamant's publication of it in his 1891 textbook and his reference to formula V as Manning's formula. A careful review of the historical record by Williams (1970), however, shows that some 10 investigators proposed a formula of this type. The first suggestion of the exponent of $\frac{2}{3}$ on the hydraulic radius actually was made by the French engineer Gauckler in 1867. Gauckler's formula also was based on Darcy and Bazin's experiments but it never received wide acceptance, partly because of the widespread use of a formula proposed by Ganguillet and Kutter in 1869 for Chezy's C. This formula for C was very complex and had a dependence on the slope and a single roughness coefficient, n, called Kutter's n. This was the result of attempting to reconcile Darcy and Bazin's data on small canals of moderate slope with the observations of Humphrey and Abbott on the Mississippi River for very small slopes. Manning, in fact, eliminated Humphrey and Abbott's data from his 170 observations because of the difficulty he perceived in measuring such small slopes and showed no small disdain for the complexity of Kutter's formula in his 1889 paper.

The final ironic twist in the development of what is now known as *Manning's formula* was the suggestion by Flamant that C_1 in Manning's formula V could be expressed as the reciprocal of Kutter's *n* in metric units. Several subsequent texts repeated this assertion, and the American hydraulician King (1918) advocated this step while referring to *n* as "Manning's *n*." What we now know as Manning's formula, which Williams (1970) suggests really should be called the *Gauckler-Manning formula,* is written today as

$$V = \frac{K_n}{n} R^{2/3} S^{1/2} \tag{4.9}$$

in which V is velocity; R is hydraulic radius; and S is bed slope. The value of $K_n = 1.0$ with R in m and V in m/s, and $K_n = 1.49$ for R in ft and V in ft/s. The latter value comes from a conversion in which Manning's n maintains the same value in either SI or English units, so that the dimensional units of K_n/n, originally m$^{1/3}$/s, have to be converted to ft$^{1/3}$/s by the factor $(3.28 \text{ ft/m})^{1/3} = 1.49$. That Equation 4.9 has endured for more than a century as a uniform flow formula would seem to indicate that Manning's labors were not in vain, although the formula that bears his name probably would be surprising to him.

4.5   TURBULENCE AND FLOW RESISTANCE

Our primary interest in this chapter is flow resistance in turbulent open channel flow. Some preliminary concepts are important in laying the groundwork for the logarithmic velocity distribution in turbulent wall-bounded flows and the resulting flow resistance laws that follow. Many investigators such as Prandtl, Taylor, von Karman, Richardson, and Kolmogorov tackled the problem of turbulence in the 20th century leading to some important physical insights and statistical analyses, but turbulence remains an ongoing and vital area of research in fluid mechanics that depends on experimental results as well as numerical analysis.

Concepts of Turbulence

As defined in Chapter 1, turbulence is an inherent instability in the flow that can be visualized as eddies or swirls existing from the largest scale of the order of the channel flow depth to the smallest scale of the order of tenths of a mm at which energy is dissipated. An essential characteristic of turbulence is its efficient mixing of momentum effected by the fluctuating vorticity associated with the eddies. This momentum exchange results in the creation of surface drag force at a solid boundary, which is the open channel flow resistance problem of interest to us in this chapter.

Turbulence manifests itself as small fluctuations with time and space in flow variables such as velocity as illustrated in Figure 4.2. A snapshot of the time record of instantaneous velocity measurements at a point as shown in Figure 4.2b

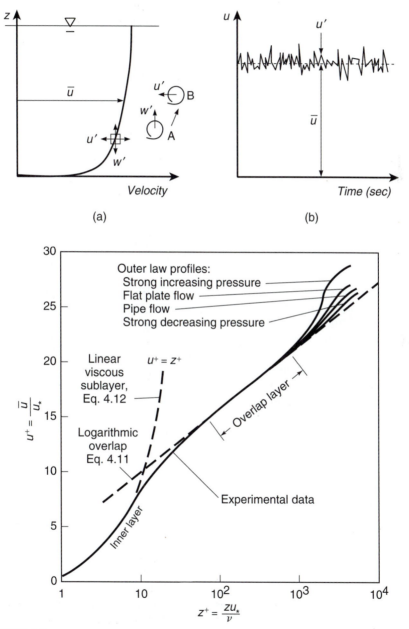

(a)   (b)

FIGURE 4.2

(a) Generation of Reynolds stress in turbulent flow; (b) time record of turbulent velocity fluctuations at a point; (c) velocity profiles in turbulent wall flow (White 2008).

displays an irregular rather than smooth appearance due to fluctuations that vary with time at each spatial point in the turbulent velocity profile shown in Figure 4.2a. The jagged appearance of the velocity record is a random set of fluctuations that cannot be repeated exactly except in terms of statistical means; in other words, the fluctuations themselves can be considered a random variable. In order to describe a turbulent flow, the instantaneous velocity u is decomposed into a time-averaged value denoted by an overbar and a fluctuating component u' illustrated for the x component of the velocity as $u = \bar{u} + u'$ where the time-averaged value is defined in Figure 4.2b by

$$\bar{u} = \frac{1}{T} \int_{t_0}^{t_0 + T} u(t) \, dt \qquad (4.10)$$

with the averaging period T taken to be much longer than the time scale of the fluctuations. If the time-averaged velocity is constant with time, then the flow can be treated as steady in the mean or, in statistical terms, as a stationary process.

The manner in which turbulent velocity fluctuations result in momentum exchange between layers of fluid leading to flow resistance is illustrated in Figure 4.2a. If at point A in the fluid at elevation z, a fluid particle with a mean velocity in the flow direction of $\bar{u}(z)$ is carried upward by a positive vertical fluctuation w' in the characteristic time t, it will encounter another fluid element at point B with a higher mean velocity in the flow or x direction. The result will be an effective negative fluctuation $u'(z + dz)$ associated with a positive vertical fluctuation w' leading to a net time-averaged momentum exchange per unit surface area given by $-\rho \overline{u'w'}$. Such a momentum exchange is equivalent to an apparent resisting shear stress called the Reynolds stress. Because it results from turbulent fluctuations at a scale much larger than the molecular scale at which viscous shear is spawned, it is much larger than the viscous stress over most of the flow depth except very near the wall and at the wall itself where it must become zero. In reality, the nonlinear convective momentum flux associated with the turbulent motion has been moved to the force side of the momentum equation and treated as a flow resistance term in the form of a Reynolds stress. Then the total shear stress in turbulent flow is written as the sum of the viscous and turbulent components

$$\tau = -\rho \overline{u'w'} + \mu \frac{d\bar{u}}{dz} \qquad (4.11a)$$

An alternative formulation of the Reynolds stress is to represent it as some eddy viscosity times the velocity gradient in analogy with molecular transfer of momentum. This concept attributed to Boussinesq is written as

$$-\rho \overline{u'w'} = \rho \varepsilon \frac{d\bar{u}}{dz} \qquad (4.11b)$$

in which ε is the eddy viscosity or turbulent momentum diffusivity. The difficulty with the eddy viscosity concept is that it depends on the turbulence rather than being a fluid property; nevertheless, it has useful applications in describing the turbulent diffusivity of suspended sediment as discussed in Chapter 10.

The distribution of the total shear stress can be derived for steady uniform flow by referring back to Figure 4.1 in which an element of fluid in the xz plane, on which the shear stress τ is acting, is shown. As in consideration of the larger finite control volume in Figure 4.1, there is no net pressure force on the fluid element in the flow direction resulting only in a balance between the net shear stress and the gravity force component in the flow direction per unit volume

$$\frac{d\tau}{dz} = -\gamma \sin\theta \tag{4.12}$$

Integration of Equation 4.12 for a constant channel slope produces a linear shear stress distribution across the flow depth with a maximum value of τ_0 at the wall and a value of zero at the free surface as discussed in Chapter 10 and shown in Figure 10.18.

Turbulent Velocity Distribution and Friction Factors

Once the linearity of shear stress distribution in steady uniform flow is established, regardless of whether the flow is laminar or turbulent, it is a straightforward process to combine it with the constitutive relation for the viscous stress (Eq. 4.11a with zero Reynolds stress) to derive the parabolic velocity distribution of laminar flow. In the case of turbulent flow, the turbulence is not a property of the fluid as is viscosity but rather a property of the flow. Consequently, the approach to obtaining the velocity distribution is by necessity different.

While Prandtl deduced a logarithmic velocity distribution in turbulent wall-bounded flow from a semiempirical mixing-length theory and von Karman from a similarity hypothesis, it is easily demonstrated to exist in laboratory experiments. Given that is the case, the logarithmic velocity distribution follows more readily from dimensional arguments. Very near a smooth wall, the time-averaged point velocity $(\bar{u})$ must depend on the fluid viscosity (μ), fluid density(ρ), and the wall shear stress (τ_0) in addition to the distance from the wall (z). A dimensional analysis results in

$$\frac{\bar{u}}{u_*} = f\left(\frac{u_* z}{\nu}\right) \tag{4.13a}$$

in which $u_* = (\tau_0/\rho)^{1/2}$ and is called the shear velocity because it is a representative velocity scale resulting from the turbulence. This functional relationship is referred to as the *law of the wall* because its existence depends on the assumption of proximity of the solid wall boundary.

In a more general sense, the law of the wall applies to an inner layer near the wall that can be subdivided into three regions. Very, very near the wall, only the viscous stress is important and it is approximately constant so that Equation 4.13a integrates to a linear velocity distribution given by

$$\frac{\bar{u}}{u_*} = \frac{u_* z}{\nu} \tag{4.13b}$$

From experimental data, this portion of the velocity distribution applies from the wall up to a value of $u_* z/\nu = 5$ as shown in Figure 4.2c. This lowest region in the flow is the viscous sublayer in which velocity fluctuations may exist such that the flow is not laminar but it is definitely dominated by the viscous shear stress.

Above the viscous sublayer at some relatively large distance from the wall in terms of the viscous length scale ν/u_*, yet also far from the free surface where the flow depth is an important length scale, the only length scale possible is the distance from the wall z. In this portion of the inner layer, it can be argued dimensionally that the velocity gradient $d\bar{u}/dz$ must depend only on the shear velocity and the distance from the wall such that, $d\bar{u}/dz = (u_*/\kappa z)$, in which κ is considered a universal constant and is referred to as von Karman's constant with a value of 0.40 to 0.41. Upon integration, this relationship becomes the logarithmic velocity distribution given by

$$\frac{\bar{u}}{u_*} = \frac{1}{\kappa} \ln \frac{z}{z_0} \tag{4.13c}$$

in which z_0 is a constant of integration that can be determined to be proportional to ν/u_* either on dimensional grounds or by considering the intersection of the logarithmic distribution with the linear distribution in the viscous sublayer. The result is the logarithmic velocity distribution in the inner layer next to a smooth wall as given by

$$\frac{\bar{u}}{u_*} = \frac{1}{\kappa} \ln \left(\frac{u_* z}{\nu} \right) + A_s \tag{4.13d}$$

in which A_s is a constant determined by Nikuradse's experiments on smooth pipes to a value of 5.5. In reality there is a transition region called the buffer layer between the viscous sublayer and the logarithmic layer as shown in Figure 4.2c in which both viscous and turbulent shear stresses are important. In the buffer layer the production of turbulence by the Reynolds stress is a maximum because of the large value of the velocity gradient there. The logarithmic layer extends from $u_* z/\nu = 30$ to much larger values of the order of 10^2 but theoretically does not apply beyond relative depths of about 0.2. However, experimental measurements have shown that the logarithmic law can be applied over nearly the full depth of flow except in cases of increasing or decreasing pressure gradients as shown in Figure 4.2c.

In the outer region away from the wall where the turbulent shear stress dominates, it can be shown again on dimensional grounds that the effect of the wall shear stress is to retard the flow or create a velocity deficit relative to the surface velocity that depends only on the relative distance from the wall. The result is the velocity defect law given in functional form as

$$\frac{u_{max} - \bar{u}}{u_*} = g \left(\frac{z}{y_0} \right) \tag{4.14a}$$

in which u_{max} is the maximum velocity at the water surface and y_0 is the flow depth. Now it can be argued that for both the outer law and the inner law of the wall to be valid in an overlap region, the velocity distribution must be logarithmic (Kundu and Cohen 2002). The logarithmic form of the velocity defect law is given by

$$\frac{u_{max} - \bar{u}}{u_*} = -\frac{1}{\kappa} \ln\frac{z}{y_0} + A_1 \qquad (4.14b)$$

and it is valid for both rough and smooth walls and can be extended into the logarithmic overlap layer. As a practical matter, the universality of the logarithmic law over nearly the full depth of flow for both smooth and rough walls (in the absence of strong secondary currents) means that the best-fit slope of experimental velocity data plotted as u vs. log (z) can be used to obtain a value of $u_* = (\tau_0/\rho)^{1/2}$ and thus the bed shear stress since $d\bar{u}/d(\log z) = 2.3u_*/\kappa$. In general, only data points for $z/y_0 < 0.6$ should be used to find the slope.

The combination of theory and experiment that led to the logarithmic velocity distribution for turbulent flow replaces the exact integration of the momentum equation, which is possible only in the case of laminar flow, to obtain the friction factor, f. The result for laminar flow is Poiseuille's law for the friction factor, $f = 64/\mathbf{Re}$. Our objective is to obtain a relation for the friction factor in turbulent flow based on the semiempirical logarithmic velocity distribution given by Equation 4.13d. The logarithmic velocity distribution is transformed into a resistance equation by integrating it over the flow thickness to obtain an expression for the dimensionless mean velocity, V/u_*. The relation for mean velocity can be expressed in terms of the friction factor, f, by rearranging Equation 4.3 as $V/u_* = \sqrt{8/f}$. The result for a smooth-walled pipe, as first given by Prandtl (Crowe et al. 2009), is

$$\frac{1}{\sqrt{f}} = 2.0 \log\left(\mathbf{Re}\sqrt{f}\right) - 0.8 \qquad (4.15)$$

For a rough-walled pipe or channel, the viscous sublayer is disrupted by roughness elements if they are larger than the thickness of the sublayer itself. In this case, the viscosity no longer is important, but the height of the roughness elements, k, becomes very influential in determining the velocity profile. A dimensional analysis indicates that the velocity distribution should depend on z/k and the dependence must be logarithmic to satisfy the overlap of the outer, or velocity-defect, law into the inner, or wall, region. The resulting velocity distribution for a rough wall is

$$\frac{\bar{u}}{u_*} = \frac{1}{\kappa} \ln\frac{z}{k} + A_r \qquad (4.16)$$

Nikuradse determined the value of A_r to be 8.5 for sand-grain roughened pipes in fully rough turbulent flow, for which $u_*k/\nu > 70$. If Equation 4.16 is integrated over a pipe cross section to obtain the mean velocity, we can derive the friction factor relation for fully rough turbulent pipe flow:

$$\frac{1}{\sqrt{f}} = \frac{1}{\kappa\sqrt{8}} \ln\frac{d}{k_s} + 1.14 \qquad (4.17)$$

in which d = pipe diameter and k_s = sand-grain roughness height from Nikuradse's experiments.

In between the turbulent pipe relations for friction factor for smooth and fully rough conditions given by Equations 4.15 and 4.17, respectively, is a transitional-rough regime, defined approximately by $4 < u_* k_s/\nu < 70$. The behavior of the friction-factor relation in this transition regime depends on the type of roughness. It is different, for example, for Nikuradse's sand-grain roughened pipes and commercial pipes. Colebrook (1939) fit a transition relation for commercial pipes that is asymptotic to both the smooth and fully rough friction-factor relations:

$$\frac{1}{\sqrt{f}} = -2 \log \left[\frac{k_s/d}{3.7} + \frac{2.51}{\mathbf{Re}\sqrt{f}} \right] \tag{4.18}$$

in which the commercial pipe roughness is expressed as an equivalent sand-grain roughness by determining the sand-grain roughness height that would give the same friction factor as for the commercial pipe in fully rough turbulent flow. Equation 4.18 is the basis for the Moody diagram shown in Figure 4.3 (see Rouse 1980).

Keulegan (1938) applied the logarithmic velocity distribution to flow in open channels. He proceeded to integrate the Nikuradse velocity distribution for fully rough turbulent flow (Equation 4.16) over a trapezoidal open channel cross section to obtain, for the friction factor,

$$\frac{1}{\sqrt{f}} = 2.03 \log \frac{R}{k_s} + 2.21 = 2.03 \log \frac{\xi R}{k_s} \tag{4.19a}$$

in which the value of $\xi = 12.26$ on the right hand side of (4.19a). In reality, ξ varies slightly with the channel shape, but a value of $\xi = 12$ is recommended by an ASCE Task Force (1963) and the slope of 2.03 often is rounded to 2.0. Keulegan derived the expression for ξ for rectangular channels to be

$$\xi = \exp \left[\ln \left(1 + 2\frac{y}{b} \right) - \frac{y}{b} + 2.4 \right] \tag{4.19b}$$

in which b = channel width and y = flow depth. For the aspect ratio b/y varying from 5 to 100, for example, ξ takes on values from 12.6 to 11.1, respectively.

The relationship between Manning's n and Darcy-Weisbach's f now can be obtained from their definitions to determine the applicability of Manning's equation:

$$n = \frac{K_n}{(8g)^{1/2}} f^{1/2} R^{1/6} \tag{4.20}$$

in which $K_n = 1.0$ for SI units and 1.49 for English units. If we substitute Equation 4.19a into Equation 4.20 with $\xi = 12$ and the slope of 2.03 rounded to 2.0, we have

$$\frac{n}{k_s^{1/6}} = \frac{\dfrac{K_n}{(8g)^{1/2}} \left(\dfrac{R}{k_s} \right)^{1/6}}{2.0 \log \left(12\dfrac{R}{k_s} \right)} \tag{4.21}$$

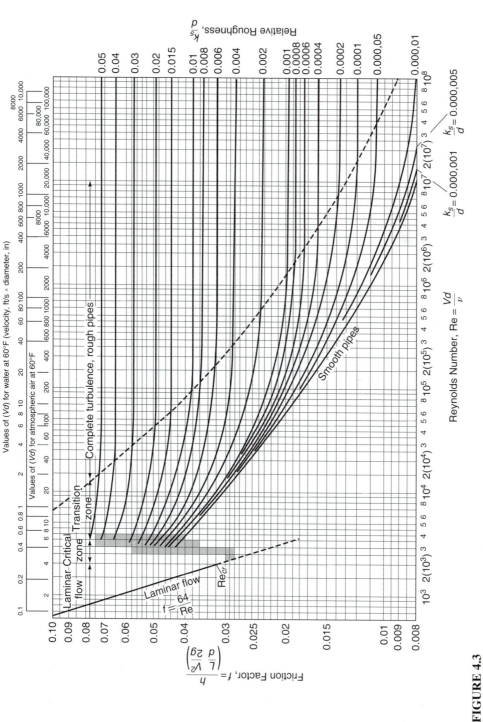

FIGURE 4.3

Moody diagram for pipe friction in pipes with smooth and rough walls (Moody 1944). (*Source: L. F. Moody, "Friction Factors for Pipe Flow,"* © *1944, ASME. Reproduced by permission of ASME.*)

which has been plotted in Figure 4.4 for both English and metric units. Over a fairly wide range of values of R/k_s, the value of $n/k_s^{1/6}$ is constant and therefore not a function of flow depth, an essential assumption of Manning's equation in which the depth dependence of the velocity for a given roughness height is assumed to be contained entirely in the $R^{2/3}$ term. The minimum value of $c_n = n/k_s^{1/6}$ in Figure 4.4 is 0.039 for metric units and $c_n = 0.032$ for English units at $R/k_s = 33.7$, although these values vary slightly with the constants assumed in the Keulegan equation (Yen 1992a). More generally, the value of $n/k_s^{1/6}$ can be shown to be within ± 5 percent of a constant value over a range of R/k_s given by $4 < R/k_s < 500$ as shown by Yen (1992a). Hager (1999) gives the limits on the constancy of n with depth and so on the range of applicability of the Manning equation to be $3.6 < R/k_s < 360$. The limitation of fully rough turbulent flow for the Manning equation also is implicit in the comparison with Keulegan's equation. This limitation requires $u_* k_s / \nu > 70$, which can be translated into the limit

$$\frac{\sqrt{gRS}}{\nu} \left[\frac{n\sqrt{g}}{K_n} \right]^6 \geq 2.3 \times 10^{-4} \tag{4.22}$$

using the minimum value of $ng^{1/2}/(K_n k_s^{1/6}) = 0.122$ from Equation 4.21 to substitute for k_s. For example, a 2 ft (0.61 m) diameter storm sewer with $n = 0.015$ flowing just full at a slope of 0.001 would exceed the limit given by the inequality in (4.22) and be in the fully rough turbulent regime.

The literature contains some disagreement about the value of $c_n = n/k_s^{1/6}$ as discussed by French (1985), partly because its value depends on whether the units of k_s are

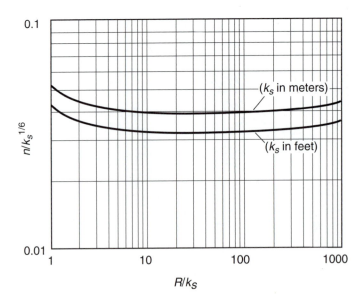

FIGURE 4.4
Manning's n variation with relative smoothness.

metric (m) or English (ft). The Strickler value of c_n is given by Henderson (1966) to be 0.034 in English units (0.041 in SI units) based on measurements made by Strickler (1923) in gravel-bed streams with $k_s = d_{50}$, the diameter for which 50 percent of the sediment particles are smaller by weight. The minimum value of c_n from Keulegan's equation is 0.032 in English units (0.039, SI) as noted earlier, which agrees well with the Strickler value considering that the effective size in the gravel-bed stream is larger than d_{50} due to bed armoring, as argued by Henderson (1966). However, several other sources, including Hager (1999), give the Strickler value of $c_n = 0.039$ in English units (0.048 in SI units). This point is considered again when discussing the resistance coefficient for rock riprap later in this chapter.

Some laboratory and field measurements of Darcy-Weisbach's f are compared with Keulegan's relation (Equation 4.19a) in Figure 4.5. The data points by Thein (1993) and Dickman (1990) were measured in a 1.07 m (3.5 ft) wide tilting flume with a coarse gravel bed in which uniform flow was set for several combinations of depth and slope. The Bathurst (1985) data in Figure 4.5 were obtained from high-gradient gravel and boulder-bed rivers in Britain. Limerinos (1970) measured the resistance coefficient in 11 gravel and cobble-bed streams in California. The data in Figure 4.5 are presented in terms of the bed friction factor, f_b, and the bed hydraulic radius, R_b, to indicate that the flume data have been corrected for the

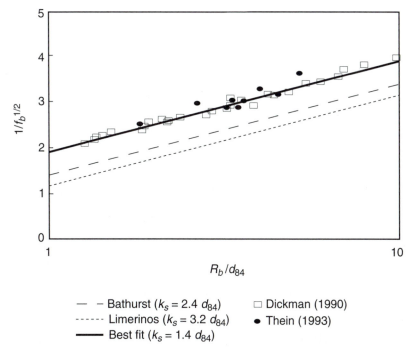

FIGURE 4.5
Friction factor in fully rough turbulent open channel flow with large roughness elements.

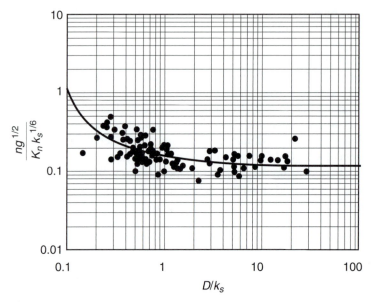

FIGURE 4.6

Comparison of field data from Blodgett (1986) on flow resistance in boulder, cobble, and gravel-bed streams with Keulegan's equation for Manning's n.

effect of smooth sidewalls. By comparing the intercepts in Figure 4.5 (where $R_b/d_{84} = 1.0$) with the Keulegan constant, a value of the equivalent sand-grain roughness, k_s, can be determined as a multiple of d_{84}, which is the sediment grain size for which 84 percent of the sediment is smaller by weight. It can be determined from Figure 4.5 that k_s/d_{84} has a value of 1.4 for the lab data, 2.4 for the Bathurst data, and 3.2 for the Limerinos data. Hey (1979) concluded that $k_s/d_{84} = 3.5$, based on data from several sources on gravel-bed streams, and suggested that wake interference losses downstream of the larger roughness elements may account for the large value of k_s.

Results for the friction factor in gravel-bed streams also can be presented in terms of Manning's n, according to Equation 4.21. Data assembled by Blodgett (1986) for boulder-, cobble-, and gravel-bed streams in the western United States are plotted in Figure 4.6 in terms of k_s, which can be determined to be 6.3 d_{50} from fitting the Keulegan equation (Equation 4.21) for Manning's n. Only the values of d_{50} were reported by Blodgett, because the interest was in obtaining a relationship for rock-riprap lined channels based on natural channel data. The data in Figure 4.6 are given in terms of the average or hydraulic depth, D, instead of hydraulic radius because Blodgett found them to be virtually identical. The Blodgett data include the data by Limerinos (1970), who reported a standard deviation in the percent difference between measured and fitted values of n to be $\pm$ 22 percent when d_{50} was used as the characteristic grain size and $\pm$ 19 percent when d_{84} was used.

4.6   DISCUSSION OF FACTORS AFFECTING f AND n

The dependence of f on the Reynolds number and relative roughness has been discussed with respect to the Moody diagram for pipe flow. The Reynolds number dependence is not as important in open channel flow, especially in large natural channels, for which the Reynolds number is quite large. If a smooth-walled conduit is flowing partly full in the smooth turbulent regime, however, Manning's equation is not directly applicable, because Manning's n can be expected to vary with the Reynolds number (Henderson 1966). In this case, the use of the Darcy-Weisbach's f is preferred, although Yen (1992a) shows that for Reynolds numbers less than the fully rough turbulent values, there is a narrower range of R/k_s, within which Manning's n still may be reasonably constant.

  The dependence of f on relative roughness in open channel flow is not as well known as in pipe flow because it is difficult to assign an equivalent sand-grain roughness for the large values of absolute roughness height typically found in open channels. Rouse (1965) discusses the importance of roughness concentration, shape, and arrangement on the equivalent sand-grain roughness height, k_s. He reports experimental results that indicate the maximum value of relative roughness occurs at a roughness concentration of 20–25 percent. Kumar and Roberson (1980) and Kumar (1992) made significant advances in obtaining an analytical relation for the variation of relative roughness with concentration and shape for randomly arranged roughness elements. This research has led to a completely general algorithm for determining rough conduit resistance (Kumar and Roberson 1980) that seems to work well for artificial roughness elements, but more limited comparisons with natural channel roughness have been made. The analytical method utilizes a drag coefficient determined for an individual roughness element, the average height of roughness elements, and an areal projection factor to describe the projection of the roughness elements on a plane perpendicular to the flow as a function of distance from the boundary. With this information, the equivalent sand-grain roughness can be estimated as a function of the concentration and shape of roughness elements. The method cannot be applied directly to in-line roughness elements, however, because it does not account for wake interference effects.

  The dependence of flow resistance on cross-sectional shape occurs as a result of changes both in the channel hydraulic radius, R, and the cross-sectional distribution of velocity and shear. Therefore, applications of the Moody diagram in open channel flow in which pipe diameter is replaced by $4R$ may not completely reflect the effects of cross-sectional shape on flow resistance. Kazemipour and Apelt (1979) suggested that two dimensionless parameters are required to characterize shape effects on the resistance coefficient f:

$$ f = F\left(\mathbf{Re}, \frac{k_s}{R}, \frac{P}{B}, \frac{B}{D} \right) \tag{4.23} $$

in which F denotes "function of"; $\mathbf{Re}$ = Reynolds number; k_s = sand-grain roughness height; R = hydraulic radius; B = channel top width; D = hydraulic depth; and P = wetted perimeter. The first shape factor, P/B, is a measure of the influence of the shear distribution on f; and the second shape factor, B/D, is a channel aspect ratio. Using the data of Shih and Grigg (1967) and Tracy and Lester (1961) for smooth rectangular channels, Kazemipour and Apelt (1979) showed that the open channel friction factor, f_c, can be obtained directly from the pipe friction factor, f_M, determined from the Moody diagram:

$$f_c = \sigma f_M \tag{4.24}$$

in which

$$\sigma = \frac{\psi_1}{\psi_2} = \frac{\left(\dfrac{P}{B}\right)^{1/2}}{\dfrac{1}{\left(\dfrac{B}{D} + 0.3\right)} + 0.9} \tag{4.25}$$

The function $\psi_2(B/D)$ is proposed as a best fit of the experimental relationship presented by Kazemipour and Apelt (1979) based on the experimental data for smooth rectangular channels that they used over a range of B/D from approximately 1 to 40. They also applied their experimental relationship successfully to limited data for fully rough turbulent flow in rectangular channels and an additional data set of their own for smooth rectangular channels (Kazemipour and Apelt 1982). Using Equation 4.25, the value of σ for a rectangular channel varies from approximately 1.04 to 1.10 as the aspect ratio (b/y) increases from 1 to 40.

Experimental research by Sturm and King (1988) on the flow resistance of horseshoe-shaped conduits flowing partly full has shown that the Kazemipour and Apelt (1982) relations for shape effects in rectangular channels cannot be extended to horseshoe conduits. The Neale and Price (1964) data for partly full flow in smooth circular conduits also show considerable scatter from Equation 4.25. For the horseshoe conduit flowing partly full, the shape effect depends on the ratio of depth to diameter, y/d, with greater influence at larger values of y/d. A summary of the results is shown in Figures 4.7 and 4.8. In Figure 4.7, the value of σ for the horseshoe conduit essentially is unity for relative depths less than 0.4, but it increases to a value of approximately 1.25 for larger relative depths. For the smooth, circular conduit data of Neale and Price (1964), an average value of $\sigma = 1.05$ occurs for $y/d > 0.2$. In Figure 4.8, the horseshoe data for the velocity ratio V/V_f, where V = partly full velocity and V_f = full flow velocity, are compared with relationships for circular and horseshoe conduits. The horseshoe data follow Manning's equation with constant n at low relative depths but then approach the Pomeroy (1967) empirical relation for circular sewers ($V/V_f = 1.093(A/A_f)^{0.316}$) at large relative depths. (The theoretical relationship for V/V_f with constant Manning's n essentially is the same for circular and horseshoe conduits.) Thus, in Figure 4.8, the reduction in velocity observed in both circular and horseshoe conduits for depths greater than half full seems to correspond to an increase in flow resistance due to the shape factor, but the velocity reduction is not as large as predicted by Camp (1946).

Nonuniformity of the open channel boundary in the direction of flow, in either plan or profile view, necessarily causes a change in the velocity distribution and

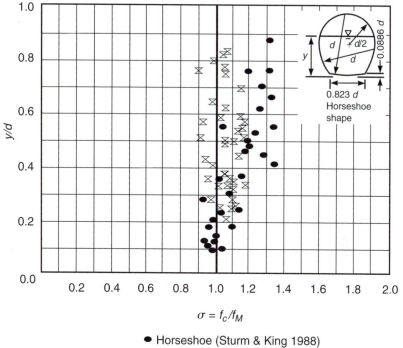

$$\sigma = f_c / f_M$$

● Horseshoe (Sturm & King 1988)
⌖ Circular (Neale & Price 1964)

FIGURE 4.7
Friction factor correction for partly full flow in horseshoe (transition) and circular (smooth) conduits (Sturm and King 1988). (*Source: T. W. Sturm and D. King, "Shape Effects on Flow Resistance in Horseshoe Conduits," J. Hydr. Engrg., © 1988, ASCE. Reproduced by permission of ASCE.*)

hydraulic resistance to flow. As an example, the development of the boundary layer in a supercritical flow discharging under a sluice gate results in a deceleration and a change in surface resistance due to the nonuniformity of the flow cross section. In gradually varied flow (e.g., a gradual nonuniformity in the flow direction), the flow resistance commonly is assumed to be the same as that obtained in a uniform flow at the same depth. The error associated with this assumption may be small in most cases, but it is essential that measured values of Manning's n for general engineering applications be obtained only for uniform flow to eliminate the effects of nonuniformity. Uniform flow in laboratory flumes is difficult to obtain unless they are very long. Tracy and Lester (1961), who measured the friction factor for a smooth channel in the 80 ft (24 m) long tilting flume at Georgia Tech, devised a procedure for determining uniform flow depth. Their technique was to establish two control gate positions, one of which provided a water surface profile that asymptotically approached uniform (normal) depth from above and the other one from below. Only in this way were they able to obtain an accurate value of uniform flow depth, even in a relatively long flume.

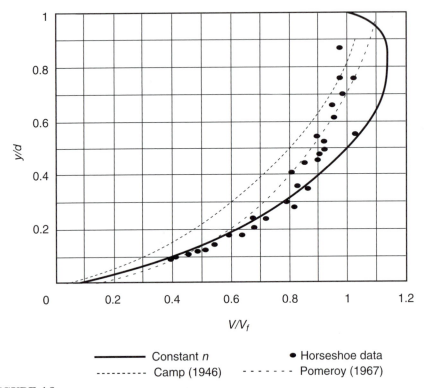

FIGURE 4.8
Relative velocity relationships for circular conduits compared to data for horseshoe conduits (Sturm and King 1988). (*Source: T. W. Sturm and D. King, "Shape Effects on Flow Resistance in Horseshoe Conduits," J. Hydr. Engrg., © 1988, ASCE. Reproduced by permission of ASCE.*)

Nonuniformities due to changes in form resistance as a result of cross-sectional changes may be considered to be Froude number dependent. Thus, flow around bridge piers or flow over a boundary that has large, widely spaced roughness elements experiences wave formation, and Froude number effects on the resistance coefficient are introduced. Supercritical flow around bends or in contractions is another case in which the resistance is Froude number dependent.

Unsteadiness of open channel flow also brings with it changes in velocity distribution and resistance to flow. The occurrence of surface instabilities in supercritical flow, commonly called *roll waves,* is an example of the unsteadiness effect. Rouse (1965) suggested that the increase in resistance due to roll waves could be related to the ratio of the flow Froude number to the critical value of Froude number above which instability occurs (= 1.5 to 2. for wide channels). Berlamont and Vanderstappen (1981) confirmed this formulation and further asserted that these resistance effects are more likely to occur in wide channels. They indicated that Froude number effects in supercritical flow may have been overlooked by some investigators because these effects are small and independent of Reynolds number when it is large.

In summary, the effects of unsteadiness, Froude number, nonuniformity, cross-sectional shape, and roughness element concentration and arrangement, as well as the usual Reynolds number and relative roughness effects, all can be expected to affect open channel flow resistance. Continued use of the Manning's n means simply that we lump all of our ignorance about flow resistance into a single coefficient. For example, it is difficult to establish the physical significance of observed changes in Manning's n with river stage because of the many factors that affect it. Engineering experience will continue to dictate the choice of Manning's n values, but they should be verified by field measurements as much as possible. In the case of turbulent, partly full flow in smooth conduits, the Darcy-Weisbach f may be the preferred resistance coefficient; however, the constancy of Manning's n over a wide range of flow conditions for a given boundary roughness, particularly in natural channels, make it a valuable tool for assessing the effects of open-channel flow resistance.

4.7   SELECTION OF MANNING'S n IN NATURAL CHANNELS

As mentioned previously, there is no substitute for experience in the selection of Manning's n for natural channels. Table 4.1 from Chow (1959) gives an idea of the variability to be expected in Manning's n. The pictures of channels with measured values of Manning's n as given by Arcement and Schneider (1984), Barnes (1967), and Chow (1959) are very useful for developing preliminary values of Manning's n. Some of these photographs are given at the end of this chapter. In addition, for those channels outside the engineer's previous experience, the more regimented procedure suggested by Cowan (1956) is helpful:

$$n = (n_b + n_1 + n_2 + n_3 + n_4)m \qquad (4.26)$$

in which n_b = the base value for a straight, uniform channel; n_1 = adjustment for surface irregularities due to bank erosion and exposed tree roots; n_2 = adjustment for variations in the shape and size of cross section in the streamwise direction; n_3 = adjustment for channel obstructions; n_4 = adjustment for type of vegetation and degree of submergence; and m = adjustment factor for channel meandering. Values for each of these adjustments are suggested by Arcement and Schneider (1984) for both natural channels and floodplains. The base channel value can be taken from Table 4.1 or, for sand and gravel-bed streams, from Keulegan's equation (4.19a). Values of n_1 vary from 0–0.02; n_2 = 0–0.015; n_3 = 0–0.05; and n_4 = 0–0.10. The meandering adjustment varies from 1.0 for minor meandering to 1.3 for severe meandering defined as a ratio of channel length to valley length exceeding 1.5.

TABLE 4.1 Values of the Manning's Roughness Coefficient n

Type of channel and description	Minimum	Normal	Maximum
A. Closed Conduits Flowing Partly Full			
A–1. Metal			
a. Brass, smooth	0.009	0.010	0.013
b. Steel			
1. Lockbar and welded	0.010	0.012	0.014
2. Riveted and spiral	0.013	0.016	0.017
c. Cast iron			
1. Coated	0.010	0.013	0.014
2. Uncoated	0.011	0.014	0.016
d. Wrought iron			
1. Black	0.012	0.014	0.015
2. Galvanized	0.013	0.016	0.017
e. Corrugated metal			
1. Subdrain	0.017	0.019	0.021
2. Storm drain	0.021	0.024	0.030
A–2. Nonmetal			
a. Lucite	0.008	0.009	0.010
b. Glass	0.009	0.010	0.013
c. Cement			
1. Neat, surface	0.010	0.011	0.013
2. Mortar	0.011	0.013	0.015
d. Concrete			
1. Culvert, straight and free of debris	0.010	0.011	0.013
2. Culvert with bends, connections, and some debris	0.011	0.013	0.014
3. Finished	0.011	0.012	0.014
4. Sewer with manholes, inlet, etc., straight	0.013	0.015	0.017
5. Unfinished, steel form	0.012	0.013	0.014
6. Unfinished, smooth wood form	0.012	0.014	0.016
7. Unfinished, rough wood form	0.015	0.017	0.020
e. Wood			
1. Stave	0.010	0.012	0.014
2. Laminated, treated	0.015	0.017	0.020
f. Clay			
1. Common drainage tile	0.011	0.013	0.017
2. Vitrified sewer	0.011	0.014	0.017
3. Vitrified sewer with manholes, inlet, etc.	0.013	0.015	0.017
4. Vitrified subdrain with open joint	0.014	0.016	0.018
g. Brickwork			
1. Glazed	0.011	0.013	0.015

Type of channel and description	Minimum	Normal	Maximum
2. Lined with cement mortar	0.012	0.015	0.017
h. Sanitary sewers coated with sewage			
slimes, with bends and connections	0.012	0.013	0.016
i. Paved invert, sewer, smooth bottom	0.016	0.019	0.020
j. Rubble masonry, cemented	0.018	0.025	0.030
B. Lined or Built-up Channels			
B–1. Metal			
a. Smooth steel surface			
1. Unpainted	0.011	0.012	0.014
2. Painted	0.012	0.013	0.017
b. Corrugated	0.021	0.025	0.030
B–2. Nonmetal			
a. Cement			
1. Neat, surface	0.010	0.011	0.013
2. Mortar	0.011	0.013	0.015
b. Wood			
1. Planed, untreated	0.010	0.012	0.014
2. Planed, creosoted	0.011	0.012	0.015
3. Unplaned	0.011	0.013	0.015
4. Plank with battens	0.012	0.015	0.018
5. Lined with roofing paper	0.010	0.014	0.017
c. Concrete			
1. Trowel finish	0.011	0.013	0.015
2. Float finish	0.013	0.015	0.016
3. Finished, with gravel on bottom	0.015	0.017	0.020
4. Unfinished	0.014	0.017	0.020
5. Gunite, good section	0.016	0.019	0.023
6. Gunite, wavy section	0.018	0.022	0.025
7. On good excavated rock	0.017	0.020	
8. On irregular excavated rock	0.022	0.027	
d. Concrete bottom float finished with			
sides of			
1. Dressed stone in mortar	0.015	0.017	0.020
2. Random stone in mortar	0.017	0.020	0.024
3. Cement rubble masonry, plastered	0.016	0.020	0.024
4. Cement rubble masonry	0.020	0.025	0.030
5. Dry rubble or riprap	0.020	0.030	0.035
e. Gravel bottom with sides of			
1. Formed concrete	0.017	0.020	0.025
2. Random stone in mortar	0.020	0.023	0.026
3. Dry rubble or riprap	0.023	0.033	0.036
f. Brick			
1. Glazed	0.011	0.013	0.015

(continued)

TABLE 4.1   (Continued)

Type of channel and description	Minimum	Normal	Maximum
2. In cement mortar	0.012	0.015	0.018
g. Masonry			
1. Cemented rubble	0.017	0.025	0.030
2. Dry rubble	0.023	0.032	0.035
h. Dressed ashlar	0.013	0.015	0.017
i. Asphalt			
1. Smooth	0.013	0.013	
2. Rough	0.016	0.016	
j. Vegetal lining	0.030	. . .	0.500
C. Excavated or Dredged			
a. Earth, straight and uniform			
1. Clean, recently completed	0.016	0.018	0.020
2. Clean, after weathering	0.018	0.022	0.025
3. Gravel, uniform section, clean	0.022	0.025	0.030
4. With short grass, few weeds	0.022	0.027	0.033
b. Earth, winding and sluggish			
1. No vegetation	0.023	0.025	0.030
2. Grass, some weeds	0.025	0.030	0.033
3. Dense weeds or aquatic plants in deep channels	0.030	0.035	0.040
4. Earth bottom and rubble sides	0.028	0.030	0.035
5. Stony bottom and weedy banks	0.025	0.035	0.040
6. Cobble bottom and clean sides	0.030	0.040	0.050
c. Dragline excavated or dredged			
1. No vegetation	0.025	0.028	0.033
2. Light brush on banks	0.035	0.050	0.060
d. Rock cuts			
1. Smooth and uniform	0.025	0.035	0.040
2. Jagged and irregular	0.035	0.040	0.050
e. Channels not maintained, weeds and brush uncut			
1. Dense weeds, high as flow depth	0.050	0.080	0.120
2. Clean bottom, brush on sides	0.040	0.050	0.080
3. Same, highest stage of flow	0.045	0.070	0.110
4. Dense brush, high stage	0.080	0.100	0.140
D. Natural Streams			
D–1. Minor streams (top width at flood stage < 100 ft)			
a. Streams on plain			
1. Clean, straight, full stage, no rifts or deep pools	0.025	0.030	0.033
2. Same as above, but more stones and weeds	0.030	0.035	0.040

Type of channel and description	Minimum	Normal	Maximum
3. Clean, winding, some pools and shoals	0.033	0.040	0.045
4. Same as above, but some weeds and stones	0.035	0.045	0.050
5. Same as above, lower stages, more ineffective slopes and sections	0.040	0.048	0.055
6. Same as 4, but more stones	0.045	0.050	0.060
7. Sluggish reaches, weedy, deep pools	0.050	0.070	0.080
8. Very weedy reaches, deep pools, or floodways with heavy stand of timber and underbrush	0.075	0.100	0.150
b. Mountain streams, no vegetation in channel, banks usually steep, trees and brush along banks submerged at high stages			
1. Bottom: gravels, cobbles, and few boulders	0.030	0.040	0.050
2. Bottom: cobbles with large boulders	0.040	0.050	0.070
D–2. Flood plains			
a. Pasture, no brush			
1. Short grass	0.025	0.030	0.035
2. High grass	0.030	0.035	0.050
b. Cultivated areas			
1. No crop	0.020	0.030	0.040
2. Mature row crops	0.025	0.035	0.045
3. Mature field crops	0.030	0.040	0.050
c. Brush			
1. Scattered brush, heavy weeds	0.035	0.050	0.070
2. Light brush and trees, in winter	0.035	0.050	0.060
3. Light brush and trees, in summer	0.040	0.060	0.080
4. Medium to dense brush, in winter	0.045	0.070	0.110
5. Medium to dense brush, in summer	0.070	0.100	0.160
d. Trees			
1. Dense willows, summer, straight	0.110	0.150	0.200
2. Cleared land with tree stumps, no sprouts	0.030	0.040	0.050
3. Same as above, but with heavy growth of sprouts	0.050	0.060	0.080

(*continued*)

TABLE 4.1   (Continued)

Type of channel and description	Minimum	Normal	Maximum
4. Heavy stand of timber, a few down trees, little undergrowth, flood stage below branches	0.080	0.100	0.120
5. Same as above, but with flood stage reaching branches	0.100	0.120	0.160
D–3. Major streams (top width at flood stage > 100 ft). The n value is less than that for minor streams of similar description, because banks offer less effective resistance.			
a. Regular section with no boulders or brush	0.025	. . .	0.060
b. Irregular and rough section	0.035	. . .	0.100

Source: Chow 1959. Used with permission of Chow estate.

4.8   CHANNELS WITH COMPOSITE ROUGHNESS

Under some circumstances, a natural or artificial channel may have varying roughness across its wetted perimeter; for example, with different lining materials on the bed and banks or vegetated banks with an unvegetated bed. The methodologies presented in this section are not meant for compound channels in which the geometry and the roughness are significantly different on the floodplains compared to the main channel. For compound channels, it is more appropriate to divide the channel into main channel and floodplain subsections with different values of the roughness coefficient to obtain the total conveyance, as will be discussed soon.

Chow (1959) presented methods by Horton, Einstein and Banks, and Lotter for obtaining a composite value of Manning's n for a single channel; that is, for the main channel only of a compound channel or a canal with laterally varying roughness. The Horton method is based on the assumption that the velocities in each wetted-perimeter subsection are equal to one another as well as equal to the mean velocity of the whole cross section. The resulting composite value of Manning's n, denoted n_c, is given by

$$n_c = \left[\frac{\sum_{i=1}^{N} P_i n_i^{3/2}}{P} \right]^{2/3} \tag{4.27}$$

in which P_i, n_i = wetted perimeter and Manning's n of any section i; P = wetted perimeter of the entire cross section; and N = total number of sections into which the wetted perimeter is divided. The Einstein and Banks method assumes that the total resisting force is equal to the sum of the resisting forces in each subsection and the hydraulic radius of each subsection is equal to the hydraulic radius of the whole section. The result is given by

$$n_c = \left[\frac{\sum\limits_{i=1}^{N} P_i n_i^2}{P} \right]^{1/2} \tag{4.28}$$

Lotter's formula is based on writing the total discharge as the sum of the discharges in the subsections:

$$n_c = \frac{PR^{5/3}}{\sum\limits_{i=1}^{N} \dfrac{P_i R_i^{5/3}}{n_i}} \tag{4.29}$$

Finally, Krishnamurthy and Christensen (1972) derived another formula based on the logarithmic velocity distribution, which gives n_c as

$$\ln n_c = \frac{\sum\limits_{i=1}^{N} P_i y_i^{3/2} \ln n_i}{\sum\limits_{i=1}^{N} P_i y_i^{3/2}} \tag{4.30}$$

in which y_i = flow depth in the ith section. Motayed and Krishnamurthy (1980) used cross-sectional data from 36 streams in Maryland, Georgia, Pennsylvania, and Oregon at U.S. Geological Survey gauging stations to test the four formulas just given. An average value of the slope of the energy grade line obtained from the measured depth and velocity distribution at a cross section was used to obtain a "measured" composite value of Manning's n to compare with the formulas. The results showed that the mean error between the computed n_c and the measured n_c was by far smallest for the Lotter formula.

4.9   UNIFORM FLOW COMPUTATIONS

Whether the Manning or Chezy equation is used, there exists a unique value of the uniform flow depth for a given channel geometry, discharge, roughness, and slope. This depth is called the *normal depth,* and its magnitude relative to the critical depth determines whether or not uniform flow is supercritical or subcritical for a given set of channel conditions. If the normal depth is greater than critical, then the uniform flow is subcritical and the slope is classified as mild. For a steep slope the normal depth is less than critical depth. The actual classification of a given channel slope can change with the discharge as the relative magnitudes of normal and critical depth change.

    The computation of normal depth using Manning's equation proceeds by rearranging the equation as

$$AR^{2/3} = \frac{A^{5/3}}{P^{2/3}} = \frac{nQ}{K_n S^{1/2}} \tag{4.31}$$

in which the right hand side is completely specified by design conditions. The design discharge may be set by flood frequency considerations; the roughness often depends on the choice of a stable lining; and the slope is a function of the topography. Equation 4.31 can be solved by trial or by a nonlinear algebraic equation solver for a known geometry. In the case of a trapezoidal channel, for example, the equation in nondimensional form becomes

$$\frac{\left[\dfrac{y_0}{b}\left(1 + \dfrac{my_0}{b}\right)\right]^{5/3}}{\left[1 + \dfrac{2y_0}{b}(1 + m^2)^{1/2}\right]^{2/3}} = \frac{AR^{2/3}}{b^{8/3}} = \frac{nQ}{K_n S^{1/2} b^{8/3}} \tag{4.32}$$

in which b = channel bottom width; m = sideslope ratio; and y_0 = normal depth. As presented, the equation can be used in SI or English units simply by substituting the appropriate value of K_n and units for Q and b consistent with K_n; that is, K_n = 1.49 for Q in cfs and b in ft while K_n = 1.0 for Q in cubic meters per second and b in meters. Equation 4.32 is shown as a graphical solution for normal depth in Figure 4.9 (Chow 1959). A similar solution can be developed for a circular channel, and it is included in the figure with the diameter as the nondimensionalizing length scale.

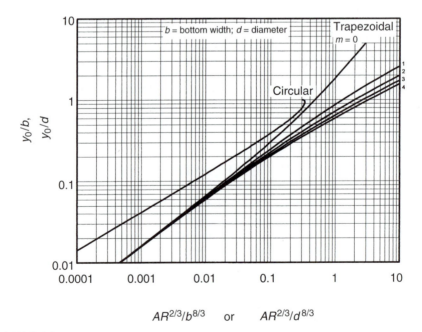

FIGURE 4.9
Curves for calculating normal depth in circular, rectangular, and trapezoidal channels (Chow 1959). (*Source: Used with permission of Chow estate.*)

When the flow is in the fully rough turbulent regime, Manning's equation is appropriate for computation of normal depth, but for the transitional and smooth turbulent regimes, the Chezy equation should be used:

$$AR^{1/2} = \frac{Qf^{1/2}}{(8gS)^{1/2}} \qquad (4.33)$$

in which f = the Darcy-Weisbach friction factor. It has been placed on the right hand side of the equation, although it depends on the Reynolds number and relative roughness, which in turn are functions of the unknown normal depth. Equation 4.33 can be solved for normal depth by assuming a value of f and iterating with the Moody diagram or Equation 4.18 (the Colebrook-White equation) with the pipe diameter replaced by $4R$ and the constant 3.7 replaced by 3.0, so that the first term on the right hand side reflects the Keulegan constant as $k_s/12R$ (Henderson 1966). The iteration required to solve Equation 4.33 may have discouraged its use in the past, so that Manning's equation often has been used without consideration of the unknown variability of Manning's n outside the fully rough flow regime. An alternative formulation of the Chezy equation for the smooth turbulent case is considered in the next section.

4.10   PARTLY FULL FLOW IN SMOOTH, CIRCULAR CONDUITS

In the case of PVC plastic pipe used for gravity sewers and detention basin outlets, the Chezy equation with the Darcy-Weisbach f rather than Manning's n is preferred. Experimental work by Neale and Price (1964) has shown that PVC pipe can be considered smooth. Furthermore, their results indicate a relatively small effect due to shape. The relation for f in smooth pipes is given by

$$\frac{1}{\sqrt{f}} = 2.0 \log \left(\mathbf{Re} \sqrt{f} \right) - 0.8 \qquad (4.34)$$

in which $\mathbf{Re} = Vd/\nu$ is the Reynolds number; d = pipe diameter; and ν = kinematic viscosity. If we replace d by $4R$ in the Reynolds number, where R is the hydraulic radius, and f by $8gA^2RS/Q^2$ from the Chezy equation, then Equation 4.34 can be recast into one with a more useful set of dimensionless variables:

$$\mathbf{Q}^* = Q/[d^2(gdS)^{1/2}];$$

$$\mathbf{Re}^* = d(gdS)^{1/2}/\nu;$$

and y/d, the relative depth.

The results of plotting (4.34) in terms of these dimensionless variables is shown in Figure 4.10. This figure can be used to find the partly full flow depth in a smooth pipe without trial and error.

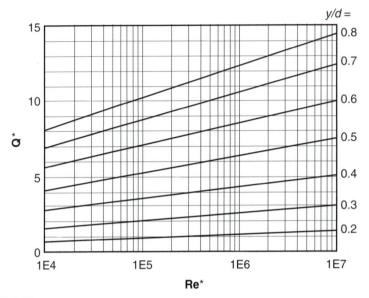

FIGURE 4.10
Discharge capacity of smooth, circular conduits flowing partly full.

4.11 GRAVITY SEWER DESIGN

The design of storm and sanitary sewers involves the determination of partly full flow capacity for a given design depth or normal depth for a given discharge in circular conduits. The design is based on discharges determined either by population estimates and corresponding wastewater rates per capita or by hydrologic calculations of peak runoff rates due to storm events. Because pressurized flow is avoided, especially in sanitary sewers, the design problem is to select a conduit size that will flow partly full for the design discharge. Even in storm sewers, undesirable flow conditions can develop as full flow is approached. When the relative depth or filling ratio, y/d, nears 1.0, air access to the free surface is reduced with intermittent opening and closing of the section (Hager 1999). Such a condition, referred to as *slugging* in culvert hydraulics, results in streaming air pockets at the crown of the pipe and pulsations that could damage pipe joints or cause undesirable fluctuations in discharge. The only practical way of avoiding these difficulties in sewers is to design for partly full flow.

A further complication of the circular cross section occurs due to changes in geometry as the pipe fills. The wetted perimeter increases more rapidly than the cross-sectional area near the crown of the pipe with the result that the discharge capacity decreases as the crown of the pipe is approached. This can be seen in Figure 4.9, as the curve for normal depth reaches a maximum in $AR^{2/3}$ and then decreases as y/d approaches 1.0. In effect, there are two possible normal depths near the crown of the pipe, and the upper one is unlikely to occur without slugging or filling the pipe.

It is sound practice to avoid these difficulties by designing the pipe for a filling ratio of about 0.8 or less at maximum design flow. Older design criteria may have specified $y/d = 0.5$ as the design filling ratio, but this does not make efficient use of the pipe capacity. The initial part of the design is to calculate a pipe diameter that will carry the maximum design discharge at, say, $y/d = 0.8$. This corresponds to a value of $nQ/K_n S^{1/2} d^{8/3} = 0.305$ from Manning's equation, as can be verified from Figure 4.9. The initial diameter then is calculated from

$$d = 1.56 \left[\frac{nQ}{K_n S^{1/2}} \right]^{3/8} \tag{4.35}$$

assuming fully rough turbulent flow, which can be checked as described previously. If Manning's equation is not applicable, then the Chezy equation with the Colebrook-White expression for the friction factor can be used. The initial diameter usually is rounded up to the next commercial pipe size, and the actual flow depth is computed for the commercial diameter. The uniform flow equation can be solved by trial and error, with a computer program, or graphically using Figure 4.9 or Figure 4.10, as appropriate, to find the normal depth.

The second part of the design is to check for the occurrence of self-cleansing velocities to prevent the build-up of deposits in the sewer. It is desirable to have a minimum velocity of at least 0.61 m/s (2.0 ft/s) to scour sand and grit from the pipe at maximum discharge, although a value of 0.91 m/s (3.0 ft/s) is preferred (ASCE 1982). Velocities as low as 0.30 m/s (1.0 ft/s) at low flows are sufficient only to prevent deposition of the lighter sewage solids, according to the ASCE manual. Hager (1999) recommends a minimum velocity of 0.60 to 0.70 m/s (2.0 to 2.3 ft/s). Once the normal depth has been determined for the selected commercial pipe diameter, the actual velocity follows from Q_{design}/A, where A is the cross-sectional area corresponding to the normal depth; and Q_{design} is the design discharge.

An alternative approach to self-cleansing velocities is the notion of equal self-cleansing, so that nearly the same average boundary shear stress occurs at both maximum and minimum flows. This may not always be possible without increasing the slope of the pipe (ASCE 1982). Hager (1999) suggests a critical shear stress τ_c of about 2.0 Pa (0.042 lbs/ft^2) for self-cleansing in separate sewer systems. The corresponding critical velocity and its variation with filling ratio are obtained by setting the shear stress $\tau_0 = \tau_c$ so that the slope $S = \tau_c/\gamma R$ in Manning's equation. Solving for the critical velocity, V_c, the result in dimensionless form is

$$V_c^* = \frac{V_c n \sqrt{g}}{K_n u_{*c} d^{1/6}} = \left[\frac{R}{d} \right]^{1/6} \tag{4.36}$$

in which u_{*c} = critical value of shear velocity = $(\tau_c/\rho)^{1/2}$. The dimensionless cleansing velocity is a unique function of the filling ratio, y/d, as shown in Figure 4.11. However, its value does not change significantly from about 0.8 for $y/d \geq 0.4$, although clearly, from Equation 4.36, the critical velocity itself depends on pipe diameter and roughness. Also shown in Figure 4.11 as a design aid to assist in determining the actual flow velocity is a plot of A/A_f, in which A = partly full flow area and A_f = full pipe flow area = $\pi d^2/4$. For $V_c^* = 0.8$, $n = 0.015$, and $\tau_c = 2.0$ Pa, the critical velocity increases from 0.68 m/s (2.2 ft/s) to 0.86 m/s (2.8 ft/s) as the diameter increases from 0.5 m (1.6 ft) to 2.0 m (6.6 ft).

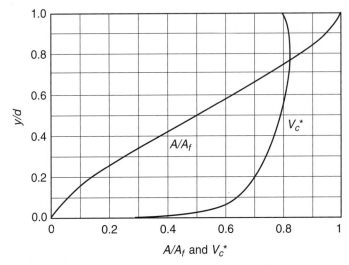

FIGURE 4.11
Dimensionless critical velocity for self-cleansing of circular sewers.

Example 4.1

Find the discharge capacity of a 24 in. (61 cm) diameter PVC storm sewer flowing at 80 percent relative depth if the slope of the sewer is 0.003. Assume that it is smooth.

Solution.   First find the geometric properties of the sewer at $y/d = 0.8$. The angle θ is

$$\theta = 2 \cos^{-1}\left(1 - 2\frac{y}{d}\right) = 2 \cos^{-1}(1 - 2 \times 0.8) = 4.4286 \text{ rad}$$

Then the area and wetted perimeter can be determined from the formulas given in Table 2.1:

$$A = (\theta - \sin\theta)\frac{d^2}{8} = [4.4286 - \sin(4.4286)]\frac{2^2}{8} = 2.69 \text{ ft}^2(0.25 \text{ m}^2)$$

$$P = \theta\frac{d}{2} = 4.4286 \times \frac{2}{2} = 4.43 \text{ ft } (1.35 \text{ m})$$

so that $R = A/P = 2.69/4.43 = 0.607$ ft (0.185 m). The friction factor comes from Equation 4.34 for smooth surfaces with $4R$ as the length scale in the Reynolds number, **Re.** This requires trial and error with Chezy's equation beginnning with an assumed value of f. For example, assume $f = 0.015$, then solve for Q and **Re:**

$$Q = \sqrt{\frac{8g}{f}}A\sqrt{RS} = \sqrt{\frac{8 \times 32.2}{0.015}} \times 2.694 \times \sqrt{0.607 \times 0.003}$$

$$= 15.1 \text{ cfs } (0.428 \text{ m}^3/\text{s})$$

$$\mathbf{Re} = \frac{(Q/A)4R}{v} = \frac{(15.1/2.694) \times 4 \times .607}{1.2 \times 10^{-5}} = 1.13 \times 10^6$$

For this value of the Reynolds number, Equation 4.34 gives $f = 0.0114$ by trial, which is used in the next iteration. In the next iteration, $Q = 17.3$ cfs (0.490 m³/s), $\mathbf{Re} = 1.30 \times 10^6$, and $f = 0.0111$. In the final iteration, $Q = 17.5$ cfs (0.496 m³/s), $\mathbf{Re} = 1.32 \times 10^6$, and $f = 0.0111$, which is the same as the previous value. Check with Figure 4.10 by computing $\mathbf{Re^*} = 2 \times (32.2 \times 2 \times 0.003)^{0.5}/1.2 \times 10^{-5} = 7.3 \times 10^4$. From Figure 4.10, read $\mathbf{Q^*} = 10.0$ and therefore $Q = 17.6$ cfs (0.499 m³/s), which is acceptable considering the graphical error. The final answer is $Q = 17.5$ cfs (0.496 m³/s). Note that the equivalent value of Manning's n from Equation 4.20 is 0.0090, but this will vary with the Reynolds number.

Example 4.2

Find the concrete sewer ($n = 0.015$) diameter required to carry a maximum design discharge of 10.0 cfs (0.283 m³/s) on a slope of 0.003. The minimum expected discharge is 2.5 cfs (0.071 m³/s). Check the velocity for self-cleansing.

Solution.   First, estimate the diameter from Equation 4.35:

$$d = 1.56 \left[\frac{nQ}{K_n S^{1/2}} \right]^{3/8} = 1.56 \times [0.015 \times 10/(1.49 \times 0.003^{1/2})]^{3/8}$$

$$= 1.96 \text{ ft } (0.597 \text{ m})$$

Round the diameter up to the next commercial pipe size of 2.0 ft (0.61 m) and solve for the normal depth of flow. First, compute the right hand side of Equation 4.31:

$$\frac{nQ}{1.49 S^{1/2}} = \frac{0.015 \times 10}{1.49 \times 0.003^{1/2}} = 1.838$$

Then set up a table as follows with assumed values of y/d from which θ, A, and R can be computed using Table 2.1. Iterate on y/d until $AR^{2/3} = 1.838$.

y/d	θ, rad	A, ft²	P, ft	R, ft	$AR^{2/3}$
0.6	3.544	1.968	3.544	0.555	1.329
0.8	4.429	2.694	4.429	0.608	1.933
0.76	4.235	2.562	4.235	0.605	1.833
0.762	4.245	2.569	4.245	0.605	1.838

This last iteration is considered acceptable; therefore, $y_0 = 0.762 \times 2 = 1.52$ ft (0.463 m) and $V = 10/2.569 = 3.89$ ft/s (1.18 m/s). This is considered more than adequate for self-cleansing at a maximum discharge. At the minimum discharge of 2.5 cfs (0.071 m³/s), calculate the normal depth using Figure 4.9:

$$\frac{AR^{2/3}}{d^{8/3}} = \frac{nQ}{1.49 S^{1/2} d^{8/3}} = \frac{.015 \times 2.5}{1.49 \times 0.003^{1/2} \times 2^{8/3}} = 0.072$$

from which y_0/d is approximately 0.33 and $y_0 = 0.66$ ft. Now, from Figure 4.11, $A/A_f = 0.29$ and $A = 0.29 \times \pi \times 2^2/4 = 0.91$ ft². Then $V = Q/A = 2.5/0.91 = 2.8$ ft/s. Obtain the critical velocity from Figure 4.11 in which $V_c^* = 0.75$ and, from Equation 4.36,

$$V_c = 0.75 \frac{K_n u_{*c} d^{1/6}}{n\sqrt{g}} = 0.75 \times \frac{1.49 \times \sqrt{0.0418/1.94} \times 2^{1/6}}{0.015 \times \sqrt{32.2}} = 2.2 \text{ ft/s } (0.67 \text{ m/s})$$

in which $\tau_c = 0.0418$ lbs/ft² (2.0 Pa). The actual velocity is well above the critical value, so this is a satisfactory design.

4.12   COMPOUND CHANNELS

A compound channel consists of a main channel, which carries base flow and frequently occurring runoff up to bank-full conditions, and a floodplain on one or both sides that carries overbank flow during times of flooding. The Manning's equation is written for compound channels in terms of the total conveyance, K, defined by $Q/S^{1/2}$, in which Q is the total discharge and S is the slope of the energy grade line, which is equal to the bed slope in uniform flow. Because of the significant difference in geometry and roughness of the floodplains compared to the main channel, the compound channel usually is divided into subsections that include the main channel and the left and right floodplains, although the floodplains may have additional subsections for varying roughness across the floodplain. If it is assumed that the energy grade line is horizontal across the cross-section for one-dimensional flow, then the slope of the energy grade line must be the same for each subsection of the compound channel as well as for the whole cross section. From continuity, $Q = \Sigma Q_i$, so it follows from equality of the slopes that $K = \Sigma k_i$, in which Q_i and k_i represent the discharge and conveyance in the ith subsection, respectively. Therefore, the total conveyance for a cross section is computed as the sum of the conveyances of the subsections. For Manning's equation, for example, the subsection conveyance is $k_i = (K_n/n_i)A_i R_i^{2/3}$, so that conveyance represents both geometric effects and roughness effects on the total conveyance and total discharge. As discussed by Cunge, Holly, and Verwey (1980), it is misleading to calculate, for a compound channel, a series of composite values of Manning's n from Manning's equation for increasing values of depth and discharge. The result is likely to be a composite n value that varies in an unexpected manner as depth increases, because this approach lumps both roughness and geometric effects into Manning's n. What is sought instead is a smooth function of increasing conveyance with increasing depth and discharge obtained by defining the total conveyance as the sum of conveyances in individual subsections. This is referred to as the *divided-channel method*.

Some difficulty arises in the divided-channel method when the hydraulic radius and wetted perimeter are defined for the floodplain and main channel subsections. The customary division into subsections, as shown in Figure 4.12, utilizes a vertical

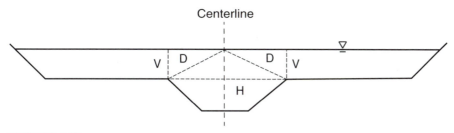

FIGURE 4.12
Compound channel with different subdivisions (H = horizontal; V = vertical; D = diagonal).

line between the subsections along which the wetted perimeter often is neglected. This is tantamount to assuming no shear stress between the main channel and floodplain flows. In fact, significant momentum exchange occurs between the faster moving main channel flow and the floodplain flow, so that the total discharge is less than what would be expected by adding the discharges of the main channel and floodplains as though they acted independently (Zheleznyakov 1971). Myers (1978) and Knight and Demetriou (1983) measured the apparent shear force on the vertical interface between the main channel and floodplain and found it to be significant. Furthermore, the mean velocity for the whole cross section actually decreases with increasing depth for overbank flow until it reaches a minimum and then begins increasing again as demonstrated by field measurements on the Sangamon River and Salt Creek in Illinois by Bhowmik and Demissie (1982). The minimum in the mean velocity for the total cross section occurred at an average floodplain depth that was 35 percent of the average main channel depth.

Several attempts have been made at quantifying the momentum transfer at the main channel-floodplain interface using concepts of imaginary interfaces included or excluded as wetted perimeter and defined at varying locations, with or without the consideration of an apparent shear stress acting on the interface. Wright and Carstens (1970) proposed that the interface be included in the wetted perimeter of the main channel and a shear force equal to the mean boundary shear stress in the main channel be applied to the floodplain interface. Yen and Overton (1973), on the other hand, suggested the idea of choosing an interface on which shear stress is in fact nearly zero. This led to several methods of choosing an interface, including a diagonal interface from the top of the main channel bank to the channel centerline at the free surface and a horizontal interface from bank to bank of the main channel, as shown in Figure 4.12. Wormleaton and Hadjipanos (1985) compared the accuracy of the vertical, diagonal, and horizontal interfaces in predicting the separate main channel and floodplain discharges measured in an experimental flume of width 1.21 m (3.97 ft) and having a fixed ratio of floodplain width to main channel half-width of 3.2. The wetted perimeter of the interface was either fully included or excluded in the calculation of wetted perimeter of the main channel. The results showed that, even though a particular choice of interface might provide a satisfactory estimate of total channel discharge, nearly all the choices tended to overpredict the separate main channel discharge and underpredict the floodplain discharge. It was further shown that these errors were magnified in the calculation of the kinetic energy flux correction coefficient.

Several empirical methods for determining discharge distribution have been developed, based on experimental data collected in the flood channel facility at Hydraulics Research, Wallingford, England, as described by Wormleaton and Merrett (1990). The channel is 56 m (184 ft) long by 10 m (33 ft) wide with a total flow capacity of 1.1 m³/s (39 cfs). In the experiments, the ratio of floodplain width to main channel half-width varied from 1 to 5.5, and the relative depth (floodplain depth/main channel depth) varied from 0.05 to 0.50. Two of the methods developed from this data include a correction to the separate main channel and floodplain discharges computed by Manning's equation. Wormleaton and Merrett (1990) applied a correction factor called the Φ *index* to the main channel and floodplain discharges calculated by a particular choice of interface (vertical, diagonal, or horizontal), which was either included or excluded from wetted perimeter. The Φ index was defined as the ratio of boundary shear force to the streamwise component of fluid weight as a measure of apparent shear force. The calculated main channel and floodplain discharges, when multiplied by the square root of the Φ index for each subsection, showed considerable improvement when compared to measured discharges; and the best performance was obtained for the diagonal interface. A regression equation was proposed for estimation of the Φ index as a function of velocity difference between main channel and floodplain, floodplain depth, and floodplain width. Ackers (1993) also proposed a discharge calculation method for compound channels using the Wallingford data. He suggested a discharge adjustment factor that depends on *coherence,* defined as the ratio of the full-channel conveyance (with the channel treated as a single unit with perimeter weighting of boundary friction factors) to the total conveyance calculated by summing the subsection conveyances. Four different zones were defined as a function of relative depth (ratio of floodplain to total depth) with a different empirical equation for discharge adjustment for each zone. In both methods, the regression equations are limited to the range of experimental variables observed in the laboratory.

An alternative approach to obtaining the discharge distribution has been the use of numerical analysis to solve the governing equations. Wark, Samuels, and Ervine (1990) and Shiono and Knight (1991) used the depth-averaged Navier-Stokes equations for steady uniform flow in a prismatic channel to solve for the lateral distribution of velocity. Their approach requires specifying the lateral distribution of eddy viscosity. Pezzinga (1994) applied a *k*-ε turbulence closure model to the three-dimensional (3D) Navier-Stokes equations for steady, uniform flow to predict secondary currents and the lateral velocity distribution. He showed that using the diagonal interface illustrated in Figure 4.12 to compute the total conveyance gave the least error in the discharge distribution in comparison with the numerical model.

Other methods for compound channel discharge distribution have appeared in the literature. Bousmar and Zech (1999) proposed a lateral momentum exchange model based on the one-dimensional momentum equation applied to the main channel with lateral inflow and outflow. They derived an additional head loss term corresponding to the exchange discharges at the interface, but it has to be obtained from the simultaneous solution of three nonlinear algebraic equations for the main channel and left and right floodplains with specification of two empirical coefficients. Myers and Lyness (1997) suggested two empirical power relations: (1) the ratio of total discharge/bank-full discharge as a function of the ratio of total depth/bank-full depth, and (2) the ratio of main channel discharge/floodplain discharge as a function of the ratio of floodplain depth/total depth. Sturm and Sadiq (1996) measured an increase in the main channel value of Manning's n of approximately 20 percent for overbank flow in comparison to the bank-full value for two different laboratory compound-channel geometries.

While it should be apparent that much research effort has been expended on the problem of discharge and its distribution in compound channels, a final solution remains elusive. The methods based on laboratory data are limited to a specific range of compound channel geometries. The 3D numerical approach of Pezzinga (1994), with a more advanced turbulence model and more extensive verification by experimental data, holds some promise for solving the problem. In the interim, either the divided channel method, using a vertical interface with the wetted perimeter included for the main channel but not the floodplain (Samuels 1989), or the divided channel method with the diagonal interface that is excluded from wetted perimeter seems to give the best results.

4.13   DESIGN OF CHANNELS WITH FLEXIBLE LININGS

When an open channel is designed for flood control, drainage, or irrigation, an important consideration is whether or not the channel bed and banks can withstand the applied flow shear stress without eroding. When the natural soil material in which a channel is constructed cannot resist erosion, then rigid or flexible linings may be considered for erosion protection. Rigid linings such as concrete may provide excellent resistance to erosion, but they may also experience other types of failures associated with frost heave, soil expansion, and high porewater pressures in the underlying soil. On the other hand, flexible linings such as rock riprap or vegetation relieve the buildup of underlying soil pressure and can even experience minor local erosion or movement of individual elements without failure because they can adjust to a changing cross-sectional shape and still provide overall channel stability.

The focus of this section is on the design of rock riprap and vegetative linings for small drainage channels such as roadside drainage ditches that must convey pavement runoff safely to a stream crossing or other drainage outlet. The design methods given here are restricted to uniform flow in prismatic channels. Typical depths are of the order of a few meters and slopes are less than about 10%. Freeboard, which is the vertical distance from the water surface to the top of the channel at the design discharge, is recommended to be about 0.15 m (0.5 ft) for such channels, and the Froude number is recommended to be less than about 0.8 for channels in subcritical flow (Kilgore and Cotton 2005).

Riprap-Lined Channels

As an application of uniform flow principles, the design procedure for riprap-lined channels as developed in NCHRP Report 108 (Anderson, Paintal, and Davenport 1970) is given in this section. It is an extension of the method of tractive force developed by the Bureau of Reclamation for stable channel design (Chow 1959). Further modifications of the procedure by Chen and Cotton (1988) are discussed.

In contrast to a fixed channel lining such as concrete, rock riprap forms a flexible channel lining that has the advantage of adjusting to minor erosion without failure and continuing to provide channel stability. The design philosophy is to choose the channel dimensions and riprap size such that the maximum boundary shear stress does not exceed the critical shear stress for erosion. As a part of the design procedure, the flow resistance of the riprap is estimated.

Experimental data on the resistance of rock riprap are summarized in Report 108, and Manning's n is taken as

$$n = 0.04d_{50}^{1/6} \tag{4.37}$$

in which d_{50} = median particle size in feet. This equation is of the same form as Strickler's equation for sand with the constant $c_n = 0.04$ in English units.

The critical shear stress relation, also based on rock riprap data, is of the same form as the Shields relation, which is described in detail in Chapter 10:

$$\tau_{0c} = 4d_{50} \tag{4.38}$$

in which τ_{0c} = critical shear stress required for initiation of motion in lbs/ft^2 and d_{50} = median particle size in feet. Equation 4.38 implicitly assumes that the particle Reynolds number is large enough that viscous effects are unimportant (i.e., Shields τ_{*c} = constant; see Chapter 10).

Shear stress distributions are analyzed on both the bed and sides of trapezoidal channels and the following relations are adopted in NCHRP Report 108:

$$(\tau_0)_{max} = 1.5\gamma RS \tag{4.39}$$

$$(\tau_0^w)_{max} = 1.2\gamma RS \tag{4.40}$$

in which $(\tau_0)_{max}$ = maximum bed shear stress, and $(\tau_0^w)_{max}$ = maximum sidewall shear stress. Also shown from the theory of stable channels (see Chapter 10) is that the tractive force ratio, K_r, at impending motion is given by

$$K_r = \frac{\tau_{0c}^w}{\tau_{0c}} = \left[1 - \frac{\sin^2\theta}{\sin^2\phi}\right]^{1/2} \tag{4.41}$$

in which θ = side slope angle; φ = angle of repose of riprap; τ_{0c}^w = critical shear stress on the sidewall; and τ_{0c} = critical shear stress for initiation of motion on the bed. The tractive force ratio, K_r, is less than 1 in value because a smaller critical shear stress is required to initiate motion on the side slope due to the gravity force component down the slope. Angles of repose and suggested side slopes, chosen such that the ratios of maximum shear stress to critical shear stress are approximately equal on the bed and banks, are summarized in Figure 4.13.

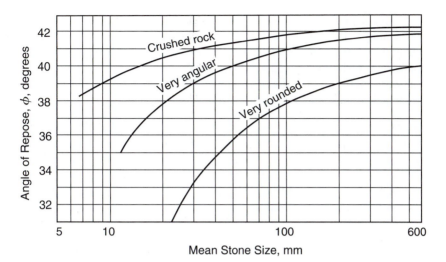

(a) Angle of Repose of Riprap

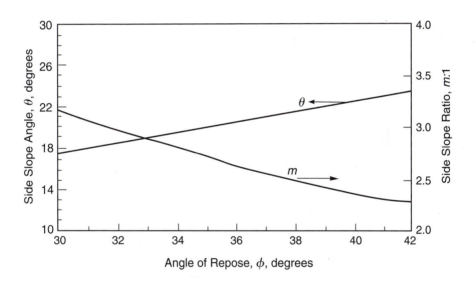

(b) Recommended Side Slopes of Trapezoidal Channels

FIGURE 4.13
Angle of repose and recommended channel side slopes for rock riprap (Anderson et al. 1970).

The riprap design procedure can be summarized as follows:

1. Choose a riprap diameter and obtain φ and θ from Figure 4.13.
2. Calculate the critical bed and wall shear stresses from Equations 4.38 and 4.41.
3. Determine Manning's n from Equation 4.37.

4. For a given channel bottom width, discharge, and slope, find the normal depth from Manning's equation.
5. Calculate maximum bed and shear stresses from Equations 4.39 and 4.40 and compare them with critical values.
6. Repeat with another riprap diameter and/or bottom width until the maximum shear stresses are just smaller than the critical values.

This procedure is simplified by Chen and Cotton (1988) in FHWA publication HEC-15 for the special case of channel side slopes that are 3:1 or flatter. In this case, the riprap on the side slopes remains stable, and failure occurs first on the channel bed. The maximum shear stress on the bed is taken to be $\tau_{max} = \gamma y_0 S$ in which y_0 = normal depth. In addition, Manning's n is computed from the relationship developed by Blodgett (1986) for the data shown previously in Figure 4.6. Blodgett (1986) obtained a best fit of the data given by

$$\frac{n}{d_{50}^{1/6}} = \frac{\dfrac{K_n}{\sqrt{8g}}(R/d_{50})^{1/6}}{0.794 + 1.85 \log(R/d_{50})} \tag{4.42}$$

which is slightly different than the Keulegan equation with a constant of 1.85 multiplying the logarithmic term rather than 2.0. In addition, Blodgett substituted the hydraulic depth for the hydraulic radius because they were nearly equal. This equation gives values of $c_n = n/d_{50}^{1/6}$ of approximately 0.046 to 0.044 in English units (0.056 to 0.054 in SI) for $30 < R/d_{50} < 185$, which is the upper limit of applicability. Therefore, Equation 4.42 gives slightly higher values of Manning's n than the Anderson et al. equation (4.37), for which $c_n = 0.04$ in English units. Recall that the Strickler value for c_n is 0.039 in English units using the value given by Hager (1999). Furthermore, Maynord (1991) determined $c_n = 0.038$ in English units from flume experiments using rock riprap in the intermediate scale of roughness ($5 < R/d_{50} < 15$) and suggested that this value also could apply in the lower range of small-scale roughness ($15 < R/d_{50} < 45$). Therefore, Equation 4.42 should give conservative estimates of Manning's n for riprap design.

The simplified procedure given in HEC-15 can be summarized as follows:

1. Choose a riprap diameter.
2. Calculate the critical bottom shear stress from Equation 4.38.
3. Estimate Manning's n from Equation 4.42 with an assumed depth.
4. Calculate the normal depth y_0 from Manning's equation and iterate on Manning's n from Equation 4.42.
5. Calculate the maximum shear stress on the bottom as $\gamma y_0 S$ and compare it with the critical stress.

Example 4.3

Determine the size of "very angular" stones required to protect a trapezoidal roadside drainage channel from erosion if the bottom width of the channel is 3.0 m (9.8 ft), the bed slope is 0.010, and the design discharge is 2.0 m³/s (70.6 cfs).

Solution.   Initially choose a median rock size of d_{50} = 50 mm (0.16 ft). Then from Figure 4.13, the angle of repose, ϕ = 40 degrees, and choose m = 2.5 so that $\theta = \tan^{-1}(1/2.5) = 21.8$ degrees. Calculate the tractive force ratio from Equation 4.41:

$$K_r = \sqrt{1 - \frac{\sin^2 \theta}{\sin^2 \phi}} = \sqrt{1 - \frac{\sin^2(21.8)}{\sin^2(40)}} = 0.816$$

The critical shear stress on the bed from Equation 4.38 is $\tau_{0c} = 4d_{50} = 4 \times (50/305) = 0.66$ lb/ft^2 (31.6 N/m^2), in which the conversion 1.0 lb/ft^2 = 47.9 N/m^2 has been used, and the critical shear stress on the side slopes is $\tau_{0c}^w = K_r\tau_{0c} = 0.816 \times 0.66 = 0.54$ lb/ft^2 (25.9 N/m^2). Manning's n from Equation 4.37 is $n = 0.04d_{50}^{1/6} = 0.04 \times (50/305)^{1/6} = 0.03$. Now solve Manning's equation for normal depth

$$\frac{A^{5/3}}{P^{2/3}} = \frac{(y_0(3.0 + 2.5y_0))^{5/3}}{(3.0 + 2\sqrt{1 + 2.5^2}y_0)^{2/3}} = \frac{nQ}{S^{1/2}} = \frac{0.03(2.0)}{0.01^{1/2}} = 0.60$$

Solving by trial, we have y_0 = 0.358 m (1.17 ft) and $R = A/P = 1.39/4.93 = 0.282$ m (0.93 ft). The maximum shear stresses on the bed and side walls are

$$(\tau_0)_{max} = 1.5\gamma RS = 1.5 \times 9810 \times 0.282 \times 0.01 = 41.5 > 31.6 \text{ N/m}^2$$

$$(\tau_0^w)_{max} = 1.2\gamma RS = 1.2 \times 9810 \times 0.282 \times 0.01 = 33.2 > 25.9 \text{ N/m}^2$$

Both maximum shear stresses exceed the corresponding critical values, so the initial riprap size of 50 mm is too small. By trying several additional riprap sizes but staying with the initial 2.5:1 side slopes (a computer program or spreadsheet facilitates the computations), the final result is d_{50} = 70 mm (0.23 ft). In this final iteration, τ_{0c} = 44.0 N/m^2 (0.92 lb/ft^2), K_r = 0.83, and $\tau_{0c}^w = K_r\tau_{0c} = 0.83 \times 44.0 = 36.5$ N/m^2 (0.76 lb/ft^2). Manning's $n = 0.04 \times (70/305)^{1/6} = 0.031$. The normal depth is 0.364 m (1.19 ft) and $R = A/P = 1.42/4.96 = 0.286$ m (0.94 ft). Finally, $(\tau_0)_{max} = 42.1 < 44.0$ N/m^2 and $(\tau_0^w)_{max} = 33.7 < 36.5$ N/m^2. These comparisons with critical shear stresses are satisfactory with some factor of safety included. A factor of safety can be decided in advance, and iteration continued until that factor of safety is reached. Note that failure is slightly more imminent on the bed than the side slopes.

Grass-Lined Channels

Channels also can be designed for stability with vegetative linings, which have been used successfully to prevent erosion of agricultural waterways, roadside drainage ditches, and emergency spillways for small lake impoundments. Vegetative linings have been classified according to their degree of vegetal retardance as Class A, B, C, D, or E by the Soil Conservation Service. Descriptive examples of types of grass, stem length, and stand density are given in Table 4.2 for each vegetal retardance class.

Both the estimation of flow resistance and establishment of a criterion for stability of grass linings are challenging design issues because of the fact that the grass stems are flexible and can bend by varying degrees during design flows. Recent turbulence measurements by Jarvela (2005) have shown that the Reynolds stresses are maximum in a shear layer just above the deflected vegetation (wheat), and that the

TABLE 4.2   Classification of vegetal cover as to degree of retardance (SCS-TP-61)

Vegetal Retardance Class	Cover	Condition
A	Weeping lovegrass	Excellent stand, tall (average 30 in.) (76 cm)
	Yellow bluestem	
	Ischaemum	Excellent stand, tall (average 36 in.) (91 cm)
B	Kudzu	Very dense growth, uncut
	Bermuda grass	Good stand, tall (average 12 in.) (30 cm)
	Native grass mixture (little bluestem, bluestem, blue gamma, and other long and short Midwest grasses)	Good stand, unmowed
	Weeping lovegrass	Good stand, tall (average 24 in.) (61 cm)
	Lespedeza sericea	Good stand, not woody, tall (average 19 in.) (48 cm)
	Alfalfa	Good stand, uncut (average 11 in.) (28 cm)
	Weeping lovegrass	Good stand, unmowed (average 13 in.) (33 cm)
	Kudzu	Dense growth, uncut
	Blue gamma	Good stand, uncut (average 13 in.) (28 cm)
C	Crabgrass	Fair stand, uncut (10 to 48 in.) (25 to 120 cm)
	Bermuda grass	Good stand, mowed (average 6 in.) (15 cm)
	Common lespedeza	Good stand, uncut (average 11 in.) (28 cm)
	Grass-legume mixture—summer (orchard grass, redtop, Italian ryegrass, and common lespedeza)	Good stand, uncut (6 to 8 in.) (15 to 20 cm)
	Centipedegrass	Very dense cover (average 6 in.) (15 cm)
	Kentucky bluegrass	Good stand, headed (6 to 12 in.) (15 to 30 cm)
D	Bermuda grass	Good stand, cut to 2.5 in. height (6 cm)
	Common lespedeza	Excellent stand, uncut (average 4.5 in.) (11 cm)
	Buffalo grass	Good stand, uncut (3 to 6 in.) (8 to 15 cm)
	Grass-legume mixture—fall, spring (orchard grass, redtop, Italian ryegrass, and common lespedeza)	Good stand, uncut (4 to 5 in.) (10 to 13 cm)
	Lespedeza sericea	After cutting to 2 in. height (5 cm)
		Very good stand before cutting
E	Bermuda grass	Good stand, cut to 1.5 in. height (4 cm)
	Bermuda grass	Burned stubble

Note: Covers classified have been tested in experimental channels. Covers were green and generally uniform.

velocity profile above the vegetation is logarithmic with respect to the ratio of distance from the bed to roughness height measured as the deflected vegetation height. Measurements by Wilson et al. (2003) confirm the existence of the shear layer above vegetation of different types and show that Reynolds stress maxima as well as values inside the vegetated layer are reduced for frond vegetation in comparison to rod vegetation (stems only).

The work of Kouwen (1992) summarizes the results of extensive experimental studies, which are based on relative roughness of vegetative linings and a logarithmic velocity profile to establish flow resistance coefficients for the types of vegetation described in Table 4.2. The biomechanical properties of the vegetation are used as a means of assessing the roughness height estimated as the deflected vegetal stem height. The method of Kouwen is introduced here as a methodology for designing stable channels with vegetative linings. Several criteria for stability are discussed including those reported by Chen and Cotton (1988).

In the early experiments (Kouwen and Unny 1973), flexible plastic elements were utilized as a model of vegetation in flumes to develop a measure of resistance to flow-induced bending that was later related to natural grasses (Kouwen and Li 1980). The stiffness parameter MEI was defined as the product of M = the number of stems per unit surface area; E = modulus of elasticity of the flexible stems; and I = moment of inertia of the stem cross-sectional area. The relative deflection of the vegetation was formulated in terms of the ratio of MEI to the applied flow shear stress resulting in a dimensionless experimental relationship given by

$$\frac{k}{h_s} = 0.14 \left[\frac{(MEI/\tau_0)^{0.25}}{h_s} \right]^{1.59} \tag{4.43}$$

in which k = roughness height of deflected stems; h_s = upright height of stems; τ_0 = applied shear stress in uniform flow = $\gamma y_0 S$ in a wide channel; y_0 = normal flow depth; and S = channel slope. To obtain an average relative roughness height for a trapezoidal channel, the mean boundary shear stress can be applied (Cotton 1999).

Assuming a logarithmic velocity profile and converting the Darcy-Weisbach f to Manning's n in a Keulegan type of resistance equation for fully rough turbulent flow (see Equations 4.19a and 4.20), we have

$$\frac{n}{k^{1/6}} = \frac{K_n}{\sqrt{8g}} \frac{(R/k)^{1/6}}{\left(a + b \log \dfrac{R}{k} \right)} \tag{4.44}$$

in which R = hydraulic radius; and k = roughness height from Equation 4.43. Although the coefficients a and b were measured for the wide rectangular case, the hydraulic radius has been used in Equation 4.44 to generalize it to other channel shapes as with the Keulegan equation. Values of a and b, which are given in Table 4.3, depend on u_*/u_{*c} = the ratio of the actual shear velocity to the vegetal critical shear velocity as an index of the flow properties required to bend the grass over from an erect waving boundary to one that is flattened. The critical shear velocity, as might be expected, is a function of the vegetative stiffness parameter, MEI, and is given by the minimum of two expressions:

$$u_{*c} = \min(0.028 + 6.33MEI^2, 0.23MEI^{0.106}) \tag{4.45}$$

Table 4.3   Values of a and b in Equation 4.44

Condition	Classification	Criteria	a	b
1	Erect	$(u_*/u_{*_c}) \leq 1.0$	0.15	1.85
2	Prone	$1.0 < (u_*/u_{*_c}) \leq 1.5$	0.20	2.70
3	Prone	$1.5 < (u_*/u_{*_c}) \leq 2.5$	0.28	3.08
4	Prone	$2.5 < (u_*/u_{*_c})$	0.29	3.50

Source: Koluven and Li (1980).

in which *MEI* is in N-m^2 and u_{*_c} is in m/s. The values of u_{*_c} are the same at *MEI* approximately equal to 0.16 N-m^2, so the first expression will be the minimum of the two for *MEI* $\leq$ 0.16 N-m^2 (vegetal retardance classes D and E), and the second expression will govern for *MEI* > 0.16 N-m^2 (vegetal retardance classes A, B, and C). The value of the uniform flow shear velocity is calculated from $u_* = (gy_0S)^{1/2}$ for wide channels or $(gRS)^{1/2}$ for trapezoidal channels since $u_* = (\tau_0/\rho)^{1/2}$ by definition. As an alternative to using only the fixed coefficient values a and b in Table 4.3 for conditions 2 and 3, values can be linearly interpolated between conditions 1 and 4 if $1.0 < u_*/u_{*_c} < 2.5$.

The dimensionless curves given by Equation 4.44 for Manning's n with vegetation conditions 1, 2, 3, and 4 (see Table 4.3) are shown in Figure 4.14. The increase in Manning's n for shallow depths relative to vegetation height is consistent with Keulegan's equation for large roughness elements such as rock riprap at shallow depths, but the coefficients a and b have different values. It also can be seen from Figure 4.14 that for a given relative roughness, the value of $(n/k^{1/6})$ decreases

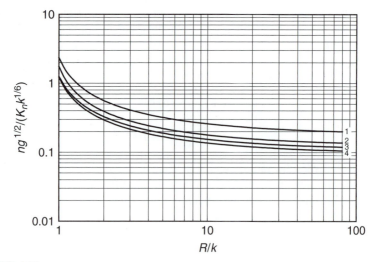

FIGURE 4.14

Manning's n for vegetated channels with vegetal conditions 1, 2, 3, or 4 as defined in Table 4.3 (after Kouwen and Li 1980).

Table 4.4   Properties of vegetal retardance classes

Vegetal retardance class	Average height, h_s, cm (ft)	Stiffness MEI, N-m^2 (lb-ft^2)	Permissible shear stress τ_p, N/m^2 (lb/ft^2)	Calculated lining failure criterion Eq. 4.43 $(k/h_s)_{min}$
A	91 (3.0)	300 (725)	177 (3.70)	0.2
B	61 (2.0)	20 (50)	100 (2.10)	0.2
C	20 (0.66)	0.5 (1.2)	48 (1.00)	0.3
D	10 (0.33)	0.05 (0.12)	29 (0.60)	0.4
E	4 (0.13	0.005 (0.012)	17 (0.35)	0.9

Source: Chen and Cotton (1988).

as the vegetation becomes more prone from condition 1 to 4, which indicates that the dominant contribution to flow resistance gradually changes from form resistance to surface resistance.

The remaining relationship needed for determination of Manning's n is the value of MEI for vegetal retardance conditions A, B, C, D, and E, which were defined in Table 4.2. These have been given by Chen and Cotton (1988) and Kilgore and Cotton (2005) and are shown in Table 4.4.

In order to find the discharge Q for a given depth of flow and channel slope, the shear stress is calculated from $\tau_0 = \gamma RS$, and for the selected vegetal retardance class, MEI is determined. Then the roughness height k is obtained from Equation 4.43. The critical shear velocity is calculated from Equation 4.45 so that the values of a and b can be selected from Table 4.3. Finally, Equation 4.44 is solved for Manning's n value, and Q can be calculated from Manning's equation. If the discharge has been specified and the depth is unknown, then iteration on the depth is necessary until the assumed depth produces a discharge equal to the specified value.

Several methods have been developed for assessing the stability of a channel lined with vegetation once the flow resistance and the depth-discharge pair have been calculated. Early designs of grass-lined channels relied on a permissible velocity for channel stability (Chow 1959). Chen and Cotton (1988) have suggested a permissible shear stress as given for each vegetal retardance class in Table 4.4. Then it is a matter of comparing the maximum shear stress with the permissible value to decide if the vegetated lining will fail or not. Permissible shear stress has the advantage that it depends only on the erodibility properties of the channel lining and boundary, while permissible velocity also depends on the depth of flow. Kouwen (1992) suggested that lining failure might be evaluated based on a minimum value of k/h_s below which field and laboratory data show a significant increase in erosion. In other words, at some critical degree of bending, the vegetation no longer shields the soil and plant roots from the high shear stress zone, and the soil begins to erode at an unacceptable rate. Temple (1980) proposed apportioning the shear stress between the soil surface and the vegetation, and evaluating failure based on exceedance of the critical shear stress of the soil itself.

This approach is described by Kilgore and Cotton (2005), and the reader is referred to their publication for details.

In Table 4.4, the minimum value of k/h_s for failure of the vegetative lining has been calculated for each vegetal retardance class from MEI and permissible shear stress values given by Chen and Cotton (1988) using Equation 4.43. The calculated minimum values of k/h_s range from 0.2 for vegetal retardance class A to 0.9 for class E as the undeflected stem height becomes shorter. In other words, the longer grass can be deflected more before exposing the soil to erosive stresses. In an independent set of experiments, Samani and Kouwen (2002) measured erosion rates of sand placed at the base of artificial grass, and found that the minimum value of k/h_s at which failure occurred for coarse sand was 0.4 for the long model grass (10.5–12.7 cm) and 0.9 for the short model grass (6.2–7.5 cm). Essentially the same results were obtained for fine sand. These minimum values of k/h_s are consistent with those calculated in Table 4.4 based on permissible shear stresses of natural grasses; however, the minimum value of k/h_s can be quite sensitive as a failure criterion.

The design procedure can be summarized as follows:

1. Choose a vegetal retardance class A, B, C, D, or E and determine the permissible shear stress, τ_p, from Table 4.4.
2. Estimate a flow depth for given bottom width b and side slopes (m:1; m $\geq$ 3) and calculate the hydraulic radius, R.
3. Obtain Manning's n value from Equations 4.43, 4.44, and 4.45 for the appropriate vegetal retardance class and the given channel slope with values of MEI, h_s, a, and b from Tables 4.3 and 4.4.
4. Calculate the normal depth from Manning's equation for the design discharge and compare with the assumed depth. Iterate on the depth until the correct Manning's n and depth y_0 have been determined.
5. Calculate the maximum bottom shear stress as $\tau_{max} = \gamma y_0 S$ and compare it with the permissible shear stress, τ_p. Adjust the channel bottom width, slope, or vegetal retardance class until $\tau_{max} \leq \tau_p$.

This design procedure can be used to design temporary linings such as jute, fiberglass roving, straw with net, and synthetic mats that are useful for stabilizing channels immediately after construction before a stand of grass develops. The permissible shear stresses and roughness values for temporary linings are given by Kilgore and Cotton (2005) and Cotton (1999).

Example 4.4

A trapezoidal roadside ditch has a bottom width of 1.5 m and side slopes of 3:1. The channel slope is 0.012, and the proposed channel lining is a grass-legume mixture that has a height of 15 to 20 cm. What is the maximum allowable discharge for this lining?

Solution. This example illustrates an alternate design procedure from the one just given that is useful for selecting the initial lining. From Table 4.2, the vegetal retardance class is C, and the permissible shear stress $\tau_p = 48$ Pa from Table 4.4. Then the maximum allowable depth comes from setting $\tau_{max} = \gamma y_0 S = \tau_p$ and solving for y_0:

$$y_0 = \frac{\tau_p}{\gamma S} = \frac{48}{9810 \times 0.012} = 0.408 \text{ m } (1.34 \text{ ft})$$

The geometric properties of area and wetted perimeter of the cross section for this depth are

$$A = y_0(b + my_0) = 0.408 \times (1.5 + 3 \times 0.408) = 1.11 \text{ m}^2 \ (12.0 \text{ ft}^2)$$

$$P = b + 2y_0\sqrt{1 + m^2} = 1.5 + 2 \times 0.408 \times \sqrt{1 + 3^2} = 4.08 \text{ m } (13.4 \text{ ft})$$

Then the hydraulic radius $R = A/P = 1.11/4.08 = 0.272$ m (0.892 ft), and $\tau_0 = \gamma RS = 9810 \times 0.272 \times 0.012 = 32.0$ N/m^2. From Table 4.4, $MEI = 0.5$ N-m^2 and the erect stem height, $h_s = 0.2$ m for Class C. The roughness height comes from Equation 4.43:

$$\frac{k}{h_s} = 0.14\left[\frac{(MEI/\tau_0)^{0.25}}{h_s}\right]^{1.59} = 0.14\left[\frac{(0.5/32.0)^{0.25}}{0.2}\right]^{1.59} = 0.346$$

so the value of $k = 0.346 \times 0.2 = 0.0693$ m. To find the values of a and b, we need the shear velocity, $u_* = (gRS)^{1/2} = (9.81 \times 0.272 \times 0.012)^{1/2} = 0.179$ m/s, and the critical vegetal shear velocity, $u_{*_c} = 0.23 \, MEI^{0.106} = 0.23(0.5)^{0.106} = 0.214$ m/s from Equation 4.45. Now because $u_*/u_{*_c} < 1$, the values of a and b from Table 4.3 are $a = 0.15$ and $b = 1.85$. The value of Manning's n is obtained from Equation 4.44 or Figure 4.14:

$$\frac{n}{k^{1/6}} = \frac{K_n}{\sqrt{8g}\left(a + b\log\dfrac{R}{k}\right)} = \frac{1.0}{\sqrt{8 \times 9.81}\left(0.15 + 1.85\log\dfrac{0.272}{0.0693}\right)} = \frac{(0.272/0.0693)^{1/6}}{} = 0.114$$

with the result that $n = 0.114(0.0693)^{1/6} = 0.073$. Finally, the allowable discharge from Manning's equation is

$$Q = \frac{K_n}{n}AR^{2/3}S^{1/2} = \frac{1.0}{0.073} \times 1.11 \times (0.272)^{2/3} \times (0.012)^{1/2}$$

$$= 0.70 \text{ m}^3/\text{s } (25 \text{ cfs})$$

The allowable discharge is compared with the design discharge to decide if this lining is suitable. The final design depth is determined from the procedure given previously.

4.14   SLOPE CLASSIFICATION

Aside from its primary use in channel design, the normal depth used in conjunction with the critical depth of flow is a useful concept in classifying slopes as mild or steep and ultimately in classifying gradually varied flow profiles. A mild slope is defined as a slope on which the uniform flow depth is subcritical; that is, normal depth, y_0, is greater than critical depth y_c. For a steep slope, the uniform flow depth is supercritical ($y_0 < y_c$). At the boundary between these two cases, it is obvious that $y_0 = y_c$, so that it is useful to define a critical slope as that value of bed slope for which uniform flow would occur at critical depth. Using Manning's equation, the critical slope, S_c, becomes

$$S_c = \frac{n^2 Q^2}{K_n^2 A_c^2 R_c^{4/3}} \tag{4.46}$$

in which A_c and R_c represent the area and hydraulic radius evaluated at critical depth. A mild slope can be defined as having a bed slope, S_0, less than the critical slope, S_c, while for a steep slope, $S_0 > S_c$. The critical slope is understood to be a calculated quantity to be used only as a criterion for classification of a slope as mild or steep.

The critical slope is a function of the discharge, so that a particular bed slope may be mild at some discharges and steep at others. This point is illustrated easily with a very wide, rectangular channel. For this shape, the hydraulic radius may be approximated as the depth of flow, and the Manning's equation simplifies considerably so that the critical slope becomes

$$S_c = \left[\frac{g^{(10/9)} n^2}{K_n^2} \right] q^{-(2/9)} \tag{4.47}$$

from which the critical slope decreases with increasing discharge. For example, a wide rectangular channel with a Manning's n value of 0.015 has a bed slope of 0.004 as shown in Figure 4.15. At this value of bed slope, the slope changes from

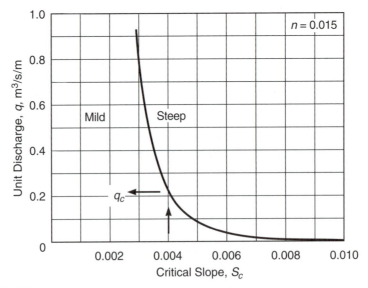

FIGURE 4.15
Critical slope for a wide, rectangular channel.

mild to steep at a discharge of 0.216 m³/s/m (2.32 cfs/ft), which is called the criti-cal discharge, q_c. The minimum possible value of the critical slope for the wide rectangular channel asymptotically approaches zero, and this is called the *limit slope*.

The limit slope for a rectangular channel that cannot be classified as very wide is finite (Rao and Sridharan 1970). If the expressions for area and hydraulic radius for a rectangular channel are substituted into Equation 4.46 and the discharge is elimi-nated by the relation between critical depth and discharge, the critical slope becomes

$$S_c = \frac{gn^2}{K_n^2} y_c \left[\frac{b + 2y_c}{by_c} \right]^{4/3} \tag{4.48}$$

If this expression for critical slope is differentiated with respect to y_c and set to zero, the minimum critical slope, or limit slope, occurs at $y_c/b = 1/6$ and has a value of

$$S_L = \frac{2.67g}{K_n^2} \frac{n^2}{b^{1/3}} \tag{4.49}$$

The expression for the critical slope in Equation 4.48 can be nondimensionalized in terms of the limit slope to produce

$$\frac{S_c}{S_L} = 0.375 \frac{\left(1 + 2\frac{y_c}{b}\right)^{4/3}}{\left(\frac{y_c}{b}\right)^{1/3}} \tag{4.50}$$

This equation is plotted in Figure 4.16, from which we see that, for a bed slope less than the limit slope, the slope remains mild for all possible discharges.

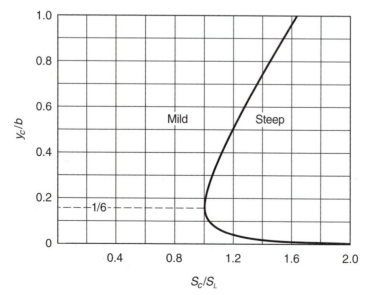

FIGURE 4.16
Critical slope for a rectangular channel in terms of the limit slope.

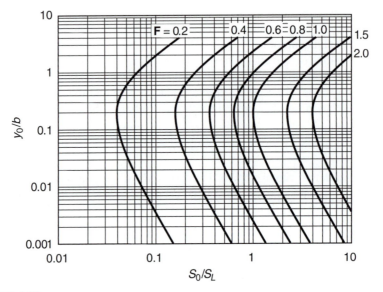

FIGURE 4.17

Normal depth vs. slope for a constant Froude number (Rao and Sridharan 1970). (*Source: N. S. L. Rao and K. Sridharan, "Limit Slope in Uniform Flow Computations," J. Hyd. Div., © 1970, ASCE. Reproduced by permission of ASCE.*)

The limit slope can be used to nondimensionalize the expression for bed slope S_0 from Manning's equation written in terms of the Froude number of the uniform flow $\mathbf{F}_0$:

$$\frac{S_0}{S_L} = 0.375\mathbf{F}_0^2 \frac{\left(1 + 2\dfrac{y_0}{b}\right)^{4/3}}{\left(\dfrac{y_0}{b}\right)^{1/3}} \qquad (4.51)$$

in which S_0 is the bed slope; S_L is the limit slope; and y_0 is the normal depth. For the case of $\mathbf{F}_0 = 1$, this equation reduces to Equation 4.50 with $S_0 = S_c$ and $y_0 = y_c$. Equation 4.51 is plotted in Figure 4.17 for different constant values of the Froude number. It can be shown that the maximum value of the Froude number occurs at $y_0/b = 1/6$ and has a value of

$$\mathbf{F}_{max} = \sqrt{\frac{S_0}{S_L}} \qquad (4.52)$$

This raises the possibility of designing a rectangular channel such that a given value of the Froude number is not exceeded for any discharge the channel may experience. It is desirable to prevent the maximum Froude number from becoming too close to unity because of the free-surface instability associated with critical flow.

Example 4.5

A concrete-lined rectangular channel has a bottom width of 3.0 m (9.8 ft) and a Manning's n of 0.015. The bed slope is 0.007, and the discharge is expected to vary from zero to 60.0 m³/s (2120 cfs). Determine if the slope is steep or mild over the full range of discharges. At what slope would the channel be mild for all discharges?

Solution.   First, find the limit slope from (4.49):

$$S_L = \frac{2.67g}{K_n^2} \frac{n^2}{b^{1/3}} = \frac{2.67 \times 9.81}{1.0^2} \frac{.015^2}{3^{1/3}} = 0.00409$$

Thus, the actual slope of 0.007 is greater than the limit slope, and discharges in both the mild and steep slope classification are possible. Set $S_c = 0.007$ in Equation 4.50 and solve for y_c. This is a trial-and-error solution with two roots, one for $y_c/b > 1/6$ and another for $y_c/b < 1/6$. The roots are $y_c/b = 1.098$ and 0.0115, from which $y_c = 3.29$ m (10.8 ft) and 0.0345 m (0.113 ft). The corresponding discharges come from the relationship between flow rate per unit of width q and y_c for rectangular channels:

$$q_c = \sqrt{gy_c^3} = \sqrt{9.81 \times 3.29^3} = 18.7 \text{ m}^2/\text{s} \ (201 \text{ ft}^2/\text{s})$$

in which only the upper value of y_c has been illustrated. The other value of q_c is 0.020 m²/s (0.22 ft²/s). The two values of critical discharge, $Q_c \ (= q_c b)$, are 56.1 m³/s (1980 cfs) and 0.060 m³/s (2.1 cfs). Between these two discharges, the slope will be steep, and for $Q > 56.1$ m³/s or $Q < 0.060$ m³/s, the slope will be mild. This can be seen in Figure 4.17 for the intersection of a vertical line, along which $S_0/S_L = 0.007/.00409 = 1.71$ and the curve for Froude number $= 1.0$. If the slope could be constructed to be less than the limit slope of 0.00409, then it would be mild for all discharges.

4.15   FLOOD CONTROL CHANNELS

Open channels designed for flood control and flood diversion may be required to carry relatively large discharges without overtopping and without becoming unstable. The design discharge is established by a frequency analysis and the assessment of flood damages associated with overtopping of the channel. The alignment of the channel and its slope and width are determined in part by existing topography, available right-of-way, and proximity of existing infrastructure such as large sewers and water mains in urban settings. The roughness coefficient depends on the choice of lining material or the natural roughness of an unlined channel. The cross-sectional shape is usually rectangular or trapezoidal and can be lined with concrete, or for low velocity channels, the channel bottom can be left natural while the banks are protected with revetment. In some cases, a small pilot channel may be constructed in the bottom of the larger channel to carry low flows effectively. Cost is an important issue and several cross sections may be investigated to determine the least-cost design that is affected by hydraulic efficiency, right-of-way costs, and construction costs.

Special design considerations include limits on the channel Froude number and specification of freeboard. Because of the instability of flows near critical depth, for which a small change in specific energy results in a large change in depth and concomitant undular surface waves, it is desirable to limit the design flow Froude number such that it is less than 0.86 or greater than 1.13 to avoid the depth range of 0.9 to 1.1 times the critical depth. Freeboard is needed to allow for uncertainties in estimates of channel roughness or backwater due to bridges, embankment settlement, accumulation of debris, and aquatic growth in the channel. Establishment of freeboard must take into account all of these factors and so there is no single best design guideline. In general, a freeboard of 2 ft (0.6 m) in rectangular channels or 2.5 ft (0.8 m) in trapezoidal channels if they are concrete lined is recommended, while the freeboard should be approximately 2.5 ft (0.8 ft) in riprap lined channels and 3 ft (0.9 m) in earthen channels (U.S. Army Corps of Engineers 1994).

The most hydraulically efficient cross section is considered next. It is followed by design guidelines for riprap protection of large flood control channels using the Corps of Engineers method (1994).

Best Hydraulic Section

From economic considerations of minimizing the flow cross-sectional area for a given design discharge, a theoretically optimum cross section can be derived, although many other factors, including channel stability and maximum Froude number, may be the overriding design criteria. Minimization of flow area implies maximization of velocity for a given discharge and, therefore, a maximization of hydraulic radius, R, for a given channel slope and roughness based on any uniform flow formula. The problem can be recast then as minimizing the wetted perimeter, P, for a fixed cross-sectional area, A, since $R = A/P$. Under this criterion, it is clear that a semicircle would provide the best hydraulic section of all. For the rectangular section, the wetted perimeter, P, is given by $P = b + 2y$ and substituting $b = A/y$, we can differentiate P with respect to y while holding A constant and set the result to zero:

$$\frac{dP}{dy} = \frac{d}{dy}\left[\frac{A}{y} + 2y\right] = -\frac{A}{y^2} + 2 = 0 \qquad (4.53)$$

Then, we see that the best rectangular section has $A = 2y^2$ and $b = 2y$, so that a semicircle can be inscribed inside it. From the same reasoning (see the Exercises), the best trapezoidal section is one for which $R = y/2$ and $m = 1/3^{0.5}$, so that the side slope angle $\theta = \tan^{-1}(1/m) = 60°$ and the shape is that of a half-hexagon inside of which a semicircle can be inscribed.

The best hydraulic section might be desirable only for a concrete-lined prismatic channel. If it is rectangular, an aspect ratio of $b/y = 2$ for the best section would mean that a subcritical Froude number would be less than its maximum value at $b/y = 6$ for a given slope. In addition, secondary currents would be much more likely in the best section because of its small aspect ratio. However, once channel stability becomes an issue, the aspect ratio is likely to greatly increase to keep the shear stress below its critical value.

Example 4.6

A flood diversion channel must be designed to carry a peak flood discharge of 150 m³/s. The topography dictates a slope of 0.0005, and the channel is to be lined with concrete. The maximum right-of-way available is 25 m. Determine the dimensions of the channel.

Solution.   First select a channel of rectangular cross section so that Manning's equation becomes

$$\frac{A^{5/3}}{P^{2/3}} = \frac{(by_0)^{5/3}}{(b + 2y_0)^{2/3}} = \frac{nQ}{K_n S^{1/2}} = \frac{0.015(150)}{1.0(0.0005)^{1/2}} = 100.6$$

in which Manning's n has been taken to be 0.015 for float-finished concrete from Table 4.1. At this point several combinations of width b and normal depth y_0 will satisfy Manning's equation. If we choose the most hydraulically efficient section, then $b = 2y_0$ and we have

$$\frac{(2y_0^2)^{5/3}}{(4y_0)^{2/3}} = 1.26y_0^{8/3} = 100.6$$

from which $y_0 = 5.17$ m (17.0 ft) and $b = 2y_0 = 10.35$ m (34.0 ft). However, this depth of excavation will encounter rock at a depth of 4.0 m (13.1 ft), which will greatly increase the excavation cost. In this case, choose a flow depth of 3.0 m (9.84 ft) and a freeboard of 0.6 m (2.0 ft) to remain above the rock, and solve Manning's equation again in the form

$$\frac{(3b)^{5/3}}{(b + 6)^{2/3}} = 100.6$$

By trial and error, $b = 19.3$ m (63.3 ft), $A = by_0 = 57.9$ m² (623 ft²), and $P = b + 2y_0 = 25.3$ m (83.0 ft). While this is a 22% increase in wetted perimeter in comparison with the most efficient hydraulic section, the width is within the allowable right-of-way and rock excavation is avoided. The mean velocity is $Q/A = 150/57.9 = 2.59$ m/s (8.50 ft/s), and the Froude number is $V/(gy_0)^{1/2} = 2.59/(9.81 \times 3.0)^{1/2} = 0.48$, which is sufficiently below critical flow. The value of $y_0/b = 0.155$, which is just below the relative depth of 0.167 at which maximum Froude number occurs, and the limit slope from Equation 4.49 is 0.0022 so that $S_0/S_L = 0.23$. By referring to Figure 4.17, it can be seen that the Froude number will be less than 0.48 for all discharges smaller than the design value of 150 m³/s (5300 cfs). The final design is a rectangular concrete channel with a flow depth of 3.0 m (9.8 ft), bottom width of 19.3 m (63.3 ft), and a freeboard of 0.6 m (2.0 ft).

Corps of Engineers (COE) Riprap Design Method

The Corps of Engineers method for riprap design is based on extensive laboratory experiments in large flumes, including a large outdoor curved channel, as well as field verification (Maynord 1988). It is well suited for side slope protection design in large flood control channels and in streams and rivers, particularly in the case of curved channels, but it can also be applied to riprap lining of straight flood control channels for both bed and bank stability. The method is limited to channels with side slopes of 1.5:1 or flatter, channel slopes less than 2%, and Froude numbers less than 1.2. The basic design equation for straight or curved channels is given in dimensionless form for the desired riprap stone size for which 30% of the distribution is lighter by weight, d_{30}:

$$\frac{d_{30}}{y} = S_f C_s C_v C_T \left[\frac{V_l}{\sqrt{K_1 (SG - 1.0)gy}}\right]^{2.5} \tag{4.54}$$

in which y = local flow depth; V_l = local depth-averaged flow velocity; SG = specific gravity of the riprap; and K_1 = side slope correction factor similar to the critical tractive force ratio as given by Equation 4.41. A series of correction factors appear in front of the bracketed term on the right-hand side of the equation, and they are defined by S_f = factor of safety = 1.1; C_s = coefficient = 0.30 for angular rock and 0.36 for rounded rock; C_v = vertical velocity distribution coefficient = 1.0 for straight channels, and as shown in Figure 4.18c for curved channels where r = radius of curvature of channel centerline and B = water-surface width; C_T = riprap thickness coefficient = 1.0 if the riprap thickness = 1 d_{100} or $1.5d_{50}$ whichever is greater.

For design of side slope riprap protection in curved channels, the local velocity is taken to be the depth-averaged local velocity over the slope at a point 20% of the slope length from the toe of the slope. The local velocity $V_l = V_{ss}$ is estimated there in ratio to the mean cross-sectional velocity, $V = Q/A$, for natural channels by the expression

$$\frac{V_{ss}}{V} = 1.74 - 0.52 \log\left(\frac{r}{B}\right) \tag{4.55}$$

in which r = radius of curvature of channel bend and B = top width of water surface; while for trapezoidal channels, the local velocity on the side slopes is given by

$$\frac{V_{ss}}{V} = 1.71 - 0.78 \log\left(\frac{r}{B}\right) \tag{4.56}$$

These empirical relationships are plotted in Figure 4.18a. The value of V_{ss} and the actual local depth at the location where V_{ss} is defined are substituted into Equation 4.54 to estimate the side slope riprap size.

The tractive force ratio, K_r, which is the ratio of critical shear stress on the bank to that on the channel bed as given by Equation 4.41, is shown in Figure 4.18b for a constant angle of repose of 40 degrees. The COE riprap design procedure considers this relationship too conservative and suggests a higher value referred to as K_1, which appears in Equation 4.54 and is given in Figure 4.18b for design.

In straight channels, side slope protection is designed for V_{ss}/V values for very large channel curvature ratios r/B, but in no case is V_{ss}/V taken to be less than 1.0.

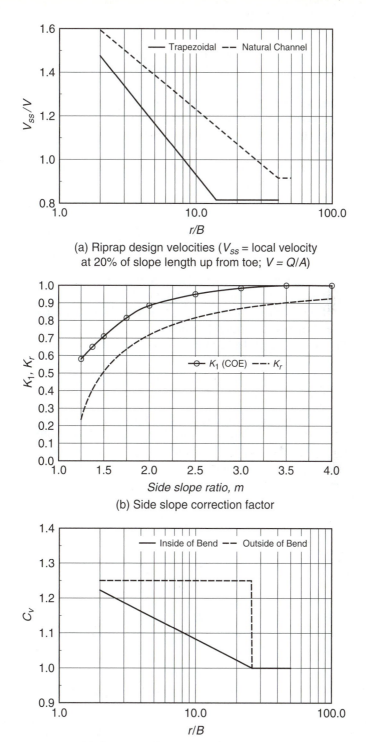

(a) Riprap design velocities (V_{ss} = local velocity at 20% of slope length up from toe; $V = Q/A$)

(b) Side slope correction factor

(c) Correction for vertical velocity distribution

FIGURE 4.18

Correction factors in Corps of Engineers riprap sizing formula.

To use Equation 4.54 for bed riprap design in a straight channel, the local velocity V_l is taken at the channel centerline to be 10–20% greater than $V = Q/A$, and the depth is the local centerline value.

In situations where the side slope riprap makes a small contribution to overall roughness such as in large rivers, Manning's n values are estimated as described previously in this chapter (Table 4.1) to determine the mean cross-sectional velocity. On the other hand, if riprap is being designed as a complete bed and bank lining, the Manning's n value is estimated from a Strickler-type relationship given by

$$n = c_n(d_{90})^{1/6} \tag{4.57}$$

in which d_{90} is the 90% finer stone size in ft; and $c_n = 0.034$ for velocity and stone size calculations, and 0.038 for capacity and freeboard calculations. In this case, iteration of Equation 4.54 and Manning's equation with n determined from (4.57) is required.

It is of interest to recognize the similarity between Equation 4.54 and the equations for critical velocity summarized in Chapter 10 (Figure 10.8). If we take a factor of safety of 1.0 with $C_s = 0.30$ for angular rock, and assume $C_v = C_T = K_1 = 1.0$ for bed riprap in a straight channel, Equation 4.54 can be rearranged in terms of the critical value of the sediment number, N_{sc}, to produce

$$N_{sc30}^2 = \frac{V_c^2}{(SG - 1)gd_{30}} = 2.2\left(\frac{d_{30}}{y}\right)^{-0.2} \tag{4.58}$$

in which the local velocity has been replaced by 1.10 times the mean cross-sectional velocity, which becomes the critical velocity V_c for initiation of sediment motion, in the definition of the sediment number. Under these conditions, Equation 4.58, which is based on the COE design equation for riprap, can be seen to be quite similar to Neill's equation given in Chapter 10 (Equation 10.21) with the sediment size taken as the 30% finer size. We shall return to this discussion in Chapter 10.

Example 4.7

Determine the stable riprap size for the outer bank of a curved natural channel at bankfull flow for which the mean channel velocity is 7.1 ft/s (2.2 m/s) and the depth is 15 ft (4.6 m). The channel is sufficiently wide that the Manning's n is unaffected by bank riprap protection. The side slope is 2:1, and the bend radius is 620 ft (189 m) with a water surface width of 200 ft (61 m). The thickness of the riprap will be $1d_{100}$ and the specific gravity of the available stone is 2.65.

Solution.   Using Equation 4.55, the maximum bend velocity ratio, V_{ss}/V, is given by

$$\frac{V_{ss}}{V} = 1.74 - 0.52 \log\left(\frac{r}{B}\right) = 1.74 - 0.52 \log\left(\frac{620}{200}\right) = 1.48$$

so that $V_{ss} = 1.48 \times 7.1 = 10.5$ ft/s (3.20 m/s) where the depth at a distance of 20% up the side slope is $0.8 \times 15 = 12$ ft (3.66 m). The value of $S_f = 1.1$, and the thickness correction $C_T = 1.0$. From Figure 4.18b, $K_1 = 0.88$. The vertical velocity correction for the outside of the bend is 1.25 from Figure 4.18c. Now the riprap size can be calculated from Equation 4.54 as

$$\frac{d_{30}}{y} = S_f\, C_s C_v C_T \left[\frac{V}{\sqrt{K_1(SG-1.0)gy}} \right]^{2.5}$$

$$= (1.1)(0.30)(1.25)(1.0)\left[\frac{10.5}{\sqrt{0.88(1.65)(32.2)(12)}} \right]^{2.5} = 0.054$$

so that $d_{30} = 0.054 \times 12. = 0.65$ ft or 8 in. (20 cm) after rounding up.

4.16 DIMENSIONALLY HOMOGENEOUS MANNING'S FORMULA

While Manning's n is firmly entrenched in engineering practice, there is a nagging desire to transform Manning's equation in some way that will make it dimensionally homogeneous, which actually was the intent of Manning when he rejected the equation that now bears his name. Equation 4.21 implies that Manning's n can be thought of as having dimensions of length to the $\frac{1}{6}$ power, with K_n then having dimensions of $L^{1/2}/T$. The nondimensionality of the equation, however, still is questionable, because K_n would have to take on a value of 1.81 ft$^{1/2}$/s compared to 1.0 m$^{1/2}$/s in the SI system if Manning's n were to be converted from m$^{1/6}$ to ft$^{1/6}$. Yen (1992b) suggested that this confusing state of affairs could be alleviated by defining Manning's equation to be

$$V = \left(\frac{K_n}{g^{1/2}} \right) \left(\frac{R^{1/6}}{n} \right) \sqrt{gRS} = \left(\frac{R^{1/6}}{n_g} \right) \sqrt{gRS} \qquad (4.59)$$

in which $n_g = ng^{1/2}/K_n$. This would allow the equation to be truly homogeneous with the capability of converting n_g from ft$^{1/6}$ to m$^{1/6}$ or vice versa with no corresponding change of coefficients in the equation, so long as the dimensional units of all other variables remained consistent. However, the current values of Manning's n would need to be converted to n_g in ft$^{1/6}$ by multiplying them by $32.2^{1/2}/1.49 = 3.81$ and to n_g in m$^{1/6}$ by multiplying them by $9.81^{1/2} = 3.13$. Yen (1992a) converted Chow's tables of Manning's n in this way. In addition, he derived values of equivalent sand-grain roughness, k_s, for these tables. Given the established nature of current values of Manning's n, the use of the tables for n_g is likely to be unpopular despite its desirability.

4.17 CHANNEL PHOTOGRAPHS

These photographs are provided by courtesy of the U.S. Geological Survey and come from the work by Barnes (1967). For each river shown, the discharge and the water surface profile over several cross sections were measured for a flood event, and Manning's n was calculated from the equation of gradually varied flow described in Chapter 5. Figures 4.19 to 4.33 give Manning's n values for main-channel flow only

FIGURE 4.19
Salt Creek at Roca, Nebraska: $n = 0.030$; depth $= 6.3$ ft. Bed consists of sand and clay. (*U.S. Geological Survey*)

FIGURE 4.20
Rio Chama near Chamita, New Mexico: $n = 0.032, 0.036$; depth $= 3.5, 3.1$ ft. Bed consists of sand and gravel. (*U.S. Geological Survey*)

FIGURE 4.21

Salt River below Stewart Mountain Dam, Arizona: $n = 0.032$; depth $= 1.8$ ft. Bed and banks consist of smooth cobbles 4 to 10 in. in diameter, average diameter about 6 in. A few boulders are as large as 18 in. in diameter. (*U.S. Geological Survey*)

FIGURE 4.22

West Fork Bitterroot River near Conner, Montana: $n = 0.036$; depth $= 4.7$ ft. Bed is gravel and boulders; $d_{50} = 172$ mm; $d_{84} = 265$ mm. (*U.S. Geological Survey*)

FIGURE 4.23

Middle Fork Vermilion River near Danville, Illinois: $n = 0.037$; depth $= 3.9$ ft. Bed is gravel and small cobbles. (*U.S. Geological Survey*)

FIGURE 4.24

Wenatchee River at Plain, Washington: $n = 0.037$; depth $= 11.1$ ft. Bed is boulders; $d_{50} = 162$ mm; $d_{84} = 320$ mm. (*U.S. Geological Survey*)

FIGURE 4.25
Etowah River near Dawsonville, Georgia: $n = 0.041, 0.039, 0.035$; depth $= 9.8, 9.0, 4.4$ ft. Bed is sand and gravel with several fallen trees in the reach. (*U.S. Geological Survey*)

FIGURE 4.26
Tobesofkee Creek near Macon, Georgia: $n = 0.043, 0.041, 0.039$; depth $= 9.2, 8.7, 6.3$ ft. Bed consists of sand, gravel, and a few rock outcrops. (*U.S. Geological Survey*)

FIGURE 4.27
Middle Fork Flathead River near Essex, Montana: $n = 0.041$; depth $= 8.4$ ft. Bed consists of boulders; $d_{50} = 142$ mm; $d_{84} = 285$ mm. (*U.S. Geological Survey*)

FIGURE 4.28
Beaver Creek near Newcastle, Wyoming: $n = 0.043$; depth $= 9.0$ ft. Bed is mostly sand and silt. (*U.S. Geological Survey*)

FIGURE 4.29
Murder Creek near Monticello, Georgia: $n = 0.045$; depth = 4.2 ft. Bed consists of sand and gravel. (*U.S. Geological Survey*)

FIGURE 4.30
South Fork Clearwater River near Grangeville, Idaho: $n = 0.051$; depth = 7.9 ft. Bed consists of rock and boulders; $d_{50} = 250$ mm; $d_{84} = 440$ mm. (*U.S. Geological Survey*)

FIGURE 4.31

Mission Creek near Cashmere, Washington: $n = 0.057$; depth $= 1.5$ ft. Bed of angular-shaped boulders as large as 1 ft in diameter. (*U.S. Geological Survey*)

FIGURE 4.32

Haw River near Benaja, North Carolina: $n = 0.059$; depth $= 4.9$ ft. Bed is composed of coarse sand and a few outcrops. (*U.S. Geological Survey*)

FIGURE 4.33
Rock Creek near Darby, Montana: $n = 0.075$; depth $= 3.1$ ft. Bed consists of boulders; $d_{50} = 220$ mm; $d_{84} = 415$ mm. (*U.S. Geological Survey*)

(Barnes 1967). The caption for each photograph shows the measured depth at the cross section along with the Manning's n value. In some cases, multiple events with different depths are shown, and the Manning's n does not necessarily remain constant. This could be due to changes in vegetation inundated for different depths, effects of large roughness elements in shallow flows, or changes in bed forms with stage, which will be discussed in more detail in Chapter 10.

REFERENCES

Ackers, P. "Stage-Discharge Functions for Two-Stage Channels: The Impact of New Research." *J. of the Inst. of Water and Environmental Management* 7, no. 1 (1993), pp. 52–61.

Anderson, A. G.; A. S. Paintal; and J. T. Davenport. *Tentative Design Procedure for Riprap-Lined Channels,* NCHRP Report 108. National Cooperative Highway Research Program, Highway Research Board, National Research Council, Washington, DC: 1970.

Arcement, G. J., and V. R. Schneider. *Guide for Selecting Manning's Roughness Coefficients for Natural Channels and Flood Plains,* Report No. FHWA-TS-84-204. Federal Highway Administration, U.S. Dept. of Transportation, National Technical Information Service, Springfield, VA: 1984.

ASCE. *Gravity Sanitary Sewer Design and Construction,* Manual No. 60. New York: ASCE, 1982.

ASCE Task Force. "Friction Factors in Open Channels, Progress Report of the Task Force on Friction Factors in Open Channels of the Committee on Hydromechanics." *J. Hyd. Div.,* ASCE 89, no. HY2 (1963), pp. 97–143.

Barnes, H. H., Jr. *Roughness Characteristics of Natural Channels,* U.S. Geological Survey Water Supply Paper 1849. Washington, DC: Government Printing Office, 1967.

Bathurst, J. C. "Flow Resistance Estimation in Mountain Rivers." *J. Hydr. Engrg.,* ASCE 111, no. HY4 (1985), pp. 625–43.

Berlamont, J. E., and N. Vanderstappen. "Unstable Turbulent Flow in Open Channels." *J. Hyd. Div.,* ASCE 107, no. HY4 (1981), pp. 427–49.

Bhowmik, N. G., and M. Demissie. "Carrying Capacity of Floodplains." *J. Hyd. Div.,* ASCE 108, no. 3 (1982), pp. 443–52.

Biswas, A. K. *History of Hydrology.* Amsterdam, the Netherlands: North-Holland Publishing Co., 1970.

Blodgett, J. C. *Rock Riprap Design for Protection of Stream Channels near Highway Structures,* Vol. 1, *Hydraulic Characteristics of Open Channels,* WRI Report 86-4127. Denver: U.S. Geological Survey, 1986.

Bousmar, D., and Y. Zech. "Momentum Transfer for Practical Flow Computation in Compound Channels." *J. Hydr. Engrg.,* ASCE 125, no. 7 (1999), pp. 696–706.

Camp, T. R. "Design of Sewers to Facilitate Flow." *Sewage Works* 18, no. 1 (1946), pp. 3–16.

Chen, Y. H., and G. K. Cotton. *Design of Roadside Channels with Flexible Linings,* FHWA-IP-87-7, Hydraulic Engineering Circular 15. Federal Highway Administration, U.S. Dept. of Transportation, National Technical Information Service, Springfield, VA: 1988.

Chow, Ven Te. *Open Channel Hydraulics.* New York: McGraw-Hill, 1959.

Colebrook, C. F. "Turbulent Flow in Pipes, with Particular Reference to the Transition Between the Smooth and Rough Pipe Laws." *J. Inst. Civ. Eng. Lond.* 111 (1939), pp. 133–56.

Cotton, G. K. "Hydraulic Design of Flood Control Channels." In *Hydraulic Design Handbook,* ed. Larry W. Mays. New York: McGraw-Hill, 1999.

Cowan, W. L. "Estimating Hydraulic Roughness Coefficients." *Agricultural Engineering* 37, no. 7 (1956), pp. 473–75.

Crowe, C. T., D. F. Elger, B. C. Williams, and J. A. Roberson. *Engineering Fluid Mechanics,* 9th ed., New York: John Wiley & Sons, 2009.

Cunge, J. A.; F. M. Holly, Jr.; and A. Verwey. *Practical Aspects of Computational Hydraulics.* Marshfield, MA: Pitman Publishing Inc., 1980.

Dickman, B. "Large Scale Roughness in Open Channel Flow," M.S. thesis, School of Civil Engineering, Georgia Institute of Technology, 1990.

Dooge, James C. I. "The Manning Formula in Context." In *Channel Flow Resistance: Centennial of Manning's Formula,* ed. B. C. Yen, pp. 136–85. Littleton, CO: Water Resources Publications, 1992.

French, R. H. *Open-Channel Hydraulics.* New York: McGraw-Hill, 1985.

Hager, W. H. *Wastewater Hydraulics.* Berlin: Springer-Verlag, 1999.

Henderson, F. M. *Open Channel Flow.* New York: The Macmillan Co., 1966.

Hey, R. D. "Flow Resistance in Gravel-Bed Rivers," *J. Hyd. Div.,* ASCE 105, no. HY4 (1979), pp. 365–79.

Jarvela, J. "Effect of Submerged Flexible Vegetation on Flow Structure and Resistance." *J. Hydrology,* Elsevier 307 (2005), pp. 233–41.

Kazemipour, A. K., and C. J. Apelt. "New Data on Shape Effects in Smooth Rectangular Channels." *J. Hydr. Res.* 20, no. 3 (1982), pp. 225–33.

Kazemipour, A. K., and C. J. Apelt. "Shape Effects on Resistance to Uniform Flow in Open Channels." *J. Hydr. Res.* 17, no. 2 (1979), pp. 129–47.

Keulegan, G. H. "Laws of Turbulent Flow in Open Channels." *J. of Res. of N.B.S.* 21 (1938), pp. 707–41.

Kilgore, R. T., and G. K. Cotton. *Design of Roadside Channels with Flexible Linings.* HEC-15, 3rd ed. FHWA, Arlington, VA, 2005.

King, H. W. *Handbook of Hydraulics,* 1st ed. New York: McGraw-Hill, 1918.

Knight, D. W., and J. D. Demetriou. "Floodplain and Main Channel Flow Interaction." *J. Hydr. Engrg.,* ASCE 109, no. 8 (1983), pp. 1073–92.

Kouwen, N. "Modern Approach to Design of Grassed Channels." *J. Irrig. And Drainage Engrg.,* ASCE 118, no. 5 (1992), pp. 733–43.

Kouwen, N., and R. M. Li. "Biomechanics of Vegetated Channel Linings." *J. Hydr. Div.,* ASCE 106, no. 6 (1980), pp. 1085–103.

Kouwen, N., and T. E. Unny. "Flexible Roughness in Open Channels." *J. Hydr. Div.,* ASCE 99, no. 5 (1973), pp. 713–28.

Krishnamurthy, M., and B. A. Christensen. "Equivalent Roughness for Shallow Channels." *J. Hyd. Div.,* ASCE 98, no. HY12 (1972), pp. 2257–63.

Kumar, S. "An Analytical Method for Computation of Rough Boundary Resistance." In *Channel Flow Resistance: Centennial of Manning's Formula,* ed. B. C. Yen, pp. 241–58. Littleton, CO: Water Resources Publications, 1992.

Kumar, S., and J. A. Roberson. "General Algorithm for Rough Conduit Resistance." *J. Hyd. Div.,* ASCE 106, no. HY11 (1980), pp. 1745–64.

Kundu, P. K., and I. M. Cohen. *Fluid Mechanics,* 2nd ed. Academic Press, London, 2002.

Limerinos, J. T. "Determination of the Manning Coefficient from Measured Bed Roughness in Natural Channels," U.S. Geological Survey Water Supply Paper 1898-B. Washington, DC: Government Printing Office, 1970.

Manning, R. "On the Flow of Water in Open Channels and Pipes." *Transactions of the Institution of Civil Engineers of Ireland* 20 (1889), pp. 166–95.

Maynord, S. T. "Flow Resistance of Riprap." *J. Hydr. Engrg.,* ASCE 117, no. 6 (1991), pp. 687–95.

Maynord, S. T. *Stable Riprap Size for Open Channel Flows.* Tech. Report HL-88-4, Waterways Experiment Station, Vicksburg, MS, 1988.

Moody, Lewis F. "Friction Factors for Pipe Flow." *Trans. ASME* 671 (November 1944).

Motayed, A. K., and M. Krishnamurthy. "Composite Roughness of Natural Channels." *J. Hyd. Div.,* ASCE 106, no. HY6 (1980), pp. 1111–16.

Myers, R. C., and J. F. Lyness. "Discharge Ratios in Smooth and Rough Compound Channels." *J. Hydr. Engr.,* ASCE 123, no. 3 (1997), pp. 182–88.

Myers, W. R. C. "Momentum Transfer in a Compound Channel." *J. Hydr. Res.* 16, no. 2 (1978), pp. 139–50.

Neale, L. C., and R. E. Price. "Flow Characteristics of PVC Sewer Pipe." *J. of Sanitary Engrg. Div.,* ASCE 90, no. SA3 (1964), pp. 109–29.

Pezzinga, G. "Velocity Distribution in Compound Channel Flows by Numerical Modeling." *J. Hydr. Engrg.,* ASCE 120, no. 10 (1994), pp. 1176–98.

Pomeroy, R. D. "Flow Velocities in Small Sewers," *J. Water Pollution Control Fed.,* ASCE 39, no. 9 (1967), pp. 1525–48.

Rao, N. S. L., and K. Sridharan. "Limit Slope in Uniform Flow Computations." *J. Hyd. Div.,* ASCE 96, no. HY1 (1970), pp. 95–102.

Rouse, Hunter. "Critical Analysis of Open-Channel Resistance." *J. Hyd. Div.,* ASCE 91, no. HY4 (1965), pp. 475–99.

Rouse, Hunter. "Some Paradoxes in the History of Hydraulics." *J. Hyd. Div.,* ASCE 106, no. HY6 (1980), pp. 1077–84.

Samani, J. M. V., and N. Kouwen. "Stability and Erosion in Grassed Channels." *J. Hydr. Engrg.,* ASCE 128, no. 1 (2002), pp. 40–45.

Samuels, P. G. "The Hydraulics of Two Stage Channels: Review of Current Knowledge," HR Paper No. 45. Wallingford, England: Hydraulics Research Limited, 1989.

SCS-TP-161. *Handbook of Channel Design for Soil and Water Conservation.* Stillwater, OK: U.S. Soil Conservation Service, 1954.

Shih, C. C., and N. S. Grigg. "A Reconsideration of the Hydraulic Radius as a Geometric Quantity in Open Channel Hydraulics." *Proc. 12th Congress, IAHR* 1 (1967), pp. 288–96.

Shiono, K., and D. W. Knight. "Turbulent Open-Channel Flows with Variable Depth Across the Channel." *J. Fluid Mech.* 222 (1991), pp. 617–46.

Strickler, A. "Beiträge zur Frage der Geschwindigkeitsformel und der Rauhigkeitszahlen für Ströme, Känale und geschlossene Leitungen" (Contributions to the question of flow roughness coefficients for rivers, channels, and conduits). *Mitteilung* **16.** Amt für Wasserwirtschaft: Bern, Switzerland, 1923 (in German).

Sturm, T. W., and D. King. "Shape Effects on Flow Resistance in Horseshoe Conduits." *J. Hydr. Engrg.,* ASCE 114, no. 11 (1988), pp. 1416–29.

Sturm, T. W., and Aftab Sadiq. "Water Surface Profiles in Compound Channel with Multiple Critical Depths." *J. Hydr. Engrg.,* ASCE 122, no. 12 (1996), pp. 703–9.

Temple, D. M. "Tractive Force Design of Vegetated Channels." *Trans., ASAE* 23, no. 4 (1980), pp. 884–90.

Thein, M. "Experimental Investigation of Flow Resistance and Velocity Distributions in a Rectangular Channel with Large Bed-roughness Elements." Ph.D. thesis, Georgia Institute of Technology, 1993.

Tracy, H. J., and C. M. Lester. "Resistance Coefficient and Velocity Distribution in Smooth Rectangular Channel," U.S. Geological Survey, Water-Supply Paper 1592-A. Washington, DC: Government Printing Office, 1961.

U.S. Army Corps of Engineers. *Design of Flood Control Channels, EM 1110-2-1601,* Dept. of the Army, Washington, DC, 1994.

Wark, J. B., P. G. Samuels, and D. A. Ervine. "A Practical Method of Estimating Velocity and Discharge in Compound Channels." In *International Conference on River Flood Hydraulics,* ed. W. R. White. New York: John Wiley & Sons, 1990.

White, F. M. *Fluid Mechanics,* 6th ed. New York: McGraw-Hill, 2008.

White, F. M. *Viscous Fluid Flow.* New York: McGraw-Hill, 1974.

Williams, G. P. "Manning Formula—A Misnomer?" *J. Hyd. Div.,* ASCE 96, no. HY1 (1970), pp. 193–99.

Wilson, C. A. M. E., T. Stoesser, P. D. Bates, and A. B. Pinzen. "Open Channel Flow through Different Forms of Submerged Flexible Vegetation." *J. Hydr. Engrg.,* ASCE 129, no. 11 (2003), pp. 847–53.

Wormleaton, P. R., and P. Hadjipanos. "Flow Distribution in Compound Channels." *J. Hydr. Engr.,* ASCE 111, no. 2 (1985), pp. 357–61.

Wormleaton, P. R., and D. J. Merrett. "An Improved Method of Calculation of Steady Uniform Flow in Prismatic Main Channel/Flood Plain Sections." *J. Hydr. Res.* 28, no. 2 (1990), pp. 157–74.

Wright, R. R., and M. R. Carstens. "Linear Momentum Flux to Overbank Sections." *J. Hyd. Div.,* ASCE 96, no. 9 (1970), pp. 1781–93.

Yen, B. C. "Hydraulic Resistance in Open Channels." In *Channel Flow Resistance: Centennial of Manning's Formula,* ed. B. C. Yen. Littleton, CO: Water Resources Publications, 1992a.

Yen, B. C. "Dimensionally Homogeneous Manning's Formula." *J. Hydr. Engrg.,* ASCE 118, no. 9 (1992b), pp. 1326–32.

Yen, C. L., and D. E. Overton. "Shape Effects on Resistance in Floodplain Channels." *J. Hyd. Div.,* ASCE 99, no. 1 (1973), pp. 219–38.

Zheleznyakov, G. V. "Interaction of Channel and Floodplain Streams." *Proceedings 14th IAHR Conference, Paris,* pp. 145–48. Delft, the Netherlands: International Association for Hydraulic Research, 1971.

EXERCISES

4.1. Determine the normal depth and critical depth in a trapezoidal channel with a bottom width of 40 ft, side slopes of 3:1, and a bed slope of 0.002 ft/ft. The Manning's n value is 0.025 and the discharge is 3,000 cfs. Is the slope steep or mild? Repeat for $n = 0.012$. Did the critical depth change? Why or why not?

4.2. Compute normal and critical depths in a concrete culvert ($n = 0.015$) with a diameter of 36 in. and a bed slope of 0.002 ft/ft if the design discharge is 15 cfs. Is the slope steep or mild? Repeat for $S = 0.02$ ft/ft.

4.3. For a discharge of 12.0 m³/s, determine the normal and critical depths in a parabolic channel that has a bank-full width of 10 m and a bank-full depth of 2.0 m. The channel has a slope of 0.005 and Manning's $n = 0.05$.

4.4. For the horseshoe conduit shape defined in Figure 4.7, derive the relationships for A/A_f, R/R_f, and Q/Q_f, where A_f, R_f, and Q_f represent the full flow values of area, hydraulic radius, and discharge, respectively. On the same graph, plot the relationships together with those for a circular conduit. Plot y/d on the vertical axis. Note that for the horseshoe conduit, $A_f = 0.8293\ d^2$ and $R_f = 0.2538\ d$.

4.5. A diversion tunnel has a horseshoe shape with a diameter of 11.8 m. The slope of the tunnel is 0.022, and it is lined with gunite ($n = 0.023$). Find the normal depth for a discharge of 950 m³/s. Is the slope steep or mild?

4.6. A trapezoidal channel has vegetated banks with Manning's $n = 0.040$ and a stable bottom with Manning's $n = 0.025$. The channel bottom width is 10 ft and the side slopes are 4:1. Find the composite value of Manning's n using the four methods given in this chapter if the flow depth is 3.0 ft.

4.7. Find the normal depth of flow in a straight and uniform triangular road ditch that has been excavated recently and is free of weeds. The side slopes are 3:1 and the longitudinal ditch slope is 0.001. The design discharge is 1.5 cfs. If the permissible shear stress to avoid erosion is 0.05 lb/ft³, is the channel stable?

4.8. Find the normal depth in a 12 in. diameter PVC storm sewer flowing at a discharge of 1.2 cfs if it has a slope of 0.001. Treat the pipe as smooth, and use the Chezy equation. Verify your solution with the graphical solution given in the text. What is the equivalent value of Manning's n for your solution? Would the value of n be the same for other pipe diameters?

4.9. A circular PVC plastic (smooth) storm sewer has a diameter of 18 in. At the design flow, it is intended to have a relative depth of 0.8. At what minimum slope can it be laid so that the velocity is at least 2 ft/s at design flow and deposited solids will be scoured out by the design flow? How would your answer for the minimum slope change for a concrete sewer?

4.10. A circular PVC plastic (smooth) storm sewer with a diameter of 24 in. has been selected to carry a peak discharge of 20 cfs on a slope of 0.005. Calculate the normal depth and the velocity. What is the value of the Froude number?

4.11. Design a concrete sewer that has a maximum design discharge of 1.0 m³/s and a minimum discharge of 0.2 m³/s if its slope is 0.0018. Check the velocity for self-cleansing.

4.12. Determine the design depth of flow in a trapezoidal roadside drainage ditch with a design discharge of 3.75 m³/s if the ditch is lined with grass having a retardance of class C. The slope of the ditch is 0.004 and it has a bottom width of 2.0 m with side slopes of 3:1. Is the channel stable?

4.13. A very wide rectangular channel is to be lined with a tall stand of Bermuda grass to prevent erosion. If the channel slope is 0.01 ft/ft, determine the maximum allowable flow rate per unit of width and velocity for channel stability.

4.14. In Example 4.4, the maximum permissible discharge for a grass-lined channel with a bottom width of 1.5 m, side slopes of 3:1, and a bed slope of 0.012 was determined for a vegetal retardance class of C. Calculate the maximum permissible discharge if the vegetal retardance class is D.

4.15. Derive a relationship between the trapezoidal channel side slope and the angle of repose of the channel riprap lining such that failure of the rock riprap occurs simultaneously on the bed and banks. Allow the angle of repose to vary between 30° and 42°. What is the minimum value of the side slope, m:1, so that failure always would occur on the bed first?

4.16. Repeat Example 4.3 to select the riprap size according to the method of Chen and Cotton (1988) in HEC-15 using the Blodgett equation (4.42) for Manning's n.

4.17. Design a riprap-lined trapezoidal channel that has a capacity of 1000 cfs and a slope of 0.0005 ft/ft. Crushed rock is to be used and the channel bottom width is not to exceed 15 ft. Determine the riprap size, the side slopes, and the design depth of flow.

4.18. A rectangular channel has a width of 10 ft and a Manning's n value of 0.020. Determine the channel slope such that uniform flow will always have a Froude number less than or equal to 0.5 regardless of the discharge.

4.19. A rectangular channel in a laboratory flume has a width of 1.25 ft and a Manning's n of 0.017. To erode a sediment sample, the shear stress needs to be 0.15 lbs/ft². A supercritical uniform flow is desired with the Froude number less than or equal to 1.5 to avoid roll waves.
 (a) Calculate the maximum and minimum slopes to satisfy the Froude number criterion.
 (b) Choose a slope in the range determined from part (a) and calculate the depth and discharge to achieve the desired shear stress.

4.20. A mountain stream has boulders with a median size (d_{50}) of 0.50 ft. The stream can be considered approximately rectangular in shape and very wide, with a slope of 0.01. You may assume that $k_s = 2d_{50}$.
 (a) For a discharge of 7.0 cfs/ft, calculate the normal depth and critical depth and classify the slope as steep or mild.
 (b) Discuss how a Manning's n that is variable with depth affects the critical slope and the slope classification.

4.21. Find the best hydraulic section for a trapezoidal channel. Express the wetted perimeter of a trapezoidal channel in terms of area, A, and depth, y, then differentiate P with respect to y, setting the result to zero to show that $R = y/2$. Also differentiate P with respect to the sideslope ratio, m, and set the result to zero. What is the best value of m and what do you conclude is the best trapezoidal shape?

4.22. Design a trapezoidal concrete-lined channel that has a flow capacity of 1500 cfs at a bed slope of 0.001. Use the most efficient trapezoidal section for which $m = 1/(3)^{1/2}$ and $b = (2/3) \times (3)^{1/2}y$.

4.23. A compound channel has symmetric floodplains, each of which is 100 m wide with Manning's $n = 0.06$, and a main channel ($n = 0.03$), which is trapezoidal with a bottom width of 10 m, side slopes of 1.5:1, and a bank-full depth of 2.5 m. If the channel slope is 0.001 and the total depth is 3.7 m, compute the uniform flow discharge using the divided channel method, first with a vertical interface both with and without wetted perimeter included for the main channel, then with a diagonal interface with wetted perimeter excluded.

4.24. A river downstream of a hydroelectric dam is subject to daily flows of 12,000 cfs at a flow depth of 6 ft during power generation by the turbines. The channel is approximately trapezoidal in shape with a bottom width of 300 ft and side slopes of 1:1 in a silty clay. At a bend in the river where the radius of curvature is 1000 ft, the outer bank is eroding but the bottom is stable. Determine the size of angular rock riprap needed to protect the banks using the COE method.

4.25. The power-law velocity distribution in a very wide open channel in uniform flow is given by

$$\frac{u}{u_*} = a\left(\frac{z}{k_s}\right)^m$$

in which u is the point velocity; u_* is the shear velocity; a is a constant; z is the distance above the channel bed; k_s is the equivalent sand-grain roughness; and m is the given fractional exponent that is constant.

(a) Find the mean velocity, V, in terms of u_{max} and m, where u_{max} is the maximum velocity at $z = y_0$; $y_0 = $ depth of flow; and $m = $ exponent in power law.

(b) Write the expression for V from part (a) in the form of a uniform flow formula and deduce the value of the exponent m that is compatible with Manning's equation.

4.26. Velocity data have been measured at the centerline of a tilting flume having a bed of crushed rock with $d_{50} = 0.060$ ft. Consider the Run 12 data that follows, for which $Q = 2.40$ cfs and $S = 0.00281$. The elevations, z_t, are given with respect to the bottom of the flume on which the rocks have been laid one layer thick. The total water depth above the flume bottom is 0.437 ft. The average bed elevation of the rocks is 0.055 ft above the flume bottom. Taking this elevation as the origin of the logarithmic velocity distribution, determine the shear stress from the velocity distribution and compare it with the value obtained from the uniform flow formula. Discuss the results.

z_i, ft	Velocity, ft/s
0.430	2.19
0.368	2.10
0.327	2.03
0.285	1.95
0.265	1.87
0.244	1.83
0.224	1.80
0.203	1.69
0.182	1.64
0.170	1.58
0.157	1.56
0.145	1.49
0.133	1.35
0.120	1.30
0.108	1.15
0.096	1.05
0.087	0.97
0.079	0.88

4.27. Write a computer program, in the language of your choice, that computes the normal depth and critical depth in a circular channel using the bisection method. The input data should include the conduit diameter, Manning's roughness coefficient, slope, and discharge. Illustrate your program with an example and verify the results for normal and critical depth.

4.28. Write a computer program to design a trapezoidal channel with a vegetative or rock riprap lining if the slope and design discharge are given. Allow the user to adjust the vegetal retardance class, or the rock size and angle of repose, and the channel bottom width and side slope interactively.

4.29. Using the computer program Ycomp on the book website, find the normal and critical depths for the following compound channel section given in a data file. The discharge is 5000 cfs, and the slope is 0.009. Plot the cross-section. Is this a subcritical or supercritical flow?

"DUCKCR",19,4
−480,796
−440,788
−420,786
−305,784

−175,782
−95,780
−50,778
−30,776
−25,774
2,772
17,772
20,774
28,780
50,780
670,780
990,782
1070,784
1120,786
1260,810
−95, .1 , 28 , .04 , 670 , .08 , 1260 , .05

5

Gradually Varied Flow

5.1 INTRODUCTION

Gradually varied flow is a steady, nonuniform flow in which the depth variation in the direction of motion is gradual enough that the transverse pressure distribution can be considered hydrostatic. This allows the flow to be treated as one dimensional with no transverse pressure gradients other than those created by gravity. The methods developed in this chapter should not be applied to regions of highly curvilinear flow, such as can be found in the vicinity of an ogee spillway crest, for example, because the centripetal acceleration in curvilinear flow alters the transverse pressure distribution so that it no longer is hydrostatic, and the pressure head no longer can be represented by the depth of flow.

Even with the assumption of gradually varied flow, an exact solution for the depth profile exists only in the case of a wide, rectangular channel. The solution of the equation of gradually varied flow in this case is called the *Bresse function*, which provides useful approximations of water surface profile lengths subject to the assumptions of a very wide channel and a constant value of Chezy's C. The solutions to all other problems, in the past, were obtained graphically or from tabulations of the varied flow function based on hydraulic exponents as developed by Bakhmeteff (1932) and Chow (1959). Currently, the use of personal computers and the application of sound numerical techniques make these older methods obsolete.

5.2 EQUATION OF GRADUALLY VARIED FLOW

In addition to the basic gradually varied flow assumption, we further assume that the flow occurs in a prismatic channel, or one that is approximately so, and that the slope of the energy grade line (EGL) can be evaluated from uniform flow formulas with uniform flow resistance coefficients, using the local depth as though the flow were locally uniform. With respect to Figure 5.1, the total head at any cross section is

$$H = z + y + \alpha \frac{V^2}{2g} \tag{5.1}$$

in which z = channel bed elevation; y = depth; and V = mean velocity. If this expression for H is differentiated with respect to x, the coordinate in the flow direction, the following equation is obtained:

$$\frac{dH}{dx} = -S_e = -S_0 + \frac{dE}{dx} \tag{5.2}$$

in which S_e is defined as the slope of the energy grade line; S_0 is the bed slope ($= -dz/dx$); and E is the specific energy. Solving for dE/dx gives the first form of the equation of gradually varied flow:

$$\frac{dE}{dx} = S_0 - S_e \tag{5.3}$$

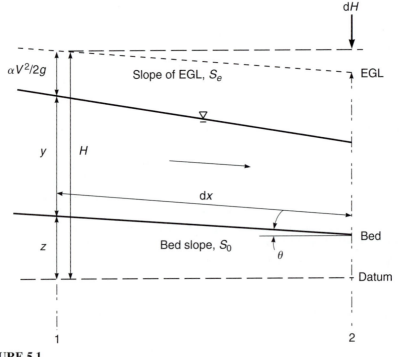

FIGURE 5.1
Definition sketch for gradually varied flow.

It is apparent from this form of the equation that the specific energy can either increase or decrease in the downstream direction, depending on the relative magnitudes of the bed slope and the slope of the energy grade line. Yen (1973) showed that, in the general case, S_e is not the same as the friction slope $S_f (= \tau_o/\gamma R)$ or the energy dissipation gradient. Nevertheless, we have no better way of evaluating this slope than uniform flow formulas such as those of Manning or Chezy. It is incorrect, however, to mix the friction slope, which clearly comes from a *momentum* analysis, with terms involving α, the kinetic *energy* correction (Martin and Wiggert 1975).

The second form of the equation of gradually varied flow can be derived if it is recognized that $dE/dx = dE/dy \cdot dy/dx$ and that, from Chapter 2, $dE/dy = 1 - \mathbf{F}^2$ provided that the Froude number is properly defined. Then, Equation 5.3 becomes

$$\frac{dy}{dx} = \frac{S_0 - S_e}{1 - \mathbf{F}^2} \tag{5.4}$$

The definition of the Froude number in Equation 5.4 depends on the channel geometry. For a compound channel, it should be the compound channel Froude number as defined in Chapter 2, while for a regular, prismatic channel, in which $d\alpha/dy$ is negligible, it assumes the conventional energy definition given by $\alpha Q^2 B/gA^3$.

5.3 CLASSIFICATION OF WATER SURFACE PROFILES

Equation 5.4 can be used to derive the expected shapes of water surface profiles for gradually varied flow on mild, steep, and horizontal slopes, for example. It is important to identify these shapes before running a water surface profile program because the location of the control, where a unique relationship exists between stage and discharge, and the direction of computation (upstream or downstream) depend on this knowledge. In effect, identification of the control for a given profile amounts to specification of the boundary condition for the numerical solution of a differential equation.

Equation 5.4 provides the tool for determining whether or not the depth is increasing or decreasing in the downstream direction and also for determining the limiting depths very far downstream and upstream for particular gradually varied flow profiles. In order to deduce the shapes of the profiles, it is sufficient to determine qualitatively the relative magnitudes of the terms on the right hand side of Equation 5.4. For this purpose and in the numerical computation of gradually varied flow profiles, we assume that the local value of the slope of the energy grade line, S_e, can be calculated from Manning's equation using the local value of depth as though the flow were uniform locally. Therefore, the following inequalities hold when comparing the magnitude of the local depth y at any point along the profile with normal depth y_0:

$$y < y_0: \quad S_e > S_0 \tag{5.5}$$

$$y > y_0: \quad S_e < S_0 \tag{5.6}$$

In addition, it is apparent that the value of the Froude number squared relative to unity is determined by the magnitude of the local depth relative to the critical depth y_c:

$$y < y_c: \qquad \mathbf{F}^2 > 1 \tag{5.7}$$

$$y > y_c: \qquad \mathbf{F}^2 < 1 \tag{5.8}$$

With these inequalities and Equation 5.4, the gradually varied flow profile shapes can be determined, as shown in Figure 5.2.

The flow profiles shown in Figure 5.2 are designated M, S, C, H, or A for mild, steep, critical, horizontal, and adverse slopes, respectively. The flow profiles are further identified numerically as 1, 2, or 3 counting from the largest depth region downward, based on the two or three regions delineated by the normal and critical depth lines. Only two regions occur for critical, horizontal, and adverse slopes, because normal depth does not exist in the latter two cases, while in the former case, normal depth equals critical depth. For those profiles having only two regions, the labeling convention of Henderson (1966) is used in which the upper profile is number one and the lower profile is number two; that is, on an adverse slope, for example, the profiles are labeled A1 and A2. Furthermore, all profiles are sketched assuming that flow is from left to right. It is important to keep in mind that the control always is downstream for subcritical flows and upstream for supercritical flows. Hence, the direction of computation of subcritical profiles is upstream, and for supercritical profiles, it is downstream.

Consider the mild slope in region 1 for which $y > y_0 > y_c$. From the inequalities in Equations 5.5 through 5.8, we can conclude that $dy/dx > 0$ so that the M1 profile always must have an increasing depth in the downstream direction. As y approaches y_0 in the upstream direction, dy/dx approaches zero asymptotically, while in the downstream direction dy/dx approaches S_0, so that a horizontal asymptote exists. The M1 profile sometimes is called the *backwater profile* and exists where a reservoir "backs up water" in the tributary stream flowing into it. In region 2 on a mild slope, where $y_c < y < y_0$, $S_e > S_0$, and $\mathbf{F} < 1$, the value of $dy/dx < 0$. As y approaches y_0 in the upstream direction, dy/dx approaches zero, so we have an asymptotic approach to normal depth from below. In the downstream direction, the M2 profile approaches critical depth where $\mathbf{F} = 1$, but the manner in which it does so is not immediately obvious. However, if we consider a mild slope followed by a steep slope, $S_e > S_0$ upstream of the slope break, where critical depth occurs, while downstream of the slope break, $S_e < S_0$ because $y > y_0$ on the steep slope. It can be reasoned then that $S_0 = S_e$ at the slope break and both the numerator and denominator of (5.4) approach zero, so that dy/dx is finite as the water surface passes through critical depth. In region 3 on a mild slope, where $y < y_c < y_0$, $S_e > S_0$ and $\mathbf{F} > 1$, we have $dy/dx > 0$. As y approaches y_c in the downstream direction, $\mathbf{F}$ approaches 1, and dy/dx approaches infinity, although a hydraulic jump would occur before that happens. In the upstream direction, both the numerator and denominator of (5.4) approach infinity as the depth approaches zero, and dy/dx approaches some positive finite limit that is of no practical interest, since there would be no flow for no depth.

It is of interest to note that both M1 and M2 profiles, which are subcritical, approach normal depth in the *upstream* direction, as controlled by the value of the downstream depth. The other profiles in Figure 5.2 can be deduced in the same way

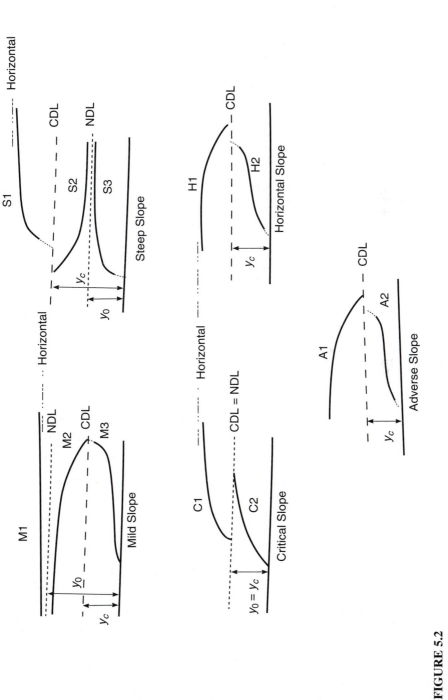

FIGURE 5.2
Gradually varied flow profile shapes on mild, steep, critical, horizontal, and adverse slopes (CDL = critical depth line; NDL = normal depth line).

as for the mild slope. In contrast to the M1 and M2 profiles, the two supercritical profiles, S2 and S3, approach normal depth in the *downstream* direction, as determined by the value of the upstream depth.

Composite flow profiles for a variety of flow situations can be sketched as shown in Figure 5.3. In Figure 5.3a, a mild slope is followed by a milder slope. If the downstream slope is very long, with uniform flow established as the control, then the depth must remain at normal depth all the way to the upstream slope. This is because the mild slope profiles cannot approach normal depth in the downstream

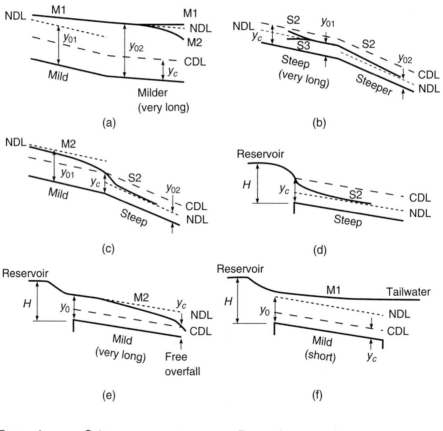

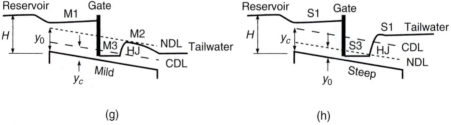

FIGURE 5.3
Composite flow profiles with various controls.

direction but only diverge from it (i.e., M1 and M2). As a result, the upstream M1 profile does not begin until the upstream slope is reached. Following the same reasoning, the steep slope followed by a steeper slope in Figure 5.3b must have an S2 or S3 profile on the upstream slope that reaches normal depth and remains there, if the slope is very long, until the break in slope is reached.

The occurrence of critical depth is a very important control, shown at the break between a mild and steep slope in Figure 5.3c. Based on the preceding reasoning, the water surface must approach some finite slope as it passes through critical depth. Critical depth also occurs at the entrance from a reservoir into a steep slope and at a free overfall, where there is a similar release or acceleration of the flow, as shown in Figures 5.3d and 5.3e, respectively. The case of the free overfall is of particular interest because in the vicinity of the brink, the streamlines gain significant curvature as the flow accelerates over the end of the channel. Strictly speaking, this violates our assumption of gradually varied flow because the pressure distribution is no longer hydrostatic at the brink. As a result, the value of critical depth calculated by setting the Froude number to unity, as we did in Chapter 2 assuming a hydrostatic pressure distribution, does not occur precisely at the brink of the overfall. Instead, the calculated critical depth occurs at a distance of 3–4 y_c upstream of the brink, and the brink depth is 0.715 y_c as measured by Rouse (Henderson 1966). As a practical matter, unless we are interested in the water surface profile in the immediate vicinity of a weir or spillway crest, we will be computing water surface profiles in relatively long channels so that assumption of critical depth at the brink does not affect the results.

The entrance from a reservoir into a mild slope is shown in Figures 5.3e and 5.3f. For the long mild channel in Figure 5.3e, the control is normal depth at the entrance, if the channel is very long (hydraulically), but switches to the tailwater depth if the channel is short as in Figure 5.3f. For a fixed reservoir elevation in Figures 5.3e and 5.3f, the maximum discharge occurs in the short channel of Figure 5.3f when the tailwater falls below critical depth at the free overfall. In that instance, flow passes through critical depth at the free overfall, which gives the maximum discharge for the available specific energy.

Flow profiles on a mild or a steep slope with a sluice gate installed midway along the channel are shown in Figures 5.3g and 5.3h, respectively. In Figure 5.3g, the sluice gate forces an M1 profile to occur upstream and an M3 profile to emerge from under the gate downstream. The M3 profile has an increasing depth until the momentum equation is satisfied for the sequent depth occurring in the downstream M2 profile. The result is a hydraulic jump (HJ). A similar situation is shown in Figure 5.3h, except that the slope is steep and there is an S3 profile upstream of the jump and an S1 profile downstream of the jump controlled by the position of the tailwater. If the upstream reservoir elevation is fixed, the sluice gate opening has no influence on the discharge unless the upstream backwater profile (M1 or S1) caused by the gate submerges the channel entrance depth. If the channel is mild and long enough, for example, the entrance depth is normal depth followed by an M1 profile, while in the case of a steep slope the entrance depth is critical depth followed by a hydraulic jump and an S1 profile. Under these circumstances, a smaller gate opening can cause the backwater profile to move farther upstream and submerge either the normal depth or the critical depth at the channel entrance such that the control moves from the channel entrance to the sluice gate.

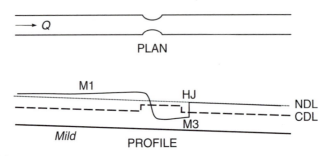

FIGURE 5.4
Water surface profiles associated with choked contraction on a mild slope.

The composite profiles shown in Figure 5.3 raise the larger issue of the location of the control, or the boundary condition, that must be applied to solve the equation of gradually varied flow. As we have just seen, the control can be critical depth, normal depth, tailwater elevation, or an obstruction such as a sluice gate or weir, for example, but the specific control can change as discharge, tailwater elevation, or sluice gate opening change. It is the interaction between the control and the water surface profile established by flow resistance that results in the appropriate solution of the equation of gradually varied flow. Within this context, we can return to our earlier discussion in Chapter 2 of choking caused by a width contraction. In particular, the channel contraction previously shown in Figure 2.11 occurs within a larger channel length in which there is a water surface profile established by critical depth and choking as the control as shown in Figure 5.4. If a contraction such as a bridge opening, for example, is situated on a mild slope and has a width such that critical depth just occurs in the contracted opening, any further decrease in width (without changes in total discharge) causes an increase in discharge per unit width, critical depth and minimum specific energy, which results in an M1 profile upstream of the contraction as shown in Figure 5.4. Effectively, the M1 profile develops the additional specific energy needed to force flow through the contracted section. Downstream of passing through critical depth in the contraction, the flow will become supercritical at a depth that satisfies the energy equation, and an M3 profile will continue downstream until the sequent depth is reached for a hydraulic jump to return the profile back to the original tailwater depth in satisfaction of the momentum equation as shown in Figure 5.4. The choked contraction on a mild slope causes additional backwater due to choking as well as a hydraulic jump, so it is to be avoided in designing bridge openings. We will return to this discussion in Chapter 6.

5.4 LAKE DISCHARGE PROBLEM

The flow situations illustrated in Figures 5.3d, 5.3e, and 5.3f lead to an important problem if the discharge is unknown, because it is unclear whether the given slope in fact is mild or steep. If the head H at the channel entrance is given, we can write the energy equation for the steep slope in Figure 5.3d between the upstream lake water surface and the channel entrance where the depth is critical (neglecting losses) to give

$$H = y_c + \frac{Q^2}{2gA_c^2} \tag{5.9}$$

For depth equal to the critical depth, the Froude number must have a value of 1 so that

$$\frac{\alpha Q^2 B_c}{gA_c^3} = 1 \tag{5.10}$$

On the other hand, if the slope is mild and the channel is very long as in Figure 5.3e, the entrance depth is normal depth and the relevant equations for solving for Q and the entrance depth are

$$H = y_0 + \frac{Q^2}{2gA_0^2} \tag{5.11}$$

$$Q = \frac{K_n}{n} A_0 R_0^{2/3} S_0^{1/2} \tag{5.12}$$

in which y_0 is the normal depth. Which of the two conditions prevails can be determined by assuming that the slope is steep and solving Equations 5.9 and 5.10 for the critical depth and critical discharge, y_c and Q_c. These values of y_c and Q_c then are substituted into Manning's equation to calculate the critical slope. If the bed slope $S_0 > S_c$, then the slope indeed is steep and the discharge is Q_c. On the other hand, if $S_0 < S_c$, then the slope is mild and Equations 5.11 and 5.12 must be solved to determine the actual Q, which will be less than Q_c. In case the slope is not very long, the normal depth, y_0 in Equations 5.11 and 5.12, must be replaced by an entrance depth, $y_e \neq y_0$, which can be determined only from water surface profile computation. In that case, Equation 5.12 is replaced by the equation of gradually varied flow, which must be solved numerically as shown in the following section.

Example 5.1

A very long rectangular channel connects two reservoirs and has a slope of 0.005. The channel has a width of 10 m (32.8 ft) and a Manning's n of 0.030. If the upstream reservoir surface is 3.50 m (11.5 ft) above the channel inlet invert and the downstream reservoir is 2.50 m (8.20 ft) above the outlet invert, determine the discharge in the channel.

Solution.   Initially assume that the slope is steep. In this case, Equations 5.9 and 5.10 are particularly simple for a rectangular channel. They become

$$y_c = \frac{2}{3} H = \frac{2}{3} \times (3.5) = 2.33 \text{ m} (7.66 \text{ ft})$$

$$q = \sqrt{gy_c^3} = \sqrt{9.81 \times 2.33^3} = 11.1 \text{ m}^2/\text{s} (120 \text{ ft}^2/\text{s})$$

in which H = upstream head of the reservoir surface relative to the channel invert and q = discharge per unit of channel width. The critical slope can be computed from

$$S_c = \frac{n^2 Q^2}{K_n^2 A_c^2 R_c^{4/3}} = \frac{0.03^2 \times (10 \times 11.14)^2}{1.0^2 \times (10 \times 2.33)^2 \times \left[\dfrac{(10 \times 2.33)}{(10 + 2 \times 2.33)}\right]^{4/3}} = 0.011$$

Now, since $S_0 < S_c$, the slope must be mild. In that case, Equations 5.11 and 5.12 must be solved simultaneously:

$$3.5 = y_0 + \frac{Q^2}{2gA_0^2} = y_0 + \frac{Q^2}{19.62 \times (10 \times y_0)^2} = y_0 + \frac{Q^2}{1962 \times y_0^2}$$

$$Q = \frac{K_n}{n} A_0 R_0^{2/3} S_0^{1/2} = \frac{1.0}{0.03} \frac{(10 \times y_0)^{5/3}}{(10 + 2 \times y_0)^{2/3}} \times (0.005)^{1/2} = 109.4 \frac{y_0^{5/3}}{(10 + 2 \times y_0)^{2/3}}$$

By trial and error, assume a value of y_0 (<3.5) and substitute it into the second equation to solve for Q. Then, substitute Q and y_0 into the first equation and iterate until the result is 3.5 m (11.5 ft) for the head. Alternatively, the second equation can be substituted into the first and a nonlinear algebraic equation solver can be applied. Solving by trial and error, the first trial gives $y_0 = 3.0$ m (9.8 ft), $Q = 108$ m³/s (3810 cfs), and $H = 3.66$ m (12.0 ft). For the second trial, $y_0 = 2.5$ m (8.2 ft), $Q = 82.8$ m³/s (2920 cfs), and $H = 3.06$ m (10.0 ft). For the third and final trial, $y_0 = 2.87$ m (9.42 ft), $Q = 101$ m³/s (3565 cfs), and $H = 3.50$ m (11.5 ft), which gives the final answer. The critical depth can be calculated to be $y_c = (q^2/g)^{1/3} = 2.18$ m (7.15 ft), so the slope indeed is mild. The Froude number of the uniform flow is 0.66. The M2 profile starts from a depth of 2.5 m (8.2 ft) at the downstream end of the channel and approaches normal depth before it reaches the upstream lake, since the channel is very long. This also can be referred to as a *hydraulically long channel*. We explore how long this really is in the next section. This example neglected the approach velocity head and the channel entrance loss, but these can be added easily without changing the solution procedure.

The difference between the hydraulically long channel in Example 5.1 and the short channel as illustrated in Figures 5.3e and 5.3f can be shown quantitatively in a canal delivery curve. For a fixed upstream water surface head, H, relative to the invert of the channel entrance (3.5 m in Example 5.1), the downstream tailwater head relative to the invert of the channel exit, H_t, is allowed to vary, and the corresponding discharges are determined for a given channel slope, S_0; length, L; roughness, n; and geometry. The results are plotted as a canal delivery curve as illustrated in Figure 5.5. The discharge for the hydraulically long channel is shown on the curve as Q_0 corresponding to uniform flow at normal depth (point N) as in Example 5.1. For this very long channel, all M2 profiles will converge to normal depth before reaching the channel entrance, so the discharge will remain at Q_0 as H_t decreases below y_0 and passes through critical depth for this discharge. If the channel is short, the maximum discharge will be Q_c corresponding to critical depth y_c (point C) for that discharge as long as $H_t \leq y_c$. This critical discharge can be found by computing the M2 profile starting at critical depth for different values of Q until the total head at the

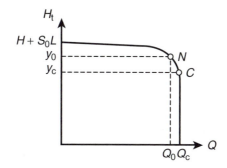

FIGURE 5.5
Canal delivery curve for fixed upstream reservoir head H and specified channel length, slope, roughness, and geometry.

channel entrance is equal to the given value of the reservoir head, H. All of the remaining points on the delivery curve for a specified channel length in Figure 5.5 can be generated using this trial and error computation of a water surface profile by iterating on Q until a particular value of H_t produces the given reservoir head, H. This canal delivery problem illustrates the importance of implementing a numerical technique for water surface profile computation as discussed in the next section.

5.5   WATER SURFACE PROFILE COMPUTATION

The computation of water surface profiles has many important applications in engineering practice. A prismatic drainage channel, storm sewer, or culvert designed for uniform flow may be checked for its performance under gradually varied flow conditions. Floodplain mapping, which is the determination of the extent of flooding for a flood of specified frequency, requires water surface profile computations in a natural channel of irregular and variable geometry, slope, and roughness.

The problem formulation in water surface profile computations usually specifies a design discharge set by frequency considerations and requires the selection of channel roughness, slope, and geometry. In the case of a natural channel, the channel roughness, slope, and geometry are measured for a series of reaches within which these parameters are relatively constant. With this information given, the mathematical problem is to solve the equation of gradually varied flow to obtain the depth as a function of distance along the channel, $y = F(x)$, subject to a boundary condition established by the channel control. The control can be a measured stage-discharge relation, normal depth, critical depth, or a depth set by a hydraulic control structure.

Two types of methods can be used to solve the equation of gradually varied flow in the form of either Equation 5.3 or Equation 5.4. In the first type, the distance is determined for a specified depth change. This approach can be classified *explicit* and sometimes is called the *direct step method*, because the solution is direct, requiring no iteration. Equation 5.4, for example, can be represented symbolically as $dy/dx = f(y)$, where $f(y)$ is the nonlinear function of y specified by the right hand side of

Equation 5.4, in which both S_e and $\mathbf{F}$ depend on the local depth y. This is an ordinary differential equation for which the variables can be separated as

$$dx = \frac{dy}{f(y)} \tag{5.13}$$

This equation can be solved by numerical integration or finite-difference approximation; in either case, a change in depth y is specified and the corresponding change in x is computed explicitly. This means no control exists over the positions x, or channel stations, where the solutions for depth are obtained, which is no problem in a prismatic channel, because the cross-sectional properties do not change with distance x. In a natural channel, on the other hand, cross-sectional properties are determined beforehand at particular locations, so that a different approach is required, in which depth is computed as a function of specified changes in distance. In this case, the variables are separated as

$$dy = f(y)\,dx \tag{5.14}$$

and it appears that the numerical solution procedure has to be iterative to compute the value of Δy for a specified Δx, because the unknown appears on both sides of the equation. If iteration is required, the approach is considered implicit. On the other hand, a class of techniques, called *predictor-corrector methods*, that essentially are explicit also can be applied to the problem posed in this way, with the depth unknown at specified locations along the channel.

Regardless of the numerical solution technique chosen for solution of the equation of gradually varied flow, we will assume that the slope of the energy grade line, S_e, can be evaluated from Manning's or Chezy's equation using the local value of depth. Essentially, the flow is assumed to behave as though it were locally uniform for the purposes of evaluating the slope of the energy grade line. Effects of nonuniformity can be lumped into the resistance coefficient, but the condition of gradually varied flow still must be satisfied.

5.6   DISTANCE DETERMINED FROM DEPTH CHANGES

Direct Step Method

In principle, the direct step method could be applied to either Equation 5.3 or 5.4 but usually is associated with the former. Equation 5.3 is placed in finite difference form by approximating the derivative dE/dx with a forward difference, as described in Appendix A, and by taking the mean value of the slope of the energy grade line over the step size $\Delta x = (x_{i+1} - x_i)$ in which distance x and the subscript i increase in the downstream direction. The result is

$$x_{i+1} - x_i = \frac{E_{i+1} - E_i}{S_0 - \bar{S}_e} \tag{5.15}$$

where $\bar{S}_e$ is the arithmetic mean slope of the energy grade line between sections i and $i + 1$, with the slope evaluated individually from Manning's equation at each cross section. The variables E_{i+1}, E_i, and $\bar{S}_e$ on the right hand side of Equation 5.15 all are functions of the depth y. The solution proceeds in a stepwise fashion in Δx by assuming values of depth y and therefore values of specific energy, E. As Equation 5.15 is written, x increases in the downstream direction. In general, upstream computations utilize Equation 5.15 multiplied by -1, so that the current value of specific energy is subtracted from the assumed value in the upstream direction and Δx becomes $(x_i - x_{i+1})$, which is negative. Therefore, if the equation is solved in the upstream direction for an M2 profile, for example, the computed values of Δx should be negative for increasing values of y. Decreasing values of y should result also in negative values of Δx for an M1 profile. For an M3 profile, which is supercritical, increasing values of depth in the downstream direction correspond to decreasing values of specific energy, and Equation 5.15 indicates positive values of Δx since $S_e > S_0$.

Although the direct step method is the easiest approach, it requires interpolation to find the final depth at the end of the profile in a channel of specified length. Some care must be taken in specifying starting depths and checking for depth limits in a computer program. In an M2 profile, for example, the starting depth should be taken slightly greater than the computed critical depth, if it is the control, because of the slight inaccuracy inherent in the numerical evaluation of critical depth. In addition, the M2 profile approaches normal depth asymptotically in the upstream direction, so that some arbitrary stopping limit must be set, such as 99 percent of normal depth.

Example 5.2

A trapezoidal channel has a bottom width, b, of 8.0 m (26.2 ft) and a sideslope ratio of 2:1. The Manning's n of the channel is 0.025, and it is laid on a slope of 0.001. If the channel ends in a free overfall, compute the water surface profile for a discharge of 30 m³/s.

Solution. First, normal depth and critical depth must be determined. From Manning's equation,

$$\frac{[y_0(8.0 + 2y_0)]^{5/3}}{[8.0 + 2y_0\sqrt{1 + 2^2}]^{2/3}} = \frac{0.025 \times 30}{1.0 \times 0.001^{1/2}} = 23.72$$

from which $y_0 = 1.754$ m (5.755 ft). Set the Froude number $(QB_c^{1/2})/(\sqrt{g}A_c^{3/2}) = 1$ and solve for critical depth:

$$\frac{[y_c(8.0 + 2y_c)]^{3/2}}{[8.0 + 4y_c]^{1/2}} = \frac{30}{\sqrt{9.81}} = 9.58$$

from which $y_c = 1.03$ m (3.38 ft). Therefore, this is a mild slope and we are computing an M2 profile that has critical depth at the free overfall as the boundary condition.

The direct step method can be programmed as shown on the book website or solved in a spreadsheet, as shown in Table 5.1. The values of y are selected in the first column; and the formulas for determining the specific energy, E, and slope of

TABLE 5.1 Water surface profile computation by the direct step method

Q,cms=	30	y0, m=	1.7538
n=	0.025	yc, m=	1.0298
S0=	0.001		
b,m=	8		
m:1	2		

(1) y m	(2) A sq. m	(3) R m	(4) V m/s	(5) E m	(6) Se	(7) Sebar	(8) Del E m	(9) Del x m	(10) Sum Del x, m
1.03	10.36	0.822	2.90	1.457	6.80E-03				0.000
1.04	10.48	0.829	2.86	1.457	6.58E-03	6.69E-03	1.62E-04	-0.028	-0.028
1.06	10.73	0.842	2.80	1.459	6.15E-03	6.36E-03	1.23E-03	-0.229	-0.257
1.08	10.97	0.855	2.73	1.461	5.75E-03	5.95E-03	2.35E-03	-0.476	-0.733
1.10	11.22	0.868	2.67	1.464	5.39E-03	5.57E-03	3.40E-03	-0.743	-1.48
1.12	11.47	0.882	2.62	1.469	5.06E-03	5.23E-03	4.36E-03	-1.03	-2.51
1.14	11.72	0.895	2.56	1.474	4.75E-03	4.90E-03	5.26E-03	-1.35	-3.85
1.16	11.97	0.908	2.51	1.480	4.47E-03	4.61E-03	6.09E-03	-1.69	-5.54
1.18	12.22	0.921	2.45	1.487	4.20E-03	4.33E-03	6.86E-03	-2.06	-7.60
1.20	12.48	0.934	2.40	1.495	3.96E-03	4.08E-03	7.58E-03	-2.46	-10.1
1.22	12.74	0.947	2.36	1.503	3.73E-03	3.84E-03	8.24E-03	-2.90	-13.0
1.24	13.00	0.959	2.31	1.512	3.52E-03	3.63E-03	8.87E-03	-3.38	-16.3
1.26	13.26	0.972	2.26	1.521	3.32E-03	3.42E-03	9.45E-03	-3.90	-20.2
1.28	13.52	0.985	2.22	1.531	3.14E-03	3.23E-03	9.99E-03	-4.47	-24.7
1.30	13.78	1.00	2.18	1.542	2.97E-03	3.06E-03	1.05E-02	-5.10	-29.8
1.32	14.04	1.01	2.14	1.553	2.81E-03	2.89E-03	1.10E-02	-5.80	-35.6

1.34	14.31	1.02	2.10	1.564	2.67E-03	2.74E-03	1.14E-02	−6.57	−42.2
1.36	14.58	1.04	2.06	1.576	2.53E-03	2.60E-03	1.18E-02	−7.42	−49.6
1.38	14.85	1.05	2.02	1.588	2.40E-03	2.46E-03	1.22E-02	−8.37	−58.0
1.40	15.12	1.06	1.98	1.601	2.28E-03	2.34E-03	1.26E-02	−9.43	−67.4
1.42	15.39	1.07	1.95	1.614	2.16E-03	2.22E-03	1.30E-02	−10.6	−78.0
1.44	15.67	1.08	1.91	1.627	2.06E-03	2.11E-03	1.33E-02	−12.0	−90
1.46	15.94	1.10	1.88	1.640	1.96E-03	2.01E-03	1.36E-02	−13.5	−104
1.48	16.22	1.11	1.85	1.654	1.86E-03	1.91E-03	1.39E-02	−15.3	−119
1.50	16.50	1.12	1.82	1.668	1.77E-03	1.82E-03	1.41E-02	−17.3	−136
1.52	16.78	1.13	1.79	1.683	1.69E-03	1.73E-03	1.44E-02	−19.7	−156
1.54	17.06	1.15	1.76	1.698	1.61E-03	1.65E-03	1.47E-02	−22.5	−178
1.56	17.35	1.16	1.73	1.712	1.54E-03	1.57E-03	1.49E-02	−25.9	−204
1.58	17.63	1.17	1.70	1.728	1.47E-03	1.50E-03	1.51E-02	−30.1	−234
1.60	17.92	1.18	1.67	1.743	1.40E-03	1.43E-03	1.53E-02	−35.3	−270
1.62	18.21	1.19	1.65	1.758	1.34E-03	1.37E-03	1.55E-02	−41.9	−312
1.64	18.50	1.21	1.62	1.774	1.28E-03	1.31E-03	1.57E-02	−50.7	−362
1.66	18.79	1.22	1.60	1.790	1.22E-03	1.25E-03	1.59E-02	−63.0	−425
1.68	19.08	1.23	1.57	1.806	1.17E-03	1.20E-03	1.60E-02	−81.0	−506
1.70	19.38	1.24	1.55	1.822	1.12E-03	1.15E-03	1.62E-02	−110	−617
1.72	19.68	1.25	1.52	1.838	1.07E-03	1.10E-03	1.63E-02	−167	−783
1.74	19.98	1.27	1.50	1.855	1.03E-03	1.05E-03	1.65E-02	−317	−1100
1.745	20.05	1.27	1.50	1.859	1.02E-03	1.02E-03	4.14E-03	−171	−1271

$A = y*(b+m*y)$

$P = b+2*y*sqrt(1+m^2)$

$R = A/P$

$V = Q/A$

$E = y + V^2/2g$

$Se = n^2*V^2/R^{4/3}$

$Sebar = (Se1 + Se2)/2$

$Del E = E2 - E1$

$Del x = Del E/(S0-Sebar)$

the energy grade line, S_e, for a given depth are shown at the bottom of the spreadsheet. The arithmetic mean of S_e (Sebar) is computed in column 7, and the change in specific energy ΔE (Del E) in the upstream direction is shown in column 8. Then, the equation of gradually varied flow in finite difference form is solved for the distance step, Δx, as

$$\Delta x = \frac{\Delta E}{S_0 - \bar{S}_e} = \frac{1.62E\text{-}04}{[0.001 - 6.69E\text{-}03]} = -0.028 \text{ m } (-0.092 \text{ ft})$$

in the first step. Note that at least three significant figures should be retained in ΔE to avoid large roundoff errors when the differences are small in comparison to the values of E. In the last column, the cumulative values of Δx are given, and these represent the distance from the starting point to the point where the specified depth y is reached. After the first step, uniform increments in depth y, with y increasing in the upstream direction, are utilized. The values of y are stopped at the finite limit of 1.745 m (5.725 ft), which is 99.5 percent of normal depth. The length required to reach this point is 1271 m (4170 ft), which is the length required for this channel to be considered hydraulically long, but that length varies, in general. The depth increments can be halved until the change in profile length becomes acceptably small. Alternatively, smaller increments in depth can be used in regions of rapidly changing depth, and larger increments may be appropriate in regions of very gradual depth changes. A portion of the computed M2 profile is shown in Figure 5.6.

Direct Numerical Integration

The direct numerical integration method is applied to Equation 5.4, which also can be solved by the direct step method, but in this case numerical integration is employed. In the integrated form, Equation 5.4 becomes

$$\int_{x_i}^{x_{i+1}} dx = x_{i+1} - x_i = \int_{y_i}^{y_{i+1}} \frac{1 - \mathbf{F}^2}{S_0 - S_e} \, dy = \int_{y_i}^{y_{i+1}} g(y) \, dy \qquad (5.16)$$

The integrand on the right hand side of Equation 5.16 is a function of y, $g(y)$, which can be integrated numerically to obtain a solution for Δx, as shown in Figure 5.7. A variety of numerical integration techniques are available, such as the trapezoidal rule and Simpson's $\frac{1}{3}$ rule, which are commonly used to find the cross-sectional area of a natural channel, for example. Simpson's rule is of higher order in accuracy than the trapezoidal rule, which simply means that the same numerical accuracy can be achieved with fewer integration steps. Application of the trapezoidal rule to the right hand side of (5.16) for a single step produces

$$x_{i+1} - x_i = \frac{g(y_{i+1}) + g(y_i)}{2} (y_{i+1} - y_i) \qquad (5.17)$$

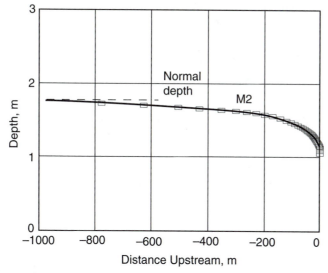

FIGURE 5.6
M2 water surface profile computed by the direct step method.

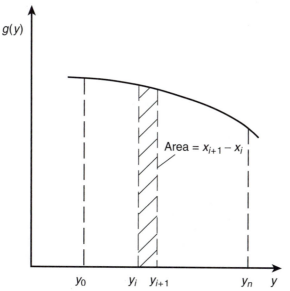

FIGURE 5.7
Water surface profile computation by direct numerical integration.

To determine the full length of a flow profile, $(x_n - x_0)$, multiple application of the trapezoidal rule results in

$$L = x_n - x_0 = \Delta y \left[\frac{g(y_0) + g(y_n) + 2 \displaystyle\sum_{i=1}^{n-1} g(y_i)}{2} \right] \quad (5.18)$$

where L = profile length and $\Delta y = (y_{i+1} - y_i)$ = uniform depth increment. Because the global truncation error in the multiple application of the trapezoidal rule is of order $(\Delta y)^2$, halving the depth increment will reduce the error in the profile length by a factor of 1/4. By successively halving the depth interval, the relative change in the profile length can be calculated with the process continuing until the relative error is less than some acceptable value.

5.7   DEPTH COMPUTED FROM DISTANCE CHANGES

The second approach to the solution of the equation of gradually varied flow is exactly opposite to the first. In this class of methods, depth changes are determined for specified changes in distance. This solution strategy is more appropriate for natural channels in which cross-sectional properties are determined by surveys at specific locations along the channel, but it can be used for artificial channels as well. If Equation 5.4 is integrated to obtain a solution for depth as a function of distance, it becomes

$$y_{i+1} - y_i = \int_{x_i}^{x_{i+1}} \frac{S_0 - S_e}{1 - \mathbf{F}^2} \, dx = \int_{x_i}^{x_{i+1}} f(y) \, dx \quad (5.19)$$

The difficulty with (5.19) is readily apparent when we recognize that the integrand itself is a function of the unknown depth, y. An alternative is to use the Taylor series expansion for y_{i+1} and drop all terms beyond the first derivative term:

$$y_{i+1} = y_i + f(y_i)\Delta x \quad (5.20)$$

where $f(y) = dy/dx$, which can be evaluated at point y_i from Equation 5.4. This method, known as *Euler's method*, simply extends the slope of the solution curve for depth y forward from x_i as a straight line to obtain the next estimate of y at x_{i+1}. The terms dropped from the Taylor series expansion make the local truncation error $O(\Delta x^2)$ as discussed in Appendix A, while the global truncation error (local plus propagated) for multiple steps is $O(\Delta x)$ (Chapra and Canale 2009). This is referred to as a *first-order method*. In general, it requires very small step sizes, and therefore considerable computational effort, to achieve acceptable accuracy.

An improved Euler's method can be formulated by evaluating the slope of the function at both x_i and x_{i+1}, then applying the arithmetic mean of the two slope estimates to move the solution forward. However, because the slope cannot be evaluated at $i + 1$, since y is unknown there, the value of y_{i+1} is first predicted by the

Euler method to evaluate the slope $f(y_{i+1})$. The value of y_{i+1} then is corrected using this estimate of the slope in the determination of the mean of the beginning and ending slopes over the interval. The resulting predictor-corrector equations, known as the *Heun method*, or *corrected Euler method*, are

$$y^0_{i+1} = y_i + f(y_i)\Delta x \tag{5.21}$$

$$y_{i+1} = y_i + \frac{[f(y_i) + f(y^0_{i+1})]}{2}\Delta x \tag{5.22}$$

in which the superscript zero is used to identify the predicted value of y_{i+1} in (5.21), which then is substituted into the corrector formula given by (5.22). This, referred to as a *one-step predictor-corrector method*, is part of a larger class of solution techniques known as *Runge-Kutta methods*. Also apparent is that Equations 5.21 and 5.22 can be iterated back and forth to improve the solution. However, the iterative approach must be used with caution because the error actually may grow rather than shrink (Beckett and Hurt 1967). If Equations 5.21 and 5.22 are not iterated, they can be shown to be a second-order Runge-Kutta method (Chapra and Canale 2009) with a global error that is $O(\Delta x^2)$.

Example 5.3

Compute the M2 profile of Example 5.2 using the corrected Euler method without iteration and compare the results.

Solution.   The solution is accomplished in the spreadsheet shown in Table 5.2, using Equations 5.21 and 5.22. First, the values of area (A1), hydraulic radius (R1), and topwidth (B1) are computed for the initial value of depth (y1), because they are needed to calculate the function $f(y)$

$$f(y) = \frac{S_0 - S_e}{1 - \mathbf{F}^2}$$

The value of $f(y1)$ is given in column 6. The predicted value of y_2 (y2:pred) at the end of the spatial interval is given in column 7, computed from Equation 5.21 using the value of $f(y_1)$ in column 6. Columns 8, 9, and 10 are needed to compute the value of f (y2:pred) in column 11. The corrected value of y_2 is computed from Equation 5.22 in column 2 of the next row for a given step size in x, and the process begins again. At a distance of 1271 m (4170 ft), the corrected Euler method gives a depth of 1.744 m (5.722 ft), while the direct step method yields a depth of 1.745 m (5.725 ft) at the same location. This is a relative difference in depth of less than 0.1 percent. If the interval size in depth y is halved in the direct step method, the resulting depth rounds to 1.744 m (5.722 ft) in agreement with the corrected Euler method. At the beginning of the computation in Table 5.2, the steps in the spatial coordinate x have been taken to be very small because of the steep, rapidly changing slope of the M2 water surface profile near critical depth.

TABLE 5.2 Water surface profile computation by the corrected Euler method

Q,cms=	30	y0, m=	1.7538
n=	0.025	yc, m=	1.0298
S0=	0.001		
b, m=	8		
m:1	2		

(1) x m	(2) y1 m	(3) A1 sq. m	(4) R1 m	(5) B1 m	(6) $f(y1)$	(7) y2:pred m	(8) A2 sq. m	(9) R2 m	(10) B2 m	(11) $f(y2{:}pred)$
0	1.0300	10.36	0.822	12.12	−1.10E+01	1.0849	11.03	0.859	12.34	−2.97E-02
−0.005	1.0575	10.70	0.840	12.23	−6.24E-02	1.0579	10.70	0.841	12.23	−6.16E-02
−0.01	1.0579	10.70	0.841	12.23	−6.16E-02	1.0585	10.71	0.841	12.23	−6.02E-02
−0.02	1.0585	10.71	0.841	12.23	−6.03E-02	1.0597	10.72	0.842	12.24	−5.77E-02
−0.04	1.0596	10.72	0.842	12.24	−5.77E-02	1.0608	10.74	0.843	12.24	−5.55E-02
−0.06	1.0608	10.74	0.842	12.24	−5.55E-02	1.0619	10.75	0.843	12.25	−5.35E-02
−0.08	1.0619	10.75	0.843	12.25	−5.35E-02	1.0629	10.76	0.844	12.25	−5.17E-02
−0.1	1.0629	10.76	0.844	12.25	−5.17E-02	1.0681	10.83	0.847	12.27	−4.42E-02
−0.2	1.0677	10.82	0.847	12.27	−4.47E-02	1.0766	10.93	0.853	12.31	−3.55E-02
−0.4	1.0757	10.92	0.852	12.30	−3.63E-02	1.0830	11.01	0.857	12.33	−3.09E-02
−0.6	1.0824	11.00	0.857	12.33	−3.12E-02	1.0887	11.08	0.861	12.35	−2.76E-02
−0.8	1.0883	11.08	0.861	12.35	−2.77E-02	1.0939	11.14	0.864	12.38	−2.51E-02
−1	1.0936	11.14	0.864	12.37	−2.52E-02	1.1188	11.45	0.881	12.48	−1.71E-02
−2	1.1147	11.40	0.878	12.46	−1.81E-02	1.1509	11.86	0.902	12.60	−1.17E-02
−4	1.1446	11.78	0.898	12.58	−1.26E-02	1.1948	12.41	0.930	12.78	−7.80E-03
−6	1.1649	12.03	0.911	12.66	−1.02E-02	1.2057	12.55	0.937	12.82	−7.13E-03

−8	1.1853	12.29	0.924	12.74	−8.46E-03	1.2022	12.51	0.935	12.81	−7.34E-03
−10	1.2011	12.49	0.934	12.80	−7.41E-03	1.2751	13.45	0.982	13.10	−4.32E-03
−20	1.2597	13.25	0.972	13.04	−4.79E-03	1.3555	14.52	1.03	13.42	−2.61E-03
−40	1.3336	14.23	1.02	13.33	−2.98E-03	1.3931	15.03	1.06	13.57	−2.08E-03
−60	1.3842	14.91	1.05	13.54	−2.19E-03	1.4281	15.50	1.08	13.71	−1.69E-03
−80	1.4230	15.43	1.07	13.69	−1.74E-03	1.4578	15.91	1.10	13.83	−1.41E-03
−100	1.4545	15.87	1.09	13.82	−1.44E-03	1.5264	16.87	1.14	14.11	−9.08E-04
−150	1.5132	16.68	1.13	14.05	−9.93E-04	1.5628	17.39	1.16	14.25	−7.01E-04
−200	1.5555	17.28	1.16	14.22	−7.40E-04	1.5925	17.81	1.18	14.37	−5.55E-04
−250	1.5879	17.75	1.18	14.35	−5.76E-04	1.6167	18.16	1.19	14.47	−4.48E-04
−300	1.6135	18.11	1.19	14.45	−4.62E-04	1.6366	18.45	1.20	14.55	−3.68E-04
−350	1.6342	18.42	1.20	14.54	−3.77E-04	1.6531	18.69	1.21	14.61	−3.06E-04
−400	1.6513	18.66	1.21	14.61	−3.12E-04	1.6669	18.89	1.22	14.67	−2.57E-04
−450	1.6655	18.87	1.22	14.66	−2.62E-04	1.6786	19.06	1.23	14.71	−2.17E-04
−500	1.6775	19.05	1.23	14.71	−2.21E-04	1.6886	19.21	1.24	14.75	−1.85E-04
−550	1.6877	19.20	1.23	14.75	−1.88E-04	1.6971	19.34	1.24	14.79	−1.59E-04
−600	1.6963	19.33	1.24	14.79	−1.61E-04	1.7044	19.44	1.24	14.82	−1.36E-04
−650	1.7038	19.44	1.24	14.82	−1.38E-04	1.7107	19.54	1.25	14.84	−1.18E-04
−700	1.7102	19.53	1.25	14.84	−1.19E-04	1.7161	19.62	1.25	14.86	−1.02E-04
−750	1.7157	19.61	1.25	14.86	−1.03E-04	1.7208	19.69	1.25	14.88	−8.83E-05
−800	1.7205	19.68	1.25	14.88	−8.94E-05	1.7249	19.75	1.26	14.90	−7.68E-05
−850	1.7246	19.75	1.26	14.90	−7.77E-05	1.7285	19.80	1.26	14.91	−6.69E-05
−900	1.7282	19.80	1.26	14.91	−6.76E-05	1.7316	19.85	1.26	14.93	−5.83E-05
−950	1.7314	19.85	1.26	14.93	−5.90E-05	1.7343	19.89	1.26	14.94	−5.10E-05
−1000	1.7341	19.89	1.26	14.94	−5.15E-05	1.7393	19.96	1.27	14.96	−3.77E-05
−1100	1.7386	19.95	1.26	14.95	−3.95E-05	1.7425	20.01	1.27	14.97	−2.91E-05
−1200	1.7420	20.01	1.27	14.97	−3.05E-05	1.7442	20.04	1.27	14.98	−2.48E-05
−1271	1.7440									

The most popular Runge-Kutta method is the fourth-order method, which requires four equations or steps to proceed from point i to point $i + 1$. The equations are recursive, in that each uses a value computed from the previous one. The method can be summarized by

$$y_{i+1} = y_i + \left[\frac{1}{6} (k_1 + 2k_2 + 2k_3 + k_4) \right] \Delta x \qquad (5.23)$$

in which

$$k_1 = f(x_i, y_i) \qquad (5.23a)$$

$$k_2 = f\left(x_i + \frac{\Delta x}{2}, y_i + \frac{\Delta x}{2} k_1 \right) \qquad (5.23b)$$

$$k_3 = f\left(x_i + \frac{\Delta x}{2}, y_i + \frac{\Delta x}{2} k_2 \right) \qquad (5.23c)$$

$$k_4 = f(x_i + \Delta x, y_i + \Delta x k_3) \qquad (5.23d)$$

The fourth-order Runge-Kutta method can be applied with adaptive step size control such that, at each step, the step size is first taken as a full step and then taken as two half steps. The difference between the two estimates of depth is used to adjust the step size so that some specified relative error criterion is met on a step-by-step basis (Chapra and Canale, 2009). In general, the direct step method often is sufficient for water surface profile computation, but the fourth-order Runge-Kutta method may be useful where a high degree of accuracy is required.

An iteration procedure for the second-order predictor-corrector method of (5.21) and (5.22) has been proposed by Prasad (1970) for water surface profile computation in rivers. His procedure is summarized by the following:

1. Calculate $f(y)$ for $y = y_i$:

$$f(y_i) = \frac{S_0 - S_e(y_i)}{1 - \mathbf{F}^2(y_i)} \qquad (5.24)$$

2. Set $f(y_{i+1}) = f(y_i)$ as an initial guess.
3. Calculate y_{i+1} for a given Δx from

$$y_{i+1} = y_i + \frac{[f(y_i) + f(y_{i+1})]}{2} \Delta x \qquad (5.25)$$

since $dy/dx = f(y)$.
4. Calculate $f(y_{i+1})$ from

$$f(y_{i+1}) = \frac{S_0 - S_e(y_{i+1})}{1 - \mathbf{F}^2(y_{i+1})} \qquad (5.26)$$

5. Check $f(y_{i+1})$ from step 4 against the previous value and repeat steps 3 through 5 until they agree within a certain error criterion.

While this method does converge, numerical problems can arise when critical depth is approached as in an M2 or M3 profile. When this happens, the denominator in $f(y_i)$ approaches zero as $\mathbf{F}^2$ approaches 1. These problems can be handled by using smaller step sizes near the critical depth and starting and stopping the profile computation within some finite interval away from critical depth. It also should be apparent that, for overbank flow, the compound channel Froude number should be used in the equation of gradually varied flow. Otherwise, incorrect values of critical depth are accepted, and the resulting profile is incorrect as well.

Example 5.4

Consider the lake discharge problem of Example 5.1, except that the mild slope ($S = 0.005$) has a length of 500 m (1640 ft), followed by a slope with a value of 0.02 and a length of 200 m (656 ft), as shown in Figure 5.8. The Manning's n of the downstream channel is 0.030 and its width is 10.0 m (32.8 ft), which are the same values as for the upstream channel. Sketch the possible water surface profiles and compute one of them for a downstream lake level of 5.0 m (16.4 ft) above the outlet invert.

Solution.   We assume at first that the mild slope length of 500 m (1640 ft) qualifies it to be hydraulically long, so the discharge is controlled by normal depth on the mild slope and it is 101 m³/s (3565 cfs), as determined in Example 5.1. This means that the critical slope of 0.011 in Example 5.1 still is valid, and therefore the downstream slope of 0.02 is steep. The critical depth of 2.18 m (7.15 ft) is the same for the steep slope, but its normal depth needs to be calculated from Manning's equation:

$$\frac{[10y_0]^{5/3}}{[10 + 2y_0]^{2/3}} = \frac{0.030 \times 101}{1.0 \times 0.02^{1/2}} = 21.42$$

from which $y_0 = 1.78$ m (5.84 ft). The downstream lake level is above both normal and critical depth on the steep slope, which means an S1 profile, as shown in Figure 5.8a. At the upstream end of the slope, critical depth will occur at the break in slope. One possibility for the composite water surface profile is an M2 on the mild slope followed by an S2 on the steep slope and a hydraulic jump to the S1 profile. Other possibilities are shown in Figure 5.9 as the downstream lake level rises. At some level, the hydraulic jump and the critical depth will be drowned out, and the S1 profile will occur along the entire steep slope and join the M1 profile on the mild slope. Which of these possibilities actually will occur can be determined only by a water surface profile computation.

The computer program WSP on the book website, which uses the direct step method, was applied to this problem with a downstream lake level of 5.0 m (16.4 ft), as the tailwater condition. First, the M2 profile was computed upstream from critical depth at the break in slope, then the S2 profile was computed downstream from the same point. Finally, the S1 profile was computed upstream from the downstream lake level. The results are shown in Figure 5.8a. The location of the hydraulic jump is determined in Figure 5.8b from the intersection of the momentum function curves computed at each step of the water surface profile computation. The length of the jump is neglected so the location is at the unique point where both the momentum equation and the equation of gradually varied flow for the S2 and S1 profiles are satisfied simultaneously.

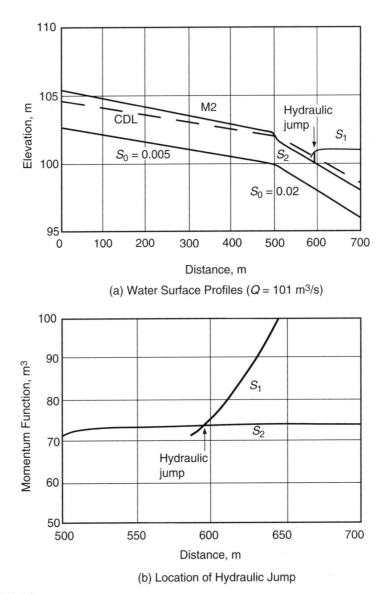

(a) Water Surface Profiles ($Q = 101$ m³/s)

(b) Location of Hydraulic Jump

FIGURE 5.8
Water surface profiles and momentum function for the location of a hydraulic jump in Example 5.4.

As a check on whether the mild slope is hydraulically long, 99.9 percent of normal depth is reached at $x = 65$ m (213 ft) downstream of the channel entrance, so the slope in fact is long enough that the control remains at the entrance to the mild slope. The S2 profile reaches normal depth within 0.1 percent at $x = 595$ m (1950 ft), which is upstream of the channel exit, so it can be considered hydraulically long as well. The hydraulic jump also occurs at $x = 595$ m (1950 ft).

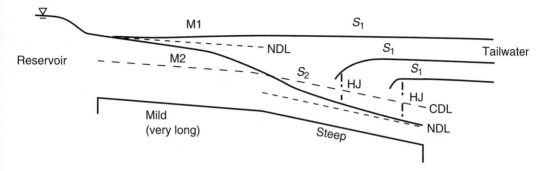

FIGURE 5.9
Possible water surface profiles for increasing tailwater with normal depth control on a mild slope.

5.8   NATURAL CHANNELS

The method of depth determined from distance is used in natural channels by solving the equation of gradually varied flow in the form of the energy equation written from one station to the next:

$$WS_2 + \alpha_2 \frac{V_2^2}{2g} = WS_1 + \alpha_1 \frac{V_1^2}{2g} + h_e \tag{5.27}$$

in which the terms are defined in Figure 5.10. This, in effect, is the integrated form of Equation 5.3, except the head loss term, h_e, includes both the boundary friction loss, $h_f = \bar{S}_e L$, and the minor loss, h_m, due to contractions and expansions from reach to reach:

$$h_e = \bar{S}_e L + K_L \left| \frac{\alpha_2 V_2^2}{2g} - \frac{\alpha_1 V_1^2}{2g} \right| \tag{5.28}$$

in which S_e = mean slope of the energy grade line; L = reach length; K_L = minor head loss coefficient; and α is evaluated by Equation 2.31. The form of Equation 5.27 is written for cross-section numbers increasing in the upstream direction. The solution is obtained by iterating on the difference between the assumed and calculated water surface elevations, using a method such as interval halving or the secant method. The programs HEC-2 and HEC-RAS (U.S. Army Corps of Engineers 2008) use the secant method for solution. When applied to natural channels, this overall solution procedure is referred to as the *standard step method* and also is used by WSPRO (Shearman 1990). Rhodes (1995) applied the Newton-Raphson technique to the iteration required in the standard step method and illustrated the method for the particular cases of prismatic rectangular and trapezoidal channels.

The default value of the minor head loss coefficient, K_L, in (5.28) is taken to be 0.0 for contractions and 0.5 for expansions by WSPRO (Shearman 1990). In HEC-2 or HEC-RAS, the recommended values of K_L are 0.1 and 0.3 for gradual contractions and expansions, respectively, and 0.6 and 0.8 for abrupt contractions and expansions.

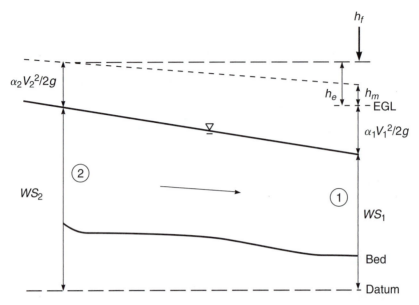

FIGURE 5.10
Definition sketch for the standard step method (U.S. Army Corps of Engineers 2008).

The computation of the mean slope of the energy grade line can be accomplished by several optional equations. In general, $S_e = (Q/K)^2$, in which K is the conveyance for any particular cross section. To obtain the mean value of S_e for two cross sections, the following options are available:

1. Average conveyance

$$\bar{S}_e = \frac{Q^2}{\left[\dfrac{K_1 + K_2}{2}\right]^2} \tag{5.29}$$

2. Average EGL slope

$$\bar{S}_e = \frac{S_{e1} + S_{e2}}{2} \tag{5.30}$$

3. Geometric mean slope

$$\bar{S}_e = \frac{Q^2}{K_1 K_2} \tag{5.31}$$

4. Harmonic mean slope

$$\bar{S}_e = \frac{2 S_{e1} S_{e2}}{S_{e1} + S_{e2}} \tag{5.32}$$

Method 1 is used as a default by HEC-2 and HEC-RAS, while method 3 is the default used by WSPRO. Method 2 has been found to be most accurate for M1 profiles, while method 4 is best for M2 profiles (U.S. Army Corps of Engineers 2008).

The computation of water surface profiles in natural channels must proceed in the upstream direction for subcritical profiles and in the downstream direction for supercritical profiles because the control is located downstream for subcritical and upstream for supercritical profiles. Whether a profile on a given slope is subcritical or supercritical depends on whether the depth is greater or less than critical depth, which is determined by the discharge and the boundary condition.

In a natural channel divided into subreaches, the normal depth changes for each subreach as the slope, roughness, and geometry change. Therefore, water surface profiles in natural channels can be viewed as transition profiles between normal depths; that is, a collection of M1 and M2 profiles on mild slopes. If the normal depth at a specific location is desired as a downstream boundary condition, several water surface profiles can be started from further downstream until asymptotic convergence to normal depth is achieved (see the M1 and M2 profiles in Figure 5.2). In reality, when the depth reached by two backwater profiles is within a specified tolerance, convergence is assumed. Davidian (1984) suggests the use of two M2 profiles to determine convergence.

Cross sections for water surface profile computation are selected to be representative of the subreaches between them, as shown in Figure 5.11. Such locations as major breaks in bed profile, minimum and maximum cross-sectional area, abrupt changes in roughness or shape, and control sections such as free overfalls always are selected for cross sections. Cross sections need to be taken at shorter intervals in bends, expansions, low-gradient streams, and where there is rapid change in conveyance (Davidian 1984).

Some cross sections may require subdivision where there are abrupt transverse changes in geometry or roughness, as in the case of overbank flows. This must be done with care, however, or unexpected results are obtained. In general, if the ratio of overbank width to depth is greater than 5 or if the ratio of main channel depth to overbank depth exceeds 2, subdivision is recommended (Davidian 1984).

The occurrence of both supercritical and subcritical depths in a river reach, referred to as a *mixed-flow regime*, requires special attention in natural channels. In a prismatic channel in which a hydraulic jump is expected, as for example in a reach with an upstream supercritical and downstream subcritical profile, the momentum function is computed for each profile and the intersection of the two momentum function profiles determines the location of a hydraulic jump, as shown in Example 5.4. In a natural channel with a slope near the critical slope, however, finding the exact location of the jump is not possible because of the continuous variation in geometric properties of the cross sections. Instead, the HEC-RAS program computes a subcritical profile in the upstream direction, starting from the downstream boundary condition, then computes a supercritical profile in the downstream direction, usually beginning from critical depth. At each cross section where both a supercritical and a subcritical solution exist, the value of the momentum function is computed and the depth with the higher value accepted. If, for example, the subcritical depth has the higher value of the momentum function, this means the jump would be submerged at this location and move upstream, so the subcritical

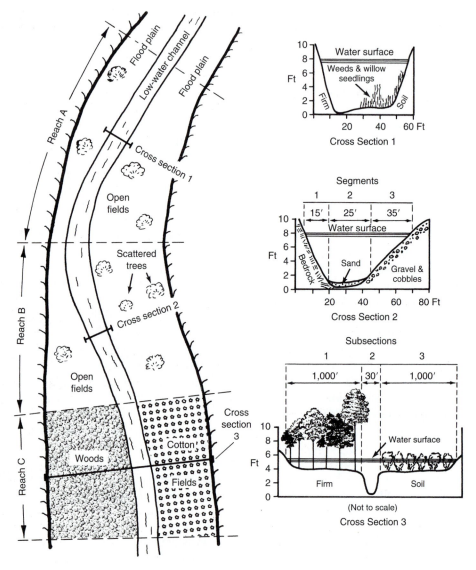

FIGURE 5.11

Hypothetical cross section showing reaches, segments, and subsections used in assigning *n* values (Arcement and Schneider 1984).

depth would be accepted. At any cross section where the HEC-RAS program or WSPRO cannot "balance" the energy equation, the critical depth is taken as the solution and computations proceed. If the depth is critical for both supercritical and subcritical profiles at a given cross section, then it is likely to be a critical control section.

Water surface profiles computed using the Prasad method and the compound channel Froude number (Sturm, Skolds, and Blalock 1985) are illustrated in Figure 5.12 for a laboratory model study. The channel is a 21.3 m (70 ft) long movable-bed model

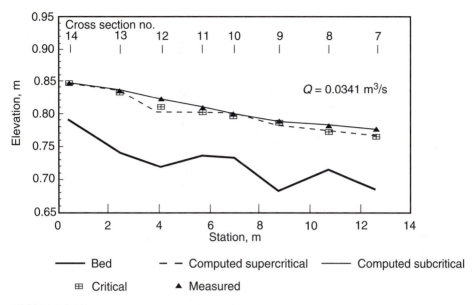

FIGURE 5.12

Measured and computed water surface profiles in a river model (Sturm, Skolds, and Blalock 1985). (*Source: T. W. Sturm, D. M. Skolds, and M. E. Blalock. "Water Surface Profiles in Compound Channels," Proc. of the ASCE Hyd. Div. Specialty Conference, Hydraulics and Hydrology in the Small Computer Age, © 1985, ASCE. Reproduced by permission of ASCE.*)

of an alluvial river. A total of eight river cross sections were used in the computations for a constant discharge of 0.0341 m³/s (1.20 cfs). Water surface profiles were measured after the sediment bed had approached equilibrium. The sediment size was uniform with $d_{50} = 3.3$ mm (0.0108 ft) and Manning's $n = 0.016$. The compound channel Froude number was used to calculate the critical depth for each cross section and to identify a particular solution of the energy equation as supercritical or subcritical. Both subcritical and supercritical profiles were computed, as shown in Figure 5.12. At cross sections 10 and 13, critical depth was returned as the solution for both profiles because neither a subcritical nor a supercritical solution could be found. The measured depths also are in close agreement with the computed values of critical depth at these two cross sections, indicating that they indeed are critical. At cross sections 12 and 9, just downstream of cross sections 13 and 10, respectively, both a supercritical and a subcritical solution exist but the measured values are subcritical. This would indicate a weak hydraulic jump or perhaps simply standing waves between cross sections 13 and 12 and between 10 and 9. Computationally, the depth with the higher value of the momentum function is chosen between the supercritical and subcritical depths at cross sections 12 and 9 (U.S. Army Corps of Engineers 2008). If, for example, the subcritical solution has the higher value of the momentum function, then the jump would be drowned out at that section and moved upstream. The importance of correctly predicting the critical depth in this example should be apparent; otherwise, an incorrect interpretation of the profile and selection of the wrong regime can occur.

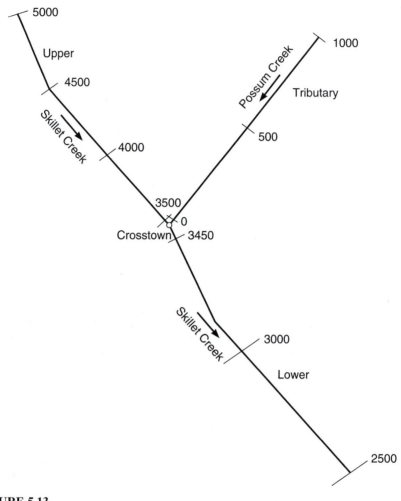

FIGURE 5.13
Stream layout schematic for HEC-RAS, Example 5.5.

EXAMPLE 5.5 (adapted from U. S. Army Corps of Engineers 2008)

Apply the HEC-RAS program to Skillet Creek, as shown in Figure 5.13, and compute
the water surface profile for a discharge of 2000 cfs (56.6 m³/s) in the upper reach, 500 cfs
(14.2 m³/s) in Possum Creek, and a total of 2500 cfs (70.8 m³/s) in the entire lower
reach of Skillet Creek. Assume a subcritical profile and use a downstream boundary
condition of the slope of the energy grade line equal to 0.0004 at Station 2500. The difference between stations indicates reach lengths in feet in Figure 5.13. Manning's n
values are 0.06 in the left floodplain, 0.035 in the main channel, and vary from 0.05 to
0.06 in the right floodplain. (All cross-section data are not shown.)

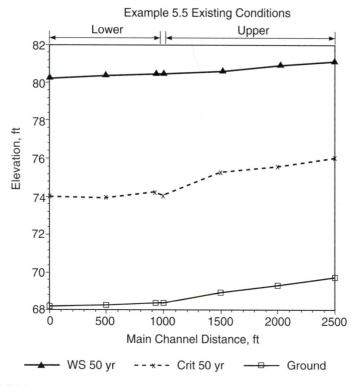

FIGURE 5.14
Computed water surface profile for HEC-RAS, Example 5.5.

Solution. The schematic layout of the river system shown in Figure 5.13 is entered graphically by the user, and the cross-section geometry data are entered and edited interactively. The user must then enter discharge data and boundary conditions before computing the profile. The computed water surface profile, along with the critical depth line, are shown in Figure 5.14 for the main stem, with the distance scale indicating distance upstream of Station 2500. The water surface profile is computed up to Station 3450 at the Crosstown junction. Then, the energy equation is applied across the junction, first from Station 3450 to Station 0 on the tributary and then from Station 3450 to Station 3500 on the main stem. A length of 50 ft was specified across the junction. Both friction losses and minor losses (contraction and expansion) are included in the energy calculation. Once the junction has been crossed, the separate profiles in the main stem and tributary can proceed. For a subcritical flow split in the downstream direction, the program requires a trial-and-error distribution of flow until the energies calculated from the two branches just downstream of the junction are equal. For supercritical and mixed flow cases, see the HEC-RAS manual (U.S. Army Corps of Engineers 2008).

Figure 5.15 illustrates the most upstream cross section at Station 5000 on the main stem. The computation determines only one critical depth, which occurs in the main channel, and the water surface elevation (WS) indicates overbank flooding. The output data for this cross section are given in Table 5.3. The water surface elevation is 81.44 ft (24.82 m)

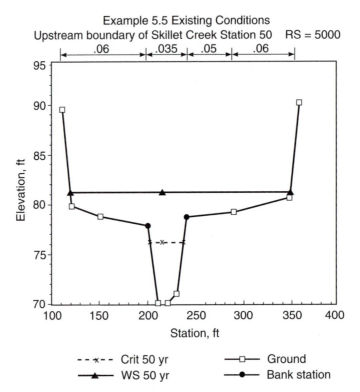

FIGURE 5.15
Upstream cross section and computed water surface elevation from HEC-RAS, Example 5.5.

TABLE 5.3   HEC-RAS cross-section output table for the upstream end of Skillet Creek

Plan: Exist River: Skillet Creek Reach: Upper Riv Sta: 5000 Profile: 50 yr

		Element	Left OB	Channel	Right OB
E.G. Elev (ft)	81.67				
Vel Head (ft)	0.23	Wt. n-Val.	0.060	0.035	0.053
W.S. Elev (ft)	81.44	Reach Len. (ft)	450.00	500.00	550.00
Crit W.S. (ft)	76.27	Flow Area (sq ft)	205.98	362.48	180.65
E.G. Slope (ft/ft)	0.000656	Area (sq ft)	205.98	362.48	180.65
Q Total (cfs)	2000.00	Flow (cfs)	241.28	1568.14	190.58
Top Width (ft)	231.87	Top Width (ft)	81.44	40.00	110.44
Vel Total (ft/s)	2.67	Avg. Vel. (ft/s)	1.17	4.33	1.05
Max Chl Dpth (ft)	11.44	Hydr. Depth (ft)	2.53	9.06	1.64
Conv. Total (cfs)	78102.3	Conv. (cfs)	9422.2	61237.9	7442.3
Length Wtd. (ft)	498.47	Wetted Per. (ft)	82.06	45.66	110.64
Min Ch El (ft)	70.00	Shear (lb/sq ft)	0.10	0.32	0.07
Alpha	2.10	Stream Power (lb/ft s)	0.12	1.41	0.07
Frctn Loss (ft)	0.32	Cum Volume (acre-ft)	6.58	14.20	6.84
C & E Loss (ft)	0.00	Cum SA (acres)	2.31	1.48	3.58

and the velocity is 2.67 ft/s (0.81 m/s). The flow is split into main channel and over-bank contributions by taking the ratio of the conveyances of each subsection to the total conveyance and multiplying times the total discharge. The main channel velocity is approximately four times greater than the overbank velocities. The geometric properties of each subsection are given in the table, leading to a value of the kinetic energy correction coefficient $\alpha = 2.10$.

5.9 FLOODWAY ENCROACHMENT ANALYSIS

Floodway boundaries are established for land-use planning and in flood insurance studies based on the amount of encroachment on the floodplain that can be allowed without exceeding some specified regulatory increase in water surface elevation. Floodway boundaries and floodway encroachment are illustrated in Figure 5.16. In the encroached areas, all floodway conveyance is assumed to be lost. In the United States, the 100-year peak flood discharge is established as the base flood for flood-way analysis, and the increase in the natural water surface elevation caused by floodway encroachment cannot exceed 1 ft.

Floodway analysis proceeds by first running a water surface profile for the base flood under natural conditions. Then, encroachments of varying amounts are added, according to certain criteria, so as not to exceed the target water surface elevation increase. The resulting boundaries from the floodway analysis usually are the result of several iterations and may have to be adjusted for undulations from cross section to cross section and for unreasonable locations when compared to existing land use and topography.

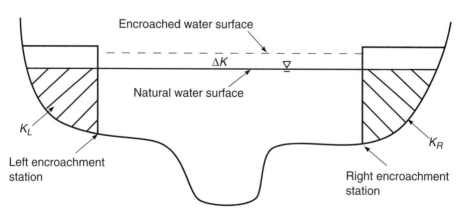

FIGURE 5.16
Floodway encroachment analysis.

Five separate methods can be selected in HEC-RAS to determine floodway boundaries. These are summarized here (Hoggan 1997; U.S. Army Corps of Engineers 2008):

1. In encroachment method 1, the exact locations and elevations of the encroachments are specified in each floodplain, as shown in Figure 5.16.
2. Encroachment method 2 specifies a fixed top width of the floodway that can be specified separately for each cross section. Each encroachment station is set at half the specified width, left and right of the channel centerline.
3. Method 3 calculates encroachment stations for a specified percent of reduction in conveyance of the natural profile for each cross section. The conveyance reduction is applied equally on each side of the cross section, but the computed encroachments are not allowed to infringe on the main channel.
4. The intent of method 4 is to specify a target for the allowable increase in the natural water surface elevation. The resulting gain in conveyance in the floodway is taken up equally on the left and right floodplains. As shown in Figure 5.16, the increase in conveyance $\Delta K = K_L + K_R$, where K_L and K_R are the blocked conveyances on the left and right floodplains and $K_L = K_R = \Delta K/2$.
5. Method 5 is an optimization technique that automatically iterates up to 20 times to achieve the target water surface elevations for all cross sections. Both a target water surface elevation increase and a target energy grade line elevation are specified. In each iteration, the entire water surface profile is computed for a set of encroachments, and then the encroachments are adjusted where the target was violated for the next iteration.

Methods 4 and 5 are most useful to establish an initial solution for the floodway boundaries. In fact, they can be run with several different target increases in water surface elevations. The final determination of the floodway boundary usually is made with method 1, which defines the specific encroachments at each cross section and allows engineering judgment to be applied to the final adjustments.

5.10   BRESSE SOLUTION

Only under very special assumptions is an analytical solution to the equation of gradually varied flow possible. This solution was first obtained by Bresse for very wide rectangular channels. The solution approach was extended by Bakhmeteff and finally fully developed by Chow (1959) into a method called the *hydraulic exponent method*. It is a numerical method in the form developed by Chow, but a very tedious one that no longer is in use.

To obtain the Bresse solution, the equation of gradually varied flow is written in the form:

$$\frac{dy}{dx} = \frac{S_0\left(1 - \dfrac{S_e}{S_0}\right)}{1 - \dfrac{Q^2 B}{g A^3}} \tag{5.33}$$

Now if Manning's equation is written in terms of conveyance, $K = Q/S^{1/2}$, the ratio of S_e/S_0 in Equation 5.33 becomes $(K_0/K)^2$, in which K_0 is the uniform flow conveyance and K is the conveyance corresponding to the local depth y. Furthermore Q^2/g in Equation 5.33 can be replaced by A_c^3/B_c for the critical condition of Froude number squared equal to 1. With these substitutions, Equation 5.33 becomes

$$\frac{dy}{dx} = \frac{S_0\left[1 - \left(\dfrac{K_0}{K}\right)^2\right]}{1 - \dfrac{A_c^3/B_c}{A^3/B}} \tag{5.34}$$

The hydraulic exponent assumptions are made at this point. We assume that the two ratio terms in the numerator and denominator on the right hand side of Equation 5.34 can be set equal to the ratio of either the normal or the critical depth to the local depth taken to a power designated M or N:

$$\frac{dy}{dx} = \frac{S_0\left[1 - \left(\dfrac{y_0}{y}\right)^N\right]}{1 - \left(\dfrac{y_c}{y}\right)^M} \tag{5.35}$$

For a rectangular channel, it is easily shown that $M = 3$. However, the value of N is a constant integer only for a wide, rectangular channel using the Chezy equation with constant C; and it, too, has the value of 3. Under these assumptions, the equation of gradually varied flow can be integrated exactly to give the solution

$$S_0 x = y - y_0\left[1 - \left(\frac{y_c}{y_0}\right)^3\right]\phi\left(\frac{y}{y_0}\right) \tag{5.36}$$

in which ϕ is a function of $y/y_0 = u$, given by

$$\phi(u) = \frac{1}{6}\ln\left[\frac{u^2 + u + 1}{(u - 1)^2}\right] - \frac{1}{3^{1/2}}\arctan\left[\frac{3^{1/2}}{2u + 1}\right] + A \tag{5.37}$$

in which A is an arbitrary constant. The value of the constant is immaterial because the function is evaluated between two points located a distance $(x_2 - x_1)$ apart, and so the constant A cancels. The Bresse varied flow function ϕ is shown graphically in Figure 5.17 for subcritical and supercritical profiles. In the cases of M1, M2, S2, and S3 profiles, the approach to normal depth is asymptotic as shown.

The determination of the downstream boundary condition for a subcritical profile in a natural channel with no critical control section requires an asymptotic method, as discussed previously. The computation is started further downstream than the reach of interest and several depths are tried successively to find an asymptotic depth as the downstream boundary condition for the reach of interest. The Bresse method for a very wide channel can be used to answer the question of how far downstream to start the process, at least in an approximate manner. The length of an M2 profile from 75 percent of normal depth downstream to 97 percent of

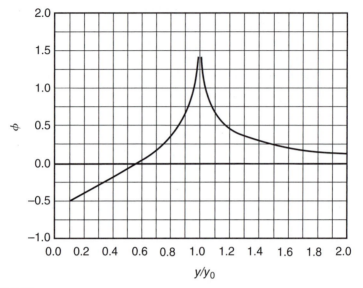

FIGURE 5.17
Bresse varied flow function ϕ for very wide channels with constant Chezy C.

normal depth upstream can be shown from the Bresse solution to be given by (Davidian 1984)

$$\frac{LS_0}{y_0} = 0.57 - 0.79\mathbf{F}^2 \qquad (\text{M2 curve}) \qquad (5.38)$$

in which L is the required total computation length; S_0 is the bed slope; y_0 is the normal depth; and $\mathbf{F}$ is the Froude number of the uniform flow. In a similar fashion, the length of an M1 profile from 125 percent of normal depth downstream to 103 percent of normal depth upstream is given by

$$\frac{LS_0}{y_0} = 0.86 - 0.64\mathbf{F}^2 \qquad (\text{M1 curve}) \qquad (5.39)$$

For example, a channel with an average slope of 0.001, normal depth of 3.0 m (9.8 ft), and a Froude number of 0.25 would have an M2 profile length of approximately 1560 m (5120 ft) while the M1 profile length would be greater, with a value of 2460 m (8070 ft).

5.11 SPATIALLY VARIED FLOW

Spatially varied flow is a gradually varied flow in which the discharge varies in the flow direction due to either a lateral inflow (Figure 5.18) or a lateral outflow. The governing equation in these two cases is different, and considerable confusion can center around which equation is appropriate for a given case.

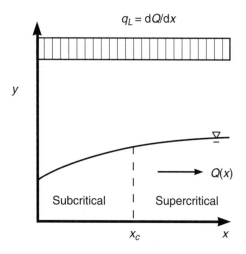

FIGURE 5.18
Spatially varied flow with lateral inflow.

For the case of lateral inflow, such as a side channel spillway, the momentum equation is more appropriate because the energy losses are not well known, while the lateral inflow momentum flux *can* be specified. If it is assumed that the momentum correction factor β is approximately unity and that the inflow enters in a direction perpendicular to the main channel flow, the general unsteady momentum equation (derived in Chapter 7) can be simplified to

$$\frac{dy}{dx} = \frac{S_0 - S_f - \dfrac{2q_L V}{gA}}{1 - \mathbf{F}^2} \tag{5.40}$$

in which S_f = friction slope = $\tau_0/\gamma R$ and q_L = lateral inflow rate per unit of channel length. In the case of the side channel spillway, q_L is a constant, such that the channel discharge $Q(x) = q_L x$, where $x = 0$ at the upstream end of the channel. Because Q varies with x, Equation 5.40 has to be solved numerically by specifying a value of x and iterating on y in a stepwise fashion along the channel.

The variation of Q with x also complicates the determination of the critical section. Critical depth can occur at any point along the channel, with subcritical flow upstream and supercritical flow downstream of the critical section. If it is assumed that critical depth occurs when the Froude number $\mathbf{F} = 1$ and the numerator of (5.40) is zero, so that $dy/dx \neq 0$, then the location of the critical section can be shown to be given by (Henderson, 1966)

$$x_c = \frac{8q_L^2}{gB^2 \left[S_0 - \dfrac{gP}{C^2 B} \right]^3} \tag{5.41}$$

in which x_c = location of critical section; q_L = lateral inflow per unit channel length; B = channel top width; S_0 = bed slope; P = wetted perimeter; and C = Chezy resistance coefficient. Equation 5.41 is solved simultaneously with the criterion that the Froude number is equal to unity at the critical section:

$$\mathbf{F}^2 = \frac{Q^2(x)B_c}{gA_c^3} = 1 \tag{5.42}$$

where $Q(x) = q_L x$. If $x_c > L$, the channel length, the control is at the downstream end of the channel with subcritical flow in the entire channel. Otherwise, the flow is subcritical upstream of x_c and supercritical downstream, as shown in Figure 5.18.

In the case of lateral outflow, such as in the side discharge weir shown in Figure 5.19, the direction of the lateral momentum flux is unknown. Furthermore, because the weir is a local disturbance, energy losses along the weir are relatively small. For these reasons, the energy approach is used more often than the momentum equation. Therefore, if we assume that $dE/dx = 0$, on differentiation of the specific energy, E, with respect to x, we have

$$\frac{dy}{dx} = \frac{Q(x)y\left(-\dfrac{dQ}{dx}\right)}{gb^2y^3 - Q^2} \tag{5.43}$$

for a rectangular channel of width b. Equation 5.43 can be placed in the form

$$\frac{dy}{dx} = \frac{\dfrac{q_L V}{gA}}{1 - \mathbf{F}^2} \tag{5.44}$$

and it only remains to specify $q_L = -dQ/dx$ from the discharge equation for a sharp-crested weir as

$$q_L = -\frac{dQ}{dx} = C_1\sqrt{2g}\,(y - P)^{3/2} \tag{5.45}$$

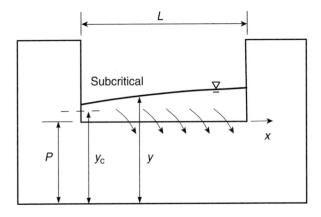

FIGURE 5.19
Spatially varied flow with lateral outflow from a side discharge weir.

in which C_1 = weir discharge coefficient, = $(2/3)C_d$ from Chapter 2. Because we assume the energy grade line to be horizontal, the energy equation gives the discharge at any section as

$$Q = by\sqrt{2g(E - y)} \qquad (5.46)$$

in which b = width of the channel and E = known constant specific energy. Substituting (5.45) and (5.46) into (5.44) and integrating, the result as obtained by De Marchi (Benefield, Judkins, and Parr 1984) is

$$\frac{xC_1}{b} = \frac{2E - 3P}{E - P}\sqrt{\frac{E - y}{y - P}} - 3\sin^{-1}\sqrt{\frac{E - y}{E - P}} + \text{constant} \qquad (5.47)$$

in which C_1 = weir discharge coefficient; E = specific energy of the flow; P = height of weir crest above channel bottom; and b = channel width. The subcritical case is shown in Figure 5.19, but it also is possible to have a supercritical profile either alone or with a hydraulic jump (see the Exercises).

Hager (1987) showed that the outflow equation used by de Marchi is exact only for small Froude numbers. He developed a generalized outflow equation for side discharge weir flow that includes the effects of lateral outflow angle and longitudinal channel width contraction. Hager (1999) gives general solutions of the free surface profile for the enhanced outflow equation.

REFERENCES

Arcement, G. J., Jr., and V. R. Schneider. *Guide for Selecting Manning's Roughness Coefficients for Natural Channels and Flood Plains.* Report No. FHWA-TS-84-204. Federal Highway Admin., U.S. Department of Transportation, National Technical Information Service, Springfield, VA: 1984.

Bakhmeteff, B. A. *Hydraulics of Open Channel Flow.* New York: McGraw-Hill, 1932.

Beckett, R., and J. Hurt. *Numerical Calculations and Algorithms.* New York: McGraw-Hill, 1967.

Benefield, L. D., J. F. Judkins, Jr., and A. D. Parr. *Treatment Plant Hydraulics for Environmental Engineers.* Englewood Cliffs, NJ: Prentice-Hall, Inc., 1984.

Chapra, S. C., and R. P. Canale. *Numerical Methods for Engineers,* 6th ed. New York: McGraw-Hill, 2009.

Chow, V. T. *Open Channel Hydraulics.* New York: McGraw-Hill, 1959.

Davidian, J. "Computation of Water Surface Profiles in Open Channels." In *Techniques of Water-Resources Investigations of the U.S. Geological Survey,* Book 3, *Applications of Hydraulics.* Washington, DC: Government Printing Office, 1984.

Hager, W. H. "Lateral Outflow Over Side Weirs." *J. Hydr. Engrg.,* ASCE 113, no. 4 (1987), pp. 491–504.

Hager, W. H. *Wastewater Hydraulics.* Berlin Heidelberg: Springer-Verlag, 1999.

Henderson, F. M. *Open Channel Flow.* New York: Macmillan, 1966.

Hoggan, D. H. *Computer-Assisted Floodplain Hydrology and Hydraulics,* 2nd ed. New York: McGraw-Hill, 1997.

Martin, C. S., and D. C. Wiggert. Discussion of "Simulation Accuracies of Gradually-Varied Flow," by J. P. Jolly and V. Yevjevich. *J. Hyd. Div.,* ASCE 101, no. HY7 (1975), pp. 1021–24.

Prasad, R. "Numerical Method of Computing Flow Profiles." *J. Hyd. Div.*, ASCE 96, no. HY1 (1970), pp. 75–86.

Rhodes, D. F. "Newton-Raphson Solution for Gradually-Varied Flow." *J. Hydr. Res.* 33, no. 2 (1995), pp. 213–18.

Shearman, J. O. *User's Manual for WSPRO—A Computer Model for Water Surface Profile Computations.* Report FHWA-IP-89-027. Federal Highway Administration, U.S. Department of Transportation, 1990.

Sturm, T. W., D. M. Skolds, and M. E. Blalock. "Water Surface Profiles in Compound Channels." *Proc. of the ASCE Hyd. Div. Specialty Conf.*, Hydraulics and Hydrology in the Small Computer Age, Lake Buena Vista, Florida, pp. 569–74, 1985.

U.S. Army Corps of Engineers. HEC-RAS Hydraulic Reference Manual, version 4. Davis, CA: U.S. Army Corps of Engineers, Hydrologic Engineering Center, 2008.

Yen, B. C. "Open Channel Flow Equations Revisited." *J. Engrg. Mech. Div.*, ASCE 99, no. EM5 (1973), pp. 979–1009.

EXERCISES

5.1. Prove from the equation of gradually varied flow that S2 and S3 profiles asymptotically approach normal depth in the downstream direction.

5.2. A reservoir discharges into a long trapezoidal channel that has a bottom width of 20 ft, side slopes of 3:1, a Manning's n of 0.025, and a bed slope of 0.001. The reservoir water surface is 10 ft above the invert of the channel entrance. Determine the channel discharge.

5.3. A reservoir discharges into a long, steep channel followed by a long channel with a mild slope. Sketch and label the possible flow profiles as the tailwater rises. Explain how you could determine if the hydraulic jump occurs on the steep or mild slope.

5.4. Compute the water surface profile of Table 5.1 in the text using the method of numerical integration with the trapezoidal rule. Use the same step sizes as in the table and determine the distance required to reach a depth of 1.74 m. Discuss the results.

5.5. A rectangular channel 6.1 m wide with $n = 0.014$ is laid on a slope of 0.001 and terminates in a free overfall. Upstream 300 m from the overfall is a sluice gate that produces a depth of 0.47 m immediately downstream. For a discharge of 17.0 m³/s, with a spreadsheet, compute the water surface profiles and the location of the hydraulic jump using the direct step method. Verify with the program WSP, or with a program that you write.

5.6. The depth of flow upstream of a sluice gate in an irrigation canal is 10.0 ft. A reservoir with a constant head of 8.0 ft above the invert of the canal entrance is located 1000 ft upstream of the gate. The tailwater depth is 15.0 ft above the invert of the canal exit at a distance of 1000 ft downstream of the gate. The canal is trapezoidal with a bottom width of 10 ft, side slopes of 3:1, Manning's n of 0.015, and a bed slope of 0.01. Plot the water surface profile upstream and downstream of the gate including the location of any hydraulic jumps. Use WSP on the book website or a program of your own. Neglect channel entrance losses.

5.7. A rectangular channel with a slope of 0.002 has a bottom width of 6 m and a Manning's n of 0.025. The channel discharge is 20 m³/s. The channel has a short width constriction to a width of 2 m followed by an expansion back to a width of 6 m.

Upstream and downstream of the constriction, the channel is very long with uniform flow acting as the downstream control. Compute and plot the complete water surface profile upstream and downstream of the constriction. Determine the location of any hydraulic jumps. Use the computer program WSP on the book website or a program of your own. Neglect any head losses in the constriction.

5.8. A very wide rectangular channel carries a discharge of 10.0 m³/s/m on a slope of 0.001 with an *n* value of 0.026. The channel ends in a free overfall. Compute the distance required for the depth to reach $0.9y_0$ using the direct step method and compare the result with that from the Bresse function.

5.9. Derive Equations 5.38 and 5.39 using the Bresse function.

5.10. For a very wide channel on a steep slope, derive a formula for the length of an S2 profile from critical depth to 1.01 y_0 using the Bresse function. What is this length in meters if the slope is 0.01, the discharge per unit of width is 2.0 m³/s/m, and Manning's *n* is 0.025?

5.11. A trapezoidal channel of bottom width 20 ft with side slopes of 2:1 is laid on a slope of 0.0005 and has an *n* value of 0.045. It drains a lake with a constant water surface level of 10 ft above the invert of the channel entrance. If the channel ends in a free overfall, calculate the discharge in the channel for channel lengths of 100 and 10,000 ft using the WSP program.

5.12. Construct and plot a delivery curve for the rectangular channel of Example 5.1 if it has a length of 100 m. The bottom width is 10 m; Manning's $n = 0.030$; upstream head is fixed at 3.5 m; and the bed slope is 0.005 as in the example. Use WSP on the book website or a program of your own. Neglect channel entrance losses.

5.13. A trapezoidal canal with a bottom width of 20 ft and 2:1 side slopes has a slope of 0.001 ft/ft and a Manning's *n* value of 0.025. The canal connects two reservoirs with the upper, very large reservoir having a constant head above the canal entrance invert of 7.0 ft. The smaller downstream reservoir has a variable head H_t relative to the invert at the canal exit. If the canal has a length of 500 ft, construct and plot a delivery curve with discharge Q on the horizontal axis and H_t on the vertical axis. Also show the case of a hydraulically long canal on the same plot. Explain the shapes of the curves in terms of the types of water surface profiles that occur. Use the computer program WSP on the book website. Neglect channel entrance losses.

5.14. A 3 ft by 3 ft box culvert that is 100 ft long is laid on a slope of 0.001 and has a Manning's *n* of 0.013. The downstream end of the culvert is a free overfall. For a discharge of 20 cfs, calculate the entrance depth using the WSP program, and the head upstream of the culvert using the energy equation with an entrance loss coefficient of 0.5 for a square-edged entrance. Compare the result with the head calculated from an assumption of a hydraulically long culvert with an entrance depth equal to normal depth.

5.15. A 6.0 ft by 6.0 ft concrete box culvert has a length of 200 ft. The upstream invert elevation is 100 ft while the downstream invert elevation is 99.6 ft. The entrance head loss coefficient is 0.5, and the Manning's $n = 0.015$. Using the computer program WSP on the book website or a program of your own, compute and plot the water surface profiles on the same graph for the five discharges and corresponding tailwater depths (relative to the downstream invert) shown in the table below. Classify each profile. Then apply the energy equation between the pool of water upstream of the culvert, where velocity head can be neglected, and a point just inside the culvert entrance where the depth and velocity are determined from the water surface profile computation. Plot a rating curve consisting of the calculated upstream head on the vertical axis vs. the discharge on the horizontal axis.

Tailwater Rating Curve

Discharge, Q, cfs	Tailwater depth, TW, ft
50	1.86
100	2.68
150	3.29
200	3.80
250	4.24

5.16. Using HEC-RAS, compute the water surface profile in Some Creek for a discharge of 10,000 cfs. Begin with a subcritical profile and a downstream water surface slope of 0.0087 as a boundary condition. Then do a mixed flow analysis with an upstream boundary condition of critical depth. The cross-section geometry, reach lengths, roughness values, and subsection breakpoints are shown in the following table. Analyze the results indicating where any hydraulic jumps may occur.

The upstream cross section for Some Creek at River Station 6000 (ft) is given by

X (ft)	Elevation (ft)	n
0	465	0.055
0	461	
23	458.8	
36	458	
45	457.8	
55	458.3	
99	458.4	0.065
110	455.9	
119	455.8	
133	455.5	
143	455.3	
150	455.4	0.040
154	454	
155	452	
160	450.3	
168	450.2	
188	450.5	
193	451.5	
200	452.7	
205	454.5	
210	455.3	0.065
229	455.6	
258	455.3	
266	456.3	
276	458	0.055
305	457.8	
344	458	
380	461	
380	465	

The left and right banks are at $X = 150$ ft and 210 ft, respectively. At subsequent stations downstream, the cross section should be adjusted with a uniform decrease in elevation from the *previous section* as follows:

River station (ft)	Decrease in elevation (ft)
4000	2.0
3000	6.0
1500	2.8
1000	6.0

Note: Interpolated cross sections may be required in Exercise 5.16 and 5.17.

5.17. Compute the water surface profile in the Red Fox River for $Q = 1000$ cfs, for which the downstream water surface elevation $WS = 3.80$, and for $Q = 10,000$ cfs with $WS = 15.05$. The stations for the four cross sections are shown here, and the elevations (Z) and n values are given in the following table (Hoggan 1997).

Cross section	Station (ft)
1	0
2	500
3	900
4	1300

Cross section 1			Cross section 2			Cross section 3			Cross section 4		
X (ft)	Z (ft)	n	X (ft)	Z (ft)	n	X (ft)	Z (ft)	n	X (ft)	Z (ft)	n
20	25	0.100	30	25	0.10	40	25	0.10	30	26	0.10
30	24		40	24		90	24		75	25	
45	22		50	22		260	22	0.05	130	24	0.05
60	20		110	20		330	20				
110	18		200	20		370	18.7	0.03	330	23	0.036
415	17	0.050	295	18		420	15		360	14	
630	16		415	17	0.05	460	11.2		370	9.5	
650	14	0.030	455	16		500	7.1		400	9.8	
655	13		505	13		530	7.5		410	13	
660	13		575	9.5	0.03	550	12		460	22	0.05
670	2		585	5		560	17.8				
675	1		596	4.2		580	19		610	22	0.10
690	0		615	4.5		600	20	0.05	650	24	
697	0.1		635	16		850	22		675	25	
700	0.8		640	18	0.10	865	24		700	26	
710	1		940	18.5		875	25				
710	13	0.050	1180	18							
940	13.5		1195	18							
1020	14	0.10	1205	20							
1215	14		1225	22							
1235	12		1245	24							
1575	12		1250	25							
1590	14										
1615	16										
1630	20										
1635	25										

5.18. The cross-section geometry for Roaring Creek follows:

X (ft)	Elevation (ft)	n
4	10.0	.050
10	9.5	
20	9.3	
30	9.4	
40	9.2	.035
42	7.0	
46	6.2	
50	6.0	
54	6.1	
58	6.2	
62	6.0	
66	7.1	
70	6.3	
72	8.3	
76	8.9	
80	9.0	.060
90	9.5	
100	9.3	.030
110	9.6	
116	10.0	

The measured water surface elevation is 9.8 ft.
(a) Manually calculate the normal discharge for a slope of 0.0008.
(b) Manually calculate the value of α and the specific energy.
(c) Is the flow subcritical or supercritical?
(d) Verify your manual calculations with the HEC-RAS program.

5.19. The data file NFPeachtreeCr.xls on the book website contains the geometry data for 12 cross sections on the North Fork of Peachtree Creek. The data for each cross section are given on a separate page of the spreadsheet and include the station (increasing upstream), reach lengths, Manning's n values, and the cross section elevations. Enter the data in HEC-RAS and carry out a steady flow analysis for two discharges: 2000 cfs (2-yr return interval) and 12,000 cfs (100-yr return interval). Use a downstream boundary condition of normal depth with a slope of 0.001. Analyze the results as follows:
(a) Give the water surface profile plots and the profile summary tables (Standard Tables 1 and 2) for each discharge. Also plot the profiles of main channel shear stress and velocity along the stream for the 2000 cfs case and identify unstable stream reaches if the critical shear stress for erosion is 0.4 lb/ft².
(b) For the 12,000 cfs case, plot profiles of total flow top width and calculate the total area flooded in acres.
(c) For the reach from River Station 23,987 to 24,963, a commercial development has been proposed with 2 ft of fill dirt in the left floodplain for the parking lot. For the steady flow 100-yr discharge of 12,000 cfs, determine the change in water surface elevations that this would cause upstream of the area of floodplain infill.

5.20. Write a computer program in the language of your choice that computes the water surface profile in a circular culvert using the method of integration by the trapezoidal rule.

5.21. Write a computer program in the language of your choice that computes a water surface profile in a trapezoidal channel using the fourth-order Runge-Kutta method. Test it with the M2 profile of Table 5.1.

5.22. For the flow over a horizontal bed with constant specific energy and discharge decreasing in the direction of flow, derive the shapes of the subcritical and supercritical profiles for a side discharge weir as shown in Figure 5.19.

5.23. Derive the energy equation for spatially varied flow in the form of Equation 5.44, but do not assume that S_0 and S_e, the bed slope and slope of the energy grade line, are equal to zero. Compare the result with Equation 5.40 and discuss.

5.24. A rectangular side discharge weir has a height of 0.35 m. It is located in a rectangular channel having a width of 0.7 m. If the downstream depth is 0.52 m for a discharge of 0.27 m^3/s, how long should the weir be for a lateral discharge of 0.21 m^3/s?

5.25. A concrete ($n = 0.013$) cooling tower collection channel is rectangular with a length of 45 ft in the flow direction and a width of 31 ft. The channel slope is 0.021 ft/ft. The addition of flow from above in the form of a continuous stream of droplets is at the rate of 0.63 cfs/ft of length. Find the location of the critical section and compute the water surface profile. How deep should the collection channel be?

6

Hydraulic Structures

6.1 INTRODUCTION

In this chapter, we consider a limited set of hydraulic structures (spillways, culverts, and bridges) that provide water conveyance to protect some other engineering structure. Spillways are used on both large and small dams to pass flood flows, thereby preventing overtopping and failure of the dam. Culverts are designed to carry peak flood discharges under roadways or other embankments to prevent embankment overflows. Finally, bridges convey vehicles over waterways, but they must accommodate through-flows of floodwaters without failure due to overtopping or foundation failure by scour.

Of primary importance for the hydraulic structures considered in this chapter is the magnitude of backwater they cause upstream of the structure for a given design discharge; that is, the head-discharge relationship for the structure. In general, this relationship can assume the form of weir flow, orifice flow, and in the case of culverts, full-pipe flow. Each type of flow has its own characteristic dependence between head and discharge. For spillways, the pressure distribution on the face of the spillway also is important, because of the possibility of cavitation and failure of the spillway surface.

Both gradually varied and rapidly varied flows are possible through these structures, but one-dimensional methods of analysis usually are sufficient and well-developed in this branch of hydraulics. Essential to the "hydraulic approach" is the specification of empirical discharge coefficients that have been well established by laboratory experiments and verified in the field. The determination of controls in the hydraulic analysis also is important, and critical depth often is the control of interest. The energy equation and the specific energy diagram are useful tools in the hydraulic analyses of this chapter.

6.2   SPILLWAYS

The concrete ogee spillway is used to transfer large flood discharges safely from a reservoir to the downstream river, usually with significant elevation changes and relatively high velocities. The characteristic ogee shape shown in Figure 6.1 is based on the shape of the underside of the nappe coming off a ventilated, sharp-crested weir. The purpose of this shape is to maintain pressure on the face of the spillway near atmospheric and well above the cavitation pressure.

As an initial departure on the task of developing the head-discharge relation-ship for ogee spillways, it is useful to use the Rehbock relationship for the discharge coefficient of a sharp-crested weir given previously in Chapter 2 as Equation 2.42. For a very high spillway, the contribution of the term involving H/P becomes small and the discharge coefficient, C_d, approaches a value of 0.611; however, this value of C_d is defined for a head of H' on a sharp-crested weir as shown in Figure 6.1. If it is converted to a value defined in terms of the head, H, which is measured rela-tive to the ogee spillway crest, then $C_d = 0.728$ because $H = 0.89H'$, as shown in Figure 6.1 (Henderson 1966). As a result, $C = Q/(LH^{3/2})$ has an equivalent value of approximately 3.9 in English units for a very high spillway.

For lower spillways, the effect of the approach velocity and the vertical con-traction of the water surface introduce an additional geometric parameter given by H/P or its inverse, in which P is the height of the spillway crest relative to the approach channel. Furthermore, the design value of the discharge coefficient is valid for one specific value of head, called the *design head*, H_d, because the pres-sure distribution changes from the ideal atmospheric pressure associated with the ogee shape whenever the head changes. As the head becomes larger than the design head, the pressures on the face of the spillway become less than atmospheric and can approach cavitation conditions. Pressures are larger than atmospheric for heads less than the design head. On the other hand, the risk of cavitation at heads higher than design head is counterbalanced by higher discharge coefficients because of the lower pressures on the face of the spillway. In other words, the spillway becomes more efficient because it passes a higher discharge for the same head with a larger

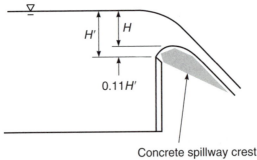

Concrete spillway crest
conforming to the underside of
nappe of sharp-crested weir

FIGURE 6.1
The ogee spillway and equivalent sharp-crested weir.

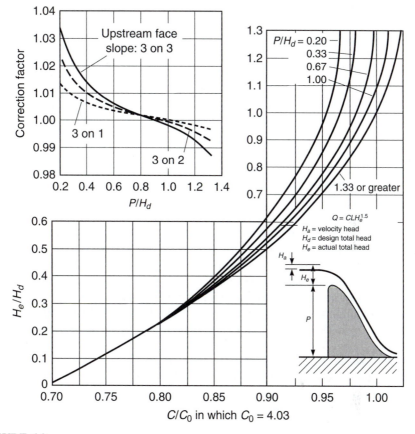

FIGURE 6.2
Discharge coefficient for the WES standard spillway shape (Chow 1959). (*Source: Used with permission of Chow estate.*)

value of the discharge coefficient. The spillway discharge coefficient is given in Figure 6.2 for the standard WES (Waterways Experiment Station) overflow spillway in terms of the influence of the spillway height relative to the design head, P/H_d, and the effect of heads other than the design head as indicated by H_e/H_d, in which H_d is the design total head and H_e is the actual total head on the spillway crest, including the approach velocity head. The discharge coefficient, C, with Q in cubic feet per second and both L and H_e in feet is defined by

$$C = \frac{Q}{LH_e^{1.5}} \tag{6.1}$$

in which L is the net effective crest length. In analogy with the sharp-crested weir equation in Chapter 2, the dimensionless coefficient of discharge, $C_d = C/[(2/3)(2g)^{1/2}]$. If C is in English units (*EN*) as in this section on spillways with head in ft and discharge in cfs, then $C_d = 0.187C$ so that C varying from 3.0 to 4.0 corresponds to $C_d = 0.56$–0.75. The inset in Figure 6.2 shows that a sloping upstream face, which can be used to prevent a separation eddy that might occur on the vertical face of a low spillway, causes an increase in the discharge coefficient for $P/H_d \leq 1.0$. The lateral contraction caused by piers and abutments tends to reduce the actual crest length, L', to its effective value, L:

$$L = L' - 2(NK_p + K_a)H_e \tag{6.2}$$

in which N = number of piers; K_p = pier contraction coefficient; and K_a = abutment contraction coefficient. For square-nosed piers, $K_p = 0.02$, while for round-nosed piers, $K_p = 0.01$, and for pointed-nose piers, $K_p = 0.0$. For square abutments with headwalls at 90° to the flow direction, $K_a = 0.20$, while for rounded abutments with the radius of curvature r in the range, $0.15 H_d \leq r \leq 0.5 H_d$, $K_a = 0.10$. Well-rounded abutments with $r > 0.5H_d$ have a value of $K_a = 0.0$ (U.S. Bureau of Reclamation 1987).

A well-established design procedure, which has been developed by the USBR (U.S. Bureau of Reclamation) and the COE (Corps of Engineers), takes advantage of the higher spillway efficiency achieved for heads greater than the design head. Essentially, the design procedure involves selecting a design head that is less than the maximum head to compute the spillway crest shape; this is called *underdesigning* the spillway crest. Tests have shown that subatmospheric pressures on the face of the spillway do not exceed about one half the design head when H_{max}/H_d does not exceed 1.33. This is shown in Figure 6.3a, in which the actual pressure distribution on a high spillway with no piers is given for H/H_d varying from 0.5 to 1.5 where $H = H_e$. At $H/H_d = 1.0$, the pressures indeed are very close to atmospheric. The minimum pressure for $H/H_d = 1.33$ is $-0.43 H_d$ at $X = -0.2H_d$, where $X = 0.0$ at the centerline of the spillway crest.

Instead of arbitrarily setting $H_e/H_d = 1.33$ at the maximum head, Cassidy (1970) suggests that a better design procedure is to establish a minimum allowable pressure on the spillway face and then determine the design head. The pressures on spillway faces are not constant but fluctuate around a mean value, so the COE now recommends a more conservative design procedure of not allowing the average pressure head to fall below -15 ft to -20 ft, even though cavitation may not be incipient until a pressure head of -25 ft is reached (Reese and Maynord 1987). In this design approach, the minimum allowable pressure head becomes the controlling feature of the design of the spillway crest, rather than a fixed value of H_e/H_d.

Once the design head is determined, the actual shape of the spillway crest downstream of the apex, in what is called the *downstream quadrant*, is given by the equation in Figure 6.3b for negligible approach velocity in which H_d = design head; and X, Y are measured from the crest axis as shown in Figure 6.3b. The *upstream quadrant* of the spillway crest is constructed from a compound circular curve, as shown in Figure 6.3b, to form the standard WES ogee spillway shape. The $0.04 H_d$ radius curve was added in the 1970s resulting in a slight increase in the spillway coefficient in Figure 6.2 for $H_e/H_d > 1.0$ and $P/H_d \geq 1.33$.

Reese and Maynord (1987) proposed, instead, a quarter of an ellipse, which is tangent to the upstream face, for the shape of the upstream quadrant as shown in Figure 6.4a. The discharge coefficients for this shape are given in Figure 6.4b for a vertical upstream face. Reese and Maynord also developed a set of cavitation safety

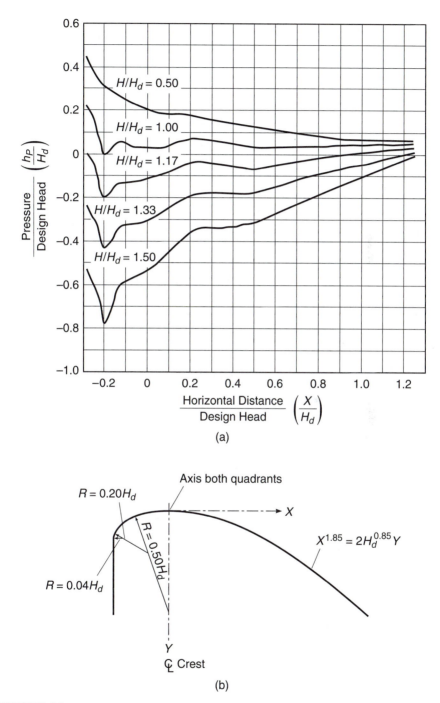

FIGURE 6.3

(a) Crest pressure on WES high-overflow spillway—no piers; (b) Standard WES ogee spill-way shape (U.S. Army Corps of Engineers 1970, Hydraulic Design Chart 111-16).

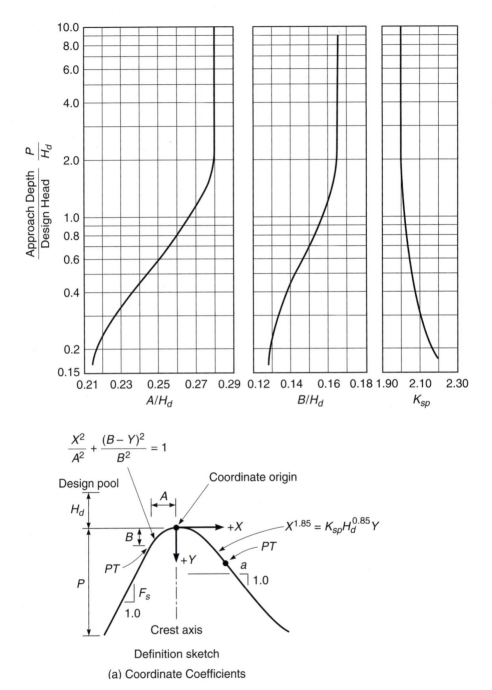

FIGURE 6.4
Elliptical crest spillway coordinate coefficients and discharge coefficients, vertical upstream face (U.S. Army Corps of Engineers 1990, Hydraulic Design Chart 111-20).

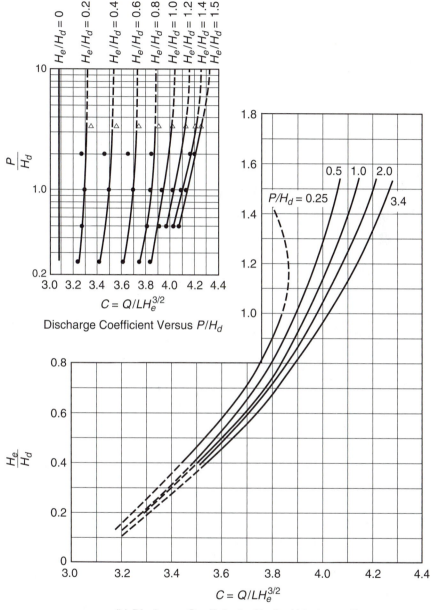

Discharge Coefficient Versus P/H_d

(b) Discharge Coefficients, Vertical Upstream Face

FIGURE 6.4 (*continued*)

curves in which the design head is determined by the allowable cavitation head. These are given in Figure 6.5 for elliptical crest spillways with and without piers. Instead of selecting H_e/H_d as 1.33, a trial design head can be chosen for a minimum pressure head of -15 ft. Then from Figure 6.5, the value of H_e/H_d and the maximum head H_e can be obtained to compare with the given value.

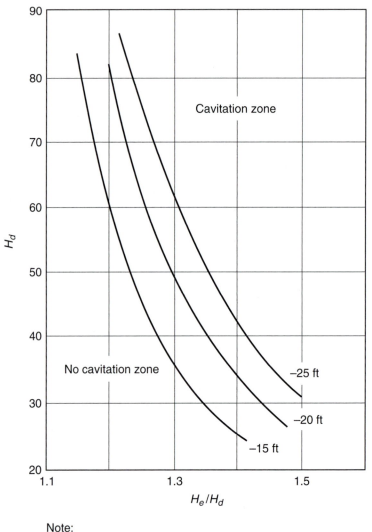

Note:
H_d = Design Total Head, ft
H_e = Actual Total Head, ft

(a) No Piers

FIGURE 6.5
Elliptical crest spillway cavitation safety curves, no piers and with piers (U.S. Army Corps of Engineers 1990, Hydraulic Design Chart 111-25).

Example 6.1

For a maximum discharge of 200,000 cfs (5666 m³/s) and a maximum total head on the spillway crest of 64 ft (19.5 m), determine the crest length with no piers, the minimum

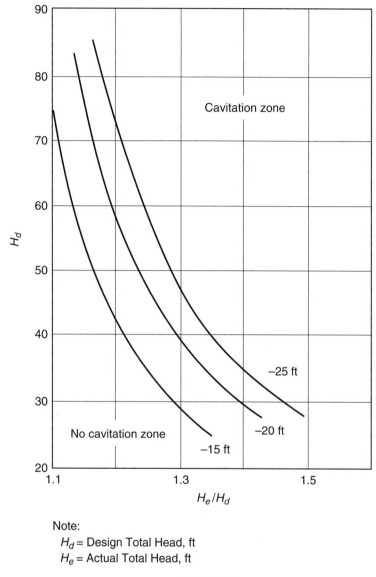

Cavitation zone

No cavitation zone

−15 ft

−20 ft

−25 ft

H_e/H_d

Note:
H_d = Design Total Head, ft
H_e = Actual Total Head, ft

(b) With Piers

FIGURE 6.5 (*continued*)

pressure on the crest, and the discharge at the design head for the standard WES ogee spillway. The height of the spillway crest, P, is 60 ft (18 m).

Solution.   For this example, use the design procedure of setting the ratio of the maximum head to design head to the value 1.33, so the design head $H_d = 64/1.33 = 48$ ft (14.6 m). Also calculate the ratio $P/H_d = 60/48 = 1.25$. Then, from Figure 6.2 for the standard WES high-overflow spillway (compound circular curve for upstream crest), the value of $C/C_0 = 1.02$ and $C = 1.02 \times 4.03 = 4.11$. Now, the required crest length is

$$L = \frac{Q_{max}}{CH_e^{3/2}} = \frac{200,000}{4.11 \times (64)^{3/2}} = 95 \text{ ft (29 m)}$$

From Figure 6.3a for $H_e/H_d = 1.33$, the minimum pressure head is $-0.43H_d$, so $p_{min}/\gamma = -20.6$ ft (-6.3 m), which is an acceptable value. However, if a less negative pressure head is desired, the value of H_e/H_d can be adjusted. Now the shape of the spillway crest is designed for the design head, H_d, of 48 ft (14.6 m). For example, the shape of the downstream portion of the crest with X and Y in feet is given by

$$X^{1.85} = 2.0H_d^{0.85}Y = (2.0 \times 48^{0.85})Y = 53.71Y$$

The discharge at the design head will have a different discharge coefficient as obtained from Figure 6.2. For $H_e/H_d = 1.0$, $C = 4.01$ and the design discharge, Q_d, is given by

$$Q_d = 4.01 \times 95 \times 48^{3/2} = 127,000 \text{ cfs } (3,600 \text{ m}^3/\text{s})$$

To design this spillway for an elliptical crest, the discharge coefficient is taken from Figure 6.4b, and the minimum pressure is determined, or specified, using Figure 6.5a.

For large reservoirs and under conditions where more control is required on the lake water surface elevation, the spillway crest overflow may be controlled by gates. The gates allow the choice of spillway crest elevation to be lower and the crest length to be shorter. For flood control, the gates can be closed to provide temporary storage of incoming flood flow and then gradually opened to release flow over the spillway after the flood at a much lower rate.

The radial or Tainter gate is widely used on ogee spillway crests as an underflow gate. As shown schematically in the inset of Figure 6.6, it is a segment of a cylinder that pivots radially on trunnions attached to the piers. A skin on the cylindrical surface is supported by a framework of structural members and radial struts that transmit the resultant hydrostatic force to the trunnions so that the required lift force on the gate is primarily the weight of the gate itself. The gate seat may be at the crest or downstream of the crest where the induced negative pressures are less. Vertical lift gates are also in use on spillways; however, they require gate slots in the piers that may induce cavitation, and they are more susceptible to hydrodynamic vibrations. In addition, the hydrostatic force is transmitted to the rollers in the gate slots, which increases the lifting force required.

As shown in the inset of Figure 6.6, the Tainter gate opening, w, gives rise to orifice flow under the gate in which the discharge is proportional to $H^{1/2}$ as opposed to $H^{3/2}$ in weir flow, where H is the head on the spillway. The geometry of the Tainter gate when the gate lip is downstream of the spillway crest makes measurement of the gate opening height more difficult. In Figure 6.6, it is defined as the minimum distance between the gate lip and the spillway surface for a given gate position. The angle β between the tangent through the point of minimum opening distance on the spillway surface and the tangent to the gate lip is used to define the gate position. In fact, the coefficient of discharge in the head-discharge relationship for orifice

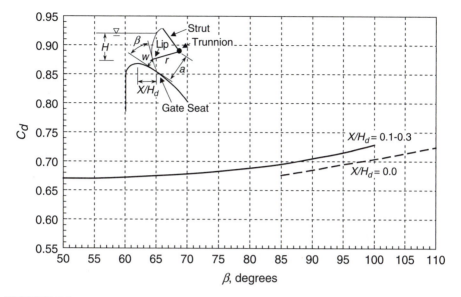

FIGURE 6.6

Coefficient of discharge for a radial gate with opening of w and head H measured to the center-line of the gate opening with H_d = spillway design head (U.S. Army Corps of Engineers 1990).

flow is given as a function of β in Figure 6.6 (U.S. Army Corps of Engineers 1990). The orifice flow rate through the gate opening, Q_G, is given by

$$Q_G = C_d L w \sqrt{2gH} \qquad (6.3a)$$

in which H = the head measured as the vertical distance from the water surface to the center of the gate opening; w = gate opening distance; L = crest length of the gate bay; and C_d = discharge coefficient from Figure 6.6 as a function of β. Once the tangent to the spillway face has been determined, the value of $\beta = \cos^{-1}[(a - w)/r]$ in which a = the perpendicular distance from the trunnion to the spillway tangent; and r = radius of the strut. If the gate seat is located on the spillway crest, then the tangent to the spillway face is just a horizontal line, and both w and a are measured perpendicular to it.

Vertical spillway gates may also have the gate seat located downstream of the spillway crest as for the radial gate in Figure 6.6, but the gate opening height, w, is measured vertically from the gate lip to the gate seat on the spillway. For this case, the U.S. Army Corps of Engineers (1990) has correlated discharge data from model studies of several large gated spillways in the form:

$$\frac{Q_G}{Q} = \frac{H_2^{3/2} - H_1^{3/2}}{H^{3/2}} \qquad (6.3b)$$

in which Q_G = discharge over the gated spillway; Q = free flow discharge without gates; H = reservoir head on the spillway crest used to calculate Q for free flow over the spillway; H_2 = reservoir head on the gate seat; and $H_1 = (H_2 - w)$ = head on the gate lip where w = gate opening height. Vischer and Hager (1998) have given a refined version of Equation 6.3b based on experiments on a vertical gate at the crest of the standard WES spillway shape (Hager and Bremen 1988). They present the results in terms of the design head H_d on the spillway:

$$\frac{Q_G}{Q_d} = \left[\left(\frac{H_2}{H_d}\right)^{3/2} - \left(\frac{H_1}{H_d}\right)^{3/2}\right]\left(\frac{1}{6} + \frac{w}{H_d}\right)^{1/9} \qquad (6.3c)$$

in which Q_d = design discharge corresponding to the design head for the ungated spillway; H_2 = head on the crest of the gated spillway; w = gate opening height; and $H_1 = H_2 - w$. Equation 6.3c is limited to values of $w/H < 0.8$ and $H/H_d < 2.5$ where H is the head on the spillway crest for calculation of the free discharge, Q. As values of w/H become very small, it can be shown that Equation 6.3c reduces to the well-known orifice discharge equation of the same form as Equation 6.3a (Hager and Bremen 1988).

Alternatively, a simple energy analysis of a vertical sluice gate in a horizontal, rectangular open channel can be adapted to a gate on the spillway crest where for unseparated flow, the spillway crest can be regarded locally as a horizontal channel (Henderson 1966; Chadwick and Morfett 1998). The energy equation is written from just upstream of the vertical gate where the flow depth is y_1 to a point downstream of the gate at the point of maximum contraction of the underflow jet, where the depth is $y_2 = C_c w$, in which C_c = the contraction coefficient and w = the vertical gate opening height. By rearranging the energy equation in terms of a dimensionless discharge coefficient, C_d, making the substitution for y_2, and solving for the gated discharge, Q_G, the result is

$$Q_G = C_d bw\sqrt{2gy_1} \qquad (6.3d)$$

which is in the form of the standard orifice flow equation. The depth y_1 corresponds to the head on the spillway, H; b = the width of the channel or the length of the spillway crest; and C_d = the coefficient of discharge defined by

$$C_d = \frac{C_c}{\sqrt{1 + C_c(w/y_1)}} \qquad (6.3e)$$

in which $C_c = 0.61$ (Henderson 1966). Equation 6.3e is valid for $E_1/y_1 = 0.1$–0.5 in which E_1 is the specific energy of the approach flow. As w/y_1 approaches zero, note that the discharge coefficient approaches the theoretical orifice contraction coefficient of 0.61.

The most significant aspect of gated flow with respect to its hydraulic characteristics is that the rate of change of head with respect to discharge abruptly increases as the flow transitions from free weir flow on the spillway to orifice flow, and the head continues rising at this faster rate as the discharge increases. The transition manifests itself as a sudden increase in head from free flow to gated flow as the gate lip catches the water surface, or a sudden decrease in head as the water surface for the gated condition drops back to free uncontrolled weir flow. This transition is of particular interest in culvert design as will be discussed next.

Example 6.2

A standard WES ogee spillway crest with compound circular curves in the upstream quadrant and vertical crest gates has a height of 6.0 m, an effective crest length of 25 m, and a design head of 9.0 m. At an operating head of 6.0 m, calculate the discharge for a gate opening height of 2.0 m using Equations 6.3c and 6.3d.

Solution.   The discharge coefficient for the design head is needed in Equation 6.3c, so we have $P/H_d = 6.0/9.0 = 0.67$, and $H_e/H_d = 1.0$. Then from Figure 6.2, $C/C_0 = 0.98$, and $C = 0.98 \times 4.03 = 3.95$. Now we can convert from the English unit value of $C = 3.95(EN)$ to the dimensionless value of C_d:

$$C_d = \frac{3.95(EN)}{\left(\frac{2}{3}\sqrt{2g}\right)} = \frac{3.95}{0.667 \times \sqrt{64.4}} = 0.738$$

The design discharge for free flow, Q_d, in SI units becomes

$$Q_d = \frac{2}{3}\sqrt{2g}\ C_d L H_d^{3/2} = 0.667 \times \sqrt{19.62} \times 0.738 \times 25 \times 9.0^{3/2} = 1472 \text{ m}^3/\text{s}$$

or 51,960 cfs. Substituting into Equation 6.3c for a head of 6.0 m, we have

$$\frac{Q_G}{Q_d} = \left[\left(\frac{H_2}{H_d}\right)^{3/2} - \left(\frac{H_1}{H_d}\right)^{3/2}\right]\left(\frac{1}{6} + \frac{w}{H_d}\right)^{1/9} = \left[\left(\frac{6.0}{9.0}\right)^{3/2} - \left(\frac{6.0-2.0}{9.0}\right)^{3/2}\right]$$

$$\left(0.167 + \frac{2.0}{9.0}\right)^{1/9} = 0.223$$

so that $Q_G = 0.223 \times 1472 = 328 \text{ m}^3/\text{s}$ (11,580 cfs). Now for Equation 6.3d, first calculate the discharge coefficient from Equation 6.3e:

$$C_d = \frac{C_c}{\sqrt{1 + C_c(w/y_1)}} = \frac{0.61}{\sqrt{1 + 0.61(2.0/6.0)}} = 0.556$$

and substitute into Equation 6.3d to obtain

$$Q_G = C_d bw\sqrt{2gy_1} = 0.556 \times 25 \times 2.0 \times \sqrt{19.62 \times 6.0} = 302 \text{ m}^3/\text{s}$$

or 10,660 cfs. The difference between the two gated discharge estimates is approximately 8%, which is not unreasonable considering that Equation 6.3d was derived for a horizontal channel rather than a spillway crest.

6.3   SPILLWAY AERATION

Even though the shape of ogee spillways can be designed to minimize the risk of damage due to cavitation, small imperfections in the spillway surface sometimes can lead to localized acceleration and corresponding pressure drops that may be unacceptable. The cost of providing a spillway surface that is smooth enough or is strengthened by surface reinforcement may become prohibitive. This has given rise to the use of artificial aeration on very high spillways to introduce air at pressures close to atmospheric pressure near the spillway face, thus preventing cavitation.

The concept of artificial aeration has stimulated interest in self-aeration, in which the natural entrainment of air at the interface with the atmosphere leads to bulking of the flow with the commonly observed white-water appearance on the face of high spillways. Early work on natural surface aeration of spillways was done by Straub and Anderson (1960) in a 50 ft (15 m) long by 1.5 ft (0.46 m) wide flume with slope angles, θ, varying from 7.5° to 75°. A sluice gate was located at the flume entrance and adjusted to achieve uniform flow and aeration conditions. The air concentration distribution was measured and shown to have two distinct regions: a lower, bubbly mixture layer and an upper layer consisting primarily of spray. Because the depth becomes ill defined in aerated flow, Straub and Anderson used a reference depth, y_0, which was the uniform flow depth of nonaerated flow. It corresponded to a measured Chezy C value of 90.5 in English units for their experiments. The effective depth of water, y_w, which was defined by $\int_0^\infty (1 - C_a)dy$, in which C_a represents the point air concentration in volume of air per unit total volume, was related to the reference depth and mean air concentration, C_m, by the relation

$$\frac{y_w}{y_0} = 1.0 - 1.3(C_m - 0.25)^2 \tag{6.4}$$

The effective depth of water also could be defined in terms of continuity as q/V, in which q = flow rate per unit of width and V = mean velocity. The mean air concentration was determined from a best fit of the experimental data in terms of the slope of the spillway, $S\ (= \sin \theta)$, and the flow rate per unit of width, q:

$$C_m = 0.743 \log_{10}\left(\frac{S}{q^{1/5}}\right) + 0.876 \tag{6.5}$$

Equation 6.5 applies for a range of air concentrations from 0.25 to 0.75, and q has units of cubic feet per second per foot. For example, for a spillway slope of 75° and a flow rate per unit of width of 600 cfs/ft (56 m³/s/m), the mean air concentration would be 0.45 (or 45 percent), defined as the ratio of volume of air to total volume. The corresponding effective depth of water from Equation 6.4 would be 95 percent of the reference depth. The effective depth of water should be used in the momentum flux term in the momentum function for the design of a stilling basin at the base of the spillway (Henderson 1966). The hydrostatic force term in the momentum function for the aerated flow becomes $(y_w)^2/[2(1 - C_m)]$.

Whether the air concentration predicted by Equation 6.5 can be achieved depends on the the length of the spillway face. In general, the point of inception of surface air entrainment would not be expected to occur until the boundary layer had grown to the point of intersection with the free surface. Keller and Rastogi (1977) solved the boundary layer equations numerically on a standard Waterways Experiment Station spillway with a vertical upstream face to obtain values of the critical distance, x_c, for the length of the boundary layer measured from the crest. Wood,

Ackers, and Loveless (1983) developed an empirical formula for x_c from a multiple regression analysis of Keller and Rastogi's results:

$$\frac{x_c}{k_s} = 13.6\left[\frac{q}{\sqrt{gk_s^3}}\right]^{0.713}\frac{1}{S^{0.277}} \tag{6.6}$$

in which S = spillway slope = $\sin\theta$; q = flow rate per unit of width; and k_s = roughness height for the spillway surface. From this equation, we can conclude that the distance required for inception of surface air entrainment depends primarily on the slope of the spillway and the flow rate per unit of width. For a concrete surface roughness height of 0.005 ft (0.0015 m), and for a spillway having q = 600 cfs/ft (56 m³/s/m) and θ = 75°, as in the previous example, the length of spillway required for self-aeration to commence would be approximately 550 ft (168 m), which corresponds to a spillway height of 531 ft (162 m).

For some spillways, even though they are high enough for self-aeration, surface air entrainment may be insufficient to prevent cavitation on the face of the spillway, especially near the crest, where it may not occur at all. Under these circumstances, aeration ramps have been used to induce an air cavity that allows entrainment of air near atmospheric pressure on the underside of the jet coming off the aeration ramp. A sketch of a typical air ramp is shown in Figure 6.7, in which the air is supplied to the air cavity from the atmosphere through lateral wedges at the edge of the spillway chute or through recesses or ducts underneath the ramp that are fed by chimneys. Turbulence causes disruption of the water surface on the underside of the nappe and air is dragged and entrained into the jet, which then is mixed with the flow downstream. The pressure in the cavity below the nappe will be slightly less than atmospheric because of head losses in the air delivery system, so that the trajectory and length of the jet will be different from that of a free jet.

With reference to Figure 6.7, a dimensional analysis of the problem leads to the following expression for the length of the jet, L, coming off the ramp:

$$\frac{L}{h} = f\left[\mathbf{F}, \frac{\Delta p_a}{\rho V^2}, \mathbf{Re}, \mathbf{We}, \text{ramp geometry}\right] \tag{6.7}$$

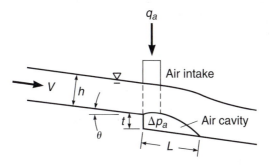

FIGURE 6.7
Definition sketch of a spillway air ramp.

in which h and V = approach flow depth and velocity, respectively, on the spillway chute; $\mathbf{F}$ = approach Froude number = $V/(gh)^{0.5}$; Δp_a = pressure drop in the air cavity relative to atmospheric pressure; $\mathbf{Re}$ = Reynolds number = Vh/ν; and $\mathbf{We}$ = $V/(\sigma/\rho L)^{0.5}$. The Reynolds number and Weber number effects tend to be small in the prototype spillway, so that for a fixed ramp geometry, the primary variables of interest are the Froude number and the subatmospheric pressure difference. It has been suggested from tests of prototype spillways that the air flow per unit of width of spillway $q_a = kVL$, where k is a constant of proportionality (de S. Pinto 1988). It follows then that

$$\frac{q_a}{q} = C_m = k\frac{L}{h} \tag{6.8}$$

The discharge ratio on the left-hand side of Equation 6.8 is equivalent to the air concentration, C_m, as shown, which should be 5–10 percent to prevent cavitation damage, based on past experience (de S. Pinto 1988). Thus, provided the constant k is known from prototype experience, the required value of L/h can be determined from Equation 6.8 for the desired air concentration. Then, from the relationship given by Equation 6.7 from physical model studies or numerical analysis of the jet trajectory for a given ramp geometry, the required underpressure Δp_a can be determined for the specified value of L/h and the known value of the Froude number. Finally, the air delivery system can be designed to provide the air flow rate with the specified pressure drop.

The value of k in Equation 6.8 has been determined to be 0.033 from the Foz do Areia prototype spillway tests (de S. Pinto 1988), but it can vary for different flow conditions and different ramp geometries. What is required is a model study with a relatively large scale (1:10 to 1:15) to eliminate Reynolds number and Weber number effects and so determine specific design values of k.

6.4   STEPPED SPILLWAYS

Stepped spillways have been used extensively around the world since antiquity, but they became very popular in the past few decades with the advent of roller-compacted concrete (RCC) and gabion construction of dams (Chanson 1994a). They provide good surface aeration but also increase the energy dissipation in the flow down the spillway in comparison to a smooth spillway. This latter feature of stepped spillways may reduce the cost of the downstream stilling basin.

Stepped spillways can operate either in a nappe flow regime or a skimming flow regime. In nappe flow, which tends to occur at lower discharges on flatter spillways, the flow consists of a series of jets that strike the floor of the succeeding steps. Each jet usually is followed by a partial hydraulic jump. In skimming flow, the jets move smoothly without breakup across the steps, which act as a series of roughness elements. A recirculating vortex forms on each step in which energy dissipates. The skimming flow regime is shown in Figure 6.8. Rajaratnam (1990) suggested that the onset of skimming flow occurs for values of y_c/h exceeding 0.8,

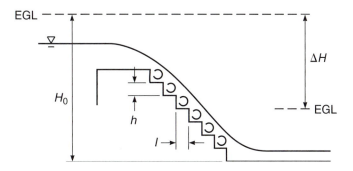

FIGURE 6.8
Definition sketch of a stepped spillway.

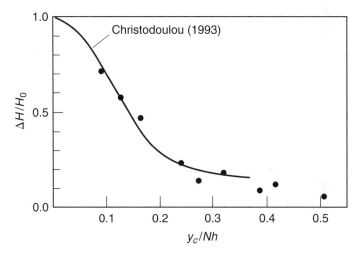

FIGURE 6.9
Model study results for head loss on a stepped spillway in skimming flow with N steps (Rice and Kadavy 1996). (*Source: C. E. Rice and K. C. Kadavy. "Model Study of a Roller Compacted Concrete Stepped Spillway," J. Hydr. Engrg., © 1996, ASCE. Reproduced by permission of ASCE.*)

where y_c is the critical depth for the flow on the spillway and h is the height of an individual step.

The amount of energy dissipation that occurs on a stepped spillway for skimming flow is one of the primary design variables. Christodoulou (1993) suggested that the energy head dissipated, ΔH, in ratio to the total head, H_0, upstream of the dam relative to the toe is related to y_c/Nh, in which y_c = critical depth; N = number of steps; and h = the height of each step, as shown in Figure 6.9. Rice and Kadavy (1996) have confirmed the validity of Figure 6.9 for a physical model of the Salado Creek spillway in Texas. Based on their data points, the Christodoulou curve in Figure 6.9 is valid for l/h values in the range of 0.7 to 2.5, $y_c/h \leq 4.5$, and $y_c/Nh \leq 0.5$.

Chanson (1994b) analyzed experimental data for stepped spillways from a large number of investigators and compared the results for relative energy loss with an analytical formulation for uniform flow conditions given by

$$\frac{\Delta H}{H_0} = 1 - \frac{C_f^{1/3} \cos \theta + 0.5 C_f^{-2/3}}{1.5 + \dfrac{H_{dam}}{y_c}} \tag{6.9}$$

in which $C_f = f/(8 \sin \theta)$; f = friction factor; $\theta = \tan^{-1} (h/l)$; H_{dam} = dam crest height above the toe; and y_c = critical flow depth. He found reasonable agreement with the experimental results, considering the degree of scatter, using $f = 1.0$ (non-aerated flow) and $\theta = 52°$ over a very wide range of H_{dam}/y_c from approximately 2 to 90. Usually, $H_{dam} = Nh$, so Equation 6.9 corresponds with the variables of Figure 6.9 except that it covers a wider range in y_c/Nh. Equation 6.9 must be used with care because of the uncertainty in the friction factor due to the effects of aeration.

Stepped spillways offer the advantage of enhanced air entrainment as well as energy dissipation. Chanson (1994b) shows that the inception of air entrainment occurs in a shorter distance on a stepped spillway than on a smooth spillway because of the more rapid rate of boundary-layer growth. However, the equilibrium air concentration is similar on stepped and smooth spillways and primarily is a function of slope. For more details on the design of stepped spillways, refer to the comprehensive treatment of the subject by Chanson (1994b).

6.5   CULVERTS

Culverts seem to be simple hydraulic structures but in fact are among the most complicated because of the wide variety of flow conditions that can occur in them. Flow can be gradually varied or rapidly varied and also a function of time. A culvert can flow full, in which case it operates under pressure-flow conditions as in pipe flow, or it can flow partly full, as an open channel. The open channel flow can be supercritical or subcritical, and its analysis may include computation of a gradually varied flow profile or a hydraulic jump. Culverts flow full when the outlet is submerged due to high tailwater but also may flow full for a very high headwater with the outlet unsubmerged. In both full and partly full flow, the submergence of the inlet or outlet is an important criterion in determining the type of flow that occurs. Perhaps the most important distinguishing characteristic of a culvert flow is whether it is under inlet or outlet control. In the case of inlet control, the head-discharge relation is determined entirely by the inlet geometry, including the inlet area, edge rounding, and shape. Tailwater conditions are immaterial for inlet control. In outlet control, on the other hand, the head-discharge relation is affected not only by the inlet but also by the barrel roughness, length, slope, shape, and area as well as the tailwater elevation. These influences on inlet and outlet control are summarized in Table 6.1. Inlet control generally occurs for short, steep culverts with a free outlet, while outlet control prevails for long, rough-barreled culverts with high tailwater conditions.

TABLE 6.1   Factors influencing culvert performance

Factor	Inlet Control	Outlet Control
Headwater elevation	x	x
Inlet area	x	x
Inlet edge configuration	x	x
Inlet shape	x	x
Barrel roughness		x
Barrel area		x
Barrel shape		x
Barrel length		x
Barrel slope		x
Tailwater elevation		x

Source: Data from Federal Highway Administration (2001).

Culvert design usually is based on the selection of a design discharge determined from frequency analysis. Interstate highway culverts, for example, may be designed to carry the 100 year peak discharge. The culvert is sized to limit the headwater resulting from the design discharge to a specified value to prevent overtopping the highway embankment. Once the design culvert size is determined, its performance may be analyzed over a wide range of discharges, including discharges that over-top the embankment. This analysis can be summarized by a plot of the complete head-discharge relation, called the *performance curve*. This step is important to accurately determine whether the culvert operates under inlet or outlet control for the design discharge. The design process is based on a selected peak discharge in steady flow, and a conservative approach is taken in which both inlet and outlet control head-discharge relationships are checked to determine the limiting control. The higher head resulting either from inlet or outlet control is compared with the allowable headwater elevation. If, at the design headwater as shown in Figure 6.10, for example, the inlet-control discharge, Q_{IC}, is less than the outlet-control discharge, Q_{OC}, then the inlet capacity is less than the barrel capacity, and the inlet controls the head-discharge relation at the design condition. This is the same as choosing the higher head for a given discharge, as can be seen in Figure 6.10. As the head increases in Figure 6.10, the culvert remains in inlet control until the intersection between the inlet-control and outlet-control curves, beyond which it is assumed to be in outlet control.

The head-discharge relationship of a culvert follows well-known hydraulic behavior. The culvert may act as a weir, an orifice, or a pipe in pressure flow. For an unsubmerged inlet, the culvert operates as a weir at the inlet and the discharge is proportional to the head to the $\frac{3}{2}$ power. If the inlet is submerged and the culvert is in inlet control, orifice flow occurs and the discharge is proportional to the head to the $\frac{1}{2}$ power. This means that the head increases more rapidly with an increase in discharge than for weir flow. In pressure flow, the head-discharge relation is determined

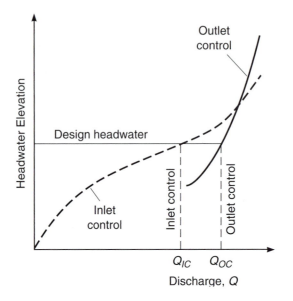

FIGURE 6.10
Culvert performance curves for the determination of inlet or outlet control (Federal Highway Administration 2001).

by the effective head, which is the difference in total head between the headwater and tailwater.

The U.S. Geological Survey (Bodhaine 1976) classifies culvert flow into six types, depending primarily on the headwater and tailwater levels and whether the slope is mild or steep. These types of flow also have been given by French (1985), but Chow (1959) used a different numbering system for the same six types of flow. Additional types of culvert flow can be identified; however, a simpler classification depends only on the type of hydraulic head-discharge relationship. In this classification, the most important criteria are whether the culvert is in inlet or outlet control and whether the inlet is submerged or unsubmerged. Submergence of the inlet occurs when the ratio of inlet head to height of the culvert, HW/d, is in the range of 1.2 to 1.5, with the latter value usually taken as the submergence criterion. Inlet head, HW, is defined as the height of the headwater above the invert of the culvert inlet, as shown in Figure 6.11.

Inlet Control

Several types of inlet control are illustrated in Figure 6.11. In Figure 6.11a, both the inlet and outlet are unsubmerged on a steep slope. Flow passes through the critical depth at the inlet and the downstream flow is supercritical (S2 curve) as it approaches normal depth. This is U.S. Geological Survey (USGS) Type 1 flow. The outlet is submerged in Figure 6.11b, which forces a hydraulic jump in the barrel. As long as the

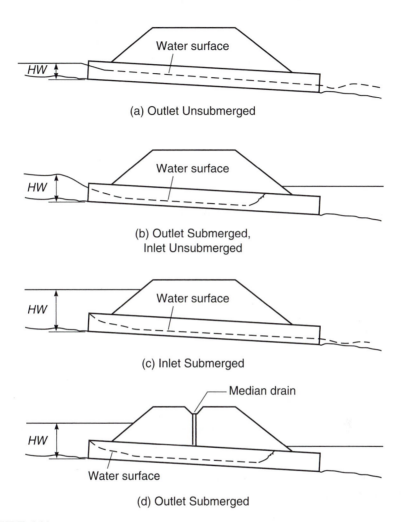

FIGURE 6.11
Types of inlet control (Federal Highway Administration 2001).

tailwater is not high enough to move the jump upstream to the inlet, the culvert remains in inlet control; that is, the head-discharge relationship does not change. In Figure 6.11c, the inlet is submerged and the outlet is unsubmerged. Critical depth occurs just downstream of the inlet, but the culvert is in orifice flow (USGS Type 5). Both the inlet and outlet are submerged in Figure 6.11d, and a vent must be provided to prevent an unstable flow situation, which oscillates between full flow and partly full flow. With the vent in place and the hydraulic jump remaining downstream of the culvert entrance, this remains inlet control with orifice flow at the entrance.

The head-discharge relationships for inlet control are based on either weir flow for an unsubmerged inlet or orifice flow for a submerged inlet. In other words, only two types of flow occur in inlet control in terms of the type of head-discharge relationship that governs: (1) inlet submerged and orifice flow, which we refer to here as Type IC-1; and inlet unsubmerged on a steep slope with weir flow, which is called Type IC-2. The head-discharge relation for weir flow (IC-2) is derived from the energy equation written from the headwater to the critical depth section, neglecting the approach velocity head:

$$HW = y_c + (1 + K_e)\frac{Q^2}{2gA_c^2} \qquad (6.10a)$$

in which HW = head above the invert of the culvert inlet; y_c = critical depth; A_c = flow area corresponding to critical depth; and K_e = entrance loss coefficient. An additional equation is needed to eliminate the critical depth, and it comes from the condition of setting the Froude number equal to unity. Equation 6.10a can be rearranged to solve for the discharge, Q:

$$Q = C_d A_c \sqrt{2g(HW - y_c)} \qquad (6.10b)$$

or it can be placed in the form of a weir equation (see Chapter 2). Note that the coefficient of discharge $C_d = 1/(1 + K_e)^{1/2}$. The USGS (Bodhaine 1976) developed values for the coefficient C_d as a function of the head to diameter ratio, HW/d, for circular culverts. For pipe culverts with a square edge in a vertical headwall, $C_d = 0.93$ for $HW/d < 0.4$, and it decreases to 0.80 at $HW/d = 1.5$, where the entrance becomes submerged. The coefficient C_d can be corrected for bevels and rounding of the entrance edge. For a standard 45° bevel with the ratio of bevel height to culvert diameter $w/d = 0.042$, the correction to the coefficient C_d is approximately 1.1. For machine tongue-and-groove reinforced concrete pipe from 18 to 36 in. in diameter, the value of $C_d = 0.95$ with no systematic variation found between C_d and HW/d. For box culverts set flush in a vertical headwall, the value of $C_d = 0.95$ for USGS Type 1 flow (IC-2).

Once the inlet is submerged (Type IC-1), the governing hydraulic equation is the orifice-flow equation given as

$$Q = C_d A_o \sqrt{2g(HW)} \qquad (6.11)$$

in which C_d = coefficient of discharge; A_o = cross-sectional area of inlet; and HW = head on the inlet invert of the culvert. Some values of C_d for orifice flow are given in Table 6.2 for various degrees of rounding with radius r and for bevels of height w as a function of HW/d. The purpose of bevels or rounding is to reduce the flow contraction at the inlet of the culvert to obtain a higher discharge coefficient. The FHWA (Federal Highway Administration) developed head-discharge relationships for inlet control using bevels of 45° or 33.7° with w/b or $w/d = 0.042$ and 0.083, respectively, where w is the height of the bevel; b is the height of a box culvert; and d is the diameter of a circular culvert. The 45° bevel is recommended for ease of construction (Federal Highway Administration 2001). From Table 6.2, we see that these two standard bevels increase the discharge coefficient by approximately 10 to 20 percent in comparison with a square-edge inlet ($r = 0$; $w = 0$). For a grooved-end concrete pipe culvert, bevels are unnecessary, because the groove gives about the same improvement in the discharge coefficient.

TABLE 6.2   Orifice discharge coefficients for culverts $[Q = C_d A_o (2g\ HW)^{1/2}]$

HW/d	r/b, r/d; w/b, w/d						
	0.0	0.02	0.04	0.06	0.08	0.10	0.14
1.4	0.44	0.46	0.49	0.50	0.50	0.51	0.51
1.5	0.46	0.49	0.52	0.53	0.53	0.54	0.54
1.6	0.47	0.51	0.54	0.55	0.55	0.56	0.56
1.7	0.48	0.52	0.55	0.57	0.57	0.57	0.57
1.8	0.49	0.54	0.57	0.58	0.58	0.58	0.58
1.9	0.50	0.55	0.58	0.59	0.60	0.60	0.60
2.0	0.51	0.56	0.59	0.60	0.61	0.61	0.62
2.5	0.54	0.59	0.62	0.64	0.64	0.65	0.66
3.0	0.55	0.61	0.64	0.66	0.67	0.69	0.70
3.5	0.57	0.62	0.65	0.67	0.69	0.70	0.71
4.0	0.58	0.63	0.66	0.68	0.70	0.71	0.72
5.0	0.59	0.64	0.67	0.69	0.71	0.72	0.73

Source: Data from Bodhaine (1976).

Between the unsubmerged and submerged portions of the inlet control head-discharge equations, a smooth transition curve connects the two. Based on extensive experimental results obtained by the National Bureau of Standards, best-fit power relationships have been obtained for both the unsubmerged and submerged portions of the inlet control head-discharge relationship. For the inlet unsubmerged, two forms of the equation are recommended:

$$\frac{HW}{d} = \frac{E_c}{d} + K\left[\frac{Q}{Ad^{0.5}}\right]^M - 0.5S \qquad (6.12a)$$

$$\frac{HW}{d} = K\left[\frac{Q}{Ad^{0.5}}\right]^M \qquad (6.12b)$$

in which HW = head above invert of culvert inlet in feet; E_c = minimum specific energy in feet; d = height of culvert inlet in feet; Q = design discharge in cubic feet per second; A = full cross-sectional area of barrel in square feet; S = culvert barrel slope in feet per foot; and K, M = constants for different types of inlets from Table 6.3. Equation 6.12a is Form 1 and preferred; Equation 6.12b is Form 2, which is used more easily. For the inlet submerged, the best-fit power relationship is of the form

$$\frac{HW}{d} = c\left[\frac{Q}{Ad^{0.5}}\right]^2 + Y - 0.5S \qquad (6.13)$$

in which c and Y are constants obtained from Table 6.3 for Q, A, and d in English units as for Equations 6.12. Equations 6.12 apply up to values of $Q/(Ad^{0.5}) = 3.5$, while Equation 6.13 is valid for $Q/(Ad^{0.5}) \geq 4.0$. Inlet control nomographs based on

TABLE 6.3 Constants for inlet control design equations

Shape and Material	Inlet Edge Description	Equation Form	Unsubmerged		Submerged	
			K	M	c	Y
Circular concrete	Square edge with headwall	1	0.0098	2.0	0.0398	0.67
	Groove end with headwall		0.0018	2.0	0.0292	0.74
	Groove end projecting		0.0045	2.0	0.0317	0.69
Circular CMP (corrugated metal pipe)	Headwall	1	0.0078	2.0	0.0379	0.69
	Mitered to slope		0.0210	1.33	0.0463	0.75
	Projecting		0.0340	1.5	0.0553	0.54
Circular	Beveled ring, 45°	1	0.0018	2.5	0.0300	0.74
	Beveled ring, 33.7°		0.0018	2.5	0.0243	0.83
Rectangular box	30–75° wingwall flares	1	0.026	1.0	0.0347	0.86
	90° and 15° wingwall flares		0.061	0.75	0.0400	0.80
	0° wingwall flares		0.061	0.75	0.0423	0.82
Rectangular box	45° wingwall flare, $w/d = 0.043$	2	0.510	0.667	0.0309	0.80
	18° to 33.7° wingwall flare, $w/d = 0.083$		0.486	0.667	0.0249	0.83
Rectangular box	90° headwall, $\frac{3}{4}$ in. chamfers	2	0.515	0.667	0.0375	0.79
	90° headwall, 45° bevels		0.495	0.667	0.0314	0.82
	90° headwall, 33.7° bevels		0.486	0.667	0.0252	0.865
Rectangular box	$\frac{3}{4}$ in. chamfers, 45° skewed headwall	2	0.545	0.667	0.04505	0.68
	$\frac{3}{4}$ in. chamfers, 30° skewed headwall		0.533	0.667	0.0425	0.705
	$\frac{3}{4}$ in. chamfers, 15° skewed headwall		0.522	0.667	0.0402	0.73
	45° bevels, 10°–45° skewed headwall		0.498	0.667	0.0327	0.75
Rectangular box, $\frac{3}{4}$ in. chamfers	45° wingwall flare, nonoffset	2	0.497	0.667	0.0339	0.803
	18.4° wingwall flare, nonoffset		0.493	0.667	0.0361	0.806
	18.4° wingwall flare, nonoffset, 30° skew		0.495	0.667	0.0368	0.71
Rectangular box, top bevels	45° wingwall flare, offset	2	0.497	0.667	0.0302	0.835
	33.7° wingwall flare, offset		0.495	0.667	0.0252	0.881
	18.4° wingwall flare, offset		0.493	0.667	0.0227	0.897
CM (corrugated metal) boxes	90° headwall	1	0.0083	2.0	0.0379	0.69
	Thick wall projecting		0.0145	1.75	0.0419	0.64
	Thin wall projecting		0.0340	1.5	0.0496	0.57

Horizontal ellipse concrete	Square edge with headwall	1	0.0100	2.0	0.0398	0.67
	Groove end with headwall		0.0018	2.5	0.0292	0.74
	Groove end projecting		0.0045	2.0	0.0317	0.69
Vertical ellipse concrete	Square edge with headwall	1	0.010	2.0	0.0398	0.67
	Groove end with headwall		0.0018	2.5	0.0292	0.74
	Groove end projecting		0.0095	2.0	0.0317	0.69
Pipe arch, 18 in. corner radius CM	90° headwall	1	0.0083	2.0	0.0379	0.69
	Mitered to slope		0.0300	1.0	0.0463	0.75
	Projecting		0.0340	1.5	0.0496	0.57
Pipe arch, 18 in. corner radius CM	Projecting	1	0.0300	1.5	0.0496	0.57
	No bevels		0.0088	2.0	0.0368	0.68
	33.7° bevels		0.0030	2.0	0.0269	0.77
Pipe arch, 31 in. corner radius CM	Projecting	1	0.0300	1.5	0.0496	0.57
	No bevels		0.0088	2.0	0.0368	0.68
	33.7° bevels		0.0030	2.0	0.0269	0.77
Arch CM	90° headwall	1	0.0083	2.0	0.0379	0.69
	Mitered to slope		0.0300	1.0	0.0463	0.75
	Thin wall projecting		0.0340	1.5	0.0496	0.57
Circular	Smooth tapered inlet throat	2	0.534	0.555	0.0196	0.90
	Rough tapered inlet throat		0.519	0.64	0.0210	0.90
Elliptical inlet face	Tapered inlet, beveled edges	2	0.536	0.622	0.0368	0.83
	Tapered inlet, square edges		0.5035	0.719	0.0478	0.80
	Tapered inlet, thin edge projecting		0.547	0.80	0.0598	0.75
Rectangular	Tapered inlet throat	2	0.475	0.667	0.0179	0.97
Rectangular concrete	Side tapered, less favorable edge	2	0.56	0.667	0.0446	0.85
	Side tapered, more favorable edge		0.56	0.667	0.0378	0.87
Rectangular concrete	Slope tapered, less favorable edge	2	0.50	0.667	0.0446	0.65
	Slope tapered, more favorable edge		0.50	0.667	0.0378	0.71

Source: Data from Federal Highway Administration (2001).

Equations 6.12 and 6.13 have been developed for manual culvert design and can be found in HDS-5 (Federal Highway Administration 2001). A full inlet control curve can be developed graphically by connecting Equations 6.12 and 6.13 with smooth curves in the transition region. For computer applications, polynomial regression has been applied to obtain best-fit relationships for the inlet control curve of the form

$$\frac{HW}{d} = A + BX + CX^2 + DX^3 + EX^4 + FX^5 - C_e S \qquad (6.14)$$

in which C_e = slope correction coefficient; S = culvert slope; and $X = Q/(Bd^{3/2})$, where Q = discharge in one barrel; B = culvert span of one barrel; and d = culvert height. The polynomial and slope correction coefficients are available in Federal Highway Administration 2008.

Outlet Control

Types of outlet control are shown in Figure 6.12. Flow condition (a) is the classic full-pipe flow, in which pressure flow occurs throughout the barrel. In flow condition (b), the outlet is submerged but the inlet is unsubmerged for low values of headwater because of the flow contraction at the inlet. The outlet is unsubmerged in flow condition (c), but the culvert still flows full due to a high headwater. In flow condition (d), the outlet not only is unsubmerged, the barrel flows partly full near the outlet and passes through critical depth there. Finally, in flow condition (e), both the inlet and outlet are unsubmerged and we have open channel flow that is subcritical on a mild slope. Flow conditions (a) and (b) are USGS flow Type 4, while conditions (c) and (d) can be considered USGS flow Type 6. Flow condition (e) for open channel flow is either USGS flow Type 2 or 3, depending on whether the downstream control is critical depth (M2 profile) or a tailwater greater than critical depth (M1 or M2 profile), respectively. As shown next, flow conditions (a), (b), (c), and (d) all can be treated as full flow with some adjustment for condition (d). Hence, we refer to these flow types here as OC-1 for outlet control with submerged inlet. Flow condition (e), on the other hand, has an unsubmerged inlet, so it is classified OC-2.

Flow conditions (a), (b), and (c) (Type OC-1) all are governed by the energy equation written from the headwater to the tailwater:

$$HW = TW - S_0 L + \left(1 + K_e + f\frac{L}{4R}\right)\frac{Q^2}{2gA^2} \qquad (6.15)$$

in which TW = tailwater depth relative to the outlet invert; S_0 = culvert slope; L = culvert length; K_e = entrance loss coefficient; f = Darcy-Weisbach friction factor; R = full-flow hydraulic radius; A = culvert cross-sectional area; and Q = culvert discharge. This equation can be rearranged and written in the form

$$Q = A\sqrt{\frac{2g(HW - TW + S_0 L)}{1 + K_e + f\frac{L}{4R}}} \qquad (6.16)$$

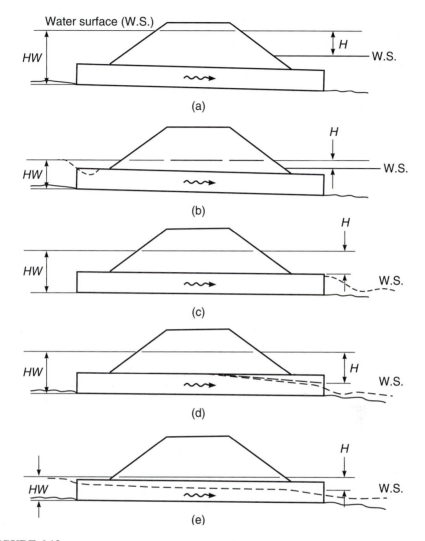

FIGURE 6.12
Types of outlet control (Federal Highway Administration 2001).

The term in parentheses in the numerator on the right hand side of Equation 6.16 is called the *effective head*, H_{eff}, because it is the difference in elevations of the headwater and tailwater. Outlet control nomographs based on H_{eff} (H in Figure 6.12) can be found in HDS-5 (Federal Highway Administration 2001). It should be emphasized that the only reason why the culvert slope appears in Equations 6.15 and 6.16 is because of the definition of the head, *HW*, relative to the invert of the culvert inlet. As long as the effective head is the same, the full-flow discharge through a culvert of specified length will be the same regardless of the barrel slope.

TABLE 6.4   Recommended Manning's n values for selected conduits

Type of conduit	Wall and joint description	Manning's n
Concrete pipe	Good joints, smooth walls	0.011–0.013
	Good joints, rough walls	0.014–0.016
	Poor joints, rough walls	0.016–0.017
Concrete box	Good joints, smooth finished walls	0.012–0.015
	Poor joints, rough, unfinished walls	0.014–0.018
Corrugated metal pipes and boxes, annular corrugations	$2\frac{2}{3}$ by $\frac{1}{2}$ in. corrugations	0.027–0.022
	6 by 1 in. corrugations	0.025–0.022
	5 by 1 in. corrugations	0.026–0.025
	3 by 1 in. corrugations	0.028–0.027
	6 by 2 in. structural plate	0.035–0.033
	9 by $2\frac{1}{2}$ in. structural plate	0.037–0.033
Corrugated metal pipes, helical corrugations, full circular flow	$2\frac{2}{3}$ by $\frac{1}{2}$ in. corrugations, 24 in. plate width	0.012–0.024
Spiral rib metal pipe	$\frac{3}{4}$ by $\frac{3}{4}$ in. recesses at 12 in. spacing, good joints	0.012–0.013

Source: Data from Federal Highway Administration (2001).

The head-loss term in Equation 6.16 sometimes is written in terms of Manning's equation instead of the Darcy-Weisbach equation, in which $fL/4R$ is replaced as follows:

$$f\frac{L}{4R} = \frac{2gn^2L}{K_n^2 R^{4/3}} \tag{6.17}$$

in which n = Manning's n value for full flow, and K_n = 1.0 for SI units and 1.49 for English units, as in Chapter 4. Typical values of Manning's n for culverts are shown in Table 6.4.

Values of the entrance loss coefficient for outlet control are given in Table 6.5. The value of K_e for a square edge in a headwall is 0.5, while for beveled edges and the groove end of concrete pipe culverts, K_e = 0.2. On box culverts with a square edge, a small reduction in K_e to a value of 0.4 is obtained for wingwalls at an angle of 30°–75° from the centerline of the barrel; otherwise, wingwalls have either no effect for concrete pipes or a detrimental effect if constructed parallel to the sides of a box culvert.

The flow condition (d) in Figure 6.12 actually requires computation of the subcritical flow profile from the outlet to the point where it intersects the crown of the culvert. Numerous backwater calculations by the FHWA, however, led to a simpler procedure for manual calculations. A full-flow hydraulic grade line is assumed to end at the outlet at a point halfway between the critical depth and the crown of the culvert, $(y_c + d)/2$, and is extended to the inlet as though full flow prevailed through the entire length of the culvert. Then the full-flow equation, Equation 6.15, can be used to calculate the head-discharge relation with TW replaced by $(y_c + d)/2$. If the tailwater is higher than $(y_c + d)/2$, then the actual tailwater depth is taken as the value

TABLE 6.5   Entrance loss coefficients: Outlet control, full or partly full entrance head loss, where $H_L = K_e \left(\dfrac{V^2}{2g} \right)$

Type of structure and design of entrance	Coefficient K_e
Pipe, concrete	
Projecting from fill, socket end (groove end)	0.2
Projecting from fill, square cut end	0.5
Headwall or headwall and wingwalls	
Socket end of pipe (groove end)	0.2
Square edge	0.5
Rounded (radius $= \frac{1}{12} d$)	0.2
Mitered to conform to fill slope	0.7
End section conforming to fill slope	0.5
Beveled edges, 33.7° or 45° bevels	0.2
Side- or slope-tapered inlet	0.2
Pipe, or pipe arch, corrugated metal	
Projecting from fill (no headwall)	0.9
Headwall or headwall and wingwalls, square edge	0.5
Mitered to conform to fill slope, paved or unpaved slope	0.7
End section conforming to fill slope	0.5
Beveled edges, 33.7° or 45° bevels	0.2
Side- or slope-tapered inlet	0.2
Box, reinforced concrete	
Headwall parallel to embankment (no wingwalls)	
Square edged on three edges	0.5
Rounded on three edges to radius of $\frac{1}{12}$ barrel dimension, or beveled	
edges on three sides	0.2
Wingwalls at 30°–75° to barrel	
Square edged at crown	0.4
Crown edge rounded to radius of $\frac{1}{12}$ barrel dimension, or beveled	
top edge	0.2
Wingwall at 10°–25° to barrel	
Square edged at crown	0.5
Wingwalls parallel (extension of sides)	
Square edged at crown	0.7
Side- or slope-tapered inlet	0.2

Source: Data from Federal Highway Administration (2001).

of *TW*. In computer programs such as HY8 (Federal Highway Administration 2008), the water surface profile for condition (d) is computed until it reaches the crown of the pipe, after which full-flow calculations are made. Thus, it is given a special Type 7 in addition to USGS Types 1 through 6, which are used in the program. Since it is a mixture of OC-1 and OC-2, as defined here, it should be given its own designation of OC-3 in HY8.

Outlet control condition (e) (Type OC-2) in Figure 6.12 requires the computation of a gradually varied flow profile from the outlet proceeding upstream to the culvert inlet. This will be either an M2 or an M1 profile. At the inlet, the velocity head and entrance losses from Table 6.5 are added to the inlet flow depth to obtain the upstream headwater, HW. The flow profile is computed in HY8 using the direct step method.

Road Overtopping

When the roadway overtops, the roadway embankment behaves like a broad-crested weir, as shown in Figure 6.13. The equation for a broad-crested weir is written for this case as

$$Q = C_w L (HW_r)^{3/2} \tag{6.18}$$

in which Q = overtopping discharge in cubic feet per second; C_w = weir discharge coefficient; L = length of roadway crest in feet; and HW_r = head on the roadway

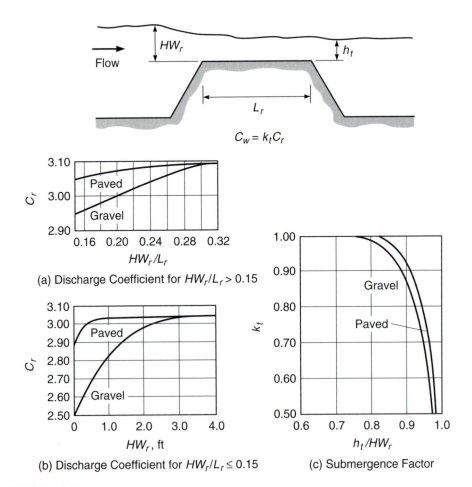

FIGURE 6.13
Discharge coefficients for roadway overtopping (Federal Highway Administration 2001).

crest in feet. Figure 6.13a gives the discharge coefficient for deep overtopping, and Figure 6.13b shows its value for shallow overtopping. The correction factor k_t in Figure 6.13c is for submergence of the weir by the tailwater. An iterative procedure has to be employed to determine the division of flow between the culvert and embankment overflow. Different headwater elevations are assumed until the sum of the culvert flow and embankment overflow equals the specified discharge.

Improved Inlets

When a culvert is in outlet control, only minimal improvements can be made to increase the discharge for a given headwater elevation. Beveling of the entrance reduces the entrance head loss, but the barrel friction loss is likely to be the dominant head loss. The barrel friction loss can be reduced by using culverts fabricated from materials having lower values of Manning's n, but this becomes an economic issue. On the other hand, a culvert that is in inlet control is amenable to considerable improvement in performance by design changes to the inlet itself.

The purpose of improved inlets is first to reduce the flow contraction, which increases the effective flow area as well as decreases the head loss that occurs in severe contractions. In addition, improved inlets can include a *fall*, or depression, that increases the head on the throat of the barrel, where the control section is located, for the same headwater elevation.

At the first level of inlet improvement, the inlet edges can be beveled. The degree of improvement can be seen in Figure 6.14, which is a set of inlet control

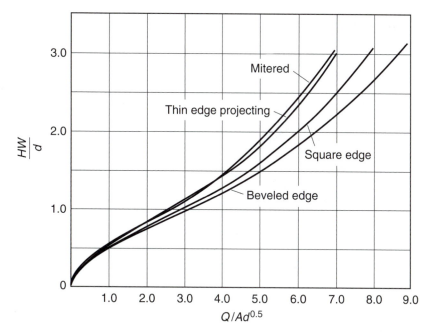

FIGURE 6.14
Inlet control curves—Circular or elliptical structural plate corrugated metal conduits with Q in cfs, A in ft^2, d in ft (Federal Highway Administration 2001).

curves for different entrance conditions constructed from Equations 6.12 and 6.13 for a circular or elliptical structural plate corrugated metal conduit. The maximum increase in discharge at $HW/d = 3.0$ due to beveling is about 20 percent in comparison to a thin edge projecting inlet.

The next level of improvement is the side-tapered inlet shown in Figure 6.15. The side-tapered inlet has an enlarged face section with a 4:1 to 6:1 side taper as a transition to the entrance to the barrel of the culvert, called the *throat*. The floor of the tapered section has the same slope as the barrel of the culvert, and the height of the face should not exceed 1.1 times the height of the barrel. The headwater height on the throat is greater than on the face due to the slope of the tapered inlet. However, an increased head on the throat can be achieved by rotating the culvert about its downstream end such that there is a fall from the natural streambed to the invert of the face. The side-tapered inlet is designed by first calculating the head on the throat, HW_t, for a given design discharge, culvert size, and allowable headwater elevation using inlet control nomographs or equations developed for this case. The elevation of

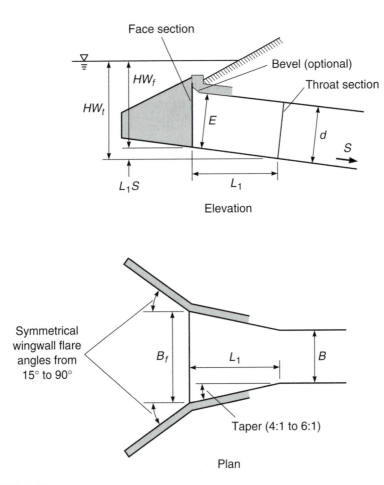

FIGURE 6.15
Side-tapered inlet, no fall (Federal Highway Administration 2001).

the throat then is set as the headwater elevation minus the head on the throat. This may require the inclusion of some fall in the throat below the normal streambed elevation. Then inlet control equations or nomographs for face control are used to obtain the minimum width of the face for the given head on the face, assuming a maximum increase in elevation of 1 ft from the throat to the face invert. The face width is rounded up slightly to be conservative, so that control will be at the throat and not the face. Once the face width is fixed, the length of the side taper is calculated from a chosen taper ratio between 4:1 and 6:1 (longitudinal:lateral), and the actual elevation of the face can be determined from the slope of the barrel. If it is more than 1 ft higher than the throat, the calculation must be repeated with a new face elevation.

The final level of inlet improvement is shown in Figure 6.16, which depicts the slope-tapered inlet. In this inlet improvement, the entire fall is concentrated from the face invert elevation at the natural streambed elevation to the throat invert elevation determined for throat control. Separate face control nomographs or equations for the slope-tapered inlet are used to find the minimum face width. The fall slope is selected to be in the range between 2:1 and 3:1 (horizontal to vertical), and the side taper remains in the range of 4:1 to 6:1 to determine the length of the tapered section. The amount of fall should be in the range between $0.25d$ and $1.5d$.

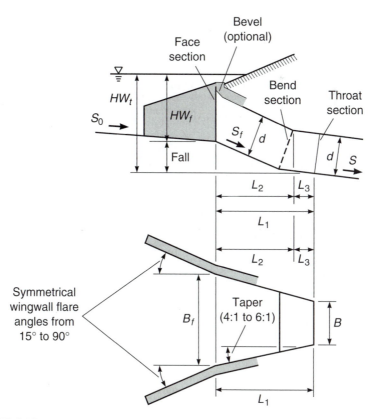

FIGURE 6.16
Slope-tapered inlet (Federal Highway Administration 2001).

Example 6.3

Design a concrete box culvert to carry a design discharge of 500 cfs (14.2 m³/s) with an allowable headwater of 10.0 ft (3.05 m) above the inlet invert. The culvert is 300 ft (91.4 m) long and has a slope of 0.02. The downstream channel is trapezoidal with a bottom width of 20 ft (6.1 m), side slopes of 2:1, $n = 0.020$, and slope $S = 0.001$ ft/ft.

Solution.   Start by choosing a 6 ft (1.8 m) by 6 ft (1.8 m) box culvert with a square edge in a headwall. Assume inlet control with the inlet submerged (IC-1), so that, from (6.11), the head for the design discharge is

$$HW = \frac{Q^2}{2g(C_d A_o)^2} = \frac{500^2}{64.4 \times 36^2 \times C_d^2} = \frac{3.00}{C_d^2}$$

Then, from Table 6.2, for $w/b = 0$, assume a value of $HW/d = 2.0$ for which $C_d = 0.51$ and $HW = 3/0.51^2 = 11.5$ ft (3.51 m). Repeat with $HW/d = 11.7/6 = 1.95$, and $C_d = 0.505$, so that $HW = 11.8$ ft (3.60 m). This is acceptable agreement, so for inlet control, the head, HW, of 11.8 ft (3.60 m) exceeds the allowable headwater. The next step could be to increase the size of the culvert, but it would be cheaper to bevel the edges. With $w/b = 0.042$, the iteration on C_d from Table 6.2 produces $C_d = 0.55$ and $HW = 9.9$ ft (3.0 m), which is just less than the allowable headwater. On the other hand, Equations 6.12 and 6.13 are somewhat more accurate for inlet control. The value of $Q/(Ad^{0.5}) = 500/(36 \times 6^{0.5}) = 5.67$, so Equation 6.13 is applicable. Table 6.3 gives $c = 0.0314$ and $Y = 0.82$ for a 45° bevel and a 90° headwall. Substituting into Equation 6.13 results in $HW = 10.9$ ft (3.32 m). This is slightly greater than the allowable headwater. For a greater factor of safety, increase the culvert size to 7 ft (2.1 m) by 6 ft (1.8 m) high but still use beveled edges. In this case, Equation 6.13 remains applicable and $HW = 9.3$ ft (2.8 m). This might be an acceptable design, but we should also check for outlet control. In fact, from Manning's equation, the normal depth in the culvert for $Q = 500$ cfs (14.2 m³/s), $n = 0.012$, $S = 0.02$, and $b = 7.0$ ft (2.1 m) is 2.97 ft (0.905 m), and critical depth $y_c = [(500/7)^2/32.2]^{1/3} = 5.41$ ft (1.65 m). Consequently, this is a steep slope and inlet control is likely to govern unless there is a high tailwater.

The tailwater for $Q = 500$ cfs (14.2 m³/s) can be calculated from Manning's equation with $n = 0.02$ and $S = 0.001$ for the given dimensions of the downstream trapezoidal channel. The result is a tailwater depth of 3.83 ft (1.17 m) above the outlet invert. Calculate $(y_c + d)/2 = (5.41 + 6)/2 = 5.7$ ft (1.7 m), which is greater than the tailwater depth of 3.83 ft (1.17 m), so use 5.7 ft (1.7 m) in the full-flow equation. Substituting into Equation 6.15 with the friction loss term evaluated by Equation 6.17 and $K_e = 0.2$ for beveled edges from Table 6.5, we have

$$HW = 5.7 - 0.02 \times 300 + \left(1 + 0.2 + \frac{64.4 \times 0.012^2 \times 300}{1.49^2 \times (42/26)^{4/3}}\right)\left(\frac{500^2}{64.4 \times 42^2}\right)$$

$$= 3.8 \text{ ft } (1.2 \text{ m})$$

Clearly the inlet control head of 9.3 ft (2.8 m) is higher, and it will control.

While this is a perfectly acceptable design, it is worthwhile to explore the effect of utilizing side-tapered and slope-tapered inlets on the original 6 ft by 6 ft box culvert design using the FHWA program HY8 (Federal Highway Administration 2008). The program HY8 allows interactive entry of culvert and inlet data and downstream channel characteristics. It then calculates the tailwater rating curve and develops a full performance curve for the selected culvert. It calculates complete water surface profiles when required and provides graphical screen results and printed output tables and files.

To design a side-tapered inlet, assume a lateral expansion of 4:1 and specify beveled edges. Then choose a face width larger than the culvert width, and the program computes the face control curve as well as the throat control performance curve. Adjust the face width until the face control curve is below the throat control curve so that the throat is the control, at least for Q greater than or equal to the design discharge. The performance curve for the 6 ft by 6 ft culvert with side-tapered inlet having a face width, B_f, of 9 ft (2.7 m) is shown in Figure 6.17 in comparison with the performance curves for a square edge and beveled edge on the 6 ft by 6 ft culvert. At the design discharge of 500 cfs (14.2 m³/s) , the head for the side-tapered inlet (SDT) is 9.11 ft (2.78 m), which is a 26 percent reduction from the head of 12.39 ft (3.78 m) for a square-edge inlet. Also shown in Figure 6.17 is the performance curve for a slope-tapered inlet (SLT) with a fall of 2 ft (0.61 m), a fall slope of 2:1, and a face width of 12.0 ft (3.66 m). The head at 500 cfs (14.2 m³/s) is 7.23 ft (2.20 m), or a 42 percent reduction from the head for a square-edge inlet. It is apparent that the culvert barrel could be reduced in size further if a side-tapered or slope-tapered inlet were used.

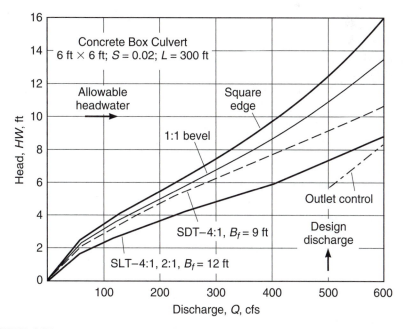

FIGURE 6.17

HY8 results showing effect of improved inlets on culvert performance curves.

Just below the performance curve for the slope-tapered inlet in Figure 6.17 is the outlet control performance curve for the slope-tapered inlet design. We see that the outlet control curve intersects the slope-tapered inlet curve at a discharge slightly greater than 600 cfs (17 m³/s). For all discharges greater than the intersection point, the culvert is in outlet control, and the head rises at a greater rate than for inlet control. One design philosophy is to use a tapered inlet with a fall such that the intersection with the outlet control curve occurs exactly at the allowable head of 10 ft (3.05 m) where Q is greater than the design value. This fully utilizes the inlet capacity of the culvert at the design head and provides a factor of safety in culvert capacity. Alternatively, the culvert with improved inlet can be designed with a fall such that the inlet control curve intersects exactly the point corresponding to the design discharge and allowable headwater. This often is acceptable, if some additional headwater can be tolerated or if road overtopping is allowed. The final possible design point is the intersection of the inlet control curve with the design discharge at the lowest possible head, which is limited by the natural water surface elevation in the stream upstream of the culvert. The final choice of design point must be made by the engineer based on local conditions and judgment.

6.6   BRIDGES

The flow constriction caused by bridge openings and bridge piers gives rise to both contraction and expansion energy losses, with a resulting rise in water surface elevation upstream of the bridge in comparison to that which would occur without the bridge. This excess water surface elevation in the bridge approach cross section, referred to as *backwater*, is shown in Figure 6.18 as h_1^*. Type I flow shown in Figure 6.18 is defined for subcritical flow throughout the approach, bridge, and exit cross sections. In Type II flow, the constriction is so severe as to produce choking and the occurrence of critical depth in the bridge opening. In Type IIA flow, the flow depth does not pass through the downstream critical depth, so a hydraulic jump does not occur. However, in the case of Type IIB flow, the flow downstream of the bridge becomes supercritical and a hydraulic jump forms immediately downstream of the bridge. Finally, Type III flow, which is not shown in Figure 6.18, occurs when an approach supercritical flow remains supercritical through the bridge opening. In Type I flow, the bridge backwater is the result of head losses, including the approach friction loss, contraction loss, and expansion loss. In the case of Type II flow, the choked condition, additional backwater is caused by the upstream depth necessary to increase the available specific energy to the minimum value in the bridge opening.

Several different methods are available for determining the bridge backwater, especially for Type I flow, which is the most common. These methods are discussed individually here and include empirical, momentum, and energy approaches to the problem.

HEC-2 and HEC-RAS

In the *normal* bridge routine in HEC-2 (U.S. Army Corps of Engineers 1991) or the energy method in HEC-RAS (U.S. Army Corps of Engineers 2008), the gradually varied flow profile calculations are continued through the bridge using the standard

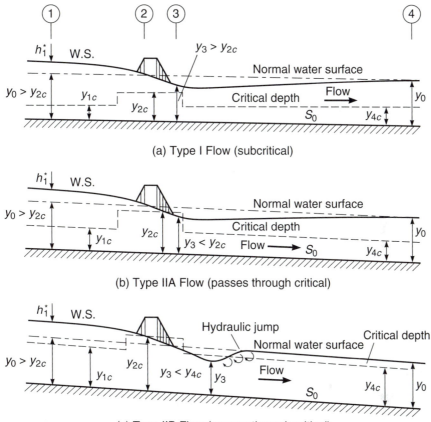

FIGURE 6.18

Flow through a bridge opening (Bradley 1978).

step method, as though the bridge opening were just another river cross section. This method usually is used when there are no piers or the head loss caused by the piers is very small. The cross sections are located as shown in Figure 6.19, numbered for consistency with other methods presented here. Cross sections 3 and 2 are located immediately downstream and upstream of the bridge opening, respectively, at a distance of only a few feet from the toe of the embankment. The approach section 1 in Figure 6.19 is in the region of parallel flow before flow contraction occurs, while the exit section 4 is located at a point where the flow has reexpanded. Traditionally, the Corps of Engineers has recommended that the length of the contraction reach from cross section 1 to 2 be taken as 1 times the average length of the side obstruction caused by the embankments (CR = 1). In addition, the expansion reach length from cross section 3 to 4 has been recommended in the past to be 4 times the average length of the side obstruction (ER = 4). However, the Corps of Engineers conducted a numerical study of the contraction and expansion reach lengths using a two-dimensional numerical model (U.S. Army Corps of Engineers 2008). The results showed that the lengths required for expansion of the flow depend on the

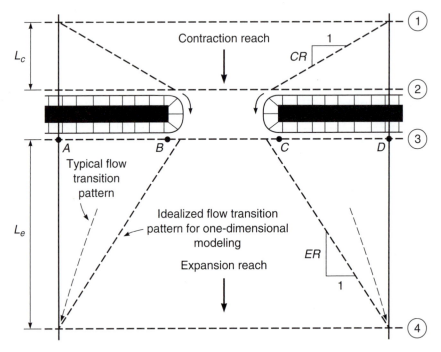

FIGURE 6.19
Cross-section locations at a bridge (U.S. Army Corps of Engineers 2008).

geometric contraction ratio, the channel slope, and the ratio of overbank to main channel values of Manning's n, while contraction reach lengths depend only on the latter two variables. In general, contraction reach lengths were in the range of 1 to 2 times the average obstruction length, and expansion reach lengths fell in the range of 1 to 2.5 times the average obstruction length. Best fits of the numerical results were obtained, but they are specific to the values of the independent variables tested in the numerical model, which included bridge opening lengths from 100 to 500 ft (30.5 to 152 m), a floodplain width of 1000 ft (305 m), overbank Manning's n values from 0.04 to 0.16, main channel Manning's n of 0.04, discharges from 5,000 to 30,000 cfs (142 to 850 m³/s), and bed slopes from 0.00019 to 0.0019 (see U.S. Army Corps of Engineers 2008).

Two additional cross sections are created by HEC-RAS inside the bridge opening. Only the effective flow area from B to C is used in the cross-section properties of cross section 3 as well as cross section 2. Standard step flow profiles are computed through this total of six cross sections with friction losses and expansion and contraction losses computed in the usual way.

In the *special* bridge method in HEC-2, the program computes a momentum balance between an upstream cross section and a section just inside the bridge section and between the bridge section and a downstream section, to determine if the flow is Type I or Type II. If the flow is Type I, then the empirical Yarnell equation (Henderson 1966) is used to determine the change in water surface elevation, ΔH_{2-3}, through the bridge opening

$$\frac{\Delta H_{2-3}}{y_3} = K_P \mathbf{F}_3^2 (K_P + 5\mathbf{F}_3^2 - 0.6)(A_r + 15A_r^4) \tag{6.19}$$

in which y_3 = downstream depth; $\mathbf{F}_3$ = downstream Froude number; K_P = pier coefficient varying from 0.9 for a pier with semicircular nose and tail to 1.25 for a pier with square nose and tail; and A_r = area ratio = obstructed area due to piers/total unobstructed area. If the flow is Type II, then HEC-2 sets the depth equal to critical depth in the bridge and determines the upstream and downstream depths from a momentum balance (Eichert and Peters 1970).

In HEC-RAS, the Yarnell method or the momentum method can be chosen as the desired bridge hydraulics analysis method for Type I flow. In the momentum method, the momentum equation is written in three steps: (1) from just upstream of the bridge to a point just inside the bridge, (2) through the bridge opening itself, and (3) from just inside the bridge exit to a point just downstream of the bridge. This provides a solution for the depth at the two cross sections inside the bridge and the cross-section immediately upstream of the bridge. The pier drag force is included in step (1).

Detection of Type II flow and calculation of the approach depth can also be accomplished using a combination of the momentum and energy approaches. First, the momentum equation is written between the bridge section and downstream section 4, with critical depth assumed in the bridge section. The result, which is given in terms of the width ratio $r = b_2/b_4$ that causes choking, is (Henderson 1966)

$$r = \frac{(2 + 1/r)^3 \mathbf{F}_4^4}{(1 + 2\mathbf{F}_4^2)^3} \tag{6.20}$$

in which b_2 = width of bridge opening; b_4 = exit channel width; and $\mathbf{F}_4$ = downstream value of the Froude number. The approach depth is obtained by writing the energy equation between the approach section 1 and the critical section inside the bridge with an appropriate head loss coefficient.

HDS-1

The Federal Highway Administration developed an energy method of bridge analysis published in the Hydraulic Design Series (HDS-1; Bradley 1978). It was used prior to the development of WSPRO (Arneson and Shearman 1998). Referring to Figure 6.20, the energy equation is applied between sections 1 and 4 to obtain

$$S_0 L_{1-4} + y_1 + \frac{\alpha_1 V_1^2}{2g} = y_4 + \frac{\alpha_4 V_4^2}{2g} + h_T \tag{6.21}$$

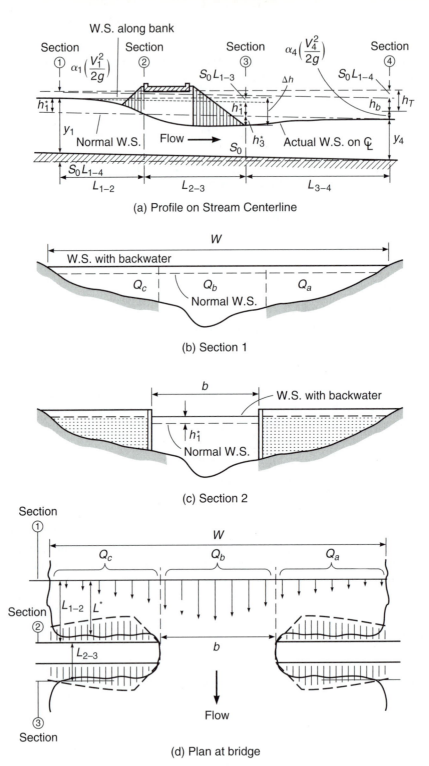

FIGURE 6.20

Normal crossing: Wingwall abutments (Bradley 1978).

268

in which h_T is the total energy loss between sections 1 and 4. With respect to the normal water surface, the uniform-flow resistance portion of h_T is just balanced by the vertical fall in the channel bottom so that from Equation 6.21, we have

$$y_1 - y_4 = \frac{\alpha_4 V_4^2}{2g} - \frac{\alpha_1 V_1^2}{2g} + h_b \tag{6.22}$$

in which h_b is the additional head loss due to the bridge constriction and can be expressed in terms of a minor loss coefficient, K^*, defined by

$$h_b = K^* \frac{\alpha_2 V_{n2}^2}{2g} \tag{6.23}$$

where V_{n2} = the mean velocity in the contracted section based on the flow area below the normal water surface inclusive of the area occupied by bridge piers. Now if $(y_1 - y_4)$ is replaced by h_1^* and Equation 6.23 is used to substitute for h_b, Equation 6.22, with the aid of continuity, becomes

$$h_1^* = K^* \frac{\alpha_2 V_{n2}^2}{2g} + \alpha_1 \left[\left(\frac{A_{n2}}{A_4} \right)^2 - \left(\frac{A_{n2}}{A_1} \right)^2 \right] \frac{V_{n2}^2}{2g} \tag{6.24}$$

The second term on the right hand side of Equation 6.24 represents the difference in velocity heads between sections 1 and 4. This term generally is much smaller than the first term, and Equation 6.24 is solved by iteration with the second term equal to zero in the first trial. It is important to note that A_{n2} is the gross water area in the contracted section measured below normal stage, and V_{n2} is a reference velocity equal to Q/A_{n2}. The value of $\alpha_2 = 1.0$, and it is assumed that $\alpha_1 = \alpha_4$.

To calculate the backwater, the value of the minor head loss coefficient, K^*, must be determined. Values of K^* have been developed from laboratory and field studies, and K^* is considered to consist of additive components

$$K^* = K_b + \Delta K_p + \Delta K_e + \Delta K_s \tag{6.25}$$

in which K_b = contraction coefficient; ΔK_p = pier coefficient; ΔK_e = eccentricity coefficient; and ΔK_s = skewness coefficient. For simplicity, we consider only a normal bridge crossing with no eccentricity or skewness effects. The values of K_b and ΔK_p can be obtained from Figures 6.21 and 6.22, respectively. The contraction coefficient K_b in Figure 6.21 depends on M_0, the bridge opening discharge ratio given as Q_b/Q and defined in Figure 6.20 for the normal water surface elevation in the approach section. For abutments exceeding 200 ft in length, the lower curve in Figure 6.21 is recommended regardless of abutment type. In Figure 6.22, the pier coefficient is given as a function of J, the ratio of area obstructed by the piers to the gross area of the bridge waterway below the normal water surface at section 2. The value of ΔK is determined first as a function of J, and then it is corrected for the value of M_0 to give $\Delta K_p = \Delta K \sigma$.

USGS Width Contraction Method

The USGS has an interest in bridges from the viewpoint of using them as flow measuring devices by measuring upstream and downstream stages. As a result, it

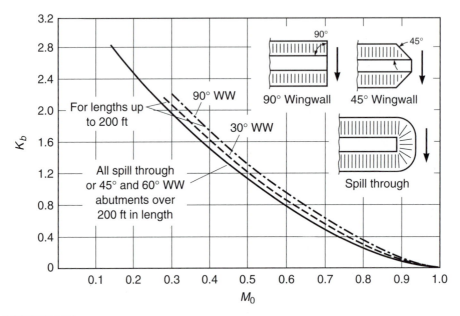

FIGURE 6.21
Backwater coefficient base curves—subcritical flow (Bradley 1978).

developed an energy approach (Kindsvater and Carter 1955; Kindsvater, Carter, and Tracy 1958; Matthai 1976) that utilizes a bridge discharge coefficient, C. First, with reference to the cross section locations in Figure 6.20, the energy equation is written between cross sections 1 and 3, but with section 3 inside the bridge, to obtain

$$\frac{\alpha_1 V_1^2}{2g} + h_1 = \frac{\alpha_3 V_3^2}{2g} + h_3 + h_e + h_f \tag{6.26}$$

in which h_1 = stage (water surface elevation) in the approach section 1; h_3 = stage in the bridge section 3; h_e = entrance head loss; and h_f = friction head loss from sections 1 to 3. If the entrance loss is expressed in terms of a minor loss coefficient as $h_e = K_e(V_3)^2/2g$ and if continuity for the bridge section is written as

$$Q = C_c b y_3 V_3 \tag{6.27}$$

in which b = bridge opening length, and C_c = the contraction coefficient, then Equation 6.26 can be solved for V_3 and expressed in terms of Q using Equation 6.27 to give

$$Q = CA_3 \sqrt{2g\left(\Delta h - h_f + \alpha_1 \frac{V_1^2}{2g}\right)} \tag{6.28}$$

In Equation 6.28, the bridge discharge coefficient, C, is defined by

$$C = \frac{C_c}{\sqrt{\alpha_3 + K_e}} \tag{6.29}$$

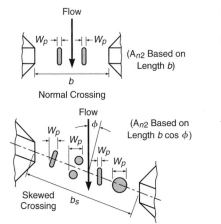

W_p = Width of pier normal to flow, ft

h_{n2} = Height of pier exposed to flow, ft

N = Number of piers

A_p = $\Sigma^N W_p h_{n2}$ = Total projected area of piers normal to flow, ft^2

A_{n2} = Gross water cross section in constriction based on normal water surface (use projected bridge length normal to flow for skew crossings)

J = $\dfrac{A_p}{A_{n2}}$

Note: Sway bracing should be included in width of pile bents.

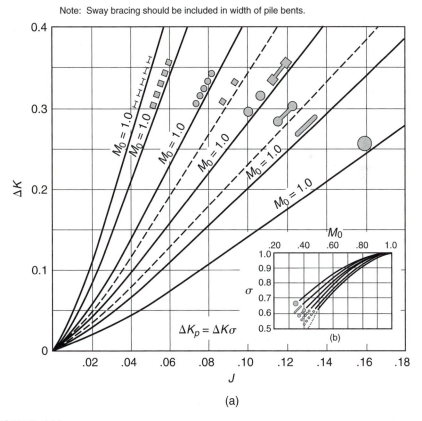

FIGURE 6.22

Incremental backwater coefficient for piers (Bradley 1978).

and $\Delta h = h_1 - h_3$ while $A_3 = by_3$. As an example, values of the bridge discharge coefficient are given in Figure 6.23 for a Type I bridge consisting of rectangular abutments with or without wingwalls. The base coefficient C' is determined from

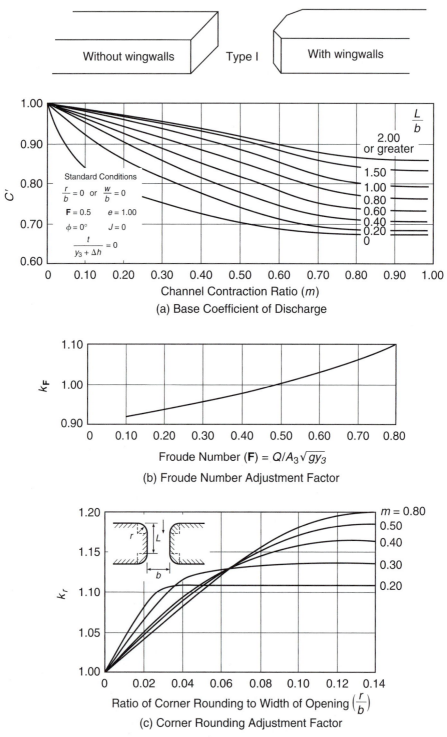

FIGURE 6.23

Bridge discharge coefficient for Type I bridge openings (Matthai 1976).

the upper graph and corrected for the Froude number and corner rounding by multiplying C' by k_F and k_r to get C. Curves of this type have been developed for three additional bridge types, discussed in the section on WSPRO. (For the complete set of curves, see Matthai 1976; French 1985; or U.S. Army Corps of Engineers 2008. The purpose here is only to show the connection between the USGS method and WSPRO.) Each bridge type has its own set of correction factors. In addition, some correction factors are common to all four bridge types, such as the correction for piers or piles as shown in Figure 6.24, and these are multiplied times the base coefficient. The pile and pier adjustment factors depend on the ratio $j = A_j/A_3$, where A_j is the submerged area of the piers projected onto the plane of cross section 3 and A_3 is the gross area of cross section 3, L/b = ratio of abutment width in the flow direction to bridge opening length, and m = channel contraction ratio. The value of the base discharge coefficient C' is a function of the channel contraction ratio, m, which is defined as the obstructed discharge in the approach channel cross section divided by the total discharge, and L/b. In terms of HDS-1, $m = (1 - M_1)$ where M_1 is the unobstructed discharge ratio, defined as in HDS-1 except that it is evaluated at the approach water surface elevation, h_1, instead of at the normal water surface elevation. To determine the backwater, Equation 6.28 can be solved for Δh, but this is only the drop in water surface from the approach to the bridge section. WSPRO, to be described next, utilizes this portion of the energy balance involving C but also the energy equation written from sections 3 to 4.

WSPRO Model

The USGS in cooperation with FHWA developed a computer program that combines step backwater analysis with bridge backwater calculations. The program, named WSPRO (Shearman et al. 1986, Arneson and Shearman 1998), is recommended by FHWA. WSPRO allows for pressure flow through the bridge, embankment overtopping, and flow through multiple bridge openings including culverts. The bridge hydraulics rely on the energy principle but have an improved technique for determining approach flow lengths and an explicit consideration of an expansion loss coefficient. The flow length improvement was found necessary when the approach flow occurs on very wide, heavily vegetated floodplains.

The cross sections necessary for the WSPRO energy analysis are shown in Figure 6.25 for a single-opening bridge with or without spur dikes having a bridge opening length of b. Cross sections 1, 3, and 4 are required for a Type I flow analysis, and they are referred to as the *approach section*, *bridge section*, and *exit section*, respectively. In addition, cross section 3F, called the *full valley section*, is needed for the water surface profile computation without the presence of the bridge contraction. Cross section 2 is used as a control point in Type II flow but requires no input data. Two more cross sections must be defined if spur dikes and a roadway profile are specified. The approach section is located a distance of b upstream of the upstream face of the bridge, while the exit section is a distance of b downstream of the downstream face.

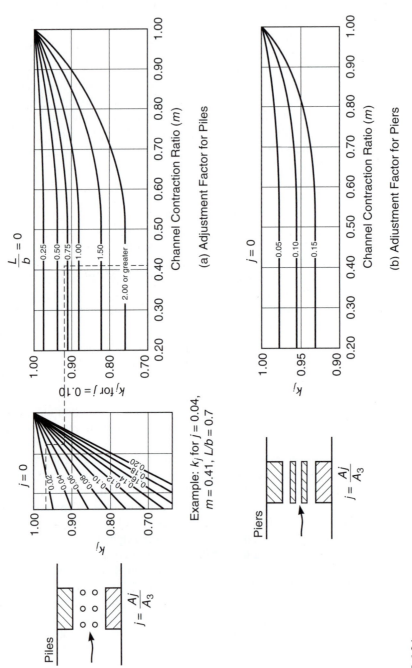

FIGURE 6.24
Bridge discharge coefficient adjustment factors for piers or piles (Matthai 1976).

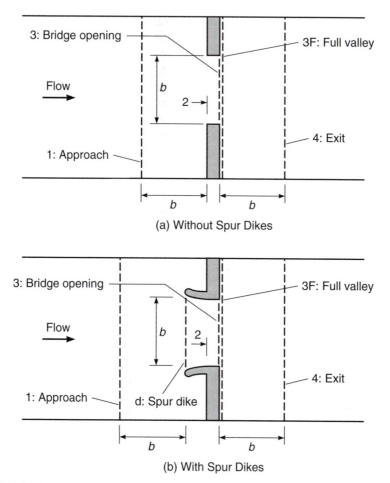

(a) Without Spur Dikes

(b) With Spur Dikes

FIGURE 6.25

WSPRO cross-section locations for stream crossing with a single waterway opening (Shearman et al. 1986).

The basic methodology for a single-opening bridge with no spur dikes and free-surface flow consists of writing the energy equation, first between cross sections 1 and 3 and then between cross sections 3 and 4, as defined in Figure 6.25:

$$h_1 = h_3 + h_{v3} + h_{f(1-2)} + h_{f(2-3)} - h_{v1} \qquad (6.30)$$

$$h_3 = h_{4n} + h_{v4n} + h_{f(3-4)} + h_e - h_{v3} \qquad (6.31)$$

in which h_i = the water surface elevation at cross section i; h_{vi} = velocity head at section i; $h_{f(i-j)}$ = the friction head loss between cross sections i and j; h_{jn} = normal water surface elevation at cross section j; and h_e = exit head loss between cross sections 3 and 4. Equations 6.30 and 6.31 are solved by assuming initial trial elevations for h_1 and h_3, which are used to compute the right hand sides of the two equations to obtain updated values on the left. Iteration is continued until the changes in h_1 and h_3 are small.

TABLE 6.6   Energy loss expressions in WSPRO. K = conveyance and C = bridge discharge coefficient

Loss between section numbers	Type of loss	Energy loss equation
1–2 (no spur dikes)	Friction	$h_{f(1-2)} = L_{av}Q^2/(K_1 K_c)$
2–3	Friction	$h_{f(2-3)} = L_{2-3}Q^2/K_3^2$
3–4	Friction	$h_{f(3-4)} = bQ^2/(K_c K_{4n})$
3–4	Expansion	$h_e = Q^2/(2gA_4^2)[2\beta_4 - \alpha_4 - 2\beta_3(A_4/A_3) + \alpha_3 (A_4/A_3)^2]$ where $\alpha_3 = 1/C^2$ and $\beta_3 = 1/C$

Source: Data from Shearman et al. (1986).

Energy loss expressions needed in Equations 6.30 and 6.31 are summarized in Table 6.6. Friction loss calculations utilize the geometric mean conveyance between any two cross sections, and the flow length from section 1 to 2 is the average length, L_{av}, as determined by the method developed by Schneider et al. (1977) and shown in Figure 6.26. The approach flow is divided into 20 streamtubes of equal conveyance, and the flow distance of each streamtube from the approach section to the bridge is averaged for the calculation of the approach friction loss. The length L_{opt} is the distance from the bridge opening to the section where the flow is nearly one dimensional, determined as a function of the geometric contraction ratio based on potential flow theory (Schneider et al. 1977). The value of L_{opt} is equal to the bridge opening length, b, at a geometric contraction ratio, $b/B = 0.26$. If L_{opt} is less than b ($b/B > 0.26$) as in Figure 6.26a, then the parallel straight-line lengths of each streamline from the approach section to the dashed line at L_{opt} plus the converging straight-line lengths to the bridge opening are averaged to obtain L_{av}. In Figure 6.26b, L_{opt} is greater than b ($b/B < 0.26$) for a very severe contraction. In this case, a parabola approximating an equipotential line is constructed from the edge of the water at the upstream distance of b. Then, the parallel streamlines are extended to intersect with the parabola before being turned to the bridge opening, if the intersection point is downstream of the dashed line located a distance of L_{opt} from the bridge opening. The approach head loss also depends on the geometric mean conveyance squared from 1 to 2, defined as the product of K_1 and K_c, where K_1 is the conveyance of the approach section, and K_c is the minimum of the conveyance at section 3 (K_3) or the conveyance K_q, defined as the conveyance of the segment of approach flow that can flow through the bridge opening with no contraction.

The friction loss through the bridge is based on the conveyance K_3 as shown in Table 6.6. The length of the expansion reach used in the friction loss calculation is one bridge opening length, b, and so the bridge exit cross-section location should not be changed. A separate expansion head loss computation is based on the approximate solution of the momentum, energy, and continuity equations for an abrupt expansion given by Henderson (1966) and discussed in Chapter 2. It depends on the coefficient of discharge for the bridge as developed by Matthai (1976). By comparing Equations 6.28 and 6.30, it can be shown that $\alpha_3 = 1/C^2$.

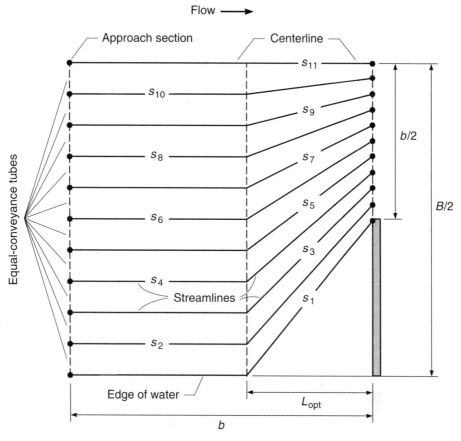

(a) For Relatively Low Degrees of Contraction

FIGURE 6.26
WSPRO definition sketches of assumed streamlines (Shearman et al. 1986).

The USGS width contraction method is used to find the bridge discharge coefficient, which then appears in the expansion head loss expression.

Pressure flow through the bridge opening is assumed to occur when the depth just upstream of the bridge opening exceeds 1.1 times the opening hydraulic depth. The flow then is calculated as orifice flow with the discharge proportional to the square root of the effective head. Unsubmerged orifice flow is illustrated in Figure 6.27 with the orifice discharge, Q_o, computed by

$$Q_o = C_d A_{3net} \sqrt{2g(Y_u - Z/2 + h_{v1})} \tag{6.32}$$

in which A_{3net} = net open area in the bridge opening, and Z = hydraulic depth = A_{3net}/b. Submerged orifice flow is treated similarly, with the head redefined as shown in Figure 6.28, and given by

$$Q_o = C_d A_{3net} \sqrt{2g\Delta h} \tag{6.33}$$

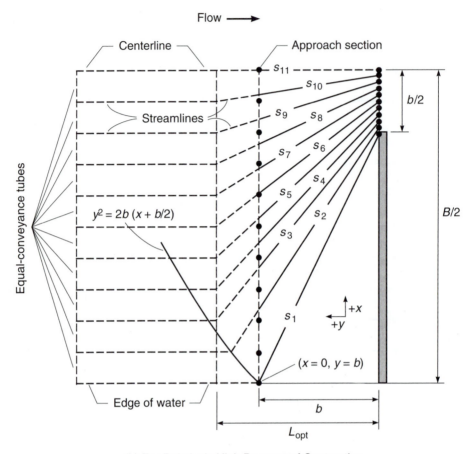

(b) For Relatively High Degrees of Contraction

FIGURE 6.26 *(continued)*

In unsubmerged orifice flow, the discharge coefficient is 0.5 over a wide range of Y_u/Z, while it is equal to 0.8 for the submerged orifice case.

WSPRO also can consider flow through the bridge opening simultaneously with embankment overflow, which is computed as a weir discharge with discharge proportional to head to the $\frac{3}{2}$ power (see Figure 6.13). This leads to classification of flow classes 1 through 6 (Shearman et al. 1986), as shown in Table 6.7. In free-surface flow, there is no contact between the water surface and the low steel elevation of the bridge. In orifice flow, only the upstream girder is submerged, while in submerged orifice flow both the upstream and downstream girders are submerged.

A total of four different bridge types can be treated by WSPRO as described in Table 6.8. Further details are given by Shearman et al. (1986).

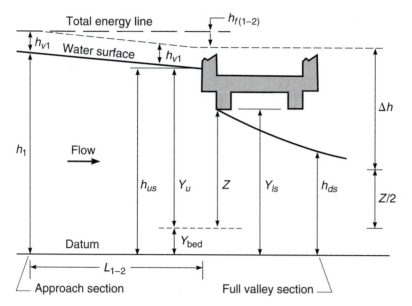

FIGURE 6.27
WSPRO definition sketch for unsubmerged orifice flow computations (Shearman et al. 1986).

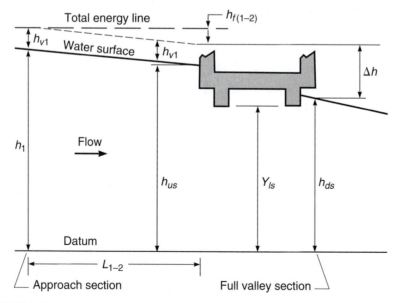

FIGURE 6.28
WSPRO definition sketch for submerged orifice flow computations (Shearman et al. 1986).

TABLE 6.7   Bridge flow classification according to submergence conditions

Flow through bridge opening only	Flow through bridge opening and over road
Class 1. Free surface flow	Class 4. Free surface flow
Class 2. Orifice flow	Class 5. Orifice flow
Class 3. Submerged orifice flow	Class 6. Submerged orifice flow

TABLE 6.8   Bridge type classification

Type	Embankments	Abutments	Wingwalls
1	Vertical	Vertical	With or without
2	Sloping	Vertical	None
3	Sloping	Sloping	None
4	Sloping	Vertical	Yes

Comparisons of WSPRO results with several other models and field measurements of water surface profiles through several bridges are given in Figure 6.29 (Shearman et al. 1986). The methods HEC-2(N) and HEC-2(S) are the normal and special bridge routines, while E431 is an older USGS method. WSPRO compares very well with the observed water surface profiles. Maximum errors are 0.3 ft for Buckhorn and Cypress Creek, and 0.4 ft and 0.6 ft for the higher and lower discharges, respectively, on Poley Creek. The results from WSPRO and HEC-2 are comparable for the entire Cypress Creek profile as well as for the profiles upstream of the bridge for Buckhorn Creek and the low discharge on Poley Creek. The water surface profile through the bridge, however, is not reproduced very well by HEC-2.

Kaatz and James (1997) compared backwater values computed by WSPRO, HEC-2, and the modified Bradley method with measured backwater values for 13 flood events at nine bridges in Louisiana, Alabama, and Mississippi. The modified Bradley method used essentially was the HDS-1 method given in this chapter, except that the contraction reach length was taken to be one bridge opening length, as in WSPRO. The bridge opening lengths varied from 40 to 130 m (130 to 430 ft) and the discharge contraction ratio, m, defined in the USGS method, varied from 54 to 79 percent of total flow obstructed in the approach section. Both the normal bridge method and the special bridge method were used in HEC-2, in which the latter method simply is an application of the Yarnell equation to determine the water surface drop through the bridge. The downstream expansion reach length for the HEC-2 methods was taken to be one bridge opening length, as in WSPRO, but the HEC-2 recommended value (4 times the average obstruction length) also was tried. When using the expansion reach length of one opening length, the HEC-2 normal bridge method gave the most consistent results, with computed backwater values showing an overall average of 2 percent less than measured backwater values, while WSPRO gave computed values with an overall average of 31 percent greater than measured backwater values. The HEC-2 special bridge method (Yarnell) and the

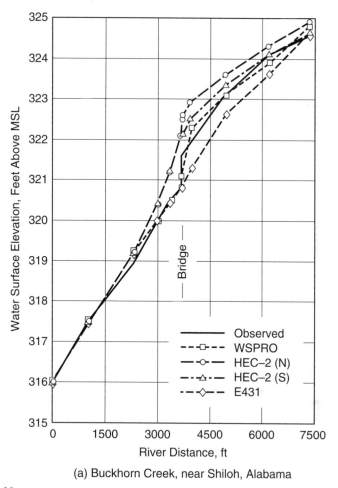

(a) Buckhorn Creek, near Shiloh, Alabama

FIGURE 6.29
Comparison of water surface profiles (Shearman et al. 1986).

modified Bradley method both gave consistently low values of computed backwater, which is not too surprising, since neither method was developed for bridges in wide, heavily vegetated floodplains. When the expansion ratio of 4:1 was applied in the HEC-2 normal bridge method, the overall average of computed backwater values was 36 percent higher than measured values and the computed water surface elevations downstream of the bridge were significantly higher. It was concluded that, although the WSPRO model gave backwater values that were somewhat high, it provided an accurate representation of the downstream water surface elevations and the water surface elevations in the immediate vicinity of the bridge.

In laboratory experiments conducted at Georgia Tech, water surface profiles were measured in a large compound channel (4.3 m (14 ft) wide by 18.3 m (60 ft) long) for which the Manning's n values were determined in uniform flow experiments

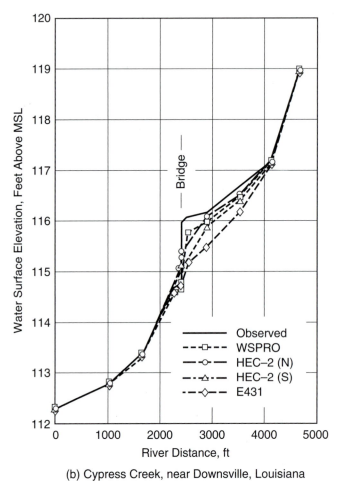

(b) Cypress Creek, near Downsville, Louisiana

FIGURE 6.29 (*continued*)

to be 0.0155 and 0.019 in the floodplain and main channel, respectively, for compound channel flow. The compound channel was asymmetric with a floodplain width of 3.66 m (12.0 ft) and a trapezoidal main channel bank-full width of 0.55 m (1.8 ft). The measured water surface profiles for a bridge abutment in place are compared with WSPRO results in Figure 6.30, in which the total depths relative to the bottom of the main channel are given. The bank-full depth is 0.15 m (0.5 ft). The abutment/embankment length for this case is 44 percent of the floodplain width ($L_a/B_f = 0.44$). Almost exact agreement is found between the WSPRO depth and the measured depth at the downstream face of the bridge, while WSPRO depths upstream of the bridge are approximately 2 to 3 percent high. Measured and computed velocity distributions are superimposed on the shape of the compound channel at the bridge approach section in Figure 6.31. The WSPRO velocities are computed from the discharges in each of 20 streamtubes having equal conveyance divided by the flow area of each streamtube. Relatively good agreement between measured and computed depth-averaged velocities is shown both in the floodplain and main

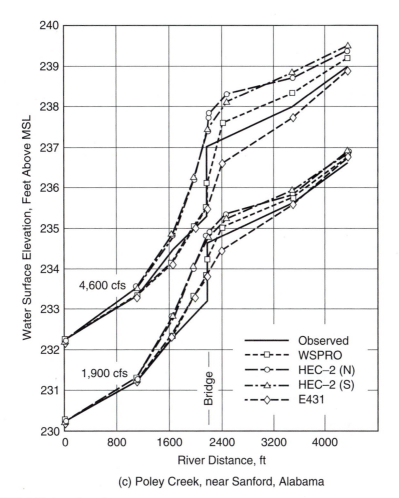

(c) Poley Creek, near Sanford, Alabama

FIGURE 6.29 (*continued*)

channel. However, WSPRO velocities computed in this way did not agree at all with measured resultant velocities near the face of the abutment, where the flow was not one-dimensional (Sturm and Chrisochoides 1998).

HEC-RAS Bridge Modeling Implementation

The computer program HEC-RAS (U.S. Army Corps of Engineers 2008) allows the computation of the water surface profile through a bridge using one or all of the following methods for Type I low flow as defined in Figure 6.18:

- Energy
- Yarnell
- Momentum
- WSPRO

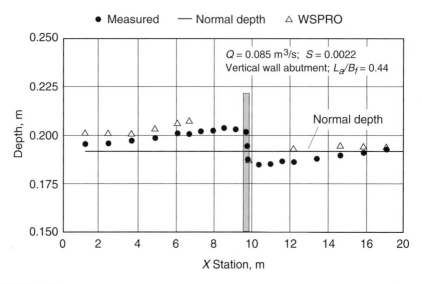

FIGURE 6.30

Comparison of measured depths and WSPRO computed depths in a laboratory compound channel (Sturm and Chrisochoides 1998). (*Source: Terry W. Sturm and Antonis Chrisochoides. One-Dimensional and Two-Dimensional Estimates of Abutment Scour Prediction Variables. In Transportation Research Record 1647, Transportation Research Board, National Research Council, Washington, D.C., 1998. Reproduced by permission of Transportation Research Board.*)

In the energy method, the water surface profile computation is carried through the bridge using the standard step method as in an ordinary river reach paying special attention to the Manning's n coefficients and the contraction and expansion coefficients for the bridge. In addition to the four cross sections numbered 1–4 in Figure 6.19, two cross sections are created inside the bridge and are referred to as BD (bridge downstream) and BU (bridge upstream) for the downstream and upstream internal bridge cross sections, respectively. These cross sections are included in the water surface profile computation through the bridge. In the Yarnell method, the change in water surface elevation through the bridge is determined directly by the Yarnell equation given previously as Equation 6.19, and no water surface elevations are computed for the BD and BU cross sections inside the bridge. The momentum method uses the four cross sections designated as 2, BU, BD, and 3 as described previously with three separate applications of the momentum equation to define the water surface profile. Implementation of the WSPRO method within HEC-RAS includes writing of the energy equation between cross sections 1–2, 2–BU, BU–BD, BD–3, and 3–4 with reference to Figure 6.19. In contrast, the original WSPRO methodology includes the friction head losses from 1–BU and BU–BD in the first writing of the energy equation given as Equation 6.30, and then incorporates the friction and exit head losses from BD–4 in the second energy equation given as Equation 6.31. (Note that section BD corresponds with the bridge opening section 3 in Figure 6.25.)

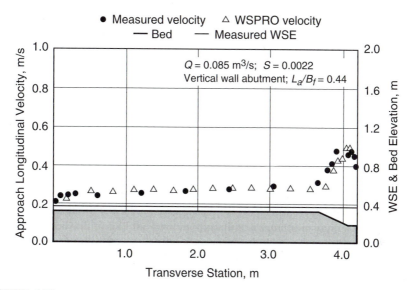

FIGURE 6.31

Comparison of measured velocity and WSPRO computed velocity in a laboratory compound channel (Sturm and Chrisochoides 1998). (*Source: Terry W. Sturm and Antonis Chrisochoides. One-Dimensional and Two-Dimensional Estimates of Abutment Scour Prediction Variables. In Transportation Research Record 1647, Transportation Research Board, National Research Council, Washington, D.C., 1998. Reproduced by permission of Transportation Research Board.*)

Because of the definition of cross sections 2 and 3 in HEC-RAS, which are just outside the bridge (usually within a few feet of the toe of the embankment), it is necessary to define *ineffective flow areas* at these sections. As long as the water surface elevations are below the low chord of the bridge and in free flow, or in orifice flow without overtopping of the bridge or roadway embankment, the floodplains need to be designated as ineffective flow areas at sections 2 and 3 because of the flow contraction and flow expansion occurring at these locations. The elevation of the ineffective flow area designation also has to be specified so that the full floodplain areas can become active once overflow of the bridge or embankments begins. The specified elevation of the ineffective flow areas is usually between the low chord and minimum top-of-road elevation on the downstream side of the bridge, while the minimum top-of-road elevation can be used on the upstream side of the bridge.

When flow is Type II (subcritical) with critical depth in the bridge cross section as shown in Figure 6.18, the momentum equation is written between the critical section and the upstream subcritical flow in section 2, and between the critical section and the downstream supercritical flow in section 3. For a Type III flow that is supercritical through all cross sections, either the momentum equation or energy equation can be utilized.

High flow situations include orifice flow or submerged orifice flow with or without overtopping of the bridge and roadway as classified in Table 6.7. The unsubmerged orifice flow relationship given by Equation 6.32 and the submerged orifice flow equation given by (6.33) are used by HEC-RAS to compute pressure flow through the bridge. Pressure flow is detected when the low flow energy head is computed to be above the maximum low chord on the upstream side of the bridge. Then the orifice flow equation is invoked to compute the orifice flow energy grade line for comparison with the free-surface flow energy grade line. The program uses the higher of the two energy grade lines. If overtopping occurs, the weir flow equation given by (6.18) is utilized with the weir coefficients shown in Figure 6.13. To determine the division in the flow rate going through the bridge and over the top of the bridge, HEC-RAS iterates on the upstream energy grade line elevation until the sum of the orifice and weir flow discharges equals the specified total discharge. The program also checks for submergence of the weir flow due to high tailwater downstream and reduces the weir discharge coefficient according to Figure 6.13c. If the ratio of the tailwater energy head to the upstream energy head exceeds 0.95 (*submergence factor*), the weir and pressure flow computation reverts to the standard step energy method because the bridge has a much smaller influence on the overall water surface profile in this instance.

The HEC-RAS user can specify that all or only one of the four low flow methods be used to compute the water surface profile. If all four methods are selected, the user can choose to have only one printed as the final result, or have the program choose the method that gives the highest energy loss through the bridge. For low flow computations, the Yarnell method is really appropriate only when almost all of the energy loss is due to the piers. For Type II flow with the occurrence of critical depth, either the momentum or energy method is applicable. The WSPRO method should be used especially when there are very wide, heavily vegetated floodplains that contribute a large friction loss to the overall head loss due to the bridge.

Some WSPRO variables have to be specified in addition to the parameters needed for the other three bridge modeling methods. The elevations of the top of the embankment and the toe of the abutment have to be entered in the *WSPRO variables* screen for both the left and right abutments. In addition, the abutment side slope and the embankment top width must be entered. The *abutment type* is selected from the choices defined in Table 6.8. The centroid station of the projected bridge opening at the approach section (*Kq* section) is another unique WSPRO parameter. The *Kq* section refers to the conveyance of that portion of the approach section that is partitioned by extending lines from the abutments to the approach section. It was discussed previously in the USGS width contraction method with respect to estimating the discharge contraction ratio of the bridge and the bridge discharge coefficient. Better results may be obtained from WSPRO if the centroid station corresponds to the centerline of the bridge opening rather than the default station taken at the centroid of the computed conveyance distribution of the approach cross section. The WSPRO data entry screen also allows the user to insert either wingwalls or guide banks upstream of the bridge to reduce head losses and scour. Finally, there is an option to specify additional contraction and expansion losses and the method of averaging conveyances from section to section in the water surface profile computation. In order to obtain comparable results with the WSPRO methodology implemented by the FHWA computer program, no additional losses should be specified, and the geometric mean conveyance method should be selected for computing the average slope of the energy grade line.

The reader is referred to *HEC-RAS Hydraulic Reference Manual* and the companion *Users Manual* for further details on the use of the computer program (U.S. Army Corps of Engineers 2008). In addition, the reference by Shearman et al. (1986) describes the WSPRO methodology in more detail. An example bridge modeling problem is given next highlighting the use of the WSPRO method within HEC-RAS.

Example 6.4

A normal, single-opening bridge having a deck width of 40 ft is to be constructed. The cross section data are given in Table 6.9, which shows the subsections and roughnesses. The average stream slope in the vicinity of the bridge is 0.00052 ft/ft. The bridge opening begins at Station $X = 230$ ft (70 m) and ends at Station 430 ft (131 m) for a total bridge opening length of 200 ft (61 m). It has vertical abutments and embankments (Type 1 bridge) and a bridge deck elevation of 35.0 ft (10.7 m) with a low chord (or low steel elevation)

TABLE 6.9   Cross section coordinates at Station 1200, Example 6.4

Distance X (ft)	Elevation (ft)	n Value	Comment
0	35.0	0.045	
10	28.0		
140	23.5		
200	21.5	0.070	
230	21.0		Left abutment
250	20.5		
280	20.4		Pier #1
300	20.0	0.035	Left bank
310	19.0		
330	10.0		Pier #2
360	3.0		
380	8.0		Pier #3
400	18.0		
430	21.0	0.045	Right bank &
450	20.0		abutment
475	17.0		
500	17.5		
540	18.0		
600	20.0		
730	28.0		
740	35.0		

of 32.0 ft (9.75 m). The bridge has three piers with a spacing of 50.0 ft (15.2 m) and a width of 3.0 ft (0.91 m) each. Pier locations are shown in Table 6.9. For a discharge of 20,000 cfs (567 m³/s), calculate the backwater caused by the bridge and the mean velocity at the bridge section using WSPRO.

Solution.   The river schematic is entered first in HEC-RAS as in Example 5.5. The river schematic for this example problem is shown in Figure 6.32. Stationing begins at River Station 2000 upstream and ends at Station 500 downstream where stations are given in ft. River cross sections are then entered starting at the upstream end and moving in the downstream direction. For simplicity, the cross sections in this example are simply raised or lowered uniformly by the product of the slope and the reach length relative to the reference section to create a prismatic natural channel. Cross-section data are given in Table 6.9 for the geometry at Station 1200 which is the reference section. Pier and abutment locations are also shown in the table, but they are entered separately in the bridge/culvert data editor. For each cross section, the following data are entered: (1) river station, (2) transverse distance and elevation geometric data by copying and adjusting elevations, (3) Manning's *n* values with horizontal variation in the cross section, (4) downstream reach lengths,

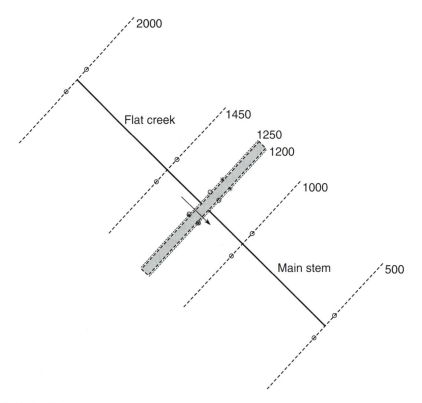

FIGURE 6.32
HEC-RAS river schematic and cross sections for Example 6.4 (Q = 20,000 cfs; S = 0.00052; bridge length = 200 ft).

and (5) main channel bank stations. The upstream bridge river station was selected to be Station 1250, which is equal to one bridge opening width downstream of the approach flow section at Station 1450 according to the WSPRO convention. The downstream bounding bridge cross section is positioned at Station 1200, which is 50 ft (15.2 m) downstream of the upstream bridge station. This distance includes the 40 ft (12.2 m) width of the bridge deck plus a 5 ft (1.5 m) allowance outside the bridge for both upstream and downstream cross sections. If the bridge has sloping embankments instead of vertical walls, then allowance is made for the distance from the edge of the road deck to the toe of the embankment to determine the locations of the bridge bounding cross sections. The exit section is located at Station 1000, which is one bridge opening width (200 ft) downstream of the downstream bounding bridge cross section. Ineffective flow areas are added at Stations 1200 and 1250.

Once the geometric cross-section data are entered, the bridge is added by entering any station between the upstream and downstream bounding cross sections. Low chord and high chord elevations are entered to block out the embankment, abutments, and bridge deck. Pier obstruction locations and widths are also entered. Finally, the bridge modeling method is chosen to be WSPRO, and the additional WSPRO variables described previously are entered.

To complete the data file, steady flow discharges and boundary conditions are entered. In this example, normal depth at a slope of 0.00052 is requested as the downstream boundary condition, which the program will calculate.

Results for the energy grade line elevation and the water surface elevation at the bridge cross section BU (inside the bridge on the upstream side) are shown in Figure 6.33. The shaded areas represent the embankment and bridge deck. Piers are also shown. The triangles and vertical lines between them indicate the location of the ineffective areas in the left and right flood plains and below the upper triangle elevations. The water surface elevation is well below the low chord of the bridge so we have free surface flow (Class 1 in Table 6.7).

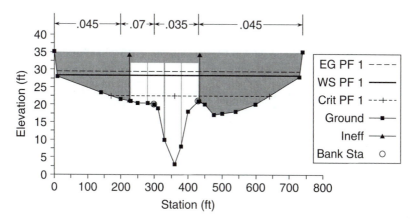

FIGURE 6.33

HEC-RAS bridge cross section (BU) with water surface elevation for Example 6.4 ($Q = 20{,}000$ cfs; $S = 0.00052$; bridge length $= 200$ ft).

The water surface profile is given in Figure 6.34, and a table of results is shown in Table 6.10. The water surface drop through the bridge is quite evident in Figure 6.34 while far upstream and downstream of the bridge, the water surface profile is that of uniform flow. The primary head loss due to the bridge can be seen in Table 6.10 at River Station 1200 representing the head loss from Station 1200 just downstream of the bridge to the exit section at Station 1000. This head loss includes 0.17 ft (0.052 m) of friction loss and 1.00 ft (0.305 m) of expansion loss (C&E). Also note that 0.14 ft (0.043 m) of friction loss occurs from the approach section to the bridge, and 0.11 ft (0.034 m) of friction loss can be attributed to flow inside the bridge. Upstream of the bridge at the approach flow section (1450), approximately 53% of the flow is in the main channel while inside the bridge, the main channel flow increases to over 90% of the total with a maximum velocity in the main channel of 10.13 ft/s (3.09 m/s). The backwater at the approach flow section (Station 1450) is the difference between the water surface elevation

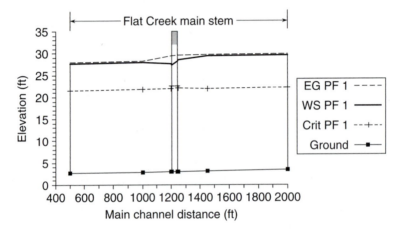

FIGURE 6.34
HEC-RAS water surface profile for Example 6.4 (Q = 20,000 cfs; S = 0.00052; bridge length = 200 ft).

TABLE 6.10   HEC-RAS output for Flat Creek (Q = 20,000 cfs; S = 0.00052; bridge length = 200 ft)

River Station (ft)	E.G. elev (ft)	W.S. elev (ft)	Crit. W.S. (ft)	Frctn loss (ft)	C&E loss (ft)	Top width (ft)	Q left (cfs)	Q chanl (cfs)	Q right (cfs)	Vel chanl (ft/s)
1450	29.71	29.48	22.01	0.14	0.00	724	2907	10528	6565	4.83
1250	29.57	28.47	22.04	0.01	0.00	721	1699	18138	162	8.80
1225 BU	29.56	28.30	22.49	0.11	0.00	191	1898	18102		9.40
1225 BD	29.45	27.31	22.46	0.01	0.00	191	1716	18284		10.13
1200	29.44	27.62	21.95	0.17	1.00	703	1488	18512		9.47
1000	28.27	27.92	21.78	0.26	0.00	720	2378	11698	5925	5.83

there of 29.48 ft (8.99 m) and the water surface elevation that would occur without the bridge in place, sometimes referred to as the unconstricted value. In general, a separate run of HEC-RAS without the bridge is required to determine the unconstricted approach water surface elevation; but in this example problem, it is simply the value at the exit section (Station 1000) plus the slope times the flow length between the exit and approach flow sections because the unconstricted case is uniform flow. Thus, the value of the backwater is $29.48 - (27.92 + 0.00052 \times 450) = 1.33$ ft (0.405 m).

REFERENCES

Arneson, L.A., and J.O. Shearman *User's Manual for WSPRO—A Computer Model for Water Surface Profile Computations.* Report FHWA-SA-98-080. Washington, DC: U.S. Dept. of Transportation, Federal Highway Administration, 1998.

Bodhaine, G. L. "Measurement of Peak Discharges at Culverts by Indirect Methods." In *Techniques of Water Resources Investigations,* Book 3, Chapter A3, Washington, DC: Government Printing Office, U.S. Geological Survey, 1976.

Bradley, J. N. *Hydraulics of Bridge Waterways (HDS-1).* Hydraulic Design Series 1. Washington, DC: Federal Highway Administration, U.S. Dept. of Transportation, 1978.

Cassidy, J. J. "Designing Spillway Crests for High Head Operation." *J. Hyd. Div.,* ASCE 96, no. HY3 (1970), pp. 745–53.

Chadwick, A., and J. Morfett. *Hydraulics in Civil and Environmental Engineering.* London and New York: Spon Press, 1998.

Chanson, H. *Hydraulic Design of Stepped Cascades, Channels, Weirs, and Spillways.* Oxford, England: Pergamon, Elsevier Science, 1994b.

Chanson, H. "Hydraulics of Skimming Flow over Stepped Channels and Spillways." *J. Hydr. Res.* 32, no. 3 (1994a), pp. 445–60.

Chow, Ven Te. *Open Channel Hydraulics.* New York: McGraw-Hill, 1959.

Christodoulou, G. C. "Energy Dissipation on Stepped Spillways." *J. Hydr. Engrg.,* ASCE 119, no. 5 (1993), pp. 644–49.

de S. Pinto, N. L. "Cavitation and Aeration." In *Advanced Dam Engineering for Design, Construction, and Rehabilitation,* ed. R. B. Jansen. New York: Van Nostrand Reinhold, 1988, pp. 620–34.

Eichert, B. S., and J. Peters. "Computer Determination of Flow Through Bridges." *J. Hyd. Div.,* ASCE 96, no. HY7 (1970), pp. 1455–68.

Federal Highway Administration. *HY-8 Culvert Hydraulic Analysis Program.* Washington, DC: Federal Highway Administration, 2008.

Federal Highway Administration. *Hydraulic Design of Culverts.* Report FHWA-IP-85-15, Hydraulic Design Series 5 (HDS-5), Zd, ed. Washington, DC: Federal Highway Administration, U.S. Dept. of Transportation, 2001.

French, R. *Open-Channel Hydraulics.* New York: McGraw-Hill, 1985.

Hager, W. H., and R. Bremen. "Plane Gate on Standard Spillway." *J. Hydr. Engrg.,* ASCE 114, no. 11(1988), pp. 1390–97.

Henderson, F. M. *Open Channel Flow.* New York: Macmillan, 1966.

Kaatz, K. J., and W. P. James. "Analysis of Alternatives for Computing Backwater at Bridges." *J. Hydr. Engrg.,* ASCE 123, no. 9 (1997), pp. 784–92.

Keller, R. J., and A. K. Rastogi. "Design Chart for Predicting Critical Point on Spillways." *J. Hyd. Div.,* ASCE 103, no. HY12 (1977), pp. 1417–29.

Kindsvater, C. E., and R. W. Carter. "Tranquil Flow Through Open-Channel Constrictions." *Transactions ASCE* 120 (1955).

Kindsvater, C. E., R. W. Carter, and H. J. Tracy. "Computation of Peak Discharge at Contractions." U.S. Geological Survey, Circular 284, Washington, DC: Government Printing Office, 1958.

Matthai, H. F. "Measurement of Peak Discharges at Width Contractions by Indirect Methods." In *Techniques of Water Resources Investigations,* Book 3, Chapter A4. U.S. Geological Survey, Washington, DC: Government Printing Office, 1976.

Rajaratnam, N. "Skimming Flow in Stepped Spillways." *J. Hydr. Engrg.,* ASCE 116, no. 4 (1990), pp. 587–91.

Reese, A. J., and S. T. Maynord. "Design of Spillway Crests." *J. Hydr. Engrg.,* ASCE 113, no. 4 (1987), pp. 476–90.

Rice, C. E., and K. C. Kadavy. "Model Study of a Roller Compacted Concrete Stepped Spillway." *J. Hydr. Engrg.,* ASCE 122, no. 6 (1996), pp. 292–97.

Schneider, V. R., J. W. Board, B. E. Colson, F. N. Lee, and L. A. Druffel. "Computation of Backwater and Discharge at Width Constrictions of Heavily Vegetated Floodplains." U.S. Geological Survey Water-Resources Investigations 76-129, Washington, DC: Government Printing Office, 1976.

Shearman, J. O., W. H. Kirby, V. R. Schneider, and H. N. Flippo. *Bridge Waterways Analysis Model: Research Report.* Report FHWA/RD-86/108. Washington, DC: Federal Highway Administration, U.S. Dept. of Transportation, 1986.

Straub, L. G., and A. G. Anderson. "Experiments on Self-Aerated Flow in Open Channels." *Trans. ASCE* (1960), p. 125.

Sturm, T. W., and A. Chrisochoides. "One-Dimensional and Two-Dimensional Estimates of Abutment Scour Prediction Variables," *Transportation Research Record 1647.* Washington, DC: National Research Council, December 1998, pp. 18–26.

U.S. Army Corps of Engineers. *Hydraulic Design Criteria.* Vicksburg, MS: U.S. Army Corps of Engineers Waterways Experiment Station, 1970.

U.S. Army Corps of Engineers. *Hydraulic Design of Spillways.* Engineer manual 1110-2-1603. Vicksburg, MS: U.S. Army Corps of Engineers, 1990.

U.S. Army Corps of Engineers. *HEC-2 User's Manual.* Davis, CA: U.S. Army Corps of Engineers, Hydrologic Engineering Center, 1991.

U.S. Army Corps of Engineers. *HEC-RAS Hydraulic Reference Manual and User's Manual, Version 4.* Davis, CA: U.S. Army Corps of Engineers, Hydrologic Engineering Center, 2008.

U.S. Bureau of Reclamation. *Design of Small Dams, a Water Resources Technical Publication,* 3rd ed. Denver: U.S. Government Printing Office, 1987.

Vischer, D. L., and W. H. Hager. *Dam Hydraulics.* Chichester: John Wiley & Sons, 1998.

Wood, I. R., P. Ackers, and J. Loveless. "General Method for Critical Point on Spillways." *J. Hydr. Engrg.,* ASCE 109, no. 2 (1983), pp. 308–12.

EXERCISES

6.1. A high overflow spillway with $P/H_d > 1.33$ has a maximum discharge of 10,000 cfs with a maximum head of 20 ft. Determine the design head, spillway crest length, and the minimum pressure on the spillway. Plot the complete spillway crest shape for a compound circular curve in the upstream quadrant of the crest.

6.2. Repeat Example 6.1 for an elliptical approach crest, using a design procedure that guarantees a minimum pressure head of -15 ft. Plot the head-discharge curve.

6.3. An ogee spillway has a crest height of 50 ft above the toe and a maximum head of 15 ft. A minimum pressure of -1.5 psi is allowed. The maximum discharge is 16,000 cfs.

    (*a*) Determine the crest length of the spillway assuming a compound circular curve for the upstream crest shape. What is the pressure at the *crest* for the maximum discharge?

    (*b*) If the spillway is designed as a stepped spillway, with each step 2 ft high by 1.5 ft long, what is the energy dissipation in feet of water at the maximum discharge?

6.4. An existing ogee spillway with an elliptical crest has a crest height of 7.0 m and a crest length of 15.2 m. A minimum gage pressure of zero (atmospheric pressure) occurs at a head of 14.0 m. What maximum head and discharge would you recommend for this spillway?

6.5 The high-head spillway designed in Example 6.1 has an effective crest length of 29.0 m and a design head of 14.6 m. The height of the spillway crest relative to the invert of the approach flow channel is $P = 18$ m. The spillway has three bays with radial gates at the crest. The gates have a radius, $r = 32.5$ m, and the height of the gate trunnion relative to the crest is $a = 12.5$ m. Construct a complete head-discharge curve for uncontrolled flow over the spillway up to the maximum discharge of 5700 m³/s with head plotted on the vertical axis and discharge on the horizontal axis. On the same graph, plot the gated head-discharge curves for gate openings of 2, 4, 6, and 8 m starting in each case at the intersection with the uncontrolled spillway curve and going up to a maximum head of 15 m relative to the spillway crest. Reference all heads to the spillway crest before plotting.

6.6 A standard WES ogee spillway with six vertical gates at the crest is used to control discharge in a large water supply and irrigation canal system. The height of the spillway crest is 11.3 ft, and the design head is 14.9 ft. The effective spillway crest length is 160 ft. The normal operating head upstream of the spillway is 12.3 ft. If a release discharge of 18,000 cfs is required, what should the gate openings be? Use Equation 6.3c.

6.7 For a head on the spillway of Exercise 6.6 of 12.3 ft and gate openings of 8 ft, calculate the gate-controlled discharge. For free flow at the same discharge, what would the head be? Use Equation 6.3c to calculate the gate-controlled discharge.

6.8 An irrigation canal is rectangular in shape with a width of 1.5 m and a maximum depth of 1.0 m. Flow is controlled by a vertical sluice gate. If it is desired to maintain the normal depth of flow of 0.80 m upstream of the gate, what should the gate opening be for an irrigation demand of 0.85 m³/s? Check to see if the hydraulic jump and the gate will be submerged on the downstream side if the tailwater depth is also the normal depth of 0.80 m.

6.9. A 0.91 m diameter corrugated metal pipe culvert ($n = 0.024$) has a length of 90 m and a slope of 0.0067. The entrance has a square edge in a headwall. At the design discharge of 1.2 m³/s, the tailwater is 0.45 m above the outlet invert. Determine the head on the culvert at the design discharge. Repeat the calculation for head if the culvert is concrete.

6.10. Show that Equation 6.10a for a box culvert in inlet control with the entrance unsubmerged can be placed in a form in which Q is proportional to the head, *HW*, to the 3/2 power.

6.11. A 3 ft by 3 ft concrete ($n = 0.012$) box culvert has a slope of 0.006 and a length of 250 ft. The entrance is a square edge in a headwall. Determine the head on the culvert for a discharge of 50 cfs and a discharge of 150 cfs. The downstream tailwater elevation is 0.5 ft above the *outlet invert* for 50 cfs and 3 ft above the *outlet invert* at 150 cfs.

6.12. Design a box culvert to carry a design discharge of 600 cfs. The culvert invert elevation is 100 ft and the allowable headwater elevation is 114 ft. The paved roadway is 500 ft long and overtops at 115 ft. The culvert length is 200 ft with a slope of 1.0 percent. The following tailwater elevations apply up to the maximum discharge of 1000 cfs:

Q, cfs	TW, ft
200	101.4
400	102.6
600	103.1
800	103.8
1000	104.1

Prepare a performance curve for the culvert design by hand and compare with the results of HY8. Also use HY8 to prepare a performance curve if the slope is 0.1 percent.

6.13. A circular concrete culvert has a diameter of 5 ft with a square-edged entrance in a headwall. The culvert is 500 ft long with a slope of 0.005 and an inlet invert elevation of 100.0 ft. The downstream channel is trapezoidal with a bottom width of 10 ft, side slopes of 2:1, slope = 0.005, and Manning's $n = 0.025$. The paved roadway has a constant elevation of 130 ft with a length of 100 ft and a width of 50 ft. The design discharge is 250 cfs and the maximum discharge is 500 cfs. Use HY8 to construct and plot the performance curve for these data, and compare this with the performance curve for a 5 ft diameter corrugated steel pipe. Also compare this with the performance curve for the 5 ft diameter concrete culvert with a side-tapered inlet.

6.14. An existing circular corrugated steel culvert has a diameter of 4 ft and a length of 250 ft at a slope of 0.018. The design discharge is 120 cfs, and the allowable headwater is 15 ft above the inlet invert of the culvert. The culvert entrance condition is a thin edge projecting pipe. The downstream channel is trapezoidal with a bottom width = 10 ft, side slopes = 2:1, bed slope = 0.004, and Manning's $n = 0.035$. Urban development upstream has increased the design discharge to 150 cfs. Use HY8 to analyze the existing culvert and plot its performance curve. Then develop a design that will convey the increase in design discharge without overtopping of the roadway.

6.15. Prove that $\alpha_3 = 1/C^2$ in the WSPRO methodology where $\alpha_3 =$ kinetic energy flux correction coefficient at section 3 and $C =$ USGS bridge discharge coefficient.

6.16. Apply the HDS-1 method to the data given in Example 6.4, and compare the backwater to that obtained from WSPRO.

6.17. Using the USGS width-contraction curves in Figures 6.23 and 6.24, determine the value of the bridge discharge coefficient and the discharge ratio m ($= M(K)$) for Example 6.4.

6.18. Change the bridge type to Type 3 for the WSPRO example (Example 6.4), and determine the backwater for Q values ranging from 5,000 to 25,000 cfs. Plot the results in a graph comparing the Type 1 and Type 3 bridges in this range of discharges for a bridge length of 200 ft.

6.19. For a bridge length of 200 ft and a Type 1 bridge, change the low chord elevation in the WSPRO example to 28 ft with a constant roadway elevation of 31 ft. Determine the backwater and the overtopping discharge, if any, for the same range of discharges as in Exercise 6.18. Plot the results in comparison with Example 6.4.

6.20. For Example 6.4, reduce the bridge length to 150 ft (with two bridge piers), and introduce a relief bridge with a length of 50 ft at a location of your choice in one of the floodplains. Plot the results for backwater over the same range of discharges as in Exercise 6.18 in comparison with the results from Example 6.4.

6.21. Analyze the existing bridge over Duck Creek using HEC-RAS (WSPRO option) and the following table of cross-section data. The bridge is Type 4 with a width of 30 ft, embankment side slopes of 2:1, embankment elevation of 790 ft, and wingwall angle of 30°. The low chord elevation is 788 ft. The design discharge is 6850 cfs with a water surface elevation of 784.66 in the exit cross section. The full valley sections just downstream and upstream of the bridge should be identical to the bridge section over the bridge opening width and the same as the exit section in both floodplains.
(a) Determine the backwater for the existing bridge.
(b) Design a new bridge to replace the old one so that the backwater is ≤0.25 ft.

Duck Creek cross sections

Exit, Station 1000

Point	Distance (ft)	Elevation (ft)	Mannings n
1	−150	792	0.08
2	−105	780	
3	−70	780	
4	−28	778	0.04
5	−24	774	
6	−20	773	
7	18	772	
8	22	772	
9	35	780	0.08
10	50	780	
11	210	779	0.05
12	600	779	0.08
13	860	780	0.05
14	1005	782	
15	1050	784	
16	1112	786	
17	1260	788	
18	1310	798	

Bridge, Station 1100

Point	Distance (ft)	Elevation (ft)	Mannings n
1	−71	788	0.04
2	−71	778.2	
3	−30	776	
4	−15	772.5	
5	16	772.5	
6	25	775	
7	25	788	

Approach, Station 1230

Point	Distance (ft)	Elevation (ft)	Mannings n
1	−480	796	0.10
2	−440	788	
3	−420	786	
4	−305	784	
5	−175	782	
6	−95	780	
7	−50	778	0.04
8	−30	776	
9	−25	774	
10	2	772	
11	17	772	
12	20	774	
13	28	780	0.08
14	50	780	
15	670	780	0.05
16	990	782	
17	1070	784	
18	1120	786	
19	1260	810	

6.22. Apply HEC-RAS to Example 6.4 using the energy, momentum, and Yarnell methods. Set up the cross sections in the schematic layout starting at the upstream station of 1450. Use a constant slope of 0.00052 and add the appropriate amount to all elevations given for station 1200 in Example 6.4. Then in the geometric data editor, copy the cross sections downstream adjusting the elevation downward according to the slope and distance between stations. Use values of 0.3 and 0.1 for the expansion and contraction loss coefficients, respectively. Establish stations at 1250, 1200, and 1000 to correspond with the WSPRO sections. Add a bridge at station 1225 using the geometric data editor. In the bridge/culvert editor, enter the deck and roadway data, pier data, and check the boxes for all three methods of computation as well as the box choosing the highest energy answer in the bridge modeling approach window. Also enter a pier drag coefficient of 2.0 and a Yarnell pier coefficient of 0.9. In the cross section data editor, add ineffective flow areas at stations 1250 and 1200 to the left and right of the bridge opening specified at elevations above the low chord elevation but below the top of the roadway. In the steady flow data menu, enter the discharge of 20,000 cfs and choose normal depth as the downstream control with a slope of 0.00052. Compare the results with WSPRO, and then make a second run with the exit section at station 800 instead of 1000. Discuss the results.

6.23. A bridge over the Rock River is 366 ft long. It has six pier bents, which are 3 ft wide and spaced at 51 ft intervals except for the two in the main river channel, which are 60 ft apart. The low chord elevation is 435 ft and the elevation of the top of the road-way is 438 ft. The embankment and abutments have side slopes of 2:1. Cross sections and bridge data are summarized in the tables below including tailwater elevations for four discharges. The bridge is located at Station 45. Assume a Manning's n of 0.05 for the channel and 0.15 for the floodplains. Use HEC-RAS with all four bridge modeling methods for low flow and the pressure/weir flow method for high flows to determine the bridge backwater for each of the four discharges shown in the table below. Then answer the following questions:

(a) What type of flow occurs for each discharge, and how much discharge overtops the bridge and embankments, if any?

(b) Which modeling method produces the highest backwater for the low flows? How much is the backwater for each of the four discharges? (Print tables of results and plotted water surface profiles.)

(c) What are the channel and left floodplain velocities inside the bridge for each discharge using the WSPRO method for low flows and the pressure/weir flow method for high flows? Where is scour most likely to occur?

Rock River cross section data

STA. -400 (Exit)		STA. 0 (D.S. toe)		STA. 90 (U.S. toe)		STA. 490 (Approach)	
Sect. dist. (ft)	Elev. (ft)	Sect. dist. (ft)	Elev. (ft)	Sect. dist. (ft)	Elev. (ft)	Sect. dist. (ft)	Elev. (ft)
2219	500.0	2000	454.6	2000	454.9	2294	460.0
2282	490.0	2100	448.6	2100	448.9	2419	450.0
2375	480.0	2200	446.2	2200	446.5	2544	440.0
2500	470.0	2300	442.7	2300	443.0	2657	430.0
2532	460.0	2400	436.9	2400	437.2	2857	425.0
2544	450.0	2500	427.3	2500	427.6	2919	426.5
2557	440.0	2600	423.1	2600	423.4	2928	424.9
2575	430.0	2700	420.9	2700	421.2	2940	416.1
2638	425.0	2800	421.5	2800	421.8	2947	414.4
2919	424.5	2862	422.2	2862	422.5	2949	408.3
2928	422.9	2900	423.9	2900	424.2	3000	400.0
2940	414.1	2919	425.1	2919	425.4	3030	405.0
2947	412.4	2928	423.5	2928	423.8	3040	419.5
2949	406.3	2940	414.7	2940	415.0	3134	420.0
3000	398.0	2947	413.0	2947	413.3	3253	430.0
3030	403.0	2949	406.9	2949	407.2	3315	440.0
3040	417.5	3000	398.6	3000	398.9	3390	450.0
3059	420.0	3030	403.6	3030	403.9	3453	460.0
3078	430.0	3040	418.1	3040	418.4	3515	470.0
3096	440.0	3050	419.4	3050	419.7	3590	480.0
3190	450.0	3058	421.6	3058	421.9	3653	490.0
3203	460.0	3100	431.9	3100	432.2	3715	500.0
3265	470.0	3141	440.1	3141	440.4	3778	510.0
3415	475.0	3200	442.5	3200	442.8		
		3300	449.8	3300	450.1		
		3400	455.0	3400	455.0		

Rock River Bridge Abutment Data

Abutment locations (ft)	Top of deck elev. (ft)	Low chord elev. (ft)
2756	438.00	435.00
3122	438.00	435.00

Rock River bridge piers

Pier locations (ft)
2807
2858
2909
2960
3020
3071

Rock River Boundary Conditions

Discharge (cfs)	Tailwater elev. @STA. -400 (ft)
4110	421.34
15000	431.91
25000	436.98
30000	439.46

7

Governing Equations of Unsteady Flow

7.1   INTRODUCTION

In unsteady flow, velocities and depths change with time at any fixed spatial position in an open channel. Open channel flow in natural channels almost always is unsteady, although it often is analyzed in a quasi-steady state for channel design or floodplain mapping. Unsteady flow in open channels by nature is nonuniform as well as unsteady because of the free surface. Mathematically, this means that the two dependent flow variables (e.g., velocity and depth or discharge and depth) are functions of both distance along the channel and time for one-dimensional applications. Problem formulation requires two partial differential equations representing the continuity and momentum principles in the two unknown dependent variables. (The differential form of the energy equation could be used in cases where the flow variables are continuous, but the momentum equation is required where they are discontinuous, as in surges or tidal bores.) The full differential forms of the two governing equations are called the *Saint-Venant equations* or the *dynamic wave equations*. Only in rather severe simplifications of the governing equations are analytical solutions available for unsteady flow. This situation has led to the extensive development of appropriate numerical techniques for the solution of the governing equations. Several of these techniques will be explored in the next chapter.

Unsteady flow problems arise in hydraulic engineering in a variety of settings, ranging from waves formed in irrigation channels by gate operation or in hydroelectric plant headraces and tailraces by turbine operation to natural flood waves and dam-break surges in rivers. The types of waves considered in these situations are called *translatory waves* because of their continuous movement along the channel as

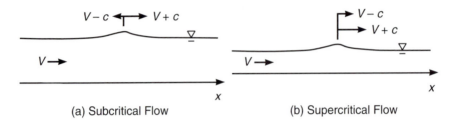

FIGURE 7.1
Wave propagation in subcritical and supercritical flow.

opposed to periodic or oscillatory ocean waves, which are not considered here. In addition, only shallow water waves are considered, in which water movement occurs over the full depth and vertical velocity and acceleration can be neglected to allow the use of one-dimensional forms of the governing equations. In all the wave problems considered, the purpose of obtaining the solution of the governing equations (referred to as *routing* in the context of flood waves) is to describe the flow velocity and depth as functions of space and time. In other words, the spatial shape and temporal development of the translatory wave are sought.

A more formal definition of the *translatory wave* describes it as a disturbance moving in the longitudinal direction that gives rise to changes in discharge, velocity, and depth with time. It propagates with an absolute speed, designated by dx/dt, which is the sum of the mean water velocity, V, and the wave celerity with respect to still water, c, as illustrated in Figure 7.1, with positive V in the positive x direction. Because the wave can move in both upstream and downstream directions, its absolute speed is given by $V \pm c$. The celerity of a long wave of small amplitude is given by $(gy)^{1/2}$, in which y is depth, so the values of dx/dt depend on the Froude number, $\mathbf{F}$, defined by V/c. In subcritical flow in which $V < c$ and $\mathbf{F} < 1$, dx/dt has two possible values given by $V + c$ in the downstream direction and $V - c$ in the upstream direction, as shown in Figure 7.1a. On the other hand, the two possible wave propagation speeds in supercritical flow, given by $(V + c)$ and $(V - c)$, are both in the downstream direction, because $V > c$ and $\mathbf{F} > 1$, as illustrated in Figure 7.1b.

The physical property of two possible wave propagation speeds is particular to hyperbolic partial differential equations, which have the mathematical property of two possible characteristic directions or paths along which discontinuities in the derivatives travel. The connection between the physical and mathematical properties of the Saint-Venant equations allows them to assume a simpler form in characteristic coordinates associated with the path of two moving observers traveling at the speeds of $(V \pm c)$. As a result, we first derive the Saint-Venant equations and then begin the study of unsteady flow, with a transformation of the equations to characteristic form to provide a deeper understanding of the physics of wave propagation as well as the initial and boundary conditions necessary to solve the Saint-Venant equations. The characteristic equations are simplified for the case of a "simple wave," with no gravity or friction effects, and applied to sluice-gate operation problems as a learning tool for understanding the characteristic form. In the next chapter, which covers numerical solution techniques for the governing equations, we also apply finite difference techniques to the solution of the governing equations in characteristic form, which has come to be called the *method of characteristics*. In addition, we consider explicit and implicit finite difference techniques applied to the untransformed Saint-Venant equations and discuss the advantages and disadvantages of each method. Applications include the problems of hydroelectric power load acceptance and rejection, dam breaks, and flood routing.

7.2   DERIVATION OF SAINT-VENANT EQUATIONS

Although the governing equations of continuity and momentum can be derived in a number of ways, we apply a control volume of small but finite length, Δx, that is reduced to zero length in the limit to obtain the final differential equation. The derivations make the following assumptions (Yevjevich 1975; Chaudhry 1993):

1. The shallow water approximations apply so that vertical accelerations are negligible, resulting in a vertical pressure distribution that is hydrostatic; and the depth, y, is small compared to the wavelength so that the wave celerity $c = (gy)^{1/2}$.
2. The channel bottom slope is small, so that $\cos^2 \theta$ in the hydrostatic pressure force formulation is approximately unity, and $\sin\theta \approx \tan\theta = S_0$, the channel bed slope, where θ is the angle of the channel bed relative to the horizontal.
3. The channel bed is stable, so that the bed elevations do not change with time.
4. The flow can be represented as one-dimensional with (a) a horizontal water surface across any cross section such that transverse velocities are negligible and (b) an average boundary shear stress that can be applied to the whole cross section.
5. The frictional bed resistance is the same in unsteady flow as in steady flow, so that the Manning or Chezy equations can be used to evaluate the mean boundary shear stress.

Additional simplifying assumptions made subsequently may be true in only certain instances. The momentum flux correction factor, β, for example, will not be assumed to be unity at first because it can be significant in river overbank flows.

Continuity Equation

First, consider the continuity equation, which will be derived from the control volume of height equal to the depth, y, and length, Δx, as shown in Figure 7.2. As in the derivation of continuity in Chapter 1, which used the Reynolds transport theorem, the basic statement of volume conservation for an incompressible fluid flowing through the control volume is Net Volume Out = $-$ Change in Storage in the time interval, Δt. This can be expressed as

$$\frac{\partial Q}{\partial x} \Delta x \Delta t - q_L \Delta x \Delta t = -\Delta x \frac{\partial A}{\partial t} \Delta t \tag{7.1}$$

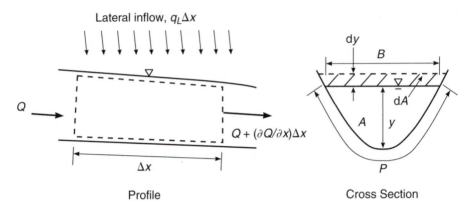

FIGURE 7.2
Control volume for derivation of unsteady continuity equation.

in which q_L = lateral inflow rate per unit length of channel and A = cross-sectional area of flow. Dividing by $\Delta x \Delta t$ and taking both the control volume length and the time interval to zero, the continuity equation is

$$\frac{\partial A}{\partial t} + \frac{\partial Q}{\partial x} = q_L \qquad (7.2)$$

Substituting $dA = Bdy$ from Figure 7.2, in which B = channel top width at the free surface, continuity becomes

$$B\frac{\partial y}{\partial t} + \frac{\partial Q}{\partial x} = q_L \qquad (7.3)$$

By definition of the discharge as $Q = AV$, in which V = mean cross-sectional velocity in the flow direction (x), the $\partial Q/\partial x$ term in (7.3) can be written as $A(\partial V/\partial x) + V(\partial A/\partial x)$, using the product rule. However, the term involving $\partial A/\partial x$ must be evaluated carefully because A can vary with both depth, y, and distance, x, if the channel width is changing:

$$\frac{\partial A}{\partial x} = \frac{\partial A}{\partial x}\bigg|_y + B\frac{\partial y}{\partial x} \qquad (7.4)$$

where the first term on the right hand side of (7.4) represents the derivative of A with respect to x while holding y constant. For prismatic channels, this term goes to zero. Finally, with these substitutions for $\partial Q/\partial x$ and then $\partial A/\partial x$, and dividing through by the top width, B, the continuity equation reduces to

$$\frac{\partial y}{\partial t} + V\frac{\partial y}{\partial x} + D\frac{\partial V}{\partial x} + \frac{V}{B}\frac{\partial A}{\partial x}\bigg|_y = \frac{q_L}{B} \qquad (7.5)$$

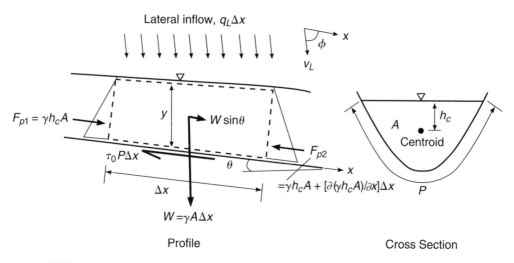

FIGURE 7.3
Control volume for derivation of unsteady momentum equation.

in which $D = A/B$ = hydraulic depth. For a prismatic channel with no lateral inflow, the fourth term on the left hand side as well as the right hand side go to zero. Furthermore, if a rectangular cross section is considered, the continuity equation becomes

$$\frac{\partial y}{\partial t} + \frac{\partial q}{\partial x} = 0 \tag{7.6}$$

in which $q = Vy$ = flow rate per unit of width. In this form, it is evident that temporal changes in depth at a point must be balanced by a longitudinal gradient in flow rate per unit of width.

Momentum Equation

The momentum equation is derived with reference to Figure 7.3, in which the forces acting on the control volume are shown. Pressure, gravity, and shear forces are considered, and these must balance the time rate of change of momentum inside the control volume and the net momentum flux out of the control volume. In the x direction, which is taken to be the flow direction, the momentum equation can be written

$$F_{px} + F_{gx} - F_{sx} = \frac{\partial}{\partial t}\left[\int_A \rho v_x \, dA\right]\Delta x + \frac{\partial}{\partial x}\left[\int_A \rho v_x^2 \, dA\right]\Delta x - \rho q_L \Delta x v_L \cos\phi \tag{7.7}$$

in which F_{px} = pressure force component in the x direction; F_{gx} = gravity force component in the x direction; F_{sx} = shear force component in the x direction; v_x = point velocity in the x direction; q_L = lateral inflow per unit of length in the flow direction; and v_L = velocity of lateral inflow inclined at angle ϕ to the x direction.

Expressions can be developed for each of the force terms. Assuming a hydrostatic pressure distribution, the pressure force, $F_{px} = F_{p1} - F_{p2}$, and is given by

$$F_{px} = -\frac{\partial}{\partial x}(\gamma h_c A)\Delta x = -\gamma A \frac{\partial y}{\partial x}\Delta x \qquad (7.8)$$

in which h_c = vertical distance below the free surface to the centroid of the flow cross-sectional area; A = cross-sectional area on which the force acts; and Ah_c = $\int_0^{y(x)}[y(x) - \eta]b(\eta)\,d\eta$, which represents the first moment of the area about the free surface with b = local width of the cross section at height η from the bottom of the channel. Note that the pressure force contribution arising from a change in cross-sectional area due to an expanding or contracting nonprismatic channel is just balanced by the component of pressure force on the channel banks in the flow direction (Liggett 1975; Cunge, Holly, and Verwey 1980). Consequently, the evaluation of the derivative shown on the far right hand side of Equation 7.8 ignores the variation in channel width with x and comes only from the integral definition of Ah_c and the Leibniz rule. The gravity force component in the x direction is given by

$$F_{gx} = \gamma A \Delta x S_0 \qquad (7.9)$$

in which S_0 = bed slope = $\tan\theta$, which has been used to approximate $\sin\theta$ for small values of slope. Finally, the boundary shear force in the x direction can be expressed as

$$F_{sx} = \tau_0 P \Delta x \qquad (7.10)$$

in which τ_0 = mean boundary shear stress; and P = boundary wetted perimeter.

On the momentum flux side of the momentum equation, the net convective flux of momentum out of the control volume can be written

$$\frac{\partial}{\partial x}\left[\int_A \rho v_x^2 dA\right]\Delta x = \frac{\partial}{\partial x}[\beta \rho V^2 A]\Delta x \qquad (7.11)$$

with β = momentum flux correction factor and V = mean cross-sectional velocity. The time rate of change of momentum inside the control volume for an incompressible fluid becomes

$$\frac{\partial}{\partial t}\left[\int_A \rho v_x dA\right]\Delta x = \rho\frac{\partial}{\partial t}[VA]\Delta x \qquad (7.12)$$

Substituting Equations 7.8 to 7.12 into Equation 7.7, dividing by $\rho\Delta x$, and letting Δx go to zero results in

$$\frac{\partial Q}{\partial t} + \frac{\partial}{\partial x}\left(\beta\frac{Q^2}{A}\right) + \frac{\partial}{\partial x}(gh_c A) = gA(S_0 - S_f) + q_L v_L \cos\phi \qquad (7.13)$$

in which $Q = AV$; S_f = friction slope = $\tau_0/(\gamma R)$; R = hydraulic radius = A/P; and q_L = lateral inflow per unit of length with velocity, v_L, and at an angle of ϕ with respect to the x direction. In order from left to right, the terms on the left-hand side of Equation 7.13 come from: (1) the time rate of change of momentum inside the control volume, (2) the net momentum flux out of the control volume, and (3) the net pressure force in the x direction. On the right-hand side, we have the contributions of: (1) the gravity force, (2) the boundary shear force, and (3) the momentum flux of the lateral inflow, all in the x direction. Equation 7.13 represents the momentum equation in conservation form for a prismatic channel. This simply means that, if the terms on the right hand side of (7.13) go to zero, the force plus momentum flux terms on the left hand side of the equation are conserved; and this may be the most appropriate form in which to apply some numerical solution schemes.

Equation 7.13 sometimes is placed in reduced form by applying the product rule of differentiation, substituting for $\partial A/\partial t$ from continuity, and dividing through by cross-sectional area, A, to yield

$$\frac{\partial V}{\partial t} + (2\beta - 1)\,V\frac{\partial V}{\partial x} + (\beta - 1)\frac{V^2}{A}\frac{\partial A}{\partial x} + V^2\frac{\partial \beta}{\partial x} + g\frac{\partial y}{\partial x}$$

$$= g(S_0 - S_f) + \frac{q_L}{A}(v_L\cos\phi - V)$$

(7.14)

Furthermore, the momentum equation often is given for the case of $\beta \approx 1$ and $\partial\beta/\partial x \approx 0$ for prismatic channels:

$$\frac{\partial V}{\partial t} + V\frac{\partial V}{\partial x} + g\frac{\partial y}{\partial x} = g(S_0 - S_f) + \frac{q_L}{A}(v_L\cos\phi - V) \qquad (7.15)$$

It is interesting to note that the two lateral inflow terms on the right hand side of (7.15) include contributions from both the convective momentum flux and the local change in momentum, respectively. The convective term goes to zero if the lateral inflow is at right angles to the main flow ($\phi = 0$), but the local contribution remains unless $q_L = 0$.

If the lateral inflow is zero, and (7.15) is rearranged as follows, the contribution of the various terms in the momentum equation with respect to different types of flow can be identified:

$$S_f = S_0 - \frac{\partial y}{\partial x} - \frac{V}{g}\frac{\partial V}{\partial x} - \frac{1}{g}\frac{\partial V}{\partial t}$$

steady, uniform flow
steady, gradually varied flow
unsteady, gradually varied flow

(7.16)

The steady, uniform flow case simply means that $\tau_0 = \gamma R S_0$, as derived previously in Chapter 4. The momentum equation for steady, gradually varied flow can be derived in a more familiar form by starting with Equations 7.2 and 7.13 with the time derivative terms set to zero. The result in terms of dy/dx, with $d\beta/dx = 0$, is given by (Yen and Wenzel 1970)

$$\frac{dy}{dx} = \frac{S_0 - S_f + \dfrac{q_L}{gA}(v_L\cos\phi - 2\beta V)}{1 - F_\beta^2}$$

(7.17)

in which $F_\beta^2 = \beta V^2/(gD) =$ momentum form of the Froude number; and $D =$ hydraulic depth $= A/B$. Equation 7.17 can be used for spatially varied flow or for steady, gradually varied flow with no lateral inflow. The differences between (7.17) and the energy form of the equation for gradually varied flow derived in Chapter 5 lie not only in the different definition of the Froude number (with β instead of α) but also in the friction slope S_f, which is defined as $\tau_0/\gamma R$ and replaces the slope of the energy grade line S_e in the energy equation. As a practical matter, both S_f and S_e are evaluated by the Manning or Chezy equations, but they have different definitions (Yen 1973).

The choice of dependent variables may depend on the numerical technique applied to solve the Saint-Venant equations. In the preferred conservation form, with discharge Q and depth y as the dependent variables, Equations 7.2 and 7.13 would be appropriate for the continuity and momentum equations, respectively. Another commonly used form is the reduced form of the continuity and momentum equations with velocity V and depth y as dependent variables, as given by Equations 7.5 and 7.15. Numerical techniques are discussed in the next chapter.

7.3   TRANSFORMATION TO CHARACTERISTIC FORM

The transformation to the characteristic form of the pair of partial differential equations given by (7.5) and (7.15) allows them to be replaced by four ordinary differential equations in the x-t plane (x represents the flow direction and t is time). Much simpler, ordinary differential equations must be satisfied along two inherent characteristic directions or paths in the x-t plane in the characteristic form. Although numerical analysis by the method of characteristics has fallen out of favor because of the difficulties involved in the supercritical case with the formation of surges, it has the advantage of being more accurate and lending a deeper understanding of the physics of shallow water wave problems as well as the mathematics of required initial and boundary conditions. In addition, the method of characteristics is essential in some explicit finite difference techniques, specifically for the evaluation of boundary conditions. Finally, the method of characteristics is useful for explaining kinematic wave routing in Chapter 9.

There are two methods of arriving at a characteristic form: (1) taking a linear combination of the momentum and continuity equations and rearranging the terms and (2) performing a matrix analysis that relies on the fundamental mathematical meaning of the characteristic form. We begin with the first approach because of its simplicity. We assume a prismatic channel without lateral inflow for the same reason.

The momentum equation (Equation 7.15) with the foregoing simplifications is multiplied alternately by the quantity $\pm(D/g)^{1/2}$ and added to the continuity equation (Equation 7.5) to give two new equations, the solution of which is the same as the original pair. The quantity D is the hydraulic depth given by A/B for a general nonrectangular cross section where A is the cross-sectional area and B is the top width. The resulting two new equations easily are shown to be given by

$$\left[\frac{\partial}{\partial t} + (V + c)\frac{\partial}{\partial x}\right]y + \frac{c}{g}\left[\frac{\partial}{\partial t} + (V + c)\frac{\partial}{\partial x}\right]V = c(S_0 - S_f) \qquad (7.18)$$

$$-\left[\frac{\partial}{\partial t} + (V - c)\frac{\partial}{\partial x}\right]y + \frac{c}{g}\left[\frac{\partial}{\partial t} + (V - c)\frac{\partial}{\partial x}\right]V = c(S_0 - S_f) \qquad (7.19)$$

in which $c = (gD)^{1/2}$ = wave celerity in a nonrectangular channel, as shown in Chapter 2. Of particular interest in (7.18) and (7.19) are the two operators appearing in brackets on the left hand sides of these two equations. First the operators are applied to depth y and then to velocity V in both equations, and they differ only in the multiplier on the x derivative, which is given by $(V + c)$ in (7.18) and $(V - c)$ in (7.19). This particular operator can be recognized as the total or material derivative D/Dt found elsewhere in fluid mechanics operating on the density, ρ, in the continuity equation or on the velocity vector to give the acceleration in the equations of motion. In general, if a function f varies with both position, x, and time, t, the total derivative is given by the chain rule to be

$$\frac{Df}{Dt} = \frac{\partial f}{\partial t} + \frac{\partial f}{\partial x}\frac{dx}{dt} \qquad (7.20)$$

Equation 7.20 can be interpreted to define the total time rate of change of the function f as seen by an observer moving through the fluid with speed dx/dt, with the first term on the right hand side of (7.20) giving the local change in f with time and the second term representing the convective change in f. Applying this interpretation to (7.18) and (7.19), it can be said that Equation 7.18 is an ordinary differential equation that must be satisfied along a path in the x-t plane described by an observer moving with the speed $(V + c)$, while Equation 7.19 must be satisfied along a path described by a second observer with speed $(V - c)$. Mathematically, the pair of governing partial differential equations has been transformed into four ordinary differential equations that have the same solution as the original system:

$$\text{along C1:} \qquad \left(\frac{Dy}{Dt}\right)_1 + \frac{c}{g}\left(\frac{DV}{Dt}\right)_1 = c(S_0 - S_f) \qquad (7.21a)$$

$$\text{C1:} \qquad \frac{dx}{dt} = (V + c) \qquad (7.21b)$$

$$\text{along C2:} \qquad -\left(\frac{Dy}{Dt}\right)_2 + \frac{c}{g}\left(\frac{DV}{Dt}\right)_2 = c(S_0 - S_f) \qquad (7.21c)$$

$$\text{C2:} \qquad \frac{dx}{dt} = (V - c) \qquad (7.21d)$$

in which the subscripts 1 and 2 refer to the two total derivative operators defined in (7.18) and (7.19) with two different speeds of moving observers, $(V + c)$ and

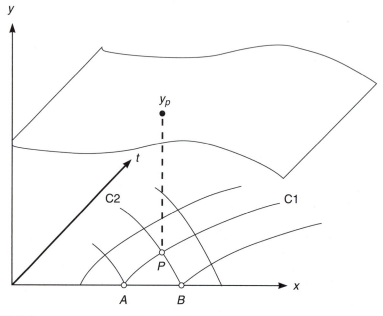

FIGURE 7.4
Characteristics in the *x-t* plane defining the solution surface for depth $y = f(x, t)$.

$(V - c)$, respectively. The family of characteristics defined by (7.21b), along which (7.21a) must be satisfied, are designated C1 characteristics, which have also been referred to as *forward characteristics* or *positive characteristics*. The C2 characteristics, also known as *backward* or *negative characteristics*, are defined by (7.21d), along which (7.21c) must be satisfied.

The two families of C1 and C2 characteristics are shown in the *x-t* plane in Figure 7.4 for a case of subcritical flow. Observer *A*, beginning at point *A*, follows the C1 characteristic path to meet observer *B*, who followed the C2 path, at point *P*. At point *P*, both observers must see the same values of depth and velocity, even though they got there by different paths and experienced different rates of change in their initial values of depth and velocity, which they picked up at the starting points, *A* and *B*. The solution for the values of depth, y_P, and velocity, V_P, at point *P* comes from the simultaneous solution of (7.21a) and (7.21c) at the position x_P and time t_P, determined by (7.21b) and (7.21d). This process can be repeated for each pair of C1 and C2 characteristics emanating from the *x* axis, along which initial conditions are specified, to determine the solutions for velocity and depth as well as positions of all points *P* at the first set of intersections or time levels. These solutions become the initial conditions for the next time levels until the solution is defined at all interior points in the *x-t* plane. The boundary conditions complete the solution for the entire *x-t* plane. In essence, the characteristic grid is a curvilinear coordinate system built as part of the solution process to define points where depth and velocity can be obtained in a simultaneous solution of all four equations given by (7.21).

Numerical solution techniques for Equations 7.21 are developed in the next chapter. The assumptions of no lateral inflow and a prismatic channel need not be made. The relaxation of these assumptions simply produces additional source terms on the right hand sides of Equations 7.21a and 7.21c. At the other end of the complexity spectrum, Equations 7.21 take on a simpler form for the special case of a rectangular prismatic channel. Because $c = (gy)^{1/2}$ for this case, it can be shown that $dy/dt = (2c/g) \, dc/dt$, from which it follows that the characteristic equations reduce to

$$\text{along C1:} \qquad \left[\frac{D(V + 2c)}{Dt} \right]_1 = g(S_0 - S_f) \qquad (7.22a)$$

$$\text{C1:} \qquad \frac{dx}{dt} = (V + c) \qquad (7.22b)$$

$$\text{along C2:} \qquad \left[\frac{D(V - 2c)}{Dt} \right]_2 = g(S_0 - S_f) \qquad (7.22c)$$

$$\text{C2:} \qquad \frac{dx}{dt} = (V - c) \qquad (7.22d)$$

From Equations 7.22, it is clear that the function subject to time variations is $(V + 2c)$ along the C1 characteristics and $(V - 2c)$ along the C2 characteristics. This suggests the interesting case, although not very practical, of the right hand sides of (7.22a) and (7.22c) becoming zero so that $(V + 2c)$ and $(V - 2c)$ become constant along the C1 and C2 characteristics, respectively. Such a simplification forms the basis of the simple wave problem for which analytical solutions exist. The simple wave problem is explored in more detail later in this chapter.

The physical connection between characteristic directions and paths of wave propagation now should be clear. The movement of elementary waves both upstream and downstream from a disturbance with speeds $(V + c)$ and $(V - c)$ in subcritical flow delineates paths along which the characteristic equations are satisfied. The complete solution describes a surface above the x-t plane that gives the values of depth and velocity for all x and t, as illustrated in Figure 7.4 for the depth.

The propagation of waves both upstream and downstream is limited to subcritical flow, while in supercritical flow the absolute speeds of $(V + c)$ and $(V - c)$ result in downstream travel only as shown in Figure 7.5, in which both the characteristics are inclined downstream.

7.4 MATHEMATICAL INTERPRETATION OF CHARACTERISTICS

As mentioned previously, a second approach for transforming the governing equations into characteristic form is a matrix analysis that arises from the mathematical interpretation of characteristics. Characteristics are defined mathematically as paths

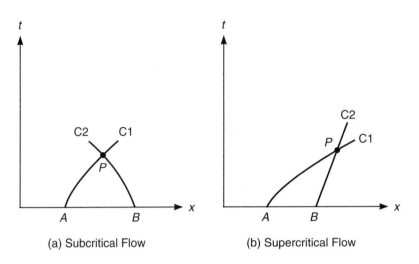

FIGURE 7.5
Characteristics in subcritical and supercritical flow.

in the x-t plane, along which discontinuities in the first- and higher-order derivatives of the dependent variables propagate. Physically, such discontinuities correspond to propagation of infinitesimally small wave disturbances in the limit.

   To translate the idea of discontinuities in derivatives into characteristic form for
the simplest case, the continuity and momentum equations for a prismatic rectangular channel without lateral inflow are written in matrix form as

$$
\begin{bmatrix}
1 & V & 0 & y \\
0 & g & 1 & V \\
dt & dx & 0 & 0 \\
0 & 0 & dt & dx
\end{bmatrix}
\begin{bmatrix}
y_t \\
y_x \\
V_t \\
V_x
\end{bmatrix}
=
\begin{bmatrix}
0 \\
g(S_0 - S_f) \\
dy \\
dV
\end{bmatrix}
\tag{7.23}
$$

in which the subscripts in the column vector on the left hand side denote partial
derivatives with respect to time, t, and distance, x. The last two equations in
(7.23) are simply the definitions of the total differentials of depth, y, and velocity,
V. If a unique solution for the derivatives exists, then from Cramer's rule, the determinant of the coefficient matrix in (7.23) must be nonzero. Therefore, a condition
for the solution to be indeterminant (and for the derivatives to be discontinuous) is
that the determinant of the coefficient matrix is exactly zero. Setting the determinant to zero results in

$$
\frac{dx}{dt} = V \pm \sqrt{gy} = V \pm c
\tag{7.24}
$$

which describes the characteristic directions. However, there is no solution of
(7.23) unless the determinant of the coefficient matrix with one column replaced
by the right hand side vector of (7.23) also is zero, in which case, the solution
has the indeterminate form 0/0 based on Cramer's rule (Lai 1986). Setting this

determinant to zero results in the characteristic equations that must be satisfied along the characteristics

$$\pm \frac{Dy}{Dt} + \frac{c}{g} \frac{DV}{Dt} = c(S_0 - S_f) \tag{7.25}$$

in which the total derivatives D/Dt appear and are defined along the C1 characteristic with the plus sign and along the C2 characteristic with the minus sign, as before. Now the transformation of variables from y to c in (7.25) yields the characteristic equations in the form given previously by Equations 7.22.

7.5   INITIAL AND BOUNDARY CONDITIONS

The dependence of the solution to the characteristic equations on initial conditions is illustrated in Figure 7.6. The solution for depth and velocity at the intersection of C1 and C2 characteristics at point P depends on knowledge of the initial conditions at A and B, as well as on all points between A and B. As observer 1 proceeds from point A, C2 characteristics emanating from the interval AB continuously intersect

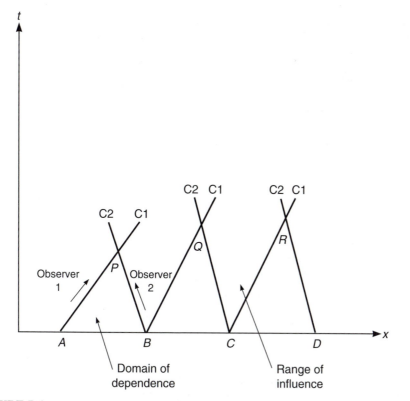

FIGURE 7.6

Domain of dependence and range of influence defined by characteristics in the x-t plane.

the path and alter the depth and velocity. In the same way, observer 2 receives information from the C1 characteristics originating from interval *AB* until meeting observer 1 at point *P*. As a result, the solution at *P* depends on the initial conditions along the interval *AB*, which is called the *interval of dependence*. In reality, an infinite number of characteristics continuously intersect *AP* and *BP* so that the region *ABP* is called the *domain of dependence* for point *P*. From a different point of view, a single point *C* on the *x* axis has initial conditions that influence the region *CQR* because all the C1 characteristics coming from the left of *CQ* and all the C2 characteristics intersecting *CR* from the right are influenced by the initial values at *C*. For this reason, the region *CQR* is called the *range of influence*.

As a consequence of wave propagation in characteristic directions, both initial conditions and boundary conditions must be specified carefully. The general rule is that the number of initial and boundary conditions must coincide with the number of characteristics entering at $t = 0$ for all *x* or at boundaries $x = 0$ and $x = L$ for all time, as shown in Figure 7.7 (Liggett and Cunge 1975; Cunge, Holly, and Verwey 1980). For the initial conditions, we see in Figure 7.7a that two conditions must be specified at point *A* to determine the initial slopes of the C1 and C2 characteristics as given by (7.24). With reference to Figure 7.6, the modification of the initial slopes at *A* and *B* comes from pairs of initial data specified on *AB* until the two characteristics intersect at point *P*. At this point, the characteristic, or compatibility, equations (7.25), are solved simultaneously for the dependent variables at *P*. As this process marches forward in time, the solutions at subsequent intersection points depend less on the initial conditions and more on information carried by characteristics coming from the boundaries and hence on the boundary conditions.

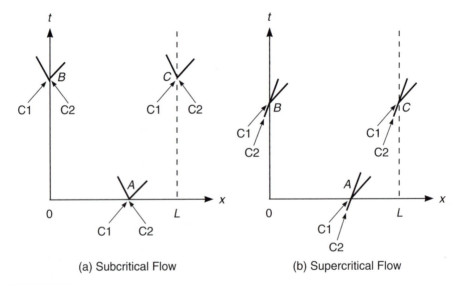

(a) Subcritical Flow   (b) Supercritical Flow

FIGURE 7.7
Specification of boundary conditions and initial conditions in subcritical and supercritical flow (after Cunge, Holly, and Verwey 1980). (*Source: Figure used courtesy of Iowa Institute of Hydraulic Research.*)

In subcritical flow, as shown in Figure 7.7a, one characteristic carries information upstream at the downstream boundary, $x = L$, and only one characteristic transmits information downstream into the solution domain from the upstream boundary at $x = 0$. In other words, only one boundary condition should be specified at both the upstream and downstream boundaries in subcritical flow. By contrast, two boundary conditions must be specified at the upstream boundary given by $x = 0$ for supercritical flow as shown in Figure 7.7b, while no boundary conditions are specified at the downstream boundary for this case.

The fact that initial conditions have less and less influence as time progresses means that, in some situations such as tidal flows in estuaries, the initial conditions need not be known very accurately, so long as a startup period is used until the solution becomes dependent only on boundary conditions. In rapid transients, such as occur in hydroelectric tailraces or headraces, on the other hand, the initial conditions must be known very well, because they will influence the early part of the solution, which is very important in the analysis of the transients that occur. In addition, if little or no friction exists, the initial conditions can continue to be reflected from upstream and downstream boundaries for a very long time, as the transient oscillates about some steady state.

The initial and boundary conditions that are specified must be independent of one another. Specifying both the value of the depth and its derivative with time, for example, as initial conditions does not satisfy the condition of independence nor does the specification of both depth, y, and $\partial Q/\partial x$, because they are related by the continuity equation. In general, a stage or discharge hydrograph, or some relation between stage and discharge as given by a rating curve, can be specified as single boundary conditions in subcritical flow. A rating curve should not be specified as an upstream boundary condition, however, because of the feedback between depth and discharge as time progresses (Cunge, Holly, and Verwey 1980).

A final consequence of characteristics to be discussed has tremendous influence on some of the numerical solution techniques described in the next chapter. By referring to Figure 7.8, we see that the characteristics define a natural coordinate system which limits the size of the time step that can be taken in a finite difference approximation. If we attempt to approximate the time derivative over a time step $\Delta t > \Delta t_c$, we are seeking a solution outside the domain of dependence established by the characteristics (Liggett and Cunge 1975). The result is instability in the numerical solution, in which a small perturbation grows without bound until it swamps the true solution. The Courant condition, which limits the time step such that the numerical solution stays inside the domain of dependence, can be stated as

$$\Delta t \leq \frac{\Delta x}{|V \pm c|} \qquad (7.26)$$

Alternatively, the Courant number C_n can be defined as the ratio of actual wave velocity to numerical wave velocity, with the result that the stability condition, also called the Courant-Friedrich-Lewy (CFL) condition, becomes (Abbott 1975)

$$C_n = \frac{|V \pm c|}{\Delta x/\Delta t} \leq 1 \qquad (7.27)$$

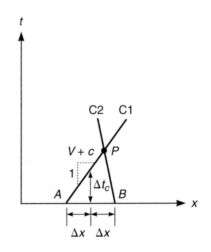

FIGURE 7.8
Limitations on the time step imposed by the Courant condition.

Because the velocity and wave celerity continuously change with time, the time step must be adjusted constantly during the numerical solution process to avoid instability.

7.6   SIMPLE WAVE

A simple wave is defined to be a wave for which $S_0 = S_f = 0$, with an initial condition of constant depth and velocity and with the water extending to infinity in at least one direction. While neglecting gravity and friction forces may not be very realistic, the simple wave assumption is useful for illustrating the solution of an unsteady flow problem in the characteristic plane. Equations 7.22a and 7.22c, the characteristic equations for a rectangular channel, assume a particularly simple form when the right hand side goes to zero. The result is

$$\text{along C1:} \qquad V + 2c = \text{constant} \qquad\qquad (7.28a)$$

$$\text{C1:} \qquad \frac{dx}{dt} = V + c \qquad\qquad (7.28b)$$

$$\text{along C2:} \qquad V - 2c = \text{constant} \qquad\qquad (7.28c)$$

$$\text{C2:} \qquad \frac{dx}{dt} = V - c \qquad\qquad (7.28d)$$

which states that $V + 2c$ is a constant along the C1 characteristics and $V - 2c$ is a constant along the C2 characteristics. The constant values in general are different

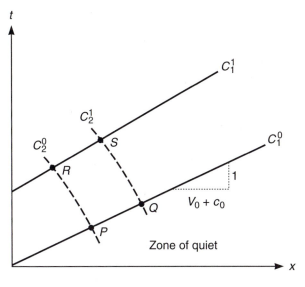

FIGURE 7.9
Straight line C1 characteristics for the simple wave.

for each characteristic and are called the *Riemann invariants* (Abbott 1975). However, the simplification is even more powerful because it can be shown that V and c are individually constant along each C1 characteristic, all of which are straight lines, and that the C2 characteristics degenerate into a constant value of $V - 2c$ everywhere in the *x-t* plane (Stoker 1957; Henderson 1966).

With reference to Figure 7.9 and following the proof by Stoker (1957), the initial C1 characteristic, C_1^0, is shown at the boundary of a constant depth region. It is a straight line inasmuch as $dx/dt = V_0 + c_0$, where V_0 and c_0 are the initial constant values of velocity and wave celerity, respectively, as required by the conditions of the simple wave problem. The constant depth region extends to the right of the initial C1 characteristic, where the initial flow is undisturbed; this region is called the *zone of quiet*, within which both C1 and C2 characteristics are straight lines, each with the same constant value of depth and velocity. We extend two C2 characteristics from the initial C1 characteristic to a second C1 characteristic, C_1^1, as shown in Figure 7.9. By definition of the characteristic equations for the C2 characteristics, we must have $V - 2c$ = constant, so that

$$V_P - 2c_P = V_R - 2c_R \tag{7.29}$$

$$V_Q - 2c_Q = V_S - 2c_S \tag{7.30}$$

but, by definition of the initial condition, we also must have $V_P = V_Q$ and $c_P = c_Q$, with the result that (7.29) and (7.30) simplify to

$$V_R - 2c_R = V_S - 2c_S \tag{7.31}$$

In addition, along the second C1 characteristic, $V + 2c$ = constant or

$$V_R + 2c_R = V_S + 2c_S \tag{7.32}$$

After adding and subtracting (7.31) and (7.32), we readily can show that $V_R = V_S$ and $c_R = c_S$, which leads to the conclusion that the second C1 characteristic also is a straight line along which the velocity and wave celerity, or depth, are constant. Generalization of (7.29), (7.30), and (7.31) to any C1 and C2 characteristic means that $V - 2c$ is a constant everywhere in the x-t plane. The C2 characteristics themselves are curved instead of straight lines, because where a single C2 characteristic crosses different C1 characteristics, there must be different values of velocity and depth, as at R and P, for example; so a different slope is given by $V - c$. The C2 characteristics in fact no longer are needed in the simple wave problem if $V - 2c$ is constant everywhere rather than just on individual C2 characteristics.

The region of C1 characteristics adjacent to the initial C1 characteristic and the zone of quiet in Figure 7.9 is referred to as the *simple wave region*. Velocities and wave celerities are determined completely in this region by the fact that $V - 2c =$ constant everywhere and that the slopes of the C1 characteristics are given by $dx/dt = V + c$. If V and c are the velocity and wave celerity at any point in the simple wave region, and if V_0 and c_0 are the constant initial conditions, the complete solution for V and c at specific locations and times defined by the slopes of the C1 characteristics is given by

$$V - 2c = V_0 - 2c_0 \tag{7.33}$$

$$\frac{dx}{dt} = V + c = V_0 - 2c_0 + 3c \tag{7.34a}$$

if the wave celerity, c (or depth), is specified as a boundary condition on the right hand side of (7.34a), or by

$$\frac{dx}{dt} = V + c = \frac{3}{2}V - \frac{V_0}{2} + c_0 \tag{7.34b}$$

if the velocity V is specified as a boundary condition. Thus, boundary conditions expressed at $x = 0$ in Figure 7.9 for all time determine the slopes of the characteristics along which both c and V are individually constant. Observers leaving from $x = 0$ at different times carry with them unique values of depth and velocity that can be located at any subsequent time in the x-t plane.

The simple wave region is applicable to negative waves, which are formed by a smaller depth propagating into a region of larger depth. Because a decreasing depth results in a smaller value of dx/dt from (7.34a), the simple wave region consists of diverging characteristics in the x-t plane. A positive wave, on the other hand, results in converging characteristics, which eventually intersect and can form a surge for which the assumption of infinitesimal wave disturbances is no longer valid. A different set of characteristics would be required upstream and downstream of the surge, across which there is an energy loss. In this case, the surge can be treated by the application of the continuity and momentum equations to a finite control volume that has been made stationary, as described in Chapter 3. Numerical solution techniques for surges are discussed in the next chapter.

Example 7.1

The initial flow conditions in an estuary are given by a velocity $V_0 = 3$ ft/s (0.91 m/s) and depth $y_0 = 8$ ft (2.4 m), as shown in Figure 7.10. The boundary condition at the mouth of the estuary where $x = 0$ is given by

$$y_L = 8 - 2\cos\left(\frac{\pi t}{6} - \frac{\pi}{2}\right) \qquad \text{(for } 0 \le t \le 3) \tag{7.35}$$

in which t is time in hours and y_L is the depth in feet at the left hand boundary. Find the depth profile at $t = 3$ hr.

Solution.   Both the physical and characteristic planes are shown in Figure 7.10. The x coordinate has been chosen positive in the direction of the advancing negative wave. The initial value of dx/dt ($= V_0 + c_0$) $= -3 + 16.05 = 13.05$ ft/s (3.98 m/s), shown as the slope of the first C1 characteristic that separates the zone of quiet from the negative wave region. Additional C1 characteristics emanate from the t axis at 0.5 hr intervals with slopes given by (7.34a), in which c is specified by the boundary condition expressed by

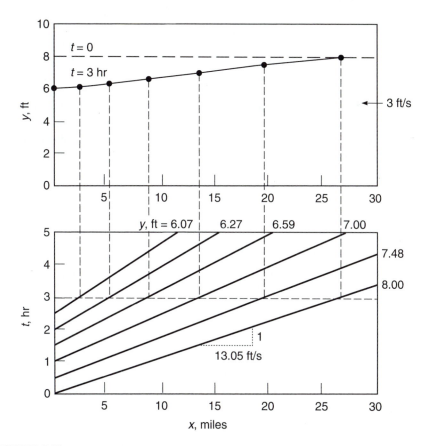

FIGURE 7.10
Simple wave solution of estuary problem.

(7.35). Along each of these characteristics, both the depth and velocity are constant, with the depth, y, specified by the boundary condition and velocity, V, determined from (7.33). The intersection of each C1 characteristic with the time line of $t_1 = 3$ hr determines the x position of the depth and velocity associated with that characteristic, and thus the depth profile as well as the velocity along the depth profile are determined. For example, the characteristic that begins at $t = 1$ hr has a depth of 7.0 ft (2.1 m) with $c = 15.0$ ft/s (4.58 m/s) from (7.35) and a slope $dx/dt = -3 - (2 \times 16.05) + (3 \times 15.0) = 9.90$ ft/s (3.02 m/s) from (7.34a). Its velocity $V = -3 - (2 \times 16.05) + (2 \times 15.0) = -5.10$ ft/s (-1.55 m/s) from (7.33). The intersection with the time line $t_1 = 3$ hr is located at $x = (dx/dt) \times (t_1 - t) = 9.90 \times 2 \times 3600/5280 = 13.5$ mi (21.7 km). So at a location of 13.5 mi (21.7 km) upstream of the estuary mouth, the depth is 7.0 ft (2.1 m) and the velocity is 5.10 ft/s (1.55 m/s) at $t = 3$ hr.

Dam-Break Problem

As another application of the method of characteristics applied to the simple wave, we consider next the sudden removal of a vertical plate behind which a known depth of water is at rest. The simple-wave solution of this problem, which Stoker (1957) referred to as the breaking of a dam, is oversimplified in comparison to the solution of a realistic dam break discussed in more detail in the next chapter. However, it illustrates the application of a velocity boundary condition and the formation of a surge in a submerged downstream river bed, and provides further insight into the unsteady development of a negative wave as interpreted by the method of characteristics. The next two examples are presented following more closely the practical approach of Henderson (1966), who related the dam-break problem to sluice gate operation and hydroelectric load acceptance in a headrace, than the mathematical treatment by Stoker (1957).

Example 7.2

A vertical plate is fixed at time $t = 0$ at $x = 0$ with a constant depth of water upstream equal to y_0 while the channel downstream of the gate is dry, as shown in Figure 7.11. The water upstream of the plate initially is at rest. At $t > 0$, the plate suddenly is accelerated to the left to a constant speed V_p. Determine the simple wave profile for this case and also for the case of the plate being removed instantaneously.

Solution.   The physical and characteristic planes are shown in Figure 7.11. The zone of quiet, denoted Region I, is established on the right hand side of the characteristic plane by a characteristic having an inverse slope of c_0 corresponding to the initial depth y_0 since the initial velocity is zero. On the left boundary, which is moving, the characteristic path is described as a straight line beginning at the origin with an inverse slope of $-V_p$. Because the water is in contact with the moving plate, it must have a constant velocity equal to that of the plate. As a result, a constant depth region is created upstream of the plate because $V - 2c$ must be a constant along the path of the plate,

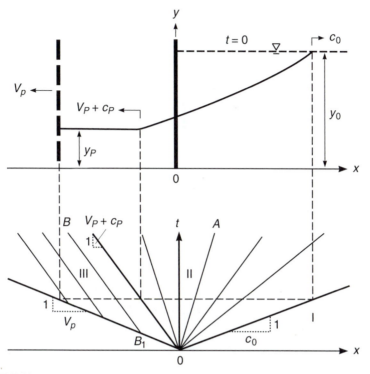

FIGURE 7.11

Simple wave solution of vertical plate removal at constant speed to left with a reservoir behind it (after Henderson 1966). (*Source: OPEN CHANNEL FLOW by Henderson, © 1966. Reprinted by permission of Prentice-Hall, Inc., Upper Saddle River, NJ.*)

which forms the left boundary in the characteristic plane. Because both V and c are constant, dx/dt also is a constant, so that the characteristics are parallel lines in Region III in Figure 7.11.

In between the zone of quiet on the right and the constant depth region on the left, the characteristics form a fan shape in Region II due to the decreasing values of $dx/dt = 3c - 2c_0$ occasioned by decreasing values of depth. For characteristic $A0$ in Region II, for example, the inverse slope of the characteristic is fixed. The depth is constant along the characteristic and determined from (7.34a) to be $c_A = (1/3) (dx/dt)_A + (2/3)c_0$. The velocity, too, is constant along the characteristic and equal to $(2c_A - 2c_0)$ from (7.33). Solving for the depth profile in Region II is only a matter of fixing a series of values of dx/dt and determining the x positions of the intersections of a fixed time line, $t = t_1$, with the characteristics. Then associated with each characteristic is a constant depth and velocity, which can be calculated from (7.34a) and (7.33), respectively.

In Region III, we must be careful to avoid an impossible situation when specifying the constant plate velocity, $-V_p$. For example, along characteristic BB_1 in the constant depth region, the wave celerity from (7.33) is $c_B = c_0 - V_p/2$, which cannot be negative. The limiting case is $c_B = 0$, for which $V_p = 2c_0$. Hence, we must have $V_p < 2c_0$ for the constant depth region to exist.

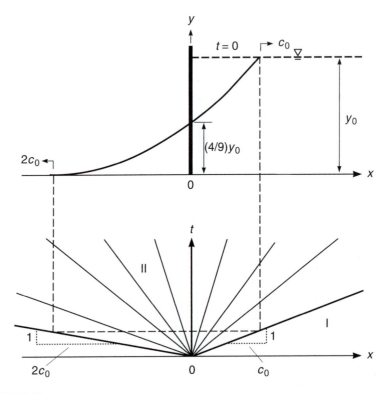

FIGURE 7.12

Simple wave solution of the instantaneous dam-break problem with a dry downstream channel (after Henderson 1966). (*Source: OPEN CHANNEL FLOW by Henderson, © 1966. Reprinted by permission of Prentice-Hall, Inc., Upper Saddle River, NJ.*)

The limiting case of $c_B = 0$ is interesting because it can be seen to have a leading feather edge of the advancing wave, which moves downstream at a speed of $2c_0$, as shown in Figure 7.12. In fact, for this case, the plate can be eliminated and imagined to be removed instantaneously because it has no real influence. For this reason, the situation shown in Figure 7.12 has come to be known as the dam-break problem but could also apply to the abrupt raising of a sluice gate. Note that the fan shaped Region II has expanded and Region III has disappeared in Figure 7.12. Furthermore, the time axis has itself become a characteristic along which velocity and depth are constant. It is easily shown from (7.34a) that, since $dx/dt = 0$ for this characteristic, we must have $c = (2/3)c_0$ or $y = (4/9)y_0 = $ constant at $x = 0$. Likewise, from (7.34b), we see that the constant velocity at $x = 0$ is $-(2/3)c_0$. The result is a constant discharge per unit of width, q, at the origin, which can be shown to be a maximum, given by

$$q_{max} = \frac{8}{27} y_0 \sqrt{g y_0} \qquad (7.36)$$

The constant discharge occurs because $|V| = c$ at the origin with the water velocity equal and opposite to the wave celerity.

Finally, the wave profile can be deduced from setting $dx/dt = x/t$ in (7.34a), since the characteristics all issue from the origin. The result for any time t_1 is

$$\frac{x}{t_1} = 3\sqrt{gy} - 2\sqrt{gy_0} \tag{7.37}$$

which can be seen to be a parabola tangent to the channel bed at the leading edge of the wave.

Example 7.3

If the initial condition in the dam-break problem of Example 7.2 includes a submerged downstream channel with depth $= y_4$ and no velocity, while the upstream depth remains constant at y_0, find the wave profile at any time t_1 if the plate suddenly is removed, or a sluice gate suddenly is raised, at $t = 0$.

Solution.   With reference to Figure 7.13, the simple wave profile cannot simply intersect the constant downstream water surface with depth $= y_4$ because this would cause a discontinuity in the velocity. The discontinuity can be resolved only by the formation of a surge with a speed V_s to the left, as shown in the figure (Henderson 1966). The intersection of the simple wave profile with the back of the surge results in a constant depth region and parallel characteristics. While $V - 2c$ still must be a constant for the simple wave, the surge must be analyzed separately by applying the momentum and continuity equations after the surge has been made stationary, as discussed in Chapter 3 and shown in Figure 7.14. The unknown values in the problem are now V_3, y_3, and V_s. First, the continuity equation can be written with reference to Figure 7.14 as

$$V_s y_4 = (V_s - V_3)y_3 \tag{7.38}$$

Second, the momentum equation also can be written for the stationary surge in Figure 7.14 to yield

$$\frac{V_s}{\sqrt{gy_4}} = -\left[\frac{1}{2}\frac{y_3}{y_4}\left(\frac{y_3}{y_4} + 1\right)\right]^{1/2} \tag{7.39}$$

in which the negative square root has been taken to agree with the sign convention in Figure 7.13 that has both V_s and V_3 in the negative x direction. Finally, the simple wave equation between points A and B in Figure 7.13 must be satisfied so that

$$V_3 = 2\sqrt{gy_3} - 2\sqrt{gy_0} \tag{7.40}$$

In principle, Equations 7.38 through 7.40 can be solved for the unknown values of V_3, y_3, and V_s. However, it is instructive to present the solution in dimensionless form, as in Figure 7.15. Equation 7.38 is divided by $c_4 = (gy_4)^{1/2}$ and solved for V_3/c_4. In the same way, Equation 7.40 is divided by c_4 and solved for y_4/y_0, keeping in mind that V_3 is inherently negative in the sign convention. Equation 7.39 is solved as a quadratic equation for y_3/y_4. The nondimensionalized solutions for V_3/c_4, y_3/y_4, and y_4/y_0 from (7.38), (7.39), and (7.40), respectively, now can be plotted as a function of V_s/c_4, as shown in Figure 7.15. For a given initial ratio of y_4/y_0, all three unknowns can be determined directly

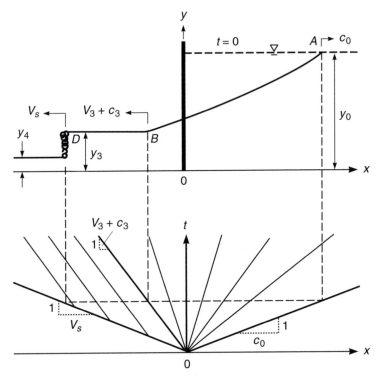

FIGURE 7.13
Simple wave solution of the instantaneous dam-break problem with a submerged down-stream channel (after Henderson 1966). (*Source: OPEN CHANNEL FLOW by Henderson, © 1966. Reprinted by permission of Prentice-Hall, Inc., Upper Saddle River, NJ.*)

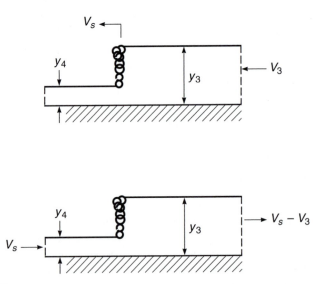

FIGURE 7.14
Making the surge in the dam-break problem stationary for momentum and continuity analysis.

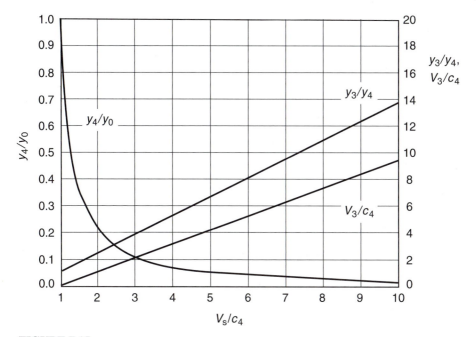

FIGURE 7.15

Variation of surge speed, depth behind the surge, and velocity behind the surge with the initial submergence ratio.

from the graph in Figure 7.15. Note that, as y_4/y_0 approaches 1, V_s approaches c_0 for a small wave disturbance. Also, V_3 is of the same sign as V_s and always is smaller in magnitude.

As drawn in Figure 7.13, it is apparent that points B and D both move downstream with the distance between them gradually increasing. However, point B could be positioned to the right of the x axis and move to the right in the upstream direction. Under these circumstances, the constant depth region extends across the x axis and submerges the constant depth of $(4/9)y_0$ that otherwise would occur, with the result that the discharge at the origin is smaller than the maximum value given by (7.36). The limiting case, of course, is for point B to lie exactly on the x axis, so that $V_3 = -(2/3)c_0$ and $y_3 = (4/9)y_0$. We can show that this limiting condition corresponds to $y_4/y_0 = 0.138$. If y_4/y_0 exceeds 0.138, then point B moves to the right, while it moves to the left for $y_4/y_0 < 0.138$. For the latter case, the Froude number, as seen by a stationary observer at $x = 0$, has a value of unity so that the depth must be critical and the discharge a maximum. Under these circumstances, the flow to the left of the point $x = 0$ is supercritical behind the surge as seen by a stationary observer and subcritical as seen by an observer moving at the speed of the surge.

The simple wave solution is not applicable generally; however, it illustrates the method of characteristics in graphical form. This will be useful in interpreting the results of the next chapter, in which the simplifying assumptions of the simple wave no longer are made and numerical solutions of the governing equations in characteristic form are sought.

REFERENCES

Abbott, M. B. "Method of Characteristics." In *Unsteady Flow in Open Channels*, ed. K. Mahmood and V. Yevjevich, vol. 1, pp. 63–88. Fort Collins, CO: Water Resources Publications, 1975.

Chaudhry, M. Hanif. *Open-Channel Flow.* Englewood Cliffs, NJ: Prentice-Hall, 1993.

Cunge, J. A., F. M. Holly, Jr., and A. Verwey. *Practical Aspects of Computational River Hydraulics.* London: Pitman Publishing Limited, 1980. (Reprinted by Iowa Institute of Hydraulic Research, Iowa City, IA, 1994.)

Henderson, F. M. *Open Channel Flow.* New York: Macmillan, 1966.

Lai, Chintu. "Numerical Modeling of Unsteady Open-Channel Flow," *Advances in Hydroscience*, vol. 14, pp. 161–333. New York: Academic Press, 1986.

Liggett, J. A. "Basic Equations of Unsteady Flow." In *Unsteady Flow in Open Channels*, ed. K. Mahmood and V. Yevjevich, vol. 1, pp. 29–62. Fort Collins, CO: Water Resources Publications, 1975.

Liggett, J. A., and J. A. Cunge. "Numerical Methods of Solution of the Unsteady Flow Equations." In *Unsteady Flow in Open Channels,* ed. K. Mahmood and V. Yevjevich, vol. 1, pp. 89–182. Fort Collins, CO: Water Resources Publications, 1975.

Stoker, J. J. *Water Waves.* New York: John Wiley & Sons, 1957.

Yen, Ben C. "Open Channel Flow Equations Revisited." *J. Engrg. Mech. Div.*, ASCE 99, no. EM5 (1973), pp. 979–1009.

Yen, Ben C., and H. G. Wenzel, Jr. "Dynamic Equations for Steady Spatially Varied Flow." *J. Hyd. Div.*, ASCE 96, no. HY3 (1970), pp. 801–14.

Yevjevich, V. "Introduction." *Unsteady Flow in Open Channels*, ed. K. Mahmood and V. Yevjevich, vol. 1, pp. 1–24. Fort Collins, CO: Water Resources Publications, 1975.

EXERCISES

7.1. Starting with Equation 7.13, derive Equation 7.14.

7.2. Derive Equations 7.18 and 7.19, the characteristic equations.

7.3. In the estuary problem given as Example 7.1, determine *algebraically* (not graphically) from the simple wave method the time in hours required for the depth to drop to 6.50 ft at a distance upstream of $x = 25,000$ ft. Plot on the same axes the depth hydrographs at $x = 0$ and $x = 25,000$ ft.

7.4. Water initially is at rest upstream and downstream of a sluice gate, which is completely closed in a rectangular channel. The upstream depth is 3.0 m and the downstream depth is 1.0 m. The gate suddenly is opened completely. Determine the speed of the surge, the depth and velocity behind the surge, and the speed of lengthening of the constant depth region. Show your results in both the physical plane and the characteristic plane.

7.5. Repeat Exercise 7.4 for the same upstream depth of 3.0 m but for a downstream depth of 0.25 m. Compare the results for the depth and velocity behind the surge.

7.6. Water flows in a rectangular channel under a sluice gate. The upstream depth is 3.0 m and the downstream depth is 1.0 m for a steady flow. If the gate is slammed shut, compute the height and speed of the surge upstream of the gate and the depth just down-stream of the gate. Under what conditions would the depth downstream of the gate go to zero? Show your results in both the physical plane and the characteristic plane.

7.7. In the steady flow of Exercise 7.6, the gate is suddenly closed partially so that the downstream depth becomes 0.5 m. Compute the height and speed of the surge upstream of the gate.

7.8. Water is initially flowing at a depth of 2.0 m with a velocity of 1.0 m/s in a rectangu-lar channel. If a landslide suddenly reduces the discharge at the downstream end of the channel to 50% of its original value, calculate the height and speed of the result-ing surge upstream of the blockage.

7.9. Water initially flows at steady state under a sluice gate. The upstream flow depth is 3.0 m and the downstream flow depth is 0.30 m. The sluice gate is raised abruptly, completely free of the flowing water.
(a) Find the depth and discharge at the gate after the gate has been raised.
(b) Determine the height and speed of the surge.

Sketch your results on the physical plane and the characteristic plane. Neglect bed slope and resistance effects.

7.10. Prove that the limiting condition of $y_4/y_0 = 0.138$ in the simple wave dam-break prob-lem determines whether the constant depth region (point B in Figure 7.13) moves upstream or downstream.

7.11. In the simple wave dam-break problem, derive an expression for the discharge per unit of width at $x = 0$, q_0, as a function of the ratio of initial depths, y_4/y_0. Nondimen-sionalize q_0 as q_0/c_0y_0 and plot the results.

7.12. Determine the maximum possible height of the surge, $(y_3 - y_4)$, in Example 7.3 in terms of y_0, and the value of y_4/y_0 for which it occurs.

7.13. The headrace for a turbine is a long rectangular canal that feeds water from a lake to the turbine penstocks. Suppose that the design discharge for a turbine is 68 m³/s. The canal width is 12 m, and the canal slope is very small with an average water depth in the canal of 3.5 m with no flow. If the turbine is brought on line suddenly (load acceptance), what will be the depth of flow at the downstream end of the canal where the penstock inlet is located? How long will it take for the negative wave to reach the reservoir if the canal is 2 km long?

7.14. Suppose that the turbine in Exercise 7.13 is operating at steady state at a discharge of 68 m³/s, and the corresponding normal depth of flow in the canal is 3.0 m. If the tur-bine is shut down suddenly (load rejection), what will be the height of the surge at the downstream end of the canal?

7.15. If a dam with a maximum water depth of 50 ft fails abruptly, estimate the time required for the surge to reach a community 5 mi downstream of the dam. What will the surge height be? Initially, the downstream river has a negligible velocity and a depth of 5 ft. What factors alter your estimates and in what direction?

8

Numerical Solution of the Unsteady Flow Equations

8.1 INTRODUCTION

Several numerical techniques have been developed to solve the partial differential equations of unsteady flow. Some of these techniques are more applicable in specific types of engineering problems than in others. The purpose of this chapter and the next one is to introduce the engineer to the more commonly used techniques, especially those found in well-known commercial and public domain codes, as well as to assist in identifying the most appropriate technique for a given problem. This chapter concentrates on solving the governing equations developed in Chapter 7 with no major simplifications. Chapter 9 considers simplified forms of the governing equations and corresponding numerical solution techniques that often are employed in some types of flood routing problems.

The methods developed in this chapter depend on approximations of the derivatives in the governing equations, either in characteristic form or in the original partial differential form. The major difference between the two approaches is that the derivatives are approximated on the characteristic grid along the characteristics themselves in the case of the characteristic form of the equations or on a fixed rectangular x-t grid in the case of the original partial differential equations. The former case is referred to as the *method of characteristics*, while the latter includes both *explicit* and *implicit finite difference* methods. Explicit finite difference methods advance the solution to the end of the time step at a single grid node, using an explicit function of the dependent variables already determined for several grid nodes at the beginning of the time step. Implicit methods, on the other hand, approximate the derivatives using values of the dependent variables both at the beginning of the time step, where they are known, and at the end of the time step for more than one grid node, where the dependent variables are unknown. In the latter case, a system of simultaneous equations must be solved to advance the solution by one time step.

The method of characteristics generally is utilized only in special cases, often as a check on some other method. However, it frequently is used in explicit finite difference techniques for a more accurate approximation of boundary conditions. The explicit finite difference method is used in problems of rapid transients such as those in the headrace or tailrace of hydroelectric turbine operations. A number of readily available codes such as BRANCH (Schaffranek, Baltzer, and Goldberg 1981), FLDWAV (Fread and Lewis 1995), and HEC-RAS (U.S. Army Corps of Engineers 2008) implement the implicit finite difference method to solve problems of flood routing and dam breaks.

In contrast to finite difference methods, finite element methods approximate the solution rather than the differential equations by using polynomial shape functions that depend on the unknown nodal values of the dependent variables. In the Galerkin approach, the residuals between the approximate solution and the exact solution are minimized by integrating the product of weighting functions and the residual over the solution domain of finite elements and setting the result to zero. The finite element method has been applied to problems of discontinuous open channel flow caused by surges (Katopedes and Wu 1986). Because it is not used as extensively as finite difference methods in widely available numerical codes for the solution of the one-dimensional unsteady flow equations, it is not discussed here.

In finite difference techniques, the continuous governing equations are approximated and transformed into discrete difference equations; therefore, it is essential that methods be chosen for which the truncation error in the approximation of the equation goes to zero as the time step, Δt, and spatial step, Δx, approach zero. Such a condition is referred to as *consistency* (Ames 1969). Without consistency, extraneous terms that are incompatible with the original differential equations may be introduced. Consistency alone, however, may not guarantee the ultimate goal of the finite difference approximation of the governing equations, which is *convergence*; that is, for the solution to approach the true solution as Δx, $\Delta t \rightarrow 0$. If small errors such as roundoff errors grow during the numerical solution process, then the solution becomes *unstable*, so that the error grows without bound, swamping the true solution and preventing convergence. The pitfall of instability is that a perfectly reasonable finite difference approximation can lead to garbage for a solution. The remedy may be to place limits on the discretization size, but in some cases it may be necessary to switch to a completely different approximation scheme. Such difficulties can cause unacceptable mistakes through uninformed use of commercial codes or casual programming of seemingly straightforward finite difference techniques, as is illustrated in this chapter.

Application of numerical techniques in hydraulics has become commonplace as computers have become more powerful. Desktop computers with parallel processors that are as fast as mainframe supercomputers of only a decade ago are very affordable. In addition, cumbersome batch processing and text-based output have been replaced by user-friendly program interfaces and colorful graphics. In this environment of readily accessible computational hydraulics, the importance of proper calibration and verification of numerical models of unsteady flow cannot be overemphasized. Application of any of the numerical techniques discussed in the next two chapters requires comparisons of computed results with measurements in the field or laboratory and appropriate adjustment of calibration factors that have physical validity. The model also should be verified with entirely different data sets from those used in the calibration procedure. It is insufficient to accept numerical model results on the basis of qualitative similarity with what might be expected to occur. Engineers must be demanding of numerical methods and never ignore the crucial link between numerical analysis and laboratory and field data.

8.2   METHOD OF CHARACTERISTICS

As discussed in Chapter 7, the transformation of the equations of unsteady flow into characteristic form gives rise to two families of characteristics that form a kind of natural coordinate system that is part of the solution. In the numerical method of characteristics, a numerical solution of the transformed equations is sought along the characteristic directions, C1 and C2. With reference to Figure 8.1, we seek a solution for velocity V and depth y at point P in the characteristic plane as a function of the values of the dependent variables at L and R. For simplicity, the rectangular

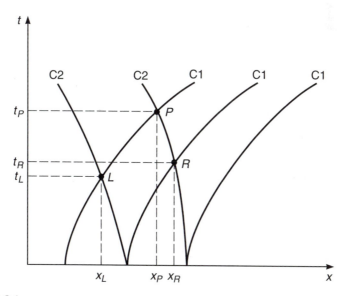

FIGURE 8.1

Numerical solution of the governing equations in characteristic form on the characteristic grid.

channel case is considered so that the characteristic equations to be solved are Equations 7.22 of Chapter 7. The equations can be written in integrated form as

$$V_P + 2c_P = V_L + 2c_L + \int_{t_L}^{t_P} g(S_0 - S_f)\, dt \tag{8.1}$$

$$x_P = x_L + \int_{t_L}^{t_P} (V + c)\, dt \tag{8.2}$$

$$V_P - 2c_P = V_R - 2c_R + \int_{t_R}^{t_P} g(S_0 - S_f)\, dt \tag{8.3}$$

$$x_P = x_R + \int_{t_R}^{t_P} (V - c)\, dt \tag{8.4}$$

in which $c = (gy)^{1/2}$. Now by evaluating the integrals approximately using the trapezoidal rule, Equations 8.1 to 8.4 in discrete form become

$$V_P + 2c_P = V_L + 2c_L + \tfrac{1}{2}(t_P - t_L)[g(S_0 - S_{fP}) + g(S_0 - S_{fL})] \tag{8.5}$$

$$x_P - x_L = \tfrac{1}{2}(t_P - t_L)(V_P + c_P + V_L + c_L) \tag{8.6}$$

$$V_P - 2c_P = V_R - 2c_R + \tfrac{1}{2}(t_P - t_R)[g(S_0 - S_{fP}) + g(S_0 - S_{fR})] \tag{8.7}$$

$$x_P - x_R = \tfrac{1}{2}(t_P - t_R)(V_P - c_P + V_R - c_R) \tag{8.8}$$

The discrete forms given in Equations 8.5 to 8.8 also could have been obtained from a forward difference approximation of the derivatives, as described in Appendix A, and taking the mean value of the integrand between points L and P and points R and P. In any case, the result is four nonlinear algebraic equations in the unknown values x_P, t_P, V_P, and c_P, which have to be solved by iteration.

One method of iteration begins by setting V_P and c_P equal to V_L and c_L, respectively, on the right hand side of (8.6) and equal to V_R and c_R, respectively, on the right hand side of (8.8) and solving for x_P and t_P. Then, by setting $S_{fP} = S_{fL}$ on the right hand side of (8.5) and $S_{fP} = S_{fR}$ on the right hand side of (8.7) and using the initial value of t_P just computed in the previous step, (8.5) and (8.7) can be solved for V_P and c_P. Thereafter, the new values of V_P and c_P are substituted on the right hand sides of (8.6) and (8.8) to obtain new values of x_P and t_P; then (8.5) and (8.7) are solved for the next values of V_P and c_P in iterative fashion. Liggett and Cunge (1975) suggested an iteration method with improved convergence properties that defines two residual functions from (8.5) and (8.7) that are driven to zero by Newton-Raphson iteration.

Regardless of the iterative procedure used, the solution is obtained on the x-t grid at irregular intervals determined by the characteristics, as shown in Figure 8.1. This requires interpolation of channel properties if they are known only at fixed grid intervals of x. In addition, the final solution must be interpolated if the water surface profile is desired at a specified time or if a stage hydrograph is required at a specified location, for example. This inconvenience is overcome by the Hartree method, also called the *method of specified time intervals*.

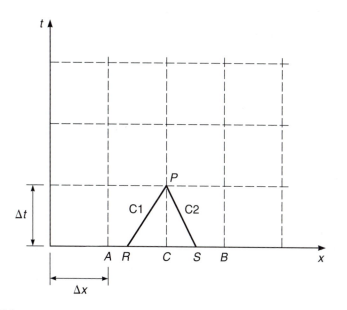

FIGURE 8.2
Numerical solution by the method of specified time intervals on the characteristic grid overlaid on a rectangular grid.

In the Hartree method, fixed time intervals and uniform spatial intervals are specified, and the solution at point P is projected backward in time to points R and S, as shown in Figure 8.2. In this case, the general nonrectangular cross section is considered, so that the equations of interest are (7.21a–d). The finite difference approximations of the derivatives along the characteristics are substituted, and the resulting discrete equations are given by

$$V_P - V_R + \frac{g}{c_R}(y_P - y_R) = g(S_0 - S_{fR})\Delta t \tag{8.9}$$

$$x_P - x_R = (V_R + c_R)\Delta t \tag{8.10}$$

$$V_P - V_S - \frac{g}{c_S}(y_P - y_S) = g(S_0 - S_{fS})\Delta t \tag{8.11}$$

$$x_P - x_S = (V_S - c_S)\Delta t \tag{8.12}$$

To avoid iteration, the values of S_f and wave celerity, c, in (8.9) and (8.11) are evaluated at points R and S, where they are known, as are the right hand sides of (8.10) and (8.12). Strictly speaking, the Hartree method uses a second-order approximation in which the mean values of S_f, c, and $V \pm c$ between points R and P and between points S and P are substituted in the finite difference approximation. The first-order method suggested by Wylie and Streeter (1978) is presented here for simplicity. In either case, the values of the dependent variables at R and S have to be determined by interpolation before they are known. If linear interpolation is utilized, for the velocity at point R with reference to Figure 8.2, we have

$$\frac{V_C - V_R}{V_C - V_A} = \frac{x_P - x_R}{\Delta x} = r(V_R + c_R) \tag{8.13}$$

in which $r = \Delta t/\Delta x$ and Equation 8.10 is used to substitute for $(x_P - x_R)$. In a similar interpolation, the value of c_R can be obtained from

$$\frac{c_C - c_R}{c_C - c_A} = r(V_R + c_R) \tag{8.14}$$

If we solve for V_R and c_R from (8.13) and (8.14), we have

$$V_R = \frac{V_C + r(-V_C c_A + c_C V_A)}{1 + r(V_C - V_A + c_C - c_A)} \tag{8.15}$$

$$c_R = \frac{c_C + rV_R(c_A - c_C)}{1 + r(c_C - c_A)} \tag{8.16}$$

In the same manner, the values of V_S and c_S can be obtained from

$$V_S = \frac{V_C + r(c_C V_B - c_B V_C)}{1 + r(-V_C + V_B + c_C - c_B)} \tag{8.17}$$

$$c_S = \frac{c_C + rV_S(c_C - c_B)}{1 + r(c_C - c_B)} \tag{8.18}$$

These interpolations assume subcritical flow, as shown in Figure 8.2, and would have to be rederived for supercritical flow. Now, with the dependent variables known at R and S, we can subtract (8.11) from (8.9) and solve for y_P to give

$$y_P = \frac{1}{(c_R + c_S)}\left[y_S c_R + y_R c_S + c_R c_S \left[\frac{(V_R - V_S)}{g} - (S_{fR} - S_{fS})\Delta t \right] \right] \tag{8.19}$$

Then, it follows from (8.9) that V_P is given by

$$V_P = V_R - g\left(\frac{y_P - y_R}{c_R} \right) - (S_{fR} - S_0)g\Delta t \tag{8.20}$$

Equations 8.19 and 8.20, together with the interpolation equations, can be used to solve explicitly for V and y at all interior points of the x-t plane, beginning with the initial conditions specified on the x axis. At the right and left boundaries for subcritical flow, (8.9) and (8.11) are solved simultaneously with the boundary conditions. The time step, however, must be chosen such that the Courant condition is satisfied for all grid points at a given time level unless the modifications suggested by Goldberg and Wylie (1983) for an implicit timeline interpolation are implemented.

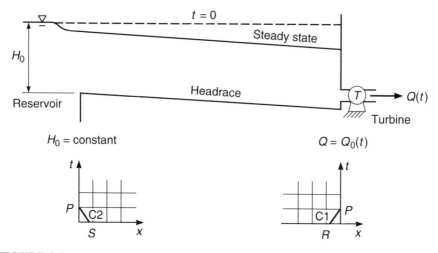

FIGURE 8.3
Boundary conditions for the hydroelectric turbine load acceptance problem.

8.3 BOUNDARY CONDITIONS

As an initial illustration of the application of boundary conditions, consider the hydroelectric turbine load acceptance/rejection problem shown in Figure 8.3. The reservoir at the upstream end supplies water to a headrace canal, which in turn conveys the discharge to the turbine shown schematically at the downstream end although in reality it is at the bottom of the penstocks. Such divided fall arrangements increase the available head on the turbines because they can be located at a lower elevation further downstream. When the turbine is brought on-line in a relatively short time, a negative wave propagates upstream from the turbine and then is reflected back as a positive wave. This is the load acceptance problem. In contrast, the load rejection problem occurs when the turbine is shut down in a finite but short time interval, causing a surge to propagate upstream. In either case, the boundary condition set by the reservoir at the left hand boundary is the maintenance of a constant value of head $H = H_0$ in the reservoir. At the downstream boundary, the turbine discharge is specified as a function of time, $Q = Q_0(t)$.

Consider first the upstream boundary condition at $x = 0$, illustrated in the characteristic plane in Figure 8.3 below the headrace entrance. Only the backward (C2) characteristic from S to P is of interest, and the equation to be satisfied along that characteristic is Equation 8.11, which can be rearranged to give

$$V_P - \frac{g y_P}{c_S} = V_S - \frac{g y_S}{c_S} - g(S_{fS} - S_0)\Delta t = K_S \qquad (8.21)$$

in which the right hand side, designated K_S for convenience, is known from the solution at the previous time step and the interpolation equations for point S given by (8.17) and (8.18). The boundary condition is applied as a condition of energy conservation at the entrance to the reservoir where, for simplicity, the entrance loss coefficient is neglected:

$$H_0 = y_P + \frac{V_P^2}{2g} \tag{8.22}$$

Equation 8.22 is solved for y_P and substituted into (8.21), which is rearranged to give a quadratic equation in V_P that is solved by the quadratic formula:

$$V_P = c_S \left[-1 + \sqrt{1 + \frac{2}{c_S} \left(\frac{gH_0}{c_S} + K_S \right)} \right] \tag{8.23}$$

At each time step, (8.23) gives the value of V_P at $x = 0$ and (8.21) or (8.22) produces the corresponding value of y_P.

At the downstream boundary, the forward (C1) characteristic is shown in Figure 8.3 where the boundary condition is given by $Q = Q_0(t)$ at $x = x_L$. The equation to be satisfied along the characteristic is (8.9), which can be rearranged as

$$V_P + \frac{gy_P}{c_R} = V_R + \frac{gy_R}{c_R} - g(S_{fR} - S_0)\Delta t = K_R \tag{8.24}$$

where, this time, the right hand side is designated as K_R and obtained from the interpolation Equations 8.15 and 8.16. From the continuity equation, the boundary condition can be stated as

$$Q_0(t) = V_P A_P \tag{8.25}$$

in which A_P is the cross-sectional area of flow that depends on the unknown depth, y_P, and on the geometric parameters for the given channel shape. On solving (8.24) for V_P and substituting into (8.25), the result is a nonlinear algebraic equation in y_P. Since this boundary condition must be applied at every time step, the Newton-Raphson technique is chosen for its solution, with the function F defined by

$$F(y_P) = Q_0(t) - A_P \left(-\frac{gy_P}{c_R} + K_R \right) \tag{8.26}$$

Then the Newton-Raphson iteration at each time step is given by

$$y_P^{k+1} = y_P^k - \frac{F(y_P^k)}{F'(y_P^k)} \tag{8.27}$$

in which the superscript k indicates the value of y_P at the kth iteration; $k + 1$ designates the value of y_P at the $(k + 1)$th iteration; and F' is the first derivative of the function F in (8.26) evaluated for y_P at the kth iteration. Equation 8.27 is iterated at each time step until there is some negligibly small change in y_P, after which V_P can be solved from Equation 8.25.

Additional boundary conditions are shown in Figure 8.4. Illustrated in Figure 8.4a is the specification of stage, h, or depth, y, as a function of time at the upstream end

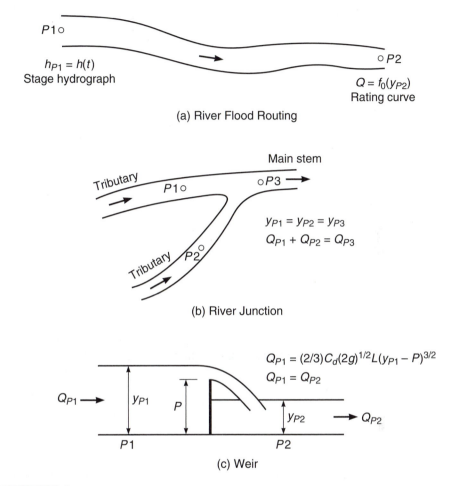

(a) River Flood Routing

(b) River Junction

(c) Weir

FIGURE 8.4
Additional types of boundary conditions for unsteady flow.

of a river reach. In this instance, Equation 8.11 is solved for V_P, given the value of y_P at the boundary. In a typical flood routing problem, the upstream boundary condition could be of this type or, alternatively, it could consist of the specification of $Q(t)$ as the inflow hydrograph to a river reach. The downstream boundary condition in a flood routing problem also might be a stage or discharge hydrograph, but in some cases, it could be a depth-discharge relationship as illustrated at the downstream end of the river reach in Figure 8.4a and given by

$$Q = A_P V_P = f_0(y_P) \tag{8.28}$$

in which $f_0(y_P)$ is a specified function determined by a gauging station, a weir, or some other control that could include uniform flow. Equation 8.24 for the right hand boundary is multiplied by A_P and rearranged to produce a function, F, that can be solved for y_P utilizing Newton-Raphson iteration:

$$F(y_P) = f_0(y_P) + \frac{g y_P A_P}{c_R} - A_P K_R \tag{8.29}$$

The value of V_P follows from (8.28).

In some situations, internal boundary conditions may be required, as illustrated in Figures 8.4b and 8.4c. Figure 8.4b shows a junction formed by two tributaries flowing into a main stem river. If no significant energy is lost and the differences in velocity head are small, the internal boundary conditions become

$$y_{P1} = y_{P2} = y_{P3} \tag{8.30}$$
$$A_{P1}V_{P1} + A_{P2}V_{P2} = A_{P3}V_{P3} \tag{8.31}$$

with unknown values of y_{P1}, y_{P2}, y_{P3}, V_{P1}, V_{P2}, and V_{P3}. The three equations represented by (8.30) and (8.31) are solved simultaneously with two forward (C1) characteristic equations written for tributaries 1 and 2 and one backward (C2) characteristic equation written for the main stem. The two forward characteristic equations are expressed in terms of two separate interpolation points for the two tributaries. The system of equations can be solved by Newton-Raphson iteration.

For a weir or spillway, as shown in Figure 8.4c, there are two grid points, P_1 and P_2, separated by a negligibly small distance with unknown values of y_{P1}, V_{P1}, y_{P2}, and V_{P2} just upstream and downstream of the weir. The solution for this internal boundary requires two boundary conditions, the forward characteristic equation upstream of the weir and the backward characteristic equation downstream of the weir for subcritical flow. The boundary conditions are written as

$$Q_{P1} = \tfrac{2}{3} C_d \sqrt{2g}\, L(y_{P1} - P)^{3/2} \tag{8.32}$$
$$Q_{P1} = Q_{P2} = A_{P2}V_{P2} \tag{8.33}$$

in which P and L = height and crest length of weir and C_d = discharge coefficient. The interpolation equations at R and S have to be written in terms of two separate points, P_1 and P_2, for which velocity and depth are determined in the previous time step.

Both the boundary conditions and the characteristic equations for the interior grid points are expressed in this section as first-order approximations. They also could be expressed as second-order approximations, as described by Liggett and Cunge (1975), but in any case the same order approximation should be used for the boundary conditions as for the interior grid points.

All the boundary conditions in Figure 8.4 are described for subcritical flow. If the flow is supercritical, both unknowns must be specified on the upstream boundary, while no boundary conditions are specified at the downstream boundary, as explained in Chapter 7. The two characteristic equations for the C1 and C2 characteristics are solved simultaneously at the downstream boundary.

Boundary conditions are specified using the method of characteristics as described in this chapter for both the numerical method of characteristics and the explicit finite difference method to be discussed next. Otherwise, instability or overspecification of the variables at the boundaries can result. In the implicit finite difference method, both internal and external boundary conditions simply become the additional compatibility equations necessary to solve the matrix equations at each time step, as explained later in this chapter.

8.4   EXPLICIT FINITE DIFFERENCE METHODS

Although explicit finite difference techniques are relatively simple to program, they are fraught with difficulties associated with instability that go beyond the satisfaction of the Courant condition. As an example, consider the computational molecule illustrated in Figure 8.5a. The computational molecule defines the grid points used in the finite difference approximations of the derivatives for a particular numerical scheme. The grid points are identified by the subscripts i and superscripts k to indicate spatial intervals and time intervals, respectively. For example, y_i^{k+1} is the discrete value of the depth at a distance of $(i\Delta x)$ from the left hand boundary, where $x = 0$, if uniform spatial intervals are used, and at a time that is $(k + 1)$ time steps from the initial time of $t = 0$, but the time steps may be nonuniform due to the requirements of the Courant condition. The time and space derivatives in the original partial differential equations are approximated on this grid and within the computational molecule, which is applied repeatedly for all the interior points at any given time step.

For the computational molecule illustrated in Figure 8.5a, an unstable finite difference scheme results if the V and y derivatives are approximated as

$$\frac{\partial V}{\partial t} = \frac{V_i^{k+1} - V_i^k}{\Delta t}; \qquad \frac{\partial y}{\partial t} = \frac{y_i^{k+1} - y_i^k}{\Delta t} \qquad (8.34)$$

$$\frac{\partial V}{\partial x} = \frac{V_{i+1}^k - V_{i-1}^k}{2\Delta x}; \qquad \frac{\partial y}{\partial x} = \frac{y_{i+1}^k - y_{i-1}^k}{2\Delta x} \qquad (8.35)$$

where the dependent variables V and y in Figure 8.5 are represented by the general function f. If these finite difference approximations are substituted into the reduced form of the continuity and momentum equations, (7.5) and (7.15), we can show that the resulting solution usually is unstable (Liggett and Cunge 1975). In some cases, stability can be achieved by artificially increasing the friction terms, but in general it is better to avoid this scheme.

Lax Diffusive Scheme

With minor modification, the unstable scheme can be made stable in the Lax diffusive scheme. The computational molecule shown in Figure 8.5b no longer uses the point (i, k) in the evaluation of the time derivative but some weighted average of the

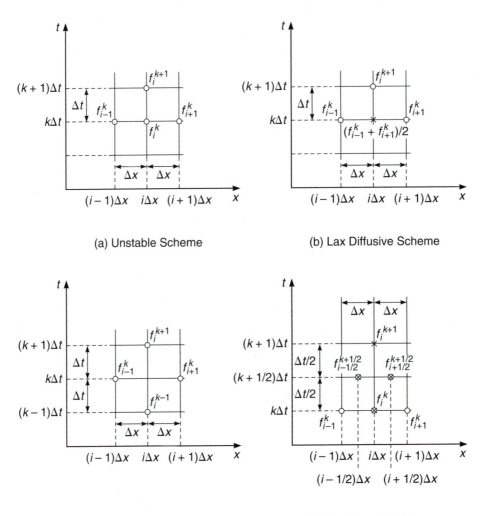

FIGURE 8.5
Computational molecules for some explicit finite difference schemes.

solution at adjacent grid points at the kth time level. Using the general function, f, to represent the dependent variables, the derivatives become

$$\frac{\partial f}{\partial t} = \frac{f_i^{k+1} - \left[\chi f_i^k + \frac{1 - \chi}{2}(f_{i+1}^k + f_{i-1}^k) \right]}{\Delta t} \tag{8.36}$$

$$\frac{\partial f}{\partial x} = \frac{f_{i+1}^k - f_{i-1}^k}{2\Delta x} \tag{8.37}$$

in which χ is a weighting factor between 0 and 1. For $\chi = 1$, we recover the unstable scheme; while for $\chi = 0$, we have a pure diffusive scheme called the Lax diffusive scheme, which is stable so long as the Courant condition is satisfied.

If the finite difference approximations suggested by (8.36) and (8.37) with $\chi = 0$ are substituted into the reduced form of the continuity and momentum equations as given by (7.5) and (7.15) for a prismatic channel without lateral inflow, the result is two difference equations that can be solved explicitly for depth and velocity at the grid point $(i, k + 1)$ with the "free variables" or coefficients evaluated as the mean of the values on either side of the grid point (i, k):

$$y_i^{k+1} = \frac{1}{2}\left(y_{i+1}^k + y_{i-1}^k\right) - \frac{\Delta t}{2\Delta x}\left(\frac{V_{i+1}^k + V_{i-1}^k}{2}\right)\left(y_{i+1}^k - y_{i-1}^k\right)$$

$$- \frac{\Delta t}{2\Delta x}\left(\frac{D_{i+1}^k + D_{i-1}^k}{2}\right)\left(V_{i+1}^k - V_{i-1}^k\right) \tag{8.38}$$

$$V_i^{k+1} = \frac{1}{2}\left(V_{i+1}^k + V_{i-1}^k\right) - \frac{\Delta t}{2\Delta x}\left(\frac{V_{i+1}^k + V_{i-1}^k}{2}\right)\left(V_{i+1}^k - V_{i-1}^k\right)$$

$$- g\frac{\Delta t}{2\Delta x}\left(y_{i+1}^k - y_{i-1}^k\right) + g\Delta t S_0 - g\Delta t\left[\frac{(S_f)_{i+1}^k + (S_f)_{i-1}^k}{2}\right] \tag{8.39}$$

in which $S_f = Q|Q|/K^2$ and $K =$ channel conveyance. The absolute value sign applied to Q in the definition of the friction slope S_f ensures the proper sign for the shear force for flows with changing directions. Liggett and Cunge (1975) show that the Lax scheme is not consistent, since the finite difference approximation introduces diffusive terms that should not appear, but it is stable provided that the Courant condition is satisfied and accurate so long as $(\Delta x)^2/\Delta t$ is small enough that the diffusive terms do not influence the solution.

The Lax diffusive scheme can also be applied to the preferred conservation form of the continuity and momentum equations as given by (7.2) and (7.13) for a prismatic channel. The difference equations are

$$A_i^{k+1} = \frac{1}{2}\left(A_{i-1}^k + A_{i+1}^k\right) - \frac{\Delta t}{2\Delta x}\left(Q_{i+1}^k - Q_{i-1}^k\right) \tag{8.40}$$

$$Q_i^{k+1} = \frac{1}{2}\left(Q_{i-1}^k + Q_{i+1}^k\right) - \frac{\Delta t}{2\Delta x}\left[\left(\frac{Q^2}{A} + gAh_c\right)_{i+1}^k - \left(\frac{Q^2}{A} + gAh_c\right)_{i-1}^k\right]$$

$$+ \Delta t\left(\frac{\phi_{i+1}^k + \phi_{i-1}^k}{2}\right) \tag{8.41}$$

in which $\phi = gA(S_0 - S_f)$. The source term ϕ has been evaluated as the mean of the values at points $(i - 1, k)$ and $(i + 1, k)$ as suggested by Terzidis and Strelkoff (1970) and Chaudhry (1993). The values of Q and A are determined at each time step, from which the values of velocity, V, and depth, y, can be calculated for the given channel geometry, and the values of Ah_c follow from its definition for the given prismatic channel shape (see Table 3.1) for use in the next time step.

Leapfrog Scheme

Another explicit method that has been used extensively is the leapfrog scheme, which has the computational molecule shown in Figure 8.5c. In terms of the general function, f, the time and space derivatives are evaluated by

$$\frac{\partial f}{\partial t} = \frac{f_i^{k+1} - f_i^{k-1}}{2\Delta t}; \qquad \frac{\partial f}{\partial x} = \frac{f_{i+1}^k - f_{i-1}^k}{2\Delta x} \tag{8.42}$$

and any coefficients are evaluated at (i, k) (Liggett and Cunge 1975). If the finite difference approximations given by (8.42) are substituted into the conservation form of the continuity and momentum equations as for the Lax scheme, the resulting difference equations for Q and A are given by

$$A_i^{k+1} = A_i^{k-1} - \frac{\Delta t}{\Delta x}(Q_{i+1}^k - Q_{i-1}^k) \tag{8.43}$$

$$Q_i^{k+1} = Q_i^{k-1} - \frac{\Delta t}{\Delta x}\left[\left(\frac{Q^2}{A} + gAh_c\right)_{i+1}^k - \left(\frac{Q^2}{A} + gAh_c\right)_{i-1}^k\right] + 2\Delta t\phi_i^k \tag{8.44}$$

in which $\phi = gA(S_0 - S_f)$ as before. An alternative expression for the source term can be developed using the grid points $(i, k + 1)$ and $(i, k - 1)$ in a weighted implicit-explicit fashion so that

$$\phi_i^k \approx gA_i^k\left[S_0 - \frac{|Q_i^k|}{(K_i^k)^2}\frac{(Q_i^{k+1} + Q_i^{k-1})}{2}\right] \tag{8.45}$$

in which K = channel conveyance (Liggett and Cunge 1975). In comparison with the Lax diffusive scheme, the leapfrog scheme is of second order rather than first order, provided that Δx and Δt are uniform, and it is nondissipative; that is, no diffusion-like numerical terms cause smearing of a wave front. This necessitates use of some artificial damping terms to simulate steep wave fronts.

Lax-Wendroff Scheme

The Lax-Wendroff scheme is developed directly from a Taylor's series expansion in the time direction in combination with the continuity and momentum equations in conservation form. Up to this point, the governing equations have been written out separately, but it sometimes is convenient to write them in the vector form:

$$\frac{\partial \mathbf{U}}{\partial t} + \frac{\partial \mathbf{F(U)}}{\partial x} = \mathbf{S(U)} \tag{8.46}$$

in which

$$\mathbf{U} = \begin{bmatrix} A \\ Q \end{bmatrix}; \qquad \mathbf{F(U)} = \begin{bmatrix} Q \\ \dfrac{Q^2}{A} + gh_cA \end{bmatrix}; \qquad \mathbf{S(U)} = \begin{bmatrix} 0 \\ gA(S_0 - S_f) \end{bmatrix} \tag{8.47}$$

The Taylor's series expansion for $\mathbf{U}^{k+1}$ is developed around the known values of $\mathbf{U}^k$ as

$$\mathbf{U}_i^{k+1} = \mathbf{U}_i^k + \Delta t \left[\frac{\partial \mathbf{U}}{\partial t} \right]_i^k + \frac{\Delta t^2}{2} \left[\frac{\partial^2 \mathbf{U}}{\partial t^2} \right]_i^k + \dots \tag{8.48}$$

in which all terms beyond the second-order term are dropped. Values for the first and second time derivatives then are expressed in terms of $\mathbf{F}(\mathbf{U})$ and its derivatives, using the original equations given by (8.46). Finally, finite difference approximations are substituted for the x derivatives of $\mathbf{F}$ (see Ames 1969). The resulting difference scheme can be simplified and is equivalent to a two-step method (Liggett and Cunge 1975; Abbott and Basco 1989) in which the Lax diffusive scheme is used in the first half of the time step at $(k + \frac{1}{2})\Delta t$ and then the leapfrog method is applied in the second half of the time step. The computational molecule is shown in Figure 8.5d, in which the circles represent the grid points involved in the first stage and the x symbols identify the computational points in the second stage of the scheme.

Applying the Lax-Wendroff scheme to the continuity and momentum equations in conservation form results in first-stage difference equations given by

$$A_{i+1/2}^{k+1/2} = \frac{1}{2} (A_{i+1}^k + A_i^k) - \frac{\Delta t}{2\Delta x} (Q_{i+1}^k - Q_i^k) \tag{8.49}$$

$$Q_{i+1/2}^{k+1/2} = \frac{1}{2} (Q_{i+1}^k + Q_i^k) - \frac{\Delta t}{2\Delta x} \left[\left(\frac{Q^2}{A} + gAh_c \right)_{i+1}^k - \left(\frac{Q^2}{A} + gAh_c \right)_i^k \right]$$

$$+ \frac{\Delta t}{2} \frac{(\phi_{i+1}^k + \phi_i^k)}{2} \tag{8.50}$$

Equations 8.49 and 8.50 are applied a second time to obtain values of A and Q at the grid point $(i - \frac{1}{2}, k + \frac{1}{2})$, as shown in Figure 8.5d. In the second stage of the scheme, the values determined at the half time step are utilized in a leapfrog type of evaluation, as given by

$$A_i^{k+1} = A_i^k - \frac{\Delta t}{\Delta x} (Q_{i+1/2}^{k+1/2} - Q_{i-1/2}^{k+1/2}) \tag{8.51}$$

$$Q_i^{k+1} = Q_i^k - \frac{\Delta t}{\Delta x} \left[\left(\frac{Q^2}{A} + gAh_c \right)_{i+1/2}^{k+1/2} - \left(\frac{Q^2}{A} + gAh_c \right)_{i-1/2}^{k+1/2} \right]$$

$$+ \Delta t \frac{(\phi_{i+1/2}^{k+1/2} + \phi_{i-1/2}^{k+1/2})}{2} \tag{8.52}$$

The Lax-Wendroff two-step scheme is of second order and dissipative (diffusive) for shorter wave components only, so that it has been used to model moving shocks (surges), as is discussed later in this chapter. This property of the method tends to smooth the wavy water surface behind the surge. However, for a hydraulic jump in steady flow, instabilities can occur at the jump, so that some additional dissipation is needed through a "dissipative interface" (Abbott and Basco 1989) or an artificial viscosity (Cunge, Holly, and Verwey 1980).

Predictor-Corrector Methods

For unsteady flow problems involving regions of both supercritical and subcritical flow that move with time (mixed-flow regimes or transcritical flow), computing through the discontinuities can introduce severe numerical difficulties. The predictor-corrector methods involve a two-step computation at each time step in which there is first a forward sweep in the spatial direction to carry the influence of upstream boundary conditions in the predictor step followed by a backward sweep in the corrector step that propagates the effect of downstream boundary conditions. The MacCormack scheme (Fennema and Chaudhry 1986) is a good example of this class of methods in which the two-step computations, with reference to the vector form of the equations in (8.46), are given by

$$\mathbf{U}_i^p = \mathbf{U}_i^k - \frac{\Delta t}{\Delta x} (\mathbf{F}_{i+1}^k - \mathbf{F}_i^k) + \Delta t \mathbf{S}_i^k \qquad (8.53)$$

$$\mathbf{U}_i^c = \mathbf{U}_i^k - \frac{\Delta t}{\Delta x} (\mathbf{F}_i^p - \mathbf{F}_{i-1}^p) + \Delta t \mathbf{S}_i^p \qquad (8.54)$$

in which the p superscript refers to the values of the variables computed in the predictor step and the c superscript refers to the values determined in the corrector step. Note that the spatial derivatives use only two grid points and they are computed as forward differences in the predictor step and backward differences in the corrector step. The order of the predictor and corrrector steps can be reversed at every other time step, but Chaudhry (1993) suggests that the predictor step should be in the direction of the advancing wave front. At the end of the predictor-corrector steps, the solution is taken as the mean of the predicted and corrected values:

$$\mathbf{U}_i^{k+1} = \tfrac{1}{2} (\mathbf{U}_i^p + \mathbf{U}_i^c) \qquad (8.55)$$

Fennema and Chaudhry (1986) and Garcia-Navarro, Alcrudo, and Saviron (1992) applied the MacCormack scheme to transcritical flow in open channels, albeit with different adaptive artificial viscosity schemes to dissipate oscillations at surge discontinuities. These dissipation methods are considered adaptive because they are applied only in regions where the water-surface gradients become large.

Meselhe, Sotiropoulos, and Holly (1997) introduced a predictor-corrector scheme derived by replacing the partial derivatives in the governing equations with Taylor series approximations centered around the grid point $(i + \tfrac{1}{2})$ for spatial derivatives and $(k + \tfrac{1}{2})$ for the time derivatives. Their MESH scheme uses only two points for evaluation of the spatial derivatives and allows for implicit evaluation of the source terms. It also employs artificial dissipation terms in the predictor-corrector equations. Simulations of choked flow over a channel bottom hump followed by a hydraulic jump as well as a jump on a steep slope downstream of a slope break agreed well with analytical solutions. They also showed satisfactory performance of the numerical scheme for supercritical flow on a steep slope followed by a hydraulic jump upstream of a weir located midway along the channel, passage to supercritical flow downstream of the weir, and another hydraulic jump downstream of the weir, which moved upstream with time due to a rising tailwater level. For further detail on these predictor-corrector methods, refer to the original papers.

Flux-Splitting Schemes

Flux-splitting schemes take advantage of the characteristic directions of the governing equations. The vector form of the equations, given by (8.46), can be rewritten in the form

$$\frac{\partial \mathbf{U}}{\partial t} + \mathbf{A} \frac{\partial \mathbf{U}}{\partial x} = \mathbf{S}(\mathbf{U}) \tag{8.56}$$

where the matrix $\mathbf{A}$ is the Jacobian of $\mathbf{F}(\mathbf{U})$, which is referred to as the *flux vector* because its components consist of the mass flux and the force plus momentum flux per unit of density. The Jacobian matrix is given by

$$\mathbf{A} = \frac{\partial \mathbf{F}}{\partial \mathbf{U}} = \begin{bmatrix} 0 & 1 \\ \dfrac{gA}{B} - \dfrac{Q^2}{A^2} & \dfrac{2Q}{A} \end{bmatrix} \tag{8.57}$$

as can be verified by the reader in the Exercises. We can show that the eigenvalues, λ, of the matrix $\mathbf{A}$ in fact are the slopes of the two characteristic directions given by $V \pm c$ by setting $\det[\mathbf{A} - \lambda\mathbf{I}] = 0$, in which $\mathbf{I}$ is the unit matrix with diagonal values of ones and zeroes for all other elements (see the Exercises to this chapter). The difference evaluation of the flux, $\Delta\mathbf{F}$, can be evaluated approximately as $\mathbf{A}\Delta\mathbf{U}$, which can be split into positive and negative parts, corresponding to the local characteristic directions. Then space derivatives involving the positive and negative components of $\mathbf{A}$ are evaluated by backward and forward finite differences to preserve the directional properties of signal propagation in subcritical and supercritical flow. Fennema and Chaudhry (1987) have applied the Beam and Warming scheme, which is of this type, to the dam-break problem described in Chapter 7. The flux splitting scheme has been modified and improved further by Jha, Akiyama, and Ura (1994, 1995). A variation of it has been introduced by Jin and Fread (1997) into the National Weather Service computer program FLDWAV for regions of mixed flow (supercritical and subcritical) near the critical state, with a moving interface between them. Such situations arise in rapid dam breaks with large differences between upstream and downstream depths.

Stability

A complete discussion of stability is beyond the scope of this introduction to numerical methods for the unsteady open channel flow equations. Stability analyses involve substitution of Fourier series terms for the solution into the finite difference scheme and determining whether the perturbations increase in amplitude with time (instability). Such classic stability analyses typically are applied to linearized forms of the equations, so that nonlinear instabilities can be found only through numerical experimentation. Suffice it to say that, for any explicit scheme, the Courant condition must be satisfied for stability:

$$\Delta t \le \frac{\Delta x}{|V \pm c|} \tag{8.58}$$

The Courant condition seems to imply that Δx can be increased to keep the time steps from becoming too small. However, the Koren condition for explicit methods, which results from the explicit treatment of the friction slope evaluation, places a limit on the spatial step size as well (Huang and Song 1985). Using numerical experiments, Huang and Song show that the Koren condition is applicable to the method of characteristics as well as to explicit schemes. The Koren condition is given by

$$\Delta t \le \frac{\sqrt{1 + 2\mathbf{F}_0} - 1}{\mathbf{F}_0 \dfrac{gS_0}{V_0}} \tag{8.59}$$

in which $\mathbf{F}_0$ = Froude number of initial steady, uniform flow of velocity, V_0, on which a disturbance is superimposed and S_0 = channel bed slope. If the Koren condition is combined with the Courant condition so that the Courant number is exactly 1, then a maximum step size, Δx_{max}, is given by

$$\frac{S_0 \Delta x_{max}}{y_0} = \left(\sqrt{1 + 2\mathbf{F}_0} - 1\right)(1 + \mathbf{F}_0) \tag{8.60}$$

in which y_0 = hydraulic depth of uniform flow. Because this limitation on Δx can become somewhat restrictive at small values of the Froude number, Huang and Song (1985) suggested several semi-implicit methods for evaluation of S_f that ease this constraint.

While several other explicit schemes have been used successfully, the ones that have been presented provide a sufficient illustration of applications in unsteady open channel flow. It is not advisable to extend an explicit scheme to the evaluation of the boundary conditions because of ambiguities and redundancies that can occur. The method of characteristics is better suited for the boundary conditions in combination with the explicit scheme for interior grid points. In general, explicit schemes may seem easier to program than other methods, but the combination of characteristics-based boundary conditions, the need for artificial dissipation, and the stability constraints on explicit methods make them more demanding to implement than may first appear. The application of the explicit method is limited to relatively short-duration transients, such as occur in hydroelectric turbine or sluice gate operations, for example, because of the limitation on the time step imposed by the Courant condition. Explicit methods ordinarily are not applied to flood routing problems in large rivers, which more often are treated by implicit methods, as described in the next section, because of their more favorable stability properties.

Example 8.1

A hydroelectric turbine increases its load linearly from 0 to 1000 cfs (28.3 m³/s) in 60 sec (load acceptance problem, see Figure 8.3). The headrace channel is trapezoidal with a length of 5000 ft (1520 m), a bottom width of 20.0 ft (6.10 m), side slopes of 1.5:1, Manning's $n = 0.015$, and a bed slope of 0.0002. Compute the depth hydrographs at $x/L = 0.0, 0.2, 0.4, 0.6, 0.8$, and 1.0, using the method of characteristics with specified time intervals and the Lax diffusive scheme.

Solution.   The channel length is divided into 50 spatial intervals, and the time step is selected so that the Courant number is ≤1 for all grid nodes at the current time level. The Lax diffusive method is applied to the conservation form of the governing equations. The method of characteristics is used for the boundary conditions for both methods. The upstream boundary is a reservoir for which the energy equation is written for flow from the lake into the entrance of the headrace channel. The downstream boundary condition is a discharge hydrograph with a linear increase in turbine discharge from zero to the steady-state value in a specified time, which is 60 sec in this example. At time $t = 0$, the water in the headrace is at rest with the same water surface elevation as the reservoir.

The results are shown in Figure 8.6a for the Lax diffusive method, and Figure 8.6b for the method of characteristics. A very rapid decrease in depth is observed at the turbines during the startup period, then we see a more gradual decrease to the minimum depth. This is followed by a gradual approach to the steady-state depth at both the upstream and downstream boundaries. The solutions are nearly indistinguishable except at the minimum depth region at the turbine ($x/L = 1.0$). The minimum depth for the Lax scheme is 4.67 ft (1.42 m), while it is 4.94 ft (1.51 m) for the method of characteristics. There also is a very slight widening or smearing of the minimum depth region by the Lax scheme, due to diffusion, which may account for the slightly smaller minimum depth.

8.5   IMPLICIT FINITE DIFFERENCE METHOD

The implicit method utilizes more than one grid value of the dependent variables at the forward time in the computational molecule, as shown in Figure 8.7. In Figure 8.7, the computational molecule is a "box" used in the Preissmann method (Cunge, Holly, and Verwey 1980). The spatial derivatives are found as weighted averages of the first-order difference approximations at the two time levels with a variable weighting factor, θ, while the time derivatives depend on the difference in the arithmetic average of the grid values at each time level (or a weighting factor of $\frac{1}{2}$). Specifically, for any function f, the spatial and time derivatives are written as

$$\frac{\partial f}{\partial x} = \frac{\theta(f_{i+1}^{k+1} - f_i^{k+1}) + (1 - \theta)(f_{i+1}^k - f_i^k)}{\Delta x} \tag{8.61}$$

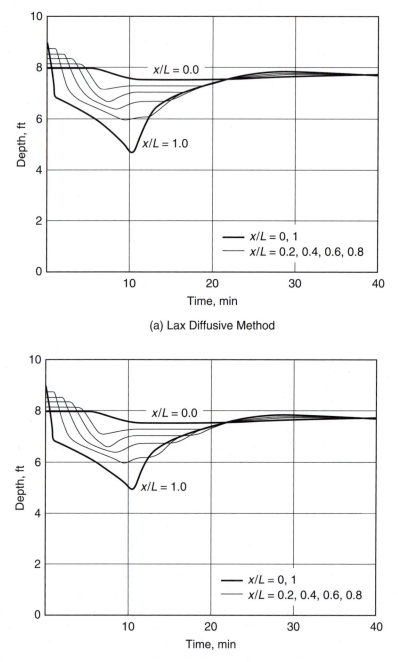

(a) Lax Diffusive Method

(b) Method of Characteristics

FIGURE 8.6
Depth hydrographs between the reservoir ($x/L = 0.0$) and the turbine ($x/L = 1.0$) for load acceptance.

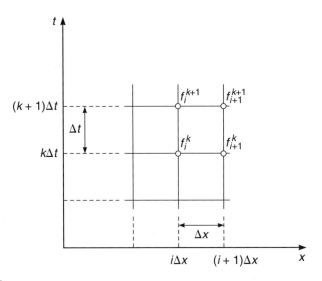

FIGURE 8.7
Preissmann implicit scheme.

$$\frac{\partial f}{\partial t} = \frac{(f_i^{k+1} + f_{i+1}^{k+1}) - (f_i^k + f_{i+1}^k)}{2\Delta t} \tag{8.62}$$

while the evaluation of the coefficients in the governing equations is given by

$$\bar{f} = \frac{\theta(f_i^{k+1} + f_{i+1}^{k+1}) + (1 - \theta)(f_i^k + f_{i+1}^k)}{2} \tag{8.63}$$

These finite difference approximations are applied to the continuity and momentum equations in the conservation form of Equations (7.2) and (7.13) for prismatic channels without lateral inflow. In the vector form of Equation 8.46, this can be written as

$$\mathbf{U}_i^{k+1} + \mathbf{U}_{i+1}^{k+1} - \mathbf{U}_i^k - \mathbf{U}_{i+1}^k + 2\frac{\Delta t}{\Delta x}[\theta(\mathbf{F}_{i+1}^{k+1} - \mathbf{F}_i^{k+1}) + (1 - \theta)(\mathbf{F}_{i+1}^k - \mathbf{F}_i^k)]$$

$$= \Delta t[\theta(\mathbf{S}_i^{k+1} + \mathbf{S}_{i+1}^{k+1}) + (1 - \theta)(\mathbf{S}_i^k + \mathbf{S}_{i+1}^k)] \tag{8.64}$$

in which the vectors were previously defined by (8.47). These equations are nonlinear, especially in the evaluation of S_f, which depends on the dependent variables Q and A as well as the conveyance K. In general, it is much easier to work with stage Z and discharge Q as the dependent variables in a natural river. Therefore, the simpler system of governing equations derived from (7.3) and (7.13) is given by

$$B\frac{\partial Z}{\partial t} + \frac{\partial Q}{\partial x} = 0 \tag{8.65}$$

$$\frac{\partial Q}{\partial t} + \frac{\partial}{\partial x}\left(\frac{Q^2}{A}\right) + gA\frac{\partial Z}{\partial x} + gA\frac{Q|Q|}{K^2} = 0 \tag{8.66}$$

in which $Z = \text{stage} = z_b + y$ and $z_b = $ bed elevation, is used more often in the implicit method, although the system is not strictly conservative. This usually is satisfactory as long as the implicit method applied to this form of the equations is not used to model very steep wave fronts, where conservation of mass and momentum must be observed more strictly (Cunge, Holly, and Verwey 1980). Note that the bed slope, S_0, does not appear explicitly in (8.66), because it has been incorporated into $\partial Z/\partial x$, since $S_0 = -\partial z_b/\partial x$ where $z_b = $ bed elevation. If the implicit approximations of (8.61), (8.62), and (8.63) are applied to Equations 8.65 and 8.66, the resulting algebraic difference equations are

$$\bar{B}\,(Z_i^{k+1} + Z_{i+1}^{k+1} - Z_i^k - Z_{i+1}^k)$$

$$+ \frac{2\Delta t}{\Delta x}[\theta(Q_{i+1}^{k+1} - Q_i^{k+1}) + (1 - \theta)(Q_{i+1}^k - Q_i^k)] = 0 \qquad (8.67)$$

$$(Q_i^{k+1} + Q_{i+1}^{k+1} - Q_i^k - Q_{i+1}^k)$$

$$+ \frac{2\Delta t}{\Delta x}\left\{\theta\left[\left(\frac{Q^2}{A}\right)_{i+1}^{k+1} - \left(\frac{Q^2}{A}\right)_i^{k+1}\right] + (1 - \theta)\left[\left(\frac{Q^2}{A}\right)_{i+1}^k - \left(\frac{Q^2}{A}\right)_i^k\right]\right\}$$

$$+ \frac{2\Delta t}{\Delta x}g\bar{A}[\theta(Z_{i+1}^{k+1} - Z_i^{k+1}) + (1 - \theta)(Z_{i+1}^k - Z_i^k)]$$

$$+ \frac{2\Delta t g\bar{A}}{\bar{K}^2}\left[\frac{\theta}{2}(Q_i^{k+1}|Q_i^{k+1}| + Q_{i+1}^{k+1}|Q_{i+1}^{k+1}|)\right.$$

$$\left. + \frac{(1 - \theta)}{2}(Q_i^k|Q_i^k| + Q_{i+1}^k|Q_{i+1}^k|)\right] = 0 \qquad (8.68)$$

in which the coefficients with an overbar are evaluated according to Equation 8.63. Equations 8.67 and 8.68 form a pair of nonlinear algebraic equations with four unknown values at the forward time level. By extension, the computational molecule will yield $2(N - 1)$ equations with $2N$ unknowns as it is applied repeatedly with overlapping across the grid in the x direction, where the total number of computational points is N and the number of reaches is $(N - 1)$. The remaining two equations must come from the boundary conditions, and the system of equations has to be solved simultaneously.

Solution of (8.67) and (8.68) is accomplished by the Newton-Raphson technique for multiple variables. If we define the left hand sides of (8.67) and (8.68) as G and H, respectively, the equations for each application of the computational molecule can be written as

$$G_i(Z_i, Q_i, Z_{i+1}, Q_{i+1}) = 0 \qquad (8.69a)$$

$$H_i(Z_i, Q_i, Z_{i+1}, Q_{i+1}) = 0 \qquad (8.69b)$$

where $i = 1, 2, \ldots, N - 1$ for $(N - 1)$ reaches. The superscripts on the dependent variables are omitted for convenience because they all are $(k + 1)$; that is, we seek the solution for these four unknowns at the $(k + 1)$th time level in terms of the known values of the dependent variables at the kth time level. Each nonlinear function has a subscript because the known values are different for each application of the equations. The two additional equations needed from the boundary conditions also can be expressed as functions set to zero. For example, an upstream specified stage hydrograph and a downstream specified stage-discharge relationship are given by

$$B_1 = Z_1 - Z_0(t) = 0 \qquad (8.69c)$$

$$B_N = Z_N - f(Q_N) = 0 \qquad (8.69d)$$

in which $Z_0(t)$ is the specified stage as a function of time and the stage-discharge relationship, or rating curve, is given by $Z = f(Q)$.

The general solution of the system of nonlinear equations represented by (8.69) can be obtained using Newton's iteration method. The solution begins with estimates of the unknown values of Z and Q that will result generally in the right hand sides of the system of equations in (8.69) being nonzero, or in other words, having residuals. At the nth iteration, this can be expressed as

$$B_1(Z_1^n, \ Q_1^n) = B_1^n \qquad (8.70a)$$

$$G_i(Z_i^n, \ Q_i^n, \ Z_{i+1}^n, \ Q_{i+1}^n) = G_i^n \qquad (8.70b)$$

$$H_i(Z_i^n, \ Q_i^n, \ Z_{i+1}^n, \ Q_{i+1}^n) = H_i^n \qquad (8.70c)$$

$$B_N(Z_N^n, \ Q_N^n) = B_N^n \qquad (8.70d)$$

in which $i = 1, 2, \ldots, N - 1$ and the superscript n refers to the present values of the unknowns and the functions at the nth iteration. Note that the residuals on the right hand sides of (8.70) simply are the evaluations of the functions with the nth estimates of the unknowns. To obtain the $(n + 1)$th estimates of the unknowns, the functions are expanded in a Taylor series while retaining only the first derivative terms. For example, for the ith continuity function, G_i, we have

$$G_i^{n+1} = G_i^n + \frac{\partial G_i^n}{\partial Z_i} \Delta Z_i + \frac{\partial G_i^n}{\partial Q_i} \Delta Q_i + \frac{\partial G_i^n}{\partial Z_{i+1}} \Delta Z_{i+1} + \frac{\partial G_i^n}{\partial Q_{i+1}} \Delta Q_{i+1} \quad (8.71)$$

in which $\Delta Z_i = (Z_i)^{n+1} - (Z_i)^n$; $\Delta Q_i = (Q_i)^{n+1} - (Q_i)^n$; $\Delta Z_{i+1} = (Z_{i+1})^{n+1} - (Z_{i+1})^n$; and $\Delta Q_{i+1} = (Q_{i+1})^{n+1} - (Q_{i+1})^n$. Equations of the form of (8.71) can be written for each of the original nonlinear equations in (8.69). Then, as in the Newton-Raphson technique for a function of one variable, we set $(G_i)^{n+1}$ and all similar functions to zero to obtain the root. The result can be rearranged as

$$\frac{\partial B_1^n}{\partial Z_1} \Delta Z_1 + \frac{\partial B_1^n}{\partial Q_1} \Delta Q_1 = -B_1^n \quad (8.72a)$$

$$\frac{\partial G_i^n}{\partial Z_i} \Delta Z_i + \frac{\partial G_i^n}{\partial Q_i} \Delta Q_i + \frac{\partial G_i^n}{\partial Z_{i+1}} \Delta Z_{i+1} + \frac{\partial G_i^n}{\partial Q_{i+1}} \Delta Q_{i+1} = -G_i^n \quad (8.72b)$$

$$\frac{\partial H_i^n}{\partial Z_i} \Delta Z_i + \frac{\partial H_i^n}{\partial Q_i} \Delta Q_i + \frac{\partial H_i^n}{\partial Z_{i+1}} \Delta Z_{i+1} + \frac{\partial H_i^n}{\partial Q_{i+1}} \Delta Q_{i+1} = -H_i^n \quad (8.72c)$$

$$\frac{\partial B_N^n}{\partial Z_N} \Delta Z_N + \frac{\partial B_N^n}{\partial Q_N} \Delta Q_N = -B_N^n \quad (8.72d)$$

in which $i = 1, 2, \ldots, N - 1$. Equations 8.72 represent a linear system of equations that can be placed in matrix form as $[\mathbf{E}] \{\Delta x\} = \{b\}$ in which $\{\Delta x\}$ = vector of changes in the unknowns at each iteration; $\{\mathbf{b}\}$ = vector of negative residuals; and $[\mathbf{E}]$ = matrix of derivatives banded along the diagonal with a maximum width of four elements. This banded property allows for more efficient solution of the system of equations. The system is solved repeatedly until the changes in the unknown values become acceptably small.

An advantage of the implicit method compared to the method of characteristics and the explicit method is its inherent stability without having to satisfy the Courant limitation of small time steps. Stability of a numerical scheme occurs when small perturbations in the solution do not grow exponentially with time. It is determined mathematically by substituting a Fourier series representation of the finite difference solution at the grid points into the difference equations and determining the conditions under which the error in the solution grows with time. The Fourier stability analysis, often attributed to von Neumann (Strelkoff 1970), generally is applied to a simpler linearized set of equations with the assumption that the results also are applicable to the more complex nonlinear system. Numerical experiments generally confirm the validity of this approach. Liggett and Cunge (1975) show, for a linearized form of the governing equations, that the condition for stability of the Preissmann scheme depends on the weighting factor θ. If $\theta = \frac{1}{2}$, then the solution is not damped with time nor does it grow with time, while for $\theta < \frac{1}{2}$ the solution grows with time and always is unstable. For $\theta > \frac{1}{2}$, the solution always is stable but some damping occurs. It is tempting then to use a value of $\theta = \frac{1}{2}$, but because of differences between the numerical wave celerity and the actual wave celerity, small undesirable oscillations in the solution can occur, although they do not grow with time. For this reason, a slightly larger value of θ is needed to damp out the oscillations. As a practical matter, Liggett and Cunge (1975) recommend $0.6 \leq \theta \leq 1.0$.

Samuels and Skeels (1990) included both the convective term and the friction slope term in their stability analysis and showed analytically that $\theta \geq \frac{1}{2}$ is required for numerical stability in agreement with previous investigators; however, they also showed that the absolute value of the Vedernikov number, $\mathbf{V}$, must be less than or equal to unity, where $\mathbf{V}$ is defined by

$$\mathbf{V} = \frac{a}{b} \frac{A}{R} \frac{dR}{dA} \mathbf{F} \quad (8.73)$$

in which a = exponent on the hydraulic radius and b = exponent on the velocity in the uniform flow evaluation of the friction slope; A = cross-sectional area of flow; R = hydraulic radius; and $\mathbf{F}$ = Froude number of the flow. The Vedernikov number arises in stability analyses of steady, uniform flow in open channels (Liggett and Cunge 1975; Chow 1959). When the Vedernikov number exceeds unity, roll waves form. The roll waves are a series of transverse ridges of high vorticity that occur in supercritical flow (Mayer 1957). The roll waves can break and resemble a succession of moving hydraulic jumps. For fully rough, turbulent flow, $b = 2$, and using the Manning equation, $a = \frac{4}{3}$, so that for a very wide channel, the Vedernikov stability limit reduces to $\mathbf{F} \leq 1.5$. What the analysis by Samuels and Skeels shows is that the Preissmann scheme must satisfy not only $\theta \geq \frac{1}{2}$ but also the physical stability limit imposed by roll waves for numerical stability to be achieved. This is the reason for the statement in some established numerical codes using the implicit method that they do not apply to supercritical flow (e.g., BRANCH).

If difficulties occur in the appplication of the Preissmann scheme, even though the stability limits on θ and the Vedernikov number are satisfied, then other sources of the difficulties must be sought. The stability analyses, for example, assume a uniform grid spacing in the flow direction, whereas the spacing is likely to be nonuniform in applications to rivers. The irregularity of the cross-section geometry, the occurrence of rapidly varied flow, and the application of the boundary conditions all could contribute to problems with the implicit method; nevertheless, it has been widely used successfully in several established codes (HEC-RAS [U.S. Army Corps of Engineers 2008], BRANCH [Schaffranek, Baltzer, and Goldberg 1981], FLDWAV [Fread and Lewis 1995]).

8.6   COMPARISON OF NUMERICAL METHODS

From the foregoing presentation of the numerical method of characteristics with specified time intervals (MOC-STI), several explicit finite difference methods, and the implicit finite difference method (Preissmann), it is apparent that an obvious advantage of the implicit method is its unconditional stability with no limits on the time step. In addition, the compactness of the Preissmann implicit scheme in particular allows it to be applied with spatial steps of variable length. As a result, the Preissmann scheme has become very popular for applications in large rivers such as routing of flood hydrographs or dam-break outflows. Reach lengths in such applications are variable because of changes in channel geometry and roughness in the flow direction. In addition, the absence of a time step limitation is advantageous for flood hydrographs that have long time bases to avoid a large computational time.

Amein and Fang (1970) applied the box (Preissmann) implicit scheme to the routing of a flood on the Neuse River from Goldsboro to Kinston, North Carolina, which is a river reach having a length of 72 km. The upstream boundary condition was specified to be the measured stage hydrograph, while the downstream condition was the measured rating curve. Initial conditions were determined from backwater calculations, starting with the measured downstream depth. For comparison of the method of characteristics (MOC), explicit, and implicit methods, a composite channel cross section was assumed, with geometric properties determined as an average over the entire reach. The computed results for all three methods were compared with the measured stage hydrograph at Kinston for two different floods each with a base time of about 15 to 20 days. The results showed similar accuracy in comparison with the observed hydrographs, but the implicit method was much more efficient. The explicit method required a time step of 0.025 hr for a subreach length of 2.4 or 4.8 km (1.5 or 3.0 mi) to maintain stability. Time steps of as large as 20 hr were possible for the implicit method with a subreach length of 4.8 km, although a somewhat shorter time step might be desirable if more rapid changes are taking place in stage or discharge. For the same subreach length of 4.8 km and a time step of 5 hr in the implicit method, the computer time was more than four times greater for the explicit method than for the implicit method.

Price (1974) compared the MOC, explicit, and implicit methods for a monoclinal wave, which is a translatory wave similar to the front of a flood wave in very long channels (see Chapter 9). It approaches a constant depth very far upstream and a smaller constant depth downstream with a wave profile in between that does not change shape as it travels downstream at a constant wave speed, c_m. The monoclinal wave is a stable, progressive wave form that results after long times when an initial constant depth is increased abruptly to a larger constant value at the upstream end of a river reach. If the wave profile is gradually varied in a wide prismatic channel, there is an analytical solution for the profile (Henderson 1966). The monoclinal wave is useful for numerical comparisons because it retains the nonlinear inertial terms in the full dynamic equations while having an analytical solution. It has a maximum speed of $(V + c)$ and a minimum speed equal to that of the "kinematic" wave, discussed in the next chapter, for which the inertial terms and the $\partial y/\partial x$ term are small in comparison to the bed slope in the momentum equation. Of interest in this chapter, however, is the comparison made by Price between specific numerical solution techniques and the analytical solution for the monoclinal wave. He selected an upstream depth of 8.0 m (26.2 ft), a downstream depth of 3.0 m (9.8 ft), and channel slopes of 0.001 and 0.00025 over a total reach length of 100 km (62 mi) having a Chezy C of 30 $m^{1/2}$/s. These data resulted in monoclinal wave speeds of 3.31 m/s (10.9 ft/s) and 1.65 m/s (5.41 ft/s) for a very wide channel with the slopes of 0.001 and 0.00025, respectively.

Price compared two explicit techniques (Lax-Wendroff and the leapfrog scheme), the method of characteristics, and the implicit scheme with the analytical solution of the monoclinal wave. Price found that the explicit and method of characteristics techniques had the least error when $\Delta x/\Delta t$ was approximately equal to the maximum Courant celerity, $V + c$; that is, a Courant number equal to 1. The implicit method exhibited the smallest error for $\Delta x/\Delta t$ approximately equal to the monoclinal wave celerity. This resulted in a larger possible time step for the implicit method than for any of the other methods and so greater computational efficiency. Furthermore, Price determined that the error in the implicit method is much less sensitive to changes in Δt for a fixed value of Δx.

Example 8.2

A natural river cross section is approximated as a trapezoidal shape with a bottom width of 100 ft (30.5 m), side slopes of 4:1, and a total in-bank flow depth of 12 ft (3.66 m). There are symmetric floodplains on either side of the main channel having a width of 300 ft (91.4 m) each, a cross slope of 150:1 toward the main channel (a drop of 2 ft or 0.61 m), and a Manning's n value of 0.08. The longitudinal bed slope of the river is 0.0005, and the Manning's n value of the main channel is 0.045. The initial condition in the river is a steady uniform flow at a rate of 200 cfs (5.67 m³/s). The length of the river reach of interest is 20,000 ft (6096 m). A flood hydrograph at the upstream boundary (River Station 20,000) is approximated as a trapezoidal shape with a time of rise of 2 hr to a peak of 5000 cfs (141.6 m³/s), followed by 1 hr at a constant discharge of 5000 cfs, and completed by a recession time of 4 hr back to the base flow of 200 cfs. This inflow hydrograph for the river reach is shown in Figure 8.8a. At the downstream boundary (River Station 0), the flow is approximately uniform. Using the HEC-RAS unsteady flow simulation module, which implements the implicit method, route the inflow hydrograph through the river reach and compare with the results of the method of characteristics program CHAR found on the book website. Also plot the flood wave profiles at 2, 4, 6, and 8 hr after the beginning of the flood hydrograph.

Solution. The river schematic and the cross sections are entered into HEC-RAS beginning at the upstream river station of 20,000 ft (6096 m). At this upstream cross section, the channel bottom elevation is arbitrarily set at 84.0 ft (25.6 m) and the bank elevation is 98.0 ft (29.9 m) with the outer boundary of the floodplain at an elevation of 100 ft (30.5 m) adjoining a vertical wall that rises to an elevation of 110 ft (33.5 m). Once this first cross section is entered, the remaining cross sections are entered at a spacing of 2000 ft (610 m) by successively copying the current cross section downstream and lowering all the elevations by 1.0 ft (0.0005 × 2000), or 0.305 m, for each new cross section. The cross-section spacing is selected here according to the guideline given by Equation 8.83 as discussed subsequently in Section 8.9. The bulk wave celerity is estimated to be approximately 3.8 ft/s or 1.2 m/s (1.5 times the uniform flow velocity corresponding to an average discharge of 2500 cfs [70.8 m³/s] — see Chapter 9), which is multiplied times the hydrograph rise time of 3 hr to the midpoint of the flat top of the inflow hydrograph and then divided by 20 to give an estimated spacing of about 2000 ft (610 m). In general, cross-section spacing in natural rivers will vary from about 100 ft (30.5 m) for steeper slopes to about 5000 ft (1500 m) for flatter slopes (US Army Corps of Engineers 2008). Boundary conditions are entered in HEC-RAS as unsteady flow data. The inflow hydrograph is entered into a table as the upstream boundary condition, and normal depth is designated as the downstream boundary condition with the given slope of 0.0005. The initial condition is specified to be a constant discharge of 200 cfs, but it is possible to enter increasing discharges downstream at flow change locations. Finally, a computation interval of 5 minutes is entered, and the implicit weighting factor, θ, is specified to be 0.6, although it can be taken to be 1.0 initially to avoid stability problems. The computation interval is selected to be less than or equal to 1/20 of the hydrograph rise time. A warm-up period of 200 time steps at constant inflow can be employed to ensure that instability in the initial condition does not cause failure of the numerical method at the very beginning of the hydrograph rise. If the hydrograph rises too fast, instability may occur anyway and a smaller computational time interval may be needed.

The routed outflow hydrograph from the HEC-RAS implicit method is shown in Figure 8.8a where it is compared with the result from the method of characteristics (MOC) obtained with the computer program CHAR, which can be found on the book website. The spatial interval for CHAR is 200 ft (61 m) based on Equation 8.60, and the total number of time steps dictated by the Courant condition results in an average time step of about 12 sec.

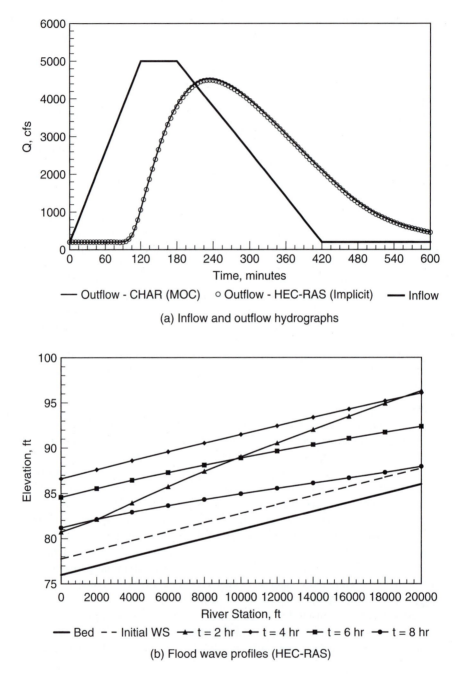

(a) Inflow and outflow hydrographs

(b) Flood wave profiles (HEC-RAS)

FIGURE 8.8
Flood routing by the method of characteristics and the implicit method.

The peak outflow rate from the implicit method is 4495 cfs (127.3 m³/s) while it is 4527 cfs (128.2 m³/s) from the method of characteristics, which is a difference of about 0.7%. In other words, comparable results are obtained from the two methods as demonstrated in Figure 8.8a, but the method of characteristics requires a spatial interval that is an order of magnitude smaller and a time step that is 25 times smaller than required by the implicit method. It is worth mentioning that if the implicit weighting factor is changed to 1.0, additional diffusion causes the peak outflow to drop to 4429 cfs (125.5 m³/s), a decrease of 1.5%.

The flood wave development is depicted in Figure 8.8b using the implicit HEC-RAS results. The flood wave rises rapidly at the upstream end of the reach forming a steep negative slope in the depth profile with respect to the x flow direction (increasing x values correspond to decreasing values of the river station), and then travels over the whole reach followed by a slower decline upstream and flatter positive slope of the depth profile as the discharge decreases on the recession side of the inflow hydrograph.

8.7   SHOCKS

In the hydraulics of unsteady open channel flow, shocks are the same as moving surges at which there is a discontinuity in depth and velocity. In the method of characteristics, the shock corresponds to an intersection of converging positive characteristics at which the methods of gradually varied flow no longer are applicable because of strong vertical accelerations and a pressure distribution that no longer is hydrostatic at the shock itself. Across the shock, both mass and the momentum function must be conserved, as discussed in Chapter 3. On either side of the shock, gradually varied unsteady flow usually exists and can be treated using any of the numerical methods in this chapter. The difficulty then is in computing the discontinuity caused by the shock itself. This important problem arises in dam-break wave fronts, rapid operation of gates in canal systems, and transients in the headrace or tailrace of a hydroelectric plant that occur upon rapid startup or shutdown of the turbines.

There are two methods of solving the problem of shock computation: shock fitting and shock capturing, also known as "computing through." In the first method, the position of the shock front at time $t + \Delta t$ is computed using the method of characteristics combined with the shock compatibility equations, which simply are the continuity and momentum equations written across the shock or surge as given previously by Equations 3.17 and 3.18. Six unknowns are found at $t + \Delta t$: the depth and velocity at the back of the surge, y_1 and V_1; depth and velocity at the front of the surge, y_2 and V_2; the speed of the surge, V_s; and the position of the surge $x_{t+\Delta t}$. However, only three equations are given by the two shock compatibility equations and the ordinary differential equation for the path of the shock, $V_s = \mathrm{d}x/\mathrm{d}t$. For a surge advancing in the positive x direction, two forward characteristics and one backward characteristic can be sketched from the unknown position and time at point P in the x-t plane backward to time level $k\Delta t$, as shown in Figure 8.9. Each of these characteristics has two equations associated with it, as described in Chapter 7, and three

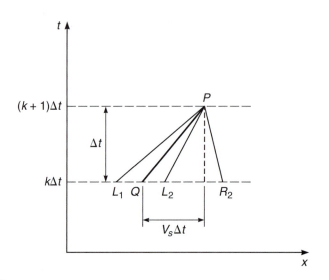

FIGURE 8.9

Shock fitting using characteristics (Lai 1986). (*Source: Figure from "Numerical Modeling of Unsteady Open Channel Flow" by Chintu Lai in ADVANCES IN HYDROSCIENCE, Volume 14, copyright © 1986 by Academic Press, reproduced by permission of the publisher.*)

more unknown values are introduced as the x positions of the intersections of these characteristics with the known time line. In all, a total of nine equations can be solved for nine unknown values to obtain not only the new position of the shock but also the depth and velocity on both sides of the shock. These latter variables then can be used as internal boundary conditions to solve the Saint-Venant equations for the gradually varied flow regions both upstream and downstream of the shock.

In the second method of computing shocks (shock capturing), the numerical solution procedure for the Saint-Venant equations is simply computed through the surge with no special treatment of the discontinuity. Martin and DeFazio (1969) applied the equivalent of the Lax diffusive scheme on a staggered grid to the problem of hydroelectric load rejection in the headrace due to shutdown of turbines and showed good agreement with measured water surface profiles of an undular surge. Martin and Zovne (1971) used the method to show reasonable agreement between computed solutions for the propagation of shocks due to an instantaneous dam break in a horizontal frictionless channel with the analytical solution of Stoker, discussed previously in Chapter 7. Terzidis and Strelkoff (1970) demonstrated the use of the Lax diffusive scheme and Lax-Wendroff scheme in computing through the propagation of a shock wave in nonuniform flow. Numerical dissipation caused by the numerical method itself tends to smooth the abrupt discontinuity in the Lax diffusive scheme, while artificial dissipation may be required for the Lax-Wendroff scheme to smooth oscillations behind the shock, although Terzidis and Strelkoff achieved similar results simply using a time step equal to eight-tenths the value required for stability. On the other hand, use of nondissipative methods such as the leapfrog scheme requires an artificial viscosity to dampen the oscillations. In the Preissmann method, taking the weighting factor $\theta > 0.5$ introduces dissipation that may avoid oscillations on the back of the shock resulting from hydroelectric load rejection in a turbine headrace; however, a value of $\theta = 1$ (fully implicit) may cause excessive damping. Wylie and Streeter (1978) showed that a value of $\theta = 0.6$ produced good agreement between the implicit method and the method of characteristics for the hydroelectric load rejection problem. For very abrupt shocks such as those that occur downstream of a very large, rapid dam break and for transcritical flow, the Preissmann method no longer may be useful, and explicit schemes have been developed for this case, as described previously (Fennema and Chaudhry 1987; Jha, Akiyama, and Ura 1995; Meselhe, Sotiropoulos, and Holly 1997). While it may not be as important in the gradually varied flow regions, it is imperative that the governing equations be written in conservation form for computing through the shock to conserve the momentum function and mass flux.

Example 8.3

A hydroelectric turbine decreases its load linearly from 1000 cfs (28.3 m³/s) to zero discharge in 10 sec. The headrace channel is trapezoidal with a length of 5000 ft (1520 m), a bottom width of 20 ft (6.1 m), side slopes of 1.5:1, Manning's $n = 0.015$, and a bed slope of 0.0002. Compute the depth hydrographs at $x/L = 0.0, 0.2, 0.4, 0.6, 0.8$, and 1.0 using the method of characteristics with specified time intervals and the Lax diffusive scheme.

Solution.   The channel length is divided into 50 spatial intervals and the time step is selected so that the Courant number is ≤1 for all grid nodes at the current time level, as in Example 8.1. The upstream boundary is a reservoir, as in Example 8.1, and the downstream boundary condition is a discharge hydrograph with a linear decrease in turbine discharge from the steady-state value of 1000 cfs (28.3 m³/s) to zero in 10 sec. At time $t = 0$, the water in the headrace is in steady uniform flow with a normal depth of 7.66 ft (2.33 m) and a critical depth of 3.85 ft (1.17 m).

The results are shown in Figure 8.10a for the Lax diffusive method, and Figure 8.10b for the method of characteristics. An abrupt increase in depth at the turbine is followed by a more gradual increase in depth due to the inertia of the flowing water. The positive wave is reflected back from the reservoir with lower depths, and it is apparent that a relatively long time is required for the water to come completely to rest, as reflections continue back and forth along the headrace. The solutions by the two methods are even closer in agreement than in Example 8.1. The only differences are a slightly more gradual rise and a very slight rounding at the peak of the depth hydrographs at intermediate points along the channel for the Lax scheme. The maximum depth for the Lax scheme is 10.34 ft (3.15 m), while it is 10.36 ft (3.16 m) for the method of characteristics.

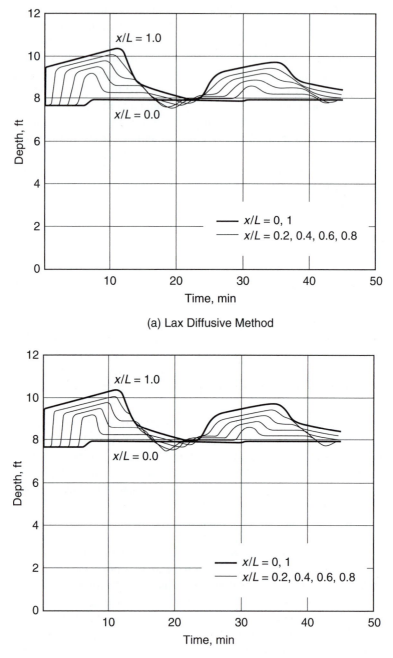

(a) Lax Diffusive Method

(b) Method of Characteristics

FIGURE 8.10
Depth hydrographs between the reservoir ($x/L = 0.0$) and the turbine ($x/L = 1.0$) for load rejection.

8.8   DAM-BREAK PROBLEM

Several large dam failures in the United States, including the Teton Dam failure on the Teton River in Idaho in 1976, have led to dam safety programs in many states and the need to predict the peak discharge and time of travel of dam-breach flood waves. In the dam-break problem, the routing of shocks in the downstream river channel depends greatly on the hydrograph created at the dam, which in turn depends on the time of failure and the geometry of the breach. The National Weather Service combined an implicit flood routing technique (Preissmann method) with a parameterization of the breach geometry to generate the outflow hydrograph resulting from a dam break and route it downstream in the program FLDWAV (Fread and Lewis 1988), which combines the formerly used programs DAMBRK and DWOPER (Chow, Maidment, and Mays 1988). The dam breach geometry is trapezoidal in shape, as shown in Figure 8.11 and given by Fread and Harbaugh (1973) and Fread (1988):

$$b_f = b_{av} - m(h_d - h_f) \tag{8.74}$$

in which b_f = final bottom width of the trapezoid; b_{av} = average breach width; m = side slope of the breach (horizontal:vertical); h_d = elevation of the top of the dam; and h_f = final elevation of the bottom of the breach. Triangular and rectangular breach shapes also can be simulated with $b_f = 0$ and $m = 0$, respectively. The instantaneous elevation, h_{bt}, of the bottom of the breach is given by

$$h_{bt} = h_d - (h_d - h_f)\left(\frac{t}{\tau}\right)^{\rho} \qquad (0 \le t \le \tau) \tag{8.75}$$

in which h_d = elevation of the top of the dam; h_f = final elevation of the bottom of the breach (taken to be the bottom of the dam unless there is an erosion retarding layer); t = time from the beginning of the breach; τ = total failure time; and

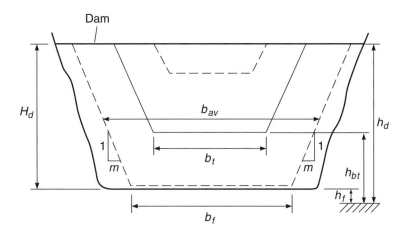

FIGURE 8.11
Definition of embankment dam breach parameters.

$\rho = 1$ to 4, with the linear rate usually assumed. Likewise, the instantaneous value of the bottom width, b_t, of the breach is

$$b_t = b_f\left(\frac{t}{\tau}\right)^{\rho} \tag{8.76}$$

in which b_f is the final bottom width of the breach. Estimates of the failure time, τ, and average breach width, b_{av}, are needed to complete the description of the time-varying geometry of the breach. Froehlich (1987) statistically analyzed 43 embankment dam failures for dams ranging in height from 12 to 285 ft (3.7 to 87 m) and proposed the following relationships:

$$b* = 0.47k_0(V*)^{0.25} \tag{8.77}$$

$$\tau* = 79(V*)^{0.47} \tag{8.78}$$

in which $b* = b_{av}/H_d$; $k_0 = 1.4$ if the failure mode is overtopping and $k_0 = 1.0$ if the failure mode is not overtopping; $V* = V_w/H_d^3$; H_d = height of dam; V_w = volume of water in reservoir at time of failure; and $\tau* = \tau(g/H_d)^{0.5}$. Equations 8.77 and 8.78 have coefficients of determination of 0.559 and 0.913, respectively. If the height of the dam, H_d, is not equal to the height of the breach, $(h_d - h_f)$, then H_d is replaced by $(h_d - h_f)$ in the definitions of the dimensionless variables. In a subsequent study, Froehlich (1995) recommended a side slope ratio $m = 1.4$ for overtopping and $m = 0.9$ otherwise.

The outflow hydrograph from the breached dam is computed either by level pool reservoir routing (see Chapter 9) or dynamic routing by the implicit numerical model with the breached dam outflow as an internal boundary condition between the upstream reservoir reach and the downstream river reach. Level pool routing is used for wide, flat reservoir surfaces with gradual changes in water surface elevation, while dynamic routing is needed for narrow valleys with significant water surface slope in the reservoir. The outflow relationship for the breach utilizes the head-discharge relationship for a trapezoidal broad-crested weir given by

$$Q_b = C_v K_s[3.1\, b_t(h_w - h_{bt})^{1.5} + 2.45m(h_w - h_{bt})^{2.5}] \tag{8.79}$$

in which Q_b = breach outflow in cubic feet per second; C_v = approach velocity correction factor; K_s = weir submergence correction factor; b_t = instantaneous bottom width of the breach in feet (Equation 8.76); h_w = elevation of the water surface in feet; and h_{bt} = instantaneous elevation of the bottom of the breach (Equation 8.75).

The FLDWAV model deals with transcritical flow and shocks or surges by two different methods (Fread and Lewis 1988; Jin and Fread 1997). The first method is an approximate approach using the implicit method in which the entire river reach is divided into supercritical and subcritical subreaches at each time step. For supercritical subreaches, two upstream boundary conditions are applied that consist of the discharge from the next upstream subreach and critical depth. For subcritical subreaches, the downstream boundary condition is critical depth and the upstream boundary condition is the discharge from the next subreach upstream. The position of surges is adjusted until the shock compatibility equations are satisfied before moving to the next time step. In the second method, an explicit, characteristics-based scheme is available, as described previously, to be used in combination with the implicit scheme for different reaches during the same routing. For transcritical or mixed-flow subreaches, the explicit scheme is applied alongside the implicit method for subreaches where near-critical flow does not occur.

To further simplify the dam-break problem and provide quicker forecasts of dam-break flood waves, dimensionless solutions have been developed; for example, Sakkas and Strelkoff (1976) for instantaneous failures and the NWS simplified dam-break method for gradual embankment failures (Wetmore and Fread 1983). These methods are based on a large number of routed hydrographs for typical values of the independent variables, with the results presented in dimensionless form.

Wurbs (1987) tested a number of dam-breach flood wave models, including simplified models, with measured field data and concluded that a dynamic routing model provides maximum accuracy although none of the methods could be considered highly accurate because of uncertainties in the breach development with time, rapid changes in downstream channel geometry, lack of one-dimensional flow conditions, and loss of flow volume. Another contributing factor to inaccuracy is that most dam-break flood waves exceed stages experienced for any historical floods so that calibration of parameters such as Manning's n is not possible. Regardless of these difficulties, dam-break flood wave propagation can be modeled to provide reasonable estimates of the consequences of a catastrophic dam failure.

8.9   PRACTICAL ASPECTS OF RIVER COMPUTATIONS

Rivers seldom are prismatic and further experience an abrupt change in cross section as the flow transitions between bank-full flow and overbank flow. The main channel may meander across the floodplain and consist of numerous branches and loops. Under these trying circumstances, the one-dimensional flow assumptions are severely strained. As long as the flow remains in the main channel or the flow completely inundates the floodplain following the general direction of the valley, one-dimensional flow is a reasonable assumption. In the transition between these two extremes, it is a question of how much lateral drop in the water surface can be tolerated as the flow moves into the floodplain on the rising side of the hydrograph and returns to the main channel on the falling side, sometimes only partially.

Several artifices have been devised for the role of floodplain storage in flood wave propagation. One possibility is to include an inactive area of flow in the floodplain in the continuity equation while using the active width in the momentum equation. In the continuity equation, the time derivative becomes $\partial(A + A_0)/\partial t$, in which A_0 represents the inactive flow area. In this way the storage effects of the floodplain are taken into account in an *ad hoc* manner, but considerable skill on the part of the modeler is necessary to designate inactive flow areas. Another approach is to treat the main channel of the river with one-dimensional methods but with storage pockets at specific nodes coming off the main channel (Cunge, Holly, and Verwey 1980). Then the challenge becomes correctly modeling the exchange of flow between the main stem and the storage pockets, usually by artificial weirs. If the storage areas are linked, then a kind of two-dimensional network of loops can be generated, and the Saint-Venant equations may be simplified in the storage reaches by neglecting the inertial terms (see Chapter 9).

If the flow is effective through a portion of the floodplains while storage alone occurs in the remaining portion, the continuity and momentum equations can be written separately for the floodplains and main channel with the discharge divided according to the ratio of active flow area conveyances. Then the equations can be combined by requiring equality of the mass and momentum exchanges between main channel and floodplains. The boundary friction forces from floodplains and main channel can be lumped into a single equivalent force term in the combined momentum equation that acts over an equivalent flow path length (U.S. Army Corps of Engineers 2008).

For meandering channels with flow in the floodplains, the flow path in the main channel may be longer than in the floodplains, and the device of a conveyance-weighted reach length can be used, in which an average length is based on the relative magnitudes of the conveyances of the left and right floodplains and the main channel. In some instances, as in the flow through multiple bridge openings, one-dimensional methods simply no longer may suffice, and two-dimensional, depth-averaged models or three-dimensional models may be required, depending on the purpose of the hydraulic modeling effort.

Calibration and verification of unsteady flow models are essential to gain confidence in their use. The selection of a particular one-dimensional model, whether the dynamic form or some simplified form as described in the next chapter, is an important consideration. Considerable time and effort are required for calibration and verification, so a simplified model should not be used if engineering river works or extensive floodplain and channel alterations are expected in the future that would require extensive recalibration.

The calibration of a one-dimensional model often begins with selection of Manning's n values based on past experience and running a steady-flow model to verify previously measured peak stages. This should be done for the entire range of discharges expected to be encountered in the unsteady flow model. Once the steady-flow values of the resistance coefficient have been established, then the unsteady flow model is implemented, with further tweaking of the resistance coefficients to reproduce measured flood hydrographs. The initial condition for the unsteady model can be the steady, gradually varied flow computation, but running the unsteady model during a startup or warmup period by maintaining steady flow may be necessary to dampen any initial instabilities.

Stage hydrographs rather than discharge hydrographs are best for calibration of the unsteady model because of the uncertainty of the stage-discharge relationship. The stage-discharge relationship, or rating curve, often is looped with higher discharges occurring on the rising limb of the hydrograph than on the recession limb. This loop, or hysteresis, in the rating curve of stage vs. discharge as shown in Figure 8.12 occurs primarily because of the depth gradient and acceleration terms in the momentum equation (see Equation 7.16). If we write the unsteady discharge Q in terms of the channel conveyance K and the friction slope S_f, we have

$$Q = KS_f^{1/2} \tag{8.80}$$

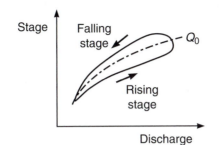

FIGURE 8.12
Loop in stage-discharge rating curve.

Similarly, the steady uniform value of discharge Q_0 is written in terms of the bed slope S_0 as

$$Q_0 = KS_0^{1/2} \qquad (8.81)$$

Then if we take the ratio of (8.80) and (8.81) and substitute for S_f from Equation 7.16, we have

$$\frac{Q}{Q_0} = \sqrt{\frac{S_f}{S_0}} = \sqrt{1 - \frac{1}{S_0}\left(\frac{\partial y}{\partial x} + \frac{V}{g}\frac{\partial V}{\partial x} + \frac{1}{g}\frac{\partial V}{\partial t}\right)} \qquad (8.82)$$

Note that Q will be different from Q_0 in magnitude only if the three dynamic terms in parentheses on the right-hand side of (8.82) are of the same order of magnitude as the bed slope; however, these three terms decrease in order of magnitude from left to right in the equation. The convective and local acceleration terms may be of the same order, but in large rivers of relatively flat slope, the $\partial y/\partial x$ term is of the same order as the bed slope and considerably larger than the acceleration terms (Henderson 1966). Under these circumstances, the $\partial y/\partial x$ term, which is negative on the rising side of the hydrograph (see Example 8.2), causes $Q > Q_0$ while the opposite is true on the falling side of the hydrograph. The result is the loop in the rating curve shown in Figure 8.12.

While use of a measured downstream stage hydrograph is a preferred downstream boundary condition, in comparison with a rating curve, such a hydrograph may not be available for future modeling runs to determine the effects of changes in the river. Therefore, it is useful to have a stage-discharge relationship that has been measured over a wide range of discharges. In order to minimize the influence of a looped rating curve, the downstream boundary should not be subject to significant backwater effects caused by a reservoir, for example. Either the downstream boundary should be moved downstream all the way to the dam or moved upstream out of the backwater influence. The establishment of a steady-state, single-valued rating curve at the downstream boundary essentially causes reflection of waves upstream that otherwise would not occur in a "free-flow" condition. This can work only if the boundary is far enough downstream not to influence the river reach of interest. One of the options offered by FLDWAV for the downstream boundary condition is a computed looped rating curve using Manning's equation with S_f determined from the implicit finite difference solution of the momentum equation for the last two spatial grid points.

The deviation of a looped rating curve for unsteady flow from the single-valued, steady-flow relationship is influenced by the rate of rise of the discharge hydrograph, roughness coefficients, and channel slope among other factors. The more rapid the hydrograph rise, the greater the deviation from the steady-state curve. For channel bed slopes in excess of approximately 0.001, loops usually do not occur, while they always exist for slopes less than 0.0001 (Cunge, Holly, and Verwey 1980). Increasing Manning's n during the calibration to reduce the computed flood peak also may cause a widening of the looped rating curve at a particular cross section.

Compound channel sections are particularly challenging during the calibration process, because the wave celerity is drastically reduced in the transition from bank-full to overbank flow due to the abrupt increase in area. The wave celerity reaches a minimum at relatively shallow depths on the floodplain, then begins to increase again. The wave celerity therefore is very sensitive to both flooded valley width and elevation of the main channel banks at which floodplain inundation begins. Calibration may require checking these particular geometric data very carefully rather than adjusting Manning's n alone to accommodate large differences between observed and computed peak travel times. Also, the location of cross sections may be deficient in that the chosen cross-section location is not representative of the river subreach of interest, especially the floodplain width.

If calibration difficulties occur, then several sources of errors should be examined. The size of the time step or distance step may be too large. Aside from stability considerations in explicit methods, the size of the time step should be small enough to adequately discretize the boundary conditions such as tidal variations or flood hydrographs. The distance step depends on the slope of the water surface and the desired accuracy. Jin and Fread (1997) recommend a distance step for the FLD-WAV model selected by

$$\Delta x \leq C_b \frac{T_r}{M} \tag{8.83}$$

in which T_r = rise time of the inflow hydrograph; C_b = bulk wave celerity of the peak discharge; and M = constant value, recommended to be about 20 for the implicit scheme. Other sources of error may be oversimplification of the basic equations (see Chapter 9 for limitations), inadequate or inaccurate flood stage data, and insufficiently detailed topographic data.

Topographic and hydraulic data needs for calibrating and verifying a dynamic flow routing model include detailed elevation data, descriptions of vegetation and other roughness elements, bridge geometry, floodplain boundaries, and stage hydrographs at several locations along the river. Existing topographic maps may be insufficient to establish variations in topography necessitating spot aerial or ground surveys to augment existing data. Bridge geometry in sufficient detail to establish a head-discharge relation as an internal boundary condition may require as-built plans or even ground truth measurements. The existence of peak stage measurements alone may require establishment of several gauging stations to obtain comprehensive, consistent data very early in the project, even if it turns out not to be data for a very large flood. While it may seem obvious, the more data that are available, the more accurately the model can predict unsteady flow variations.

For a more detailed discussion of the application of unsteady flow models to rivers along with case studies, refer to Cunge, Holly, and Verwey (1980).

REFERENCES

Abbott, M. B., and D. R. Basco. *Computational Fluid Dynamics.* New York: John Wiley & Sons, 1989.

Amein, M., and C. S. Fang. "Implicit Flood Routing in Natural Channels." *J. Hyd. Div.,* ASCE 96, no. HY12 (1970), pp. 2481–2500.

Ames, William F. *Numerical Methods for Partial Differential Equations.* New York: Barnes and Noble, 1969.

Chaudhry, M. H. *Open-Channel Flow.* Englewood Cliffs, NJ: Prentice-Hall, 1993.

Chow, Ven Te. *Open-Channel Hydraulics.* New York: McGraw-Hill, 1959.

Chow, Ven Te; D. R. Maidment; and L. W. Mays. *Applied Hydrology.* New York: McGraw-Hill, 1988.

Cunge, J. A., F. M. Holly, Jr., and A. Verwey. *Practical Aspects of Computational River Hydraulics.* London: Pitman Publishing Limited, 1980 (Reprinted by Iowa Institute of Hydraulic Research, Iowa City, IA, 1994.)

Fennema, R. J., and M. H. Chaudhry. "Explicit Numerical Schemes for Unsteady Free-Surface Flows with Shocks." *Water Resources Res.* 22, no. 13 (1986), pp. 1923–30.

Fennema, R. J., and M. H. Chaudhry. "Simulation of One-Dimensional Dam-Break Flows." *J. Hydr. Res.* 25, no. 1 (1987), pp. 41–51.

Fread, D. L. *The NWS DAMBRK Model: Theoretical Background/User Documentation.* Silver Spring, MD: Hydrologic Research Laboratory, National Weather Service, 1988.

Fread, D. L., and T. E. Harbaugh. "Transient Simulation of Breached Earth Dams." *J. Hyd. Div.,* ASCE 99, no. HY1 (1973), pp. 139–54.

Fread, D. L., and J. M. Lewis. "FLDWAV: A Generalized Flood Routing Model." In *Proceedings of the 1988 National Conference on Hydraulic Engineering,* ed. S. R. Abt and J. Gessler, pp. 668–73. Colorado Springs, CO: ASCE, 1988.

Fread, D. L., and J. M. Lewis. *The NWS FLDWAV Model Quick User's Guide.* Silver Spring, MD: Hydrological Research Laboratory, National Weather Service, 1995.

Froehlich, D. C. "Embankment-Dam Breach Parameters." In *Proceedings of the 1987 National Conference on Hydraulic Engineering,* ed. R. M. Ragan, pp. 570–75. Williamsburg, VA, ASCE, 1987.

Froehlich, D. C. "Embankment Dam Breach Parameters Revisited." In *Proceedings of the First International Conference on Water Resources Engineering,* ed. W. H. Espey, Jr. and P. G. Combs, pp. 887–91, San Antonio, TX, ASCE, 1995.

Garcia-Navarro, P.; F. Alcrudo; and J. M. Saviron. "1-D Open Channel Flow Simulation Using TVD-McCormack Scheme." *J. Hydr. Engrg.,* ASCE 118, no. 10 (1992), pp. 1359–72.

Goldberg, D. E., and E. B. Wylie. "Characteristics Method Using Time-Line Interpolations." *J. Hydr. Engrg.*, ASCE 109, no. 5 (1983), pp. 670–83.

Henderson, F. M. *Open Channel Flow.* New York: Macmillan, 1966.

Huang, J., and C. C. S. Song. "Stability of Dynamic Flood Routing Schemes." *J. Hydr. Engrg.*, ASCE 111, no. 12 (1985), pp. 1497–1505.

Jha, A. K., J. Akiyama, and M. Ura. "First- and Second-Order Flux Difference Splitting Schemes for Dam-Break Problem." *J. Hydr. Engrg.*, ASCE 121, no. 12 (1995), pp. 877–84.

Jha, A. K., J. Akiyama, and M. Ura. "Modeling Unsteady Open-Channel Flows—Modification to Beam and Warming Scheme." *J. Hydr. Engrg.*, ASCE 120, no. 4 (1994), pp. 461–76.

Jin, M., and D. L. Fread. "Dynamic Flood Routing with Explicit and Implicit Numerical Solution Schemes." *J. Hydr. Engrg.*, ASCE 123, no. 3 (1997), pp. 166–73.

Katopedes, N., and C. T. Wu. "Explicit Computation of Discontinuous Channel Flow." *J. Hydr. Engrg.*, ASCE 112, no. 6 (1986), pp. 456–75.

Lai, C. "Numerical Modeling of Unsteady Open-channel Flow." In *Advances in Hydroscience*, vol. 14, pp. 161–333. New York: Academic Press, 1986.

Liggett, J. A., and J. A. Cunge. "Numerical Methods of Solution of the Unsteady Flow Equations." In *Unsteady Flow in Open Channels*, ed. K. Mahmood and V. Yevjevich, vol. 1, pp. 89–182. Fort Collins, CO: Water Resources Publications, 1975.

Martin, C. S., and F. G. DeFazio. "Open-Channel Surge Simulation by Digital Computer." *J. Hyd. Div.*, ASCE 95, no. HY6 (1969), pp. 2049–70.

Martin, C. S., and J. J. Zovne. "Finite Difference Simulation of Bore Propagation." *J. Hyd. Div.*, ASCE 97, no. HY7 (1971), pp. 993–1007.

Mayer, P. G. W. "A Study of Roll Waves and Slug Flows in Inclined Open Channels." Ph.D. thesis, Cornell University, 1957.

Meselhe, E. A.; F. Sotiropoulos; and F. M. Holly, Jr. "Numerical Simulation of Transcritical Flow in Open Channels." *J. Hydr. Engrg.*, ASCE 123, no. 9 (1997), pp. 774–83.

Price, R. K. "A Comparison of Four Numerical Methods for Flood Routing." *J. Hyd. Div.*, ASCE 100, no. HY7 (1974), pp. 879–99.

Sakkas, J. G., and T. Strelkoff. "Dimensionless Solution of Dam-Break Flood Waves." *J. Hyd. Div.*, ASCE 102, no. HY2 (1976), pp. 171–84.

Samuels, P. G., and C. P. Skeels. "Stability Limits for Preissmann's Scheme." *J. Hydr. Engrg.*, ASCE 116, no. 8 (1990), pp. 997–1012.

Schaffranek, R. W.; R. A. Baltzer; and D. E. Goldberg. "A Model for Simulation of Flow in Singular and Interconnected Channels." Chapter C3 in *Techniques of Water Resources Investigations of the U.S. Geological Survey.* Washington, DC: Government Printing Office, 1981.

Strelkoff, T. "Numerical Solution of Saint-Venant Equations." *J. Hyd. Div.*, ASCE 96, no. HY1 (1970), pp. 223–52.

Terzidis, G., and T. Strelkoff. "Computation of Open-Channel Surges and Shocks." *J. Hyd. Div.*, ASCE 96, no. HY12 (1970), pp. 2581–2610.

U.S. Army Corps of Engineers. *HEC-RAS Hydraulic Reference Manual.* Davis, CA: U.S. Army Corps of Engineers, Hydrologic Engineering Center, 2008.

Wetmore, J. N., and D. L. Fread. *The NWS Simplified Dam-Break Model Executive Brief.* Silver Spring, MD: National Weather Service, Office of Hydrology, 1983.

Wurbs, R. A. "Dam-Breach Flood Wave Models." *J. Hydr. Engrg.*, ASCE 113, no. 1 (1987), pp. 29–46.

Wylie, E. B., and V. L. Streeter. *Fluid Transients.* New York: McGraw-Hill, 1978.

EXERCISES

8.1. Write out the complete difference equations for an internal boundary condition of a weir. What if the weir were in a submerged condition?

8.2. What causes the diffusive behavior in the Lax diffusive method?

8.3. Derive the Lax-Wendroff method using the Taylor's series expansion for $U(t + \Delta t)$ about $U(t)$. Hint: Write the second derivative of U with respect to t in terms of derivatives of F using the governing equations and substitute $A\Delta U = \Delta F$.

8.4. Verify the definition of A given by (8.57) by taking the Jacobian of F ($\partial F/\partial U$).

8.5. Show that the eigenvalues, λ, of A are given by $V + c$ and $V - c$ by taking the determinant of $(A - \lambda I)$.

8.6. Rederive the Lax diffusive scheme using the method of treating the source term given by (8.45).

8.7. Apply the simple wave method of Chapter 7 to Example 8.1. Calculate the minimum depth at the turbine and compare it with the computer results. What is the maximum discharge that can be supplied to the turbine by the negative wave?

8.8. Apply the momentum and continuity equations in finite volume form (shock compatibility equations) to the surge developed in Example 8.3 and compare the results to the numerical results.

8.9. For an earthen dam that is 80 ft in height with a volume of water in storage of 50,000 ac-ft, estimate the time of failure and the average breach width. What will be the height and bottom width of the breach at one third the time to failure?

8.10. Use the computer program CHAR (on the website) to route a triangular dam-breach hydrograph with a peak discharge of 100,000 cfs at a time to peak of 1 hr and a base time of 3 hr in a downstream rectangular prismatic channel having a width of 200 ft, a slope of 0.0003, and Manning's $n = 0.025$. The initial discharge in the channel is 2500 cfs, and it is 10 mi long. What will be the peak discharge at the downstream boundary and how much time will it take for the peak discharge to arrive?

8.11. Apply the computer program CHAR (on the website) for a hydroelectric load acceptance problem in a trapezoidal headrace having a bottom width of 10 m, side slopes of 0.5:1, Manning's n of 0.016, and a bed slope of 0.0001. The steady-state turbine discharge is 40 m³/s, which is brought on line in 2 min. The headrace is 3 km long. Repeat for a slope of 0.0004.

8.12. Apply the computer program CHAR (on the website) for the same conditions as in Exercise 8.11 except for the load rejection problem.

8.13. Apply the computer program LAX (on the website) for the same conditions as in Exercise 8.11 and compare the results with those from CHAR.

8.14. In the load acceptance problem, show that the negative wave cannot supply the steady state turbine discharge if the corresponding Froude number of the uniform flow, F_0, exceeds 0.319. Set the steady state discharge per unit of channel width, Vy_0, equal to the unit maximum discharge for the negative wave, which depends on the upstream head or specific energy, E_0, if the slope is small. Then solve for F_0.

8.15. A steady discharge of 8000 cfs is released by hydroelectric turbines at a dam into the downstream river at full load. The discharge is increased linearly from the minimum release rate of 700 cfs to 8000 cfs in 20 minutes, held steady at 8000 cfs for 2 hr, and brought back down to 700 cfs linearly in 20 minutes. The river cross section is approximately trapezoidal in shape with a bottom width of 300 ft and 1:1 side slopes. The bed slope is 0.0005 ft/ft, and Manning's $n = 0.035$. What will be the time to peak and the peak discharge at a location 25,000 ft downstream of the dam? Use the computer program LAX or CHAR.

8.16. In the dam-break problem of Exercise 8.10, repeat the computation using CHAR or LAX, but for a Manning's $n = 0.05$ instead of 0.025. Plot on the same axes the maximum stage as a function of distance downstream of the dam for both values of Manning's n.

8.17. Repeat the flood routing problem of Example 8.2 using HEC-RAS, but decrease the main channel width linearly along the channel with a 5% decrease for each successive station in the downstream direction. Use the *Adjust Station* feature of the geometric cross-section editor with a multiplying factor of 0.95 times the width of the immediate upstream section applied to the main channel only. Compare the outflow hydrograph with that from Example 8.2 and discuss.

8.18. Repeat the flood-routing problem of Example 8.2 using HEC-RAS, but double all the inflow hydrograph ordinates including the base flow and initial condition. Compare the results with Example 8.2, and discuss how they are affected by the overbank flow.

8.19. Use HEC-RAS and the NFPeachtreeCr. xls geometry file on the book website (see Exercise 5.19) to conduct a flood-routing analysis. The downstream boundary condition is normal depth for a slope of 0.001. The upstream boundary condition is the inflow hydrograph given in the table below. Plot the inflow and outflow hydrographs and the profile of maximum water surface elevation. Also plot the computed rating curve at the upstream station.

Time, hr	Inflow, cfs	Time, hr	Inflow, cfs
0	200	11	8000
1	200	12	6000
2	200	13	4500
3	200	14	3400
4	2000	15	2400
5	4000	16	1600
6	6000	17	900
7	9000	18	500
8	12000	19	200
9	11700	20	200
10	10000	21	200

Simplified Methods of Flow Routing

9.1 INTRODUCTION

While in the previous chapter, the full dynamic equations of continuity and momentum were solved numerically, this chapter presents simplified methods in which one or more terms of the governing equations are neglected. These simplified methods are presented in the context of flow routing problems, where they most often are used. By *flow routing*, we refer to the tracking in time and space of a wave characteristic such as the peak discharge or stage as it moves along the flow path but superimposed on the physical flow itself. Flow routing problems range from the routing of a flood or dam-break surge in a river to routing runoff from a parking lot or upland watershed to generate a runoff hydrograph. The general solution sought in the flow routing problem is the distribution of discharge or stage with time at the downstream end of a river reach; that is, the outflow hydrograph, given the inflow hydrograph at the upstream end of the reach and the stream geometry, slope, and roughness. In particular, the translation and attenuation of the peak discharge or stage with respect to time most often are of interest. Simplified methods of flow routing may be employed in situations such as long-term hydrologic modeling of large-scale watershed systems for which solving the full dynamic equations may not be warranted. However, it is essential to understand the limitations of simplified flow routing methods.

While flow routing methods can be classified in a number of ways, one of the most important distinctions is between *hydrologic routing* and *hydraulic routing*. In hydrologic or storage routing, the momentum equation is ignored altogether and the one-dimensional continuity equation is integrated spatially in the flow direction so

369

that it becomes a lumped system spatially, with no variation of parameters, within the resulting control volume. Hydraulic routing is a distributed system method that determines the flow as a function of both space and time (Chow, Maidment, and Mays 1988).

The continuity equation in hydrologic routing simplifies to the storage equation given as

$$\frac{dS}{dt} = I - O \tag{9.1}$$

in which S = storage in the reach (control volume); I = inflow rate to the reach; and O = outflow rate from the reach. An additional equation is required to solve for the outflow in Equation 9.1, and it is provided by a known functional relationship between storage and the inflow and outflow; that is, $S = f(I, O)$. In contrast, hydraulic routing includes the full one-dimensional, unsteady continuity equation and all or part of the momentum equation, as follows:

$$\frac{\partial Q}{\partial x} + \frac{\partial A}{\partial t} = 0 \tag{9.2}$$

$$\frac{\partial V}{\partial t} + V\frac{\partial V}{\partial x} + g\frac{\partial y}{\partial x} - g(S_0 - S_f) = 0$$

$$\begin{array}{ll} \vdash\!-\!-\!-\!-\!-\!-\!\dashv \text{ kinematic} & \\ \vdash\!-\!-\!-\!-\!-\!-\!-\!-\!-\!\dashv \text{ diffusion} & (9.3) \\ \vdash\!-\!-\!-\!-\!-\!-\!-\!-\!-\!-\!-\!-\!-\!\dashv \text{ dynamic} & \end{array}$$

As shown in Equation 9.3, dynamic routing includes all terms in the momentum equation, while diffusion routing neglects the inertia terms (local and convective acceleration), and kinematic routing includes only the gravity and flow resistance terms. In terms of spatial variation, all the hydraulic routing methods can be considered distributed models but applicable only under conditions for which the neglected terms are small relative to the remaining terms. The previous chapter considered the case of dynamic routing, while this chapter treats simplified routing methods, both lumped and distributed.

Finite difference numerical techniques are described for the simplified methods of flow routing in this chapter. Problems of stability and numerical diffusion are considered because they are just as important as in the previous chapter, when solving the dynamic routing equations. Solutions of the simplified equations also are compared with solutions of the dynamic equations so that the conditions of applicability of the simplified methods can be identified. In addition, we show that the kinematic routing method can be recast in terms of the method of characteristics with a result analogous to the simple wave problem, treated in Chapter 7. Finally, we see that a hybrid method, the Muskingum-Cunge method, can be developed by matching numerical diffusion and physical diffusion terms in the routing equations.

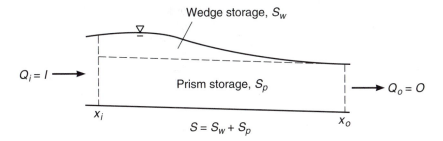

FIGURE 9.1
Inflow, outflow, and storage in hydrologic routing through a river reach.

9.2 HYDROLOGIC ROUTING

With reference to the control volume shown in Figure 9.1, the one-dimensional continuity equation given by (9.2) can be integrated along the flow path from the inflow section at x_i to the outflow section at x_o as follows:

$$\int_{x_i}^{x_o} \frac{\partial Q}{\partial x} \, dx + \int_{x_i}^{x_o} \frac{\partial A}{\partial t} \, dx = 0 \tag{9.4}$$

The first integral in (9.4) becomes the difference between the discharge evaluated at x_o and x_i, $(Q_o - Q_i)$, or simply the difference between outflow and inflow rates for the control volume. By application of the Leibniz rule, the time derivative can be brought outside the second integral so that (9.4) becomes

$$Q_o - Q_i + \frac{d}{dt} \int_{x_i}^{x_o} A \, dx = 0 \tag{9.5}$$

Now recognizing the integral in the third term as the storage volume, S, and replacing the outflow rate, Q_o, with O and the inflow rate, Q_i, with I, we recover the storage equation given by (9.1).

In the Muskingum method of river routing, the second relationship (in addition to the storage equation) required to solve the routing problem is supplied by a linear relationship between storage and a weighted function of inflow and outflow:

$$S = \theta[XI + (1 - X)O] \tag{9.6}$$

in which θ = time constant; X = weighting factor (<1); S = storage; I = inflow rate; and O = outflow rate. Equation 9.6 sometimes is justified physically by arguing that channel storage consists of prism storage and wedge storage, as illustrated in Figure 9.1; thus, it depends on both inflow and outflow. Prism storage is that portion of storage associated with a steady uniform flow profile dependent only on outflow, while wedge storage is the remaining storage that occurs during the unsteady rise and fall of stage, so it depends on the difference between inflow and outflow for the river reach. Alternatively, it can be argued that Equation 9.6 is simply a conceptual model with the parameter θ indicative of the translation of the inflow hydrograph and the parameter X considered a weighting factor related to storage and attenuation of the flood peak, with both θ and X to be determined by calibration. If the time constant, θ, indeed is held constant for all discharges, then Equation 9.6 establishes a linear relationship between storage and weighted flow in which the coefficient of proportionality can be interpreted as the wave travel time in the river reach as will be justified later.

As will be discussed subsequently, the value of the Muskingum X generally falls in the range of 0.0 to 0.5 (although not always). These limits on the range of X values establish two different behaviors with respect to the propagation of a flood wave. For $X = 0$, Equation 9.6 simplifies to the storage relationship for a linear reservoir, several of which sometimes are used in series for catchment routing. Reservoir routing can be interpreted as a special case of Muskingum river routing, in which the storage depends only on outflow. So long as the reservoir water surface can be assumed to be approximately horizontal, otherwise known as *level pool routing*, both the storage and the outflow depend solely on the reservoir water surface elevation and therefore on each other. As the reservoir inflow increases with time and a portion goes into reservoir storage, the outflow is reduced, as shown in Figure 9.2a. The result is a peak outflow attenuated in comparison to the peak inflow, which is precisely the purpose of a flood control reservoir. It is instructive to note that the peak outflow occurs at the intersection of the inflow and outflow hydrographs. This is because the maximum storage occurs at the same time as the maximum outflow when $dS/dt = 0$ and therefore $I = O$. Reservoir routing provides the limiting case of pure storage with associated attenuation and spreading in time of the outflow hydrograph which is sometimes referred to as diffusion.

At the other extreme, $X = 0.5$, the Muskingum routing technique weights equally the inflow and outflow in the storage relationship. The result, as is shown subsequently, is pure translation of the inflow hydrograph, in which each discharge is delayed in time by the wave travel time in the reach determined by θ. In this special case, the inflow hydrograph theoretically undergoes no attenuation or change in shape, as shown in Figure 9.2b. Most rivers behave between the two extremes given in Figure 9.2 and exhibit both diffusion and translation of the inflow hydrograph.

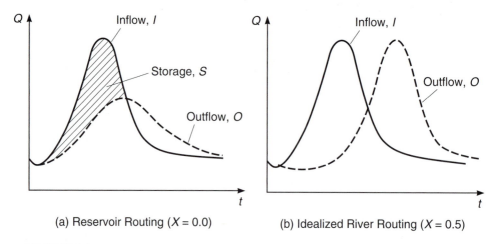

(a) Reservoir Routing ($X = 0.0$)   (b) Idealized River Routing ($X = 0.5$)

FIGURE 9.2
Routing of hydrographs with pure storage (a) and pure translation (b).

Reservoir Routing

Equations 9.1 and 9.6 can be solved numerically using a finite difference technique. If a forward difference is taken for the time derivative in Equation 9.1 with the mean values of I and O evaluated over the time interval Δt, the following results:

$$\frac{S_2 - S_1}{\Delta t} = \frac{I_1 + I_2}{2} - \frac{O_1 + O_2}{2} \tag{9.7}$$

in which the subscripts 1 and 2 refer to the beginning and end of the time interval, analogous to the indices k and $k + 1$ for the current and subsequent time levels used in Chapter 8. In the storage-indication method, as it sometimes is called, Equation 9.7 is multiplied by 2 and rearranged to yield

$$\frac{2S_2}{\Delta t} + O_2 = I_1 + I_2 + \frac{2S_1}{\Delta t} - O_1 \tag{9.8}$$

Because both storage and outflow are functions of reservoir stage, a separate relationship can be developed between the left hand side of Equation 9.8 and outflow O_2. Then the right hand side of Equation 9.8 is computed in subsequent time steps to determine the left hand side, from which O_2 follows, based on the separate relationship of $2S/\Delta t + O$ vs. O.

Example 9.1

A small water supply lake has a normal pool elevation given by $Z = 0.0$ m with an emergency spillway crest elevation given by $Z = 0.50$ m (1.64 ft). The emergency spillway is a broad-crested weir with a discharge coefficient $C_d = 0.848$ and a crest length of 20 m (66 ft). The elevation-storage-outflow relationship is given in Table 9.1. If the lake level initially is at the normal pool, route the inflow hydrograph given in Table 9.2 through the spillway and determine the peak outflow.

TABLE 9.1 Elevation-storage-outflow relationships of Example 9.1

Z, m	S, m³	O, m³/s	$2S/\Delta t + O$
0.0	6.90E+06	0.0	7668
0.5	7.72E+06	0.0	8579
1.0	8.58E+06	10.2	9546
1.5	9.49E+06	28.9	10569
2.0	1.04E+07	53.1	11642
2.5	1.14E+07	81.8	12767
3.0	1.24E+07	114.3	13942
3.5	1.35E+07	150.2	15166
4.0	1.46E+07	189.3	16440
4.5	1.58E+07	231.3	17763
5.0	1.70E+07	276.0	19136

TABLE 9.2   Storage-indication method of flow routing of Example 9.1

Time, hr	I, m³/s	$2S/\Delta t - O$	$2S/\Delta t + O$	O, m³/s
0.0	0.0	7668	7668	0.0
0.5	79.2	7748	7748	0.0
1.0	212.5	8039	8039	0.0
1.5	320.5	8572	8572	0.0
2.0	381.9	9260	9275	7.3
2.5	400.0	10003	10042	19.3
3.0	386.1	10722	10789	33.9
3.5	352.3	11362	11460	49.0
4.0	308.4	11897	12023	62.8
4.5	261.7	12319	12467	74.1
5.0	216.5	12632	12797	82.6
5.5	175.6	12846	13024	88.9
6.0	140.1	12977	13162	92.7
6.5	110.2	13038	13227	94.5
7.0	85.7	13044	13234	94.7
7.5	65.9	13009	13196	93.7
8.0	50.3	12942	13125	91.7
8.5	38.1	12852	13030	89.1
9.0	28.6	12746	12918	86.0
9.5	21.4	12631	12796	82.6
10.0	15.9	12510	12668	79.3
10.5	11.7	12386	12537	75.9
11.0	8.6	12261	12406	72.6
11.5	6.3	12137	12276	69.3
12.0	4.6	12016	12148	66.0
12.5	3.4	11898	12024	62.8
13.0	2.4	11785	11904	59.8
13.5	1.8	11675	11789	56.9
14.0	1.3	11570	11678	54.0
14.5	0.9	11469	11572	51.5
15.0	0.7	11372	11471	49.2
15.5	0.5	11279	11373	47.1
16.0	0.3	11190	11280	45.0
16.5	0.2	11105	11191	42.9
17.0	0.2	11023	11105	41.0
17.5	0.1	10945	11024	39.2
18.0	0.1	10871	10945	37.4
18.5	0.1	10799	10871	35.7
19.0	0.0	10731	10799	34.1
19.5	0.0	10666	10731	32.6
20.0	0.0	10604	10666	31.1

Solution. First, the routing interval $\Delta t = 0.5$ hr is chosen such that the rising side of the inflow hydrograph is adequately discretized. Then, in Table 9.1, calculations are shown for the quantity $2S/\Delta t + O$ to be used in the routing with careful attention being paid to using consistent units, which are cubic meters per second in this example. The routing table (Table 9.2) is developed based on Equation 9.8. To start the routing, the first outflow value is set to zero and the corresponding value of $2S/\Delta t + O$ is placed in the table corresponding to the initial condition of normal pool elevation ($Z = 0$), which is below the spillway crest. The value of $2S/\Delta t - O$ follows from the value in the $2S/\Delta t + O$ column minus twice the outflow at the same value of time. Then, Equation 9.8 is solved for $2S/\Delta t + O$ in the next time step and interpolated in Table 9.1 to obtain the corresponding outflow value. The outflow remains at zero until the lake level rises above the spillway crest, after which finite values of outflow are obtained from the interpolation in Table 9.1. This process is continued until the outflow falls to some small value. The peak outflow rate, which occurs at the point of intersection with the inflow hydrograph, is 94.7 m³/s (3340 cfs) at $t = 7.0$ hr, and it has been reduced from the peak inflow rate of 400 m³/s (14,100 cfs) at $t = 2.5$ hr. The results are shown in Figure 9.3.

River Routing

For Muskingum river routing, Equation 9.6 can be employed to evaluate S_2 and S_1 with the results substituted into (9.7) to yield

$$\theta[X(I_2 - I_1) + (1 - X)(O_2 - O_1)] = \frac{\Delta t}{2}[(I_1 + I_2) - (O_1 + O_2)] \quad (9.9)$$

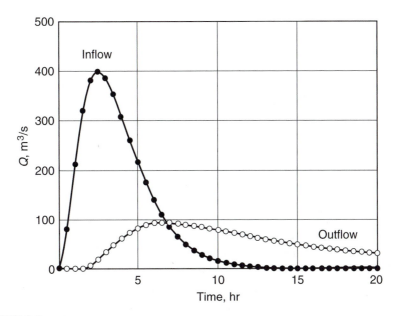

FIGURE 9.3
Inflow and outflow hydrographs for reservoir routing, Example 9.1.

Collecting terms in (9.9) and solving for O_2, we have

$$O_2 = C_0 I_2 + C_1 I_1 + C_2 O_1 \qquad (9.10)$$

in which the coefficients C_0, C_1, and C_2, called *routing coefficients*, are defined by

$$C_0 = \frac{-\theta X + 0.5\Delta t}{\theta - \theta X + 0.5\Delta t} \qquad (9.11)$$

$$C_1 = \frac{\theta X + 0.5\Delta t}{\theta - \theta X + 0.5\Delta t} \qquad (9.12)$$

$$C_2 = \frac{\theta - \theta X - 0.5\Delta t}{\theta - \theta X + 0.5\Delta t} \qquad (9.13)$$

Because the denominator is the same for all the routing coefficients, we readily see that $(C_0 + C_1 + C_2) = 1$. Thus, at least conceptually, the routing coefficients can be viewed as weighting factors applied to the inflows at the beginning and end of the time interval and to the outflow at the beginning of the time interval to solve for the outflow at the end of the time interval. The routing equation as defined by (9.10) can be applied repeatedly to obtain the outflow hydrograph at the end of the river reach, given the inflow hydrograph that enters the reach.

In the context of the finite difference notation used in Chapter 8, Equation 9.10 can be rewritten as

$$Q_{i+1}^{k+1} = C_0 Q_i^{k+1} + C_1 Q_i^k + C_2 Q_{i+1}^k \qquad (9.14)$$

in which the computational molecule is rectangular, representing only one reach and a single time step. Note from (9.14) that, for pure translation over the time interval Δt, the values of C_0 and C_2 should equal zero so that $C_1 = 1$ and $Q_{i+1}^{k+1} = Q_i^k$. These conditions are satisfied in Equations 9.11 through 9.13 for the special case of $X = 0.5$ and $\Delta t = \theta$, as stated previously. The latter requirement of $\Delta t = \theta$ is equivalent to specifying a value of unity for the Courant number, provided that θ can be interpreted as the wave travel time over the reach length Δx as is proven later.

The choice of the time step, Δt, and the spatial interval, Δx, are important, and some general limits must be considered. First, Δt should be chosen such that the rising limb of the hydrograph is approximated adequately by a series of straight lines, which usually requires $\Delta t \leq t_p/5$, where t_p is the time to peak of the inflow hydrograph. Second, it would seem that negative values of the routing coefficients are counterintuitive if they indeed represent weighting values for the inflow and outflow (Miller and Cunge 1975). However, Ponce and Theurer (1982) showed from numerical experiments that it is necessary only for $C_0 \geq 0$, while C_1 and C_2 can be negative without affecting the accuracy of the routing (defined as avoidance of negative outflows). Requiring $C_0 \geq 0$ is equivalent to developing an inequality such that the numerator of C_0 in (9.11) remains nonnegative, so that the following limit on Δt must be satisfied:

$$\Delta t \geq 2\theta X \qquad (9.15)$$

If θ is the wave travel time defined by $\Delta x/V_w$, where V_w is a representative value of the wave travel speed, then (9.15) can be viewed as a limit on Δx for a value of

Δt determined by the required discretization of the time of rise of the inflow hydrograph:

$$\Delta x \leq \frac{V_w \Delta t}{2X} \tag{9.16}$$

Weinmann and Laurenson (1979) suggest a less severe limit, in which Δt on the right hand side of the inequality in (9.16) is replaced by T_r = rise time of the inflow hydrograph. Cunge (1969) shows from a stability analysis that the condition for stability is $X \leq 0.5$ and further suggests that $X \geq 0$ for the physical interpretation of wedge and prism storage to make sense. However, Ponce and Theurer (1982) argue that negative values of X are possible. This is discussed later in more detail for an extension of the Muskingum method called the *Muskingum-Cunge* or *Muskingum diffusion scheme*.

While the Muskingum method appears complete, it depends strongly on the parameters θ and X. In general, these are taken to be constant for a given river reach, and the original method of estimating them requires measured values of inflow and outflow for the river reach under consideration. Because they essentially are calibration constants when determined in this way, there is no assurance that they will have the same values for a flood different from the calibration flood.

If it is assumed that the complete inflow and outflow hydrographs have been measured for a given river reach, then the cumulative storage can be computed from a rearrangement of Equation 9.7 as

$$S_2 = S_1 + \frac{\Delta t}{2} (I_2 + I_1 - O_2 - O_1) \tag{9.17}$$

Repeated application of (9.17) for successive values of time allows the determination of the cumulative storage, S, at any time t. The initial value of storage is usually taken as zero. Then, according to Equation 9.6, we seek a linear relationship between relative storage, S, and the weighted flow value $\{XI + (1 - X)O\}$, which also can be computed from the inflow and outflow hydrographs as a function of time. However, river relationships generally display some degree of hysteresis because of greater storage on the rising side of the hydrograph (Bras 1990). Thus, as a practical matter, the value of X that produces the best single-valued relationship, or narrowest loop, is determined by trial and error, and the slope of the best-fit straight line gives the value of θ as required by Equation 9.6 and illustrated in Figure 9.4 for Example 9.2.

As an alternative to the graphical method for estimating θ and X shown in Figure 9.4, these parameters can be determined by a least-squares parameter estimation technique. Singh and McCann (1980) show that the least-squares method of minimizing the difference between observed and estimated storage is equivalent to maximizing the correlation coefficient between S and the weighted flow in the graphical method. The least-squares technique seeks to minimize the error function, E, given by

$$E = \sum_j [AI_j + BO_j + S_1 - S_j]^2 \tag{9.18}$$

in which $A = \theta X$; $B = \theta(1 - X)$; S_1 = initial storage; and S_j = observed relative storage at the jth time step. Gill (1977) proposed such a technique, and the values of A and B are given by Aldama (1990). The summation takes place with j running

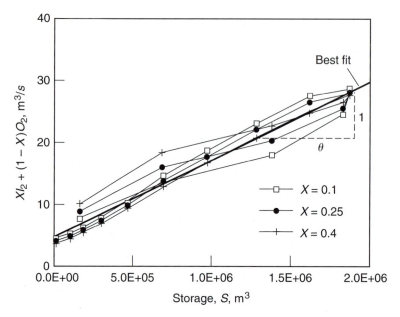

FIGURE 9.4
Graphical method for determining Muskingum θ and X, Example 9.2.

from 1 to N observed values of inflow, outflow, and relative storage for a given river reach. The error, E, is minimized by differentiating with respect to A and B and setting the results to zero. The resulting equations are solved for A and B to give the following expressions:

$$A = \frac{1}{C}[(\Sigma I_j O_j \Sigma O_j - \Sigma I_j \Sigma O_j^2)\Sigma S_j \tag{9.19}$$

$$+ (N\Sigma O_j^2 - (\Sigma O_j)^2)\Sigma I_j S_j + (\Sigma I_j \Sigma O_j - N\Sigma I_j O_j)\Sigma O_j S_j]$$

$$B = \frac{1}{C}[(\Sigma I_j \Sigma I_j O_j - \Sigma I_j^2 \Sigma O_j)\Sigma S_j \tag{9.20}$$

$$+ (\Sigma I_j \Sigma O_j - N\Sigma I_j O_j)\Sigma I_j S_j + (N\Sigma I_j^2 - (\Sigma I_j)^2)\Sigma O_j S_j]$$

$$C = N[\Sigma I_j^2 \Sigma O_j^2 - (\Sigma I_j O_j)^2] + 2\Sigma I_j \Sigma O_j \Sigma I_j O_j \tag{9.21}$$

$$- (\Sigma I_j)^2 \Sigma O_j^2 - \Sigma I_j^2 (\Sigma O_j)^2$$

Once A and B are computed, θ and X can be determined from

$$\theta = A + B, \qquad X = \frac{A}{A + B} \tag{9.22}$$

The Muskingum-Cunge technique is an extension of the Muskingum method, in which the values of θ and X can be related to the discharge and geometric properties of the channel. To achieve a better understanding of how this is accomplished, the kinematic wave and diffusion routing techniques are considered next.

Example 9.2

Utilize the observed inflows and outflows for a river reach given in Table 9.3 (Hjelmfelt and Cassidy 1975) to obtain values of θ and X using both the graphical method and the least-squares method. Then determine the routing coefficients and route the inflow hydrograph through the river reach.

Solution.   The storage is calculated from the average inflow and outflow rates over a single time step using Equation 9.17 and accumulated beginning with zero initial storage as shown in Table 9.3. Then various values of X are substituted to obtain the weighted inflow and outflow quantity, $XI + (1 - X)O$, at the end of each time step. The storage is related to this quantity by Equation 9.6, so the plot shown in Figure 9.4 allows the determination of the inverse slope, which is equal to the Muskingum time constant, θ. Figure 9.4 shows the results for values of $X = 0.10$, 0.25, and 0.40. By trial and error, the narrowest loop occurs for X approximately equal to 0.25 with intersection of the rising and falling limbs about midway along the storage axis. The best-fit line of the data for $X = 0.25$ gives a value of $\theta = 0.92$ days from Figure 9.4, and these are the results for the graphical method. Alternatively, Equations 9.19 through 9.22 can be solved for the data given in Table 9.3 to produce the values $X = 0.243$ and $\theta = 0.897$ days. These latter values are chosen as the Muskingum parameters, with $\Delta t = 0.5$ days to calculate the Muskingum routing coefficients from Equations 9.11 through 9.13, with the result

$$C_0 = 0.034, \qquad C_1 = 0.504, \qquad C_2 = 0.462$$

TABLE 9.3   Computation of storage and Muskingum parameters of Example 9.2

						$XI + (1 - X)O$		
t, days	I, m³/s	O, m³/s	I_{avg}	O_{avg}	S, m³	$X = 0.10$	0.25	0.40
0.0	2.2	2.0			0.00E+00	2.0	2.1	2.1
0.5	14.5	7.0	8.4	4.5	1.66E+05	7.8	8.9	10.0
1.0	28.4	11.7	21.5	9.4	6.89E+05	13.4	15.9	18.4
1.5	31.8	16.5	30.1	14.1	1.38E+06	18.0	20.3	22.6
2.0	29.7	24.0	30.8	20.3	1.83E+06	24.6	25.4	26.3
2.5	25.3	29.1	27.5	26.6	1.87E+06	28.7	28.2	27.6
3.0	20.4	28.4	22.9	28.8	1.62E+06	27.6	26.4	25.2
3.5	16.3	23.8	18.4	26.1	1.29E+06	23.1	21.9	20.8
4.0	12.6	19.4	14.5	21.6	9.76E+05	18.7	17.7	16.7
4.5	9.3	15.3	11.0	17.4	7.00E+05	14.7	13.8	12.9
5.0	6.7	11.2	8.0	13.3	4.73E+05	10.8	10.1	9.4
5.5	5.0	8.2	5.9	9.7	3.07E+05	7.9	7.4	6.9
6.0	4.1	6.4	4.6	7.3	1.88E+05	6.2	5.8	5.5
6.5	3.6	5.2	3.9	5.8	1.04E+05	5.0	4.8	4.6
7.0	2.4	4.6	3.0	4.9	2.16E+04	4.4	4.1	3.7

For this case, $\Delta t > 2\theta X$ so that $C_0 > 0$. Finally, the solution of the routing equation, Equation 9.10, can proceed as shown in Table 9.4. The outflow initially is assumed to be equal to the inflow, and the routing progresses from one time step to the next. The results are shown in Figure 9.5 in which the calculated and observed outflows can be

TABLE 9.4 Muskingum routing (C_0 = 0.034, C_1 = 0.504, C_2 = 0.462) of Example 9.2

t, days	I, m³/s	$C_0 \times I_2$	$C_1 \times I_1$	$C_2 \times O_1$	O_2, m³/s
0.0	2.2				2.2
0.5	14.5	0.50	1.11	1.02	2.62
1.0	28.4	0.97	7.31	1.21	9.49
1.5	31.8	1.09	14.32	4.38	19.79
2.0	29.7	1.02	16.03	9.13	26.18
2.5	25.3	0.86	14.98	12.09	27.93
3.0	20.4	0.70	12.76	12.89	26.34
3.5	16.3	0.56	10.29	12.16	23.00
4.0	12.6	0.43	8.22	10.62	19.27
4.5	9.3	0.32	6.35	8.89	15.56
5.0	6.7	0.23	4.69	7.18	12.10
5.5	5.0	0.17	3.38	5.59	9.14
6.0	4.1	0.14	2.52	4.22	6.88
6.5	3.6	0.12	2.07	3.17	5.37
7.0	2.4	0.08	1.82	2.48	4.37

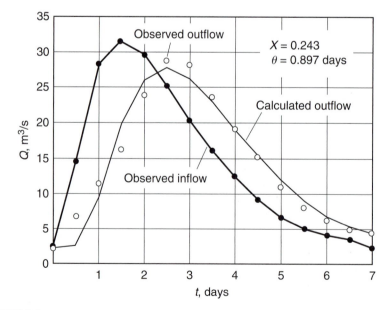

FIGURE 9.5
Inflow and outflow hydrographs for Muskingum routing, Example 9.2.

compared. The routed outflow peak is within approximately 4 percent of the observed value at the same time to peak. There generally is a better fit of the data on the falling side of the hydrograph than on the rising side.

9.3   KINEMATIC WAVE ROUTING

As shown in Equation 9.3, the momentum equation is simplified in kinematic wave routing by neglecting both the inertia terms and the pressure gradient term, so that it becomes

$$S_0 = S_f \tag{9.23}$$

Equation 9.23 is incorporated into the continuity equation given by (9.2) to obtain the kinematic wave equation. One interpretation of (9.23) is that uniform flow can be assumed in a quasisteady fashion from one time step to the next over each reach length in a finite difference numerical solution of the continuity equation. Thus, Equation 9.23 is equivalent to expressing the discharge Q in terms of a uniform flow formula such as Manning's equation, which can be rearranged as

$$Q = \left[\frac{K_n}{n} \frac{S_0^{1/2}}{P^{2/3}} \right] A^{5/3} = b_0 A^a \tag{9.24}$$

in which b_0 = constant for a wide channel of constant slope and roughness; A = cross-sectional flow area; and exponent $a = 5/3$. Under these conditions, the discharge Q is a function of cross-sectional area A alone so that

$$\frac{dQ}{dA} = ab_0 A^{a-1} = aV \tag{9.25}$$

in which $V = Q/A$ = mean flow velocity. The significance of (9.25) can be seen by writing the continuity equation given by (9.2) as

$$\frac{\partial Q}{\partial x} + \frac{\partial A}{\partial t} = \frac{\partial Q}{\partial x} + \left(\frac{dA}{dQ} \right) \frac{\partial Q}{\partial t} = 0 \tag{9.26}$$

which can be rearranged in the form

$$\frac{\partial Q}{\partial t} + \left(\frac{dQ}{dA} \right) \frac{\partial Q}{\partial x} = 0 \tag{9.27}$$

We assume a unique relationship between stage (or area) and discharge, so that dQ/dA is an ordinary derivative. The physical meaning of dQ/dA can be shown by setting the total differential of Q to zero resulting in

$$dQ = \frac{\partial Q}{\partial t} \, dt + \frac{\partial Q}{\partial x} \, dx = 0 \tag{9.28}$$

On comparing (9.27) and (9.28), it is obvious that $dQ/dA = dx/dt$, which can be interpreted as an absolute kinematic wave celerity, c_k, in the kinematic wave equation, which now can be written as

$$\frac{dQ}{dt} = \frac{\partial Q}{\partial t} + c_k \frac{\partial Q}{\partial x} = 0 \tag{9.29}$$

From the foregoing, we can conclude that Equation 9.29 has a single family of characteristics along which the discharge Q = constant in the x-t plane with a positive slope given by the kinematic wave celerity, c_k. The characteristics are straight lines because of the assumption of a unique depth-discharge relationship so that a given characteristic has a constant value of discharge and depth and, therefore, constant wave celerity. In terms of our discussion of characteristics in Chapter 7, an observer moving at the speed c_k along a particular characteristic path would see no change in the discharge Q associated with that characteristic. In other words, the partial differential equation (9.29) can be expressed in characteristic form by the following pair of equations:

$$Q = \text{constant} \tag{9.30}$$

$$\frac{dx}{dt} = c_k \tag{9.31}$$

in which the constant and the kinematic wave celerity in general are different for each characteristic. Furthermore, from (9.25), (9.27), and (9.29), the absolute kinematic wave celerity, c_k, can be expressed as

$$c_k = \frac{dQ}{dA} = \frac{1}{B}\frac{dQ}{dy} = aV \tag{9.32}$$

in which B = water surface top width. Equation 9.32 states that the kinematic wave celerity can be determined from the inverse of the flow top width times the slope of the stage-discharge relationship, which was shown to be equivalent to a constant times the mean velocity. The constant $a = \frac{5}{3}$, using Manning's equation, and $a = \frac{3}{2}$, from Chezy's equation. Implicit in this estimate of the wave celerity is the existence of a single-valued, stage-discharge relationship.

In general, c_k would be expected to vary with depth and therefore with Q; however, a simplification often is possible, in which c_k is assumed to be constant and equal to a reference value corresponding either to the peak Q or an average Q for the inflow hydrograph. Under these conditions, the characteristics become straight, parallel lines, as shown in Figure 9.6. Along these characteristics, a specific value of Q (or depth) is translated at the constant speed of c_k. Therefore, the kinematic wave equation for constant wave celerity is linear with an analytical solution represented by a pure translation of the inflow hydrograph.

When allowed to be variable, it is clear that c_k will increase with increasing Q and also with increasing depth, y. Therefore, higher discharges will move downstream at a higher speed, resulting in a steepening of the wave front, or leading edge of the kinematic wave. (The rising limb of the hydrograph also will steepen.) However, attenuation or subsidence of the kinematic wave still will not occur, because of the omission of the pressure gradient and inertia terms in the momentum equation, which are important in a "dynamic" wave.

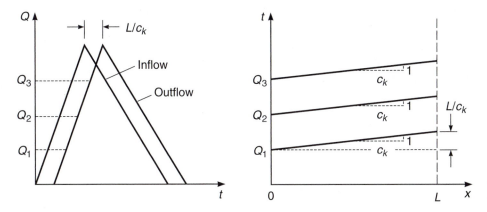

FIGURE 9.6
Pure translation of the linear kinematic wave.

While it is well known that river floods usually subside or attenuate, the question remains as to the conditions for which the kinematic wave method is applicable, since it does not allow for subsidence. If the momentum equation is solved for S_f, the following results:

$$S_f = S_0 - \frac{\partial y}{\partial x} - \frac{V}{g}\frac{\partial V}{\partial x} - \frac{1}{g}\frac{\partial V}{\partial t} \tag{9.33}$$

What is required then is to determine the relative magnitude of the bed slope, S_0, in comparison to the remaining three dynamic terms on the right hand side of (9.33). From an order of magnitude analysis, Henderson (1966) concluded that S_0 is much larger than the remaining terms for floods in steep rivers; while for very flat rivers with low Froude numbers, S_0 and $\partial y/\partial x$ are of the same order and the inertia terms are negligible. Equation 9.33 can be nondimensionalized in terms of a steady-state uniform flow velocity, V_0; a corresponding normal depth, y_0; and a reference channel reach length, L_0. Nondimensionalizing and dividing through by S_0 results in a dimensionless equation given by

$$\frac{S_f}{S_0} = 1 - \left[\frac{y_0}{L_0 S_0}\right]\frac{\partial y^o}{\partial x^o} - \left[\frac{V_0^2}{gL_0 S_0}\right]\left(V^o\frac{\partial V^o}{\partial x^o} + \frac{\partial V^o}{\partial t^o}\right) \tag{9.34}$$

in which $y^o = y/y_0$; $x^o = x/L_0$; $V^o = V/V_0$; and $t^o = tV_0/L_0$. The dimensionless numbers multiplying the inertia terms and pressure force terms, respectively, can be transformed as

$$\frac{V_0^2}{gL_0 S_0} = \frac{\mathbf{F}_0^2 y_0}{L_0 S_0} = \frac{1}{k} \tag{9.35}$$

$$\frac{y_0}{L_0 S_0} = \frac{1}{k\mathbf{F}_0^2} \tag{9.36}$$

in which $\mathbf{F}_0 = V_0/(gy_0)^{1/2}$ = a reference Froude number and k = a kinematic flow number defined by Woolhiser and Liggett (1967). For large values of k, the dynamic terms are small relative to the bed slope. If $k > 10$, the kinematic wave approximation is considered satisfactory, especially for overland flow for which k can have values in excess of 1000 (Woolhiser and Liggett 1967). Also, Miller and Cunge (1975) suggested that the Froude number should be less than 2 for the kinematic wave equation to apply, not only because of the formation of roll waves for larger values but also because this is the limit at which the kinematic wave celerity becomes equal to the dynamic wave celerity, as shown by Equation 9.38 later. Also useful to note is that, from the ratio of the coefficients in (9.35) and (9.36), the inertia terms approach the same order as the pressure force term as the Froude number approaches unity.

Ponce, Li, and Simons (1978) applied a sinusoidal perturbation to a uniform flow and examined the attenuation factor to determine the applicability of the kinematic wave model. Ponce (1989) suggested from the results that the kinematic wave model is applicable if

$$\frac{T_r V_0 S_0}{y_0} > 85 \tag{9.37}$$

in which V_0 and y_0 represent average velocity and flow depth, respectively; S_0 = bed slope; and T_r = time of rise of the inflow hydrograph. This criterion indicates that both steep slopes and long hydrograph rise times tend to favor the use of kinematic wave routing in which inertia and pressure gradient terms can be neglected in comparison to the resulting quasisteady balance between gravity and friction terms in the momentum equation.

In spite of these limitations on the use of the kinematic wave approximation, it can be shown that both kinematic wave and dynamic wave behavior occur in a river flood wave (Ferrick and Goodman 1998). Henderson (1966) argued that the bulk of a natural flood wave of small height moves at the kinematic wave speed, c_k, while the leading edge of the wave experiences dynamic effects and rapid subsidence. Because the kinematic wave moves in the downstream direction only, its speed, c_k, can be compared with the downstream speed of the dynamic wave, which has been given previously as $V + c$, by taking the ratio of the two wave speeds:

$$\frac{c_k}{V + c} = \frac{a\mathbf{F}}{\mathbf{F} + 1} \tag{9.38}$$

in which $\mathbf{F}$ = Froude number = V/c and $c = (gy)^{1/2}$. It can be shown from (9.38) for $a = 3/2$ (Chezy) that $c_k < (V + c)$ so long as $\mathbf{F} < 2$ and attenuation of the "dynamic forerunner" will result (Henderson 1966). It would seem then that the kinematic wave moves downstream more slowly than the dynamic wave forerunner unless $\mathbf{F} > 2$, at which time it may steepen to form a surge.

Kinematic Shocks and the Monoclinal Wave

The question arises as to whether the steepening of the kinematic wave can lead to some stable form, which has been called a *kinematic shock* (Henderson 1966),

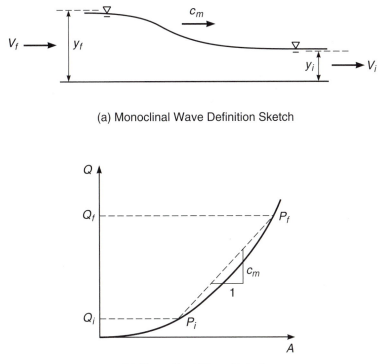

(a) Monoclinal Wave Definition Sketch

(b) Monoclinal Wave Celerity

FIGURE 9.7
Monoclinal wave definition and wave celerity from discharge-area curve.

before actually becoming an abrupt surge as the dynamic terms in the momentum equation become important. This leads to the special case of a uniformly progressive wave called the *monoclinal rising wave* shown in Figure 9.7. The monoclinal wave can be conceptualized as the result of an abrupt increase in discharge at the upstream end of a very long prismatic channel. Very far upstream, the flow is uniform with a depth of y_f and velocity, V_f, while very far downstream the flow is uniform with depth y_i such that $y_c < y_i < y_f$, where y_c = critical depth for the moving wave profile. A stationary observer would see the depth gradually increase from the initial uniform flow value, y_i, to the final value, y_f, as the wave profile moves downstream with time. The slope of the profile is relatively gentle so that it cannot be considered a surge, but because it does not change form as it moves downstream, it can be treated using the same methodology as for surges. For a monoclinal wave moving downstream at a constant absolute speed, c_m, the problem can be made stationary by superimposing a speed of c_m in the upstream direction. The continuity equation then is applied at points i and f along the wave profile, as shown in Figure 9.7a, to yield

$$(c_m - V_f)A_f = (c_m - V_i)A_i = Q_r \qquad (9.39)$$

in which Q_r is referred to as the *overrun discharge* that is seen by a moving observer with the speed c_m. If Equation 9.39 is solved for the monoclinal wave speed, the result is the so-called Kleitz-Seddon principle (Chow 1959), given by

$$c_m = \frac{Q_f - Q_i}{A_f - A_i} \tag{9.40}$$

which can be illustrated, as in Figure 9.7b, by the slope of the straight line between points P_i and P_f. The curve in Figure 9.7b is shown concave up, because the velocity generally increases with stage and flow area for flow in the main channel alone. We can deduce from the figure that c_m is greater than the flow velocity at either point i or f and that the maximum value of c_m occurs as the depth, y_i, approaches y_f. For the special case of a very wide channel, Equation 9.40 becomes

$$c_m = \frac{V_f y_f - V_i y_i}{y_f - y_i} \tag{9.41}$$

With the aid of the Chezy equation for a very wide channel, the ratio of the monoclinal wave celerity, c_m, to the kinematic wave celerity of the initial uniform flow, c_{ki}, can be determined from Equation 9.41 as

$$\frac{c_m}{c_{ki}} = \frac{2}{3} \frac{\left(\dfrac{y_f}{y_i}\right)^{3/2} - 1}{\dfrac{y_f}{y_i} - 1} \tag{9.42}$$

in which c_{ki} has been determined from the Chezy equation and Equation 9.32 as $(3/2)V_i$; that is, $a = 3/2$. It is clear from (9.42) that the monoclinal wave celerity is greater than the initial kinematic wave celerity as well as the initial uniform flow velocity, V_i, and it depends only on the specified ratio of depths y_f/y_i and V_i. As y_f approaches y_i, we can see from (9.42) that the monoclinal wave celerity approaches, as a lower limit, the kinematic wave celerity of the initial uniform flow.

Determination of the shape of the monoclinal wave profile requires the use of the momentum equation applied from the viewpoint of a moving observer with the constant absolute speed, c_m, who sees a steady, gradually varied flow profile. Under these circumstances, the equation of gradually varied flow for a very wide channel with Chezy friction becomes

$$\frac{dy}{dx} = \frac{S_0 - \dfrac{q^2}{C^2 y^3}}{1 - \dfrac{q_r^2}{g y^3}} \tag{9.43}$$

in which y = depth; x = distance along the channel; q = flow rate per unit of width = Vy; C = Chezy resistance coefficient; and q_r = overrun discharge per unit of width = $(c_m - V)y$. Note that the evaluation of the friction slope depends on the absolute velocity and discharge, while the Froude number squared term $q_r^2/(gy^3)$, which comes from convective inertia, is based on the relative velocity and overrun discharge as seen by the moving observer. Therefore, the critical depth that defines the limit of stability of the monoclinal wave is given by $y_c = (q_r^2/g)^{1/3}$, corresponding to the overrun Froude number having a value of unity. As y_i approaches y_c, the denominator of (9.43) approaches zero and the slope of the profile becomes infinite with the formation of a surge.

The surge that forms at the stability limit of the monoclinal wave defines a maximum celerity that is reached. If the continuity definition of overrun discharge from Equation 9.39 is simplified for a very wide channel and set equal to the limiting value for the overrun Froude number equal to 1 with $y_c = y_i$, the result is

$$q_r = (c_m - V_i)y_i = \sqrt{gy_i^3} \tag{9.44}$$

Equation 9.44 can be solved to obtain the maximum value of $c_m = V_i + c_i$, where $c_i = (gy_i)^{1/2}$. In other words, the monoclinal wave has a maximum celerity corresponding to the dynamic wave celerity of the initial uniform flow, while it has a minimum celerity equal to the kinematic wave celerity of the initial uniform flow, as shown by Equation 9.42. This can be placed in dimensionless form (Ferrick and Goodman 1998) and expressed as

$$1 \le \frac{c_m}{c_{ki}} \le \frac{2}{3}\left(1 + \frac{1}{\mathbf{F}_i}\right) \tag{9.45}$$

in which $\mathbf{F}_i$ = Froude number of initial uniform flow, and the kinematic wave celerity has been taken as $(3/2)V_i$ from the Chezy equation. The same conclusion for the upper limit was reached previously when comparing the kinematic wave celerity and dynamic wave celerity in Equation 9.38. Hence, for a given Froude number of the initial uniform flow, there is a maximum celerity for the monoclinal wave that decreases as the Froude number increases to a value of 2, at which time the upper and lower limits both collapse to the kinematic wave celerity. Setting the right hand side of Equation 9.42 equal to the upper limit given by (9.45), the stability limit on the initial Froude number can be defined in terms of the ratio of the final and initial normal depths (Ferrick and Goodman 1998):

$$\mathbf{F}_i = \frac{(y_r^{1/2} + 1)}{y_r} \tag{9.46}$$

in which $y_r = y_f/y_i$. For a given initial Froude number, Equation 9.46 gives the maximum value of the depth ratio, beyond which the monoclinal wave becomes unstable and remains at its maximum dynamic celerity.

Monoclinal Wave Profile

There is an analytical solution to Equation 9.43 for the profile of the stable monoclinal wave (Lighthill and Whitham 1955; Chow 1959; Henderson 1966). The details of the solution are given by Agsorn and Dooge (1991). The result is

$$\frac{S_0 x}{y_i} = \frac{y}{y_i} + a_1 \ln(y - y_i) + a_2 \ln(y_f - y) + a_3 \ln(y - Y_0) + C_I \tag{9.47}$$

in which

$$a_1 = \frac{y_i^3 - y_c^3}{y_i(y_i - y_f)(y_i - Y_0)} \tag{9.48a}$$

$$a_2 = \frac{y_f^3 - y_c^3}{y_i(y_f - y_i)(y_f - Y_0)} \tag{9.48b}$$

$$a_3 = \frac{Y_0^3 - y_c^3}{y_i(Y_0 - y_i)(Y_0 - y_f)} \tag{9.48c}$$

$$Y_0 = \frac{y_i y_f}{\left(\sqrt{y_i} + \sqrt{y_f}\right)^2} \tag{9.48d}$$

and C_I = constant of integration. To obtain the wave profile at any time t_1, the solution for the profile given by Equation 9.47 is translated a distance of $c_m t_1$ with c_m given by Equation 9.42. Because the profile is infinitely long, some depth slightly less than the final normal depth can be specified at $x = 0$ with the constant of integration, C_I, determined accordingly. Alternatively, C_I can be determined such that the mid-depth of the wave occurs at $x = 0$ when $t = 0$ and subsequently travels at the speed c_m, like all other points on the wave profile.

Agsorn and Dooge (1991) confirmed the existence of a stable monoclinal wave profile using numerical experiments. The theoretical solution was used to obtain an upstream hydrograph that then was routed downstream using the method of characteristics for the full dynamic equations. The resulting routed hydrograph propagated downstream at the speed given by the Kleitz-Seddon principle without change in shape. Furthermore, there was convergence to the theoretical monoclinal shape for a uniformly rising inflow hydrograph. Therefore, the monoclinal wave is a special case of a dynamic wave of equilibrium shape at large values of time in which kinematic steepening is balanced by dynamic smoothing effects. It has been used for testing numerical methods and evaluating the effect of various terms in the momentum equation when compared to simplified routing methods.

9.4   DIFFUSION ROUTING

Because kinematic routing cannot predict subsidence of the flood wave but only translation, it is appropriate to consider the effect of including the pressure gradient terms of the dynamic equation while still neglecting the inertia terms. With this simplification, the momentum equation becomes

$$S_f = S_0 - \frac{\partial y}{\partial x} \tag{9.49}$$

Writing $S_f = Q^2/K^2$, in which K = channel conveyance, substituting into (9.49) and differentiating with respect to time, the simplified momentum equation becomes

$$\frac{2Q}{K^2}\frac{\partial Q}{\partial t} - \frac{2Q^2}{K^3}\frac{\partial K}{\partial t} = -\frac{\partial^2 y}{\partial t\partial x} \tag{9.50}$$

The right hand side of (9.50) can be eliminated by differentiating the continuity equation (Equation 9.2) with respect to distance x to obtain

$$\frac{\partial^2 Q}{\partial x^2} + B\frac{\partial^2 y}{\partial x\partial t} = 0 \tag{9.51}$$

in which we have assumed a rectangular channel of constant width B. Substituting (9.50) into (9.51), we have

$$\frac{2Q}{K^2}\frac{\partial Q}{\partial t} - \frac{2Q^2}{K^3}\frac{\partial K}{\partial t} = \frac{1}{B}\frac{\partial^2 Q}{\partial x^2} \tag{9.52}$$

Because the conveyance K is a single-valued function of depth y and therefore of cross-sectional area A, its derivative with respect to time in the second term on the left of (9.52) can be transformed, with the aid of the continuity equation, to

$$\frac{\partial K}{\partial t} = \frac{dK}{dA}\frac{\partial A}{\partial t} = -\frac{dK}{dA}\frac{\partial Q}{\partial x} \tag{9.53}$$

If we further assume that dK/dA can be evaluated from the uniform flow formula in which $K = Q/S_0^{0.5}$ and then substitute the result from (9.53) back into (9.52), we have

$$\frac{\partial Q}{\partial t} + \frac{dQ}{dA}\frac{\partial Q}{\partial x} = \frac{Q}{2BS_0}\frac{\partial^2 Q}{\partial x^2} \tag{9.54}$$

If dQ/dA is interpreted as the kinematic wave celerity, c_k, as previously, the left hand side of (9.54) is the same as the kinematic routing equation, but the right hand side now has the appearance of a "diffusion term," with an apparent diffusion coefficient given by $D = Q/(2BS_0)$. From the behavior of the diffusion/dispersion term in river mixing problems, the diffusion analogy makes it clear that attenuation in Q will be produced by this simplified routing equation in addition to advection at the kinematic wave speed.

For constant wave celerity and diffusion coefficient, Equation 9.54 is the governing equation for linear diffusion routing for which there are exact solutions. The same equation results if depth rather than discharge is the dependent variable. For example, Henderson (1966) gives the Hayami (1951) solution for routing an upstream depth hydrograph that consists of a series of unit step changes in depth.

It is instructive to derive the linear diffusion equation for depth from a slightly different viewpoint than that used to obtain (9.54) to gain further insight into the limitations of diffusion routing. If the depth, y, and velocity, V, are written in terms of their initial uniform flow values, y_i and V_i, plus small perturbations from these values as $y = y_i + y'$ and $V = V_i + V'$, then using an order-of-magnitude analysis the continuity equation for a wide channel becomes

$$\frac{\partial y'}{\partial t} + V_i \frac{\partial y'}{\partial x} + y_i \frac{\partial V'}{\partial x} = 0 \tag{9.55}$$

Likewise, the momentum equation absent the inertia terms is

$$\frac{\partial y'}{\partial x} + S_0 \left(\frac{S_f}{S_0} - 1 \right) = 0 \tag{9.56}$$

The ratio S_f/S_0 can be evaluated using the Chezy equation for a wide channel. Neglecting the appropriate terms in this ratio from order-of-magnitude considerations, the momentum equation reduces to

$$\frac{\partial y'}{\partial x} + S_0 \left(\frac{2V'}{V_i} - \frac{y'}{y_i} \right) = 0 \tag{9.57}$$

The momentum equation is differentiated with respect to x, and the term $\partial V'/\partial x$ in the resulting equation is eliminated by substitution from the continuity equation. With some algebra and rearrangement of terms, the result is given by

$$\frac{\partial \phi}{\partial t} + c_k \frac{\partial \phi}{\partial x} = D \frac{\partial^2 \phi}{\partial x^2} \tag{9.58}$$

in which $\phi = y'$; $c_k = (3/2)V_i$ from Chezy; and $D = (V_i y_i)/(2S_0)$ for a wide channel. It is apparent that, once again, we have derived the linear diffusion routing equation, but it is strictly valid for small deviations in depth from the initial depth. Nevertheless, of particular interest is the solution to (9.58) for the upstream boundary condition of an abrupt increase in stage from the initial value ϕ_i to ϕ_f. The variable ϕ could be redefined easily in dimensionless terms as $\phi_d = (y - y_i)/(y_f - y_i)$. The solution to the linear diffusion equation then is given by Carslaw and Yeager (1959) to be

$$\phi_d = \frac{1}{2} \left[\text{erfc} \left(\frac{x - c_k t}{(4Dt)^{1/2}} \right) + \exp \left(\frac{c_k x}{D} \right) \text{erfc} \left(\frac{x + c_k t}{(4Dt)^{1/2}} \right) \right] \tag{9.59}$$

in which erfc = the complementary error function. The solution indicates that the wave will move downstream at the speed c_k while spreading or "diffusing" at a rate controlled by the apparent diffusion coefficient, D. By definition of D, more diffusion will occur for smaller slopes and larger values of depth.

Some comparisons have been made by Ferrick and Goodman (1998) to emphasize the effect of the diffusion term with respect to the inertia terms. They compared the solution of the linearized dynamic form of the momentum and continuity equations with the diffusion routing solution for a flood wave. The boundary condition consisted of an abrupt increase in depth and discharge at the upstream boundary, starting from an initial condition of steady, uniform flow. They found that the initial shock traveled downstream with the dynamic forerunner at the speed of $(V_i + c_i)$ and remained distinct from the diffusion wave profile until after the shock attenuated. Then, the profile celerities converged and approached the kinematic wave celerity.

We must point out an inconsistency in the derivation of the diffusion routing equation (9.54) that occurs because of the assumption in its derivation that dK/dA can be evaluated from the uniform flow formula; that is, by specifically invoking the bed slope in place of the friction slope. By definition, the diffusion routing equation includes the pressure gradient term as represented by dy/dx; yet to obtain the final form of the diffusion equation, dy/dx in effect is taken to be zero in the evaluation of the parameters, so that $S_0 = S_f$ as in kinematic routing. The second derivation of the linear diffusion routing equation (9.58) further implies that $y = y_i$ for the evaluation of constant values of c_k and D. Therefore, in the strictest sense, Equation 9.54 is applicable only for the case of quasiuniform flows with relatively small values of the pressure gradient, while the linear form given by Equation 9.58 further suggests the limitation of small deviations from the initial uniform flow depth.

In the variable-parameter case, in which the parameters c_k and D are allowed to vary with Q but are obtained from a uniform flow formula, Cappeleare (1997) shows that mass (or volume) is not conserved in the routing; that is, the outflow volume under the outflow hydrograph tends to be smaller than the inflow volume under the inflow hydrograph. He proposes a more accurate nonlinear diffusion routing method that properly accounts for the pressure gradient effect on the evaluation of the variable parameters, but it requires a more sophisticated numerical solution technique. The advantage of the linear approach in which the parameters are constant is that volume is conserved, and the river can be divided into a series of reaches with parameters varying from reach to reach as the physical characteristics of the channel change as described in the next section.

Example 9.3

A very wide river channel has a slope of 0.0005, and initially it is flowing at a uniform depth of 1.0 m (3.3 ft). The Chezy $C = 24$ in SI units ($n = 0.042$). If the depth of flow is abruptly increased to 1.2 m at the upstream end, where $x = 0$, compute the monoclinal wave profile and compare it with the diffusion wave solution at various values of time.

Solution.   The initial flow velocity follows from a solution of the Chezy equation

$$V_i = C y_i^{1/2} S_0^{1/2} = 24 \times (1.0)^{1/2} \times 0.0005^{1/2} = 0.537 \text{ m/s } (1.76 \text{ ft/s})$$

so that the kinematic wave celerity $c_k = (3/2)V_i = 0.805$ m/s (2.64 ft/s). The diffusion coefficient then is

$$D = \frac{q}{2S_0} = \frac{0.537 \times 1.0}{2 \times 0.0005} = 537 \text{ m}^2/\text{s } (5780 \text{ ft}^2/\text{s})$$

The monoclinal wave celerity is computed from Equation 9.42 to give

$$c_m = \frac{2}{3} \frac{(1.2)^{3/2} - 1}{1.2 - 1} \times 0.805 = 0.844 \text{ m/s } (2.77 \text{ ft/s})$$

The monoclinal wave celerity is only slightly greater than the kinematic wave celerity because the increase in depth is only 20 percent of the initial depth. Such a small increase is necessary for the solution of the linear diffusion equation to apply. Equation 9.59 is solved for a series of x values at a specified time to obtain the diffusion wave solution. The values of time are taken to be 50, 100, 200, and 400×10^3 sec as shown in Figure 9.8. The values of ϕ_d are defined by $(y - y_i)/(y_f - y_i)$, where y_i and y_f are the initial and final depths of flow, respectively. The solution for the monoclinal wave profile comes from Equations 9.47 and 9.48 with $a_1 = -6.814$, $a_2 = 9.271$, and $a_3 = 0.0160$. The integration constant is chosen such that the mid-depth point ($\phi_d = 0.5$) travels at the speed c_m from $x = 0$ at $t = 0$. As shown in Figure 9.8, the shape of the monoclinal wave profile does not change at successive times. The diffusion wave profile shows increased spreading due to diffusion as time progresses until it approaches the shape of the monoclinal wave. It lags the monoclinal wave slightly, however, because of the smaller kinematic wave speed. Both the time and corresponding distance required for the diffusion wave to approach the shape of the low-amplitude monoclinal wave are long, so that a very long river would be necessary to achieve convergence. The applicability of the diffusion wave solution for this problem depends on the time being long enough for the initial shock to have dissipated. The rate of diffusion depends on the channel slope, initial depth, and channel roughness with greater diffusion and attenuation occurring for smoother, deeper flows in channels of flatter slope.

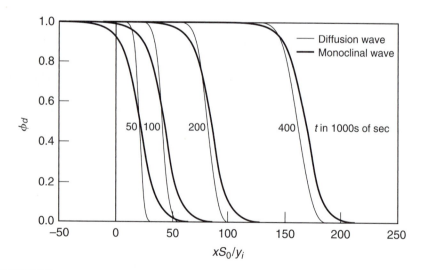

FIGURE 9.8

Comparison of diffusion wave and monoclinal wave profiles at various times.

9.5   MUSKINGUM-CUNGE METHOD

The Muskingum-Cunge method is a generalization of the Muskingum method that takes advantage of the diffusion contributions of the momentum equation by allowing for true wave attenuation through a matching of numerical and physical diffusion. First, a numerical discretization of the kinematic wave equation is developed to set the stage for quantifying numerical diffusion within the context of the Muskingum method. With reference to the computational molecule shown in Figure 9.9, the kinematic wave equation as given by Equation 9.29 is discretized with weighting factors X and Y to give

$$\frac{X(Q_i^{k+1} - Q_i^k) + (1 - X)(Q_{i+1}^{k+1} - Q_{i+1}^k)}{\Delta t}$$

$$+ c_k \frac{Y(Q_{i+1}^k - Q_i^k) + (1 - Y)(Q_{i+1}^{k+1} - Q_i^{k+1})}{\Delta x} = 0 \qquad (9.60)$$

If Equation 9.60 is rearranged in the form of the Muskingum routing equation, as given by Equation 9.14, then the routing coefficients can be expressed in terms of the weighting factors, X and Y:

$$C_0 = \frac{-X + C_n(1 - Y)}{1 - X + C_n(1 - Y)} \qquad (9.61a)$$

$$C_1 = \frac{X + C_n Y}{1 - X + C_n(1 - Y)} \qquad (9.61b)$$

$$C_2 = \frac{1 - X - C_n Y}{1 - X + C_n(1 - Y)} \qquad (9.61c)$$

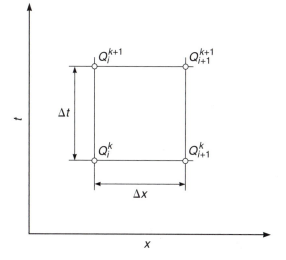

FIGURE 9.9

Computational molecule for numerical solution of kinematic wave routing problem.

in which C_n is the Courant number defined by $c_k \Delta t/\Delta x$. Treating the kinematic wave celerity as a constant, some special cases of the coefficients given by Equations 9.61a–c can be considered. For example, a centered, second-order finite difference scheme results if $X = Y = 0.5$, and the routing coefficients become $C_0 = (C_n-1)/(1 + C_n)$; $C_1 = 1$; and $C_2 = -C_1$.

Furthermore, if the Courant number is exactly 1, then the coefficients become $C_0 = 0$, $C_1 = 1$, and $C_2 = 0$; that is, $Q_{i+1}^{k+1} = Q_i^k$, so that the centered finite difference scheme produces pure translation only for the Courant number equal to 1, and thus it represents an exact solution of the kinematic wave equation for this special case.

Example 9.4

Route the triangular inflow hydrograph shown in Figure 9.10 using the finite difference approximation to the kinematic wave equation given by Equation 9.60 with $X = Y = 0.5$ and $C_n = 1.4$.

Solution.   The routing coefficients from Equations 9.61 become

$$C_0 = 0.167; \qquad C_1 = 1.0; \qquad C_2 = -0.167$$

The value of Δt is chosen to be 0.5 hr and the routing computations are carried out as shown in Table 9.5. The scheme appears to be stable, but the shape of the hydrograph changes as shown in Figure 9.10, with the numerical solution leading the analytical

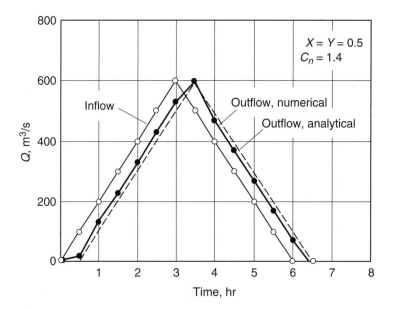

FIGURE 9.10
Comparison of numerical and analytical solutions of kinematic wave equation for $X = Y = 0.5$ and $C_n = 1.4$.

TABLE 9.5 Numerical solution of linear kinematic wave equation ($X = 0.5$; $Y = 0.5$; $C_n = 1.4$; $C_0 = 0.167$; $C_1 = 1.0$; $C_2 = -0.167$) of Example 9.4

					Numerical, Analytical	
t, hr	I, m³/s	$C_0 \times I_2$	$C_1 \times I_1$	$C_2 \times O_1$	O_2, m³/s	O_2, m³/s
0.0	0				0	0
0.5	100	16.67	0	0.00	17	0
1.0	200	33.33	100	−2.78	131	100
1.5	300	50.00	200	−21.76	228	200
2.0	400	66.67	300	−38.04	329	300
2.5	500	83.33	400	−54.77	429	400
3.0	600	100.00	500	−71.43	529	500
3.5	500	83.33	600	−88.10	595	600
4.0	400	66.67	500	−99.21	467	500
4.5	300	50.00	400	−77.91	372	400
5.0	200	33.33	300	−62.01	271	300
5.5	100	16.67	200	−45.22	171	200
6.0	0	0.00	100	−28.57	71	100
6.5	0	0.00	0	−11.90	−12	0

solution on both the rising and falling sides of the outflow hydrograph. The numerical peak outflow is only slightly smaller than the analytical value. The distortion is caused by the numerical solution itself.

Example 9.4 brings up the more general questions of numerical stability and consistency for variable values of X and Y; that is, does the solution of the finite difference equation amplify and grow without bound or not and does the solution converge to that of the original partial differential equation, respectively? To answer these questions, the remainder, R, or difference between the finite difference approximation of the differential equation and the differential equation itself can be determined from a Taylor series expansion of the function Q ($i\Delta x$, $k\Delta t$) about the point ($i\Delta x$, $k\Delta t$). As shown by Cunge (1969) and Ponce, Chen, and Simons (1979), the remainder R can be expressed as

$$R = c_k \Delta x \left[\left(\frac{1}{2} - X \right) + C_n \left(\frac{1}{2} - Y \right) \right] \frac{\partial^2 Q}{\partial x^2}$$

$$+ c_k \Delta x^2 \left\{ (1 - C_n) \left[\frac{1}{2} (X + C_n Y) - \frac{1}{3} (1 + C_n) \right] \right\} \frac{\partial^3 Q}{\partial x^3} + O(\Delta x^3) \quad (9.62)$$

The coefficient multiplying the second derivative of Q behaves like an artificial or numerical diffusion due only to the numerical approximation itself, because it does not appear in the original kinematic wave equation. It is clear that, for $X = Y = 0.5$, the numerical diffusion coefficient goes to zero (convergence) and the approximation error is of second-order $O(\Delta x^2)$ unless the value of the Courant number $C_n = 1$. In this case, the coefficient multiplying the third derivative term, which causes numerical dispersion or changes in shape, also goes to zero. For the Courant number not equal to 1, as in Example 9.4 and Figure 9.10, numerical dispersion results, even though numerical diffusion is not present. Furthermore, for $Y = 0.5$ and $X > 0.5$, we see that the numerical diffusion coefficient becomes negative, which causes numerical amplification of the solution and instability, as proven by Cunge (1969).

Example 9.5

Route the triangular inflow hydrograph of Example 9.4 using the finite difference approximation of the kinematic wave equation with $X = 0.1$ and $Y = 0.5$. Set the Courant number $C_n = 1.0$.

Solution. The routing coefficients are recalculated from Equations 9.61 for $X = 0.1$, $Y = 0.5$, and Courant number of 1.0 to yield

$$C_0 = 0.286; \qquad C_1 = 0.428; \qquad C_2 = 0.286$$

On examination of Equation 9.62 for the remainder of the numerical approximation, it is apparent that the numerical diffusion coefficient has a finite value but the dispersion term goes to zero. The routing computations are shown in Table 9.6 and plotted in Figure 9.11. The striking result in Figure 9.11 is the attenuation of the peak outflow caused by pure numerical diffusion that is a property of the numerical approximation and not of the analytical solution of the kinematic wave equation.

To generalize the computation of the routing coefficients for the specific case of $Y = 0.5$ but X variable, substitutions are made so that Equation 9.60 becomes

$$X(Q_i^{k+1} - Q_i^k) + (1 - X)(Q_{i+1}^{k+1} - Q_{i+1}^k) + \frac{C_n}{2}(Q_{i+1}^k - Q_i^k + Q_{i+1}^{k+1} - Q_i^{k+1}) = 0$$

$$(9.63)$$

Collecting terms and placing Equation 9.63 in the form of the Muskingum routing equation as given by Equation 9.14, the routing coefficients are

$$C_0 = \frac{0.5C_n - X}{1 - X + 0.5C_n} \qquad (9.64)$$

$$C_1 = \frac{0.5C_n + X}{1 - X + 0.5C_n} \qquad (9.65)$$

$$C_2 = \frac{1 - X - 0.5C_n}{1 - X + 0.5C_n} \qquad (9.66)$$

TABLE 9.6 **Numerical solution of kinematic wave equation** ($X = 0.1$; $Y = 0.5$; $C_n = 1.0$; $C_0 = 0.2857$; $C_1 = 0.4286$; $C_2 = 0.2857$) **of Example 9.5**

					Numerical, Analytical	
t, hr	I, m³/s	$C_0 \times I_2$	$C_1 \times I_1$	$C_2 \times O_1$	O_2, m³/s	O_2, m³/s
0.0	0				0	0
0.5	100	28.57	0.00	0.00	29	0
1.0	200	57.14	42.86	8.16	108	100
1.5	300	85.71	85.71	30.90	202	200
2.0	400	114.29	128.57	57.81	301	300
2.5	500	142.86	171.43	85.90	400	400
3.0	600	171.43	214.29	114.34	500	500
3.5	500	142.86	257.14	142.87	543	600
4.0	400	114.29	214.29	155.11	484	500
4.5	300	85.71	171.43	138.19	395	400
5.0	200	57.14	128.57	112.95	299	300
5.5	100	28.57	85.71	85.33	200	200
6.0	0	0.00	42.86	57.03	100	100
6.5	0	0.00	0.00	28.54	29	0

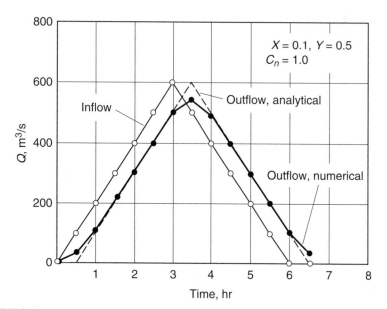

FIGURE 9.11

Comparison of numerical and analytical solutions of kinematic wave equation for $X = 0.1$, $Y = 0.5$, and $C_n = 1.0$.

Now if both the numerators and denominators of Equations 9.64 through 9.66 are multiplied by the Muskingum constant, θ, and compared with the Muskingum routing coefficient definitions given by Equations 9.11 through 9.13, it is obvious that they are identical if θ represents the kinematic wave travel time given by $\Delta x/c_k$, where c_k is the kinematic wave speed, and if X represents the Muskingum weighting factor. It follows that the Muskingum method in fact is a numerical solution of the linear kinematic wave equation that shows attenuation of the flood wave through numerical diffusion, as illustrated by Example 9.5. For the special case of $X = 0.5$ and $C_n = 1$, the Muskingum method provides the exact solution of the linear kinematic wave with pure translation, as determined from Equations 9.60 and 9.61.

Under these circumstances, it would seem that the Muskingum method is not very useful unless the numerical diffusion that it produces is related in some way to the apparent physical diffusion and wave attenuation that actually occur in a river. Cunge (1969) set the numerical diffusion coefficient from the approximation error expressed by Equation 9.62 for $Y = 0.5$ and variable X equal to the apparent physical diffusion coefficient as derived in Equation 9.54. The resulting expression can be solved for the Muskingum weighting factor X to produce

$$X = 0.5\left(1 - \frac{Q}{BS_0c_k\Delta x}\right) \tag{9.67}$$

with $\theta = \Delta x/c_k$ as before. When X in the Muskingum method is calculated from Equation 9.67, the method is referred to as the *Muskingum-Cunge method*, in which the routing parameters depend in a known way on the flow characteristics and channel properties, so that numerical and physical diffusion effects are matched. Either variable parameters or constant parameters can be used as is discussed in more detail later.

Koussis (1980) refined the Cunge approach by maintaining the time derivatives as continuous but still weighted, while discretizing only the x derivative in the kinematic wave equation. He then assumed a linear variation in the inflow hydrograph over the time interval Δt and obtained alternate expressions for the Muskingum coefficients:

$$C_0' = 1 - \frac{1 - \beta}{C_n} \tag{9.68}$$

$$C_1' = \frac{1 - \beta}{C_n} - \beta \tag{9.69}$$

$$C_2' = \beta = \exp\left(-\frac{C_n}{1 - X}\right) \tag{9.70}$$

in which C_n = Courant number = $\Delta t/\theta$ and θ = travel time in reach = $\Delta x/c_k$. Following the same procedure of matching physical and numerical diffusion, Koussis developed an expression for the Muskingum weighting factor X given by

$$X = 1 - \frac{C_n}{\ln\left(\dfrac{1 + \lambda + C_n}{1 + \lambda - C_n}\right)} \tag{9.71}$$

in which

$$\lambda = \frac{Q}{BS_0 c_k \Delta x} \tag{9.72}$$

Although this expression for Muskingum X seems more refined than (9.67), which gives $X = 0.5 \, (1 - \lambda)$, Koussis found little to recommend one formulation over the other. Later, Perumal (1989) showed that the conventional and refined Muskingum schemes were identical for $[C_n/(1 - X)] \leq 0.184$, while the conventional Muskingum-Cunge scheme was slightly better than the refined scheme when both were compared with an analytical solution given by Nash (1959) outside this range.

The issue of constant vs. variable parameters in the Muskingum-Cunge method already has been alluded to, and it really is a question of whether c_k and λ are calculated as a function of the varying discharge Q to produce variable routing coefficients or they are taken as constant as a function of a reference discharge. It should be clear from the outset, however, that allowing the coefficients to vary with Q does not remove the approximation of evaluating them based on a uniform flow formula as a function of the bed slope with the actual depth assumed to be normal. Koussis (1978) proposed the use of a constant value of X but a variable $\theta = \Delta x/c_k$ with discharge. Ponce and Yevjevich (1978) tested several methods of determining variable parameters and suggested that the best performance came from evaluating c_k and λ for each application of the computational molecule from either three-point or four-point averages of c_k and Q (to be used in the computation of λ). In the three-point method, Q and c_k are taken as the average of the values at grid points (i, k), $(i + 1, k)$ and $(i, k + 1)$ in Figure 9.9. In the four-point method, all four grid points are used in the averaging process, which necessarily requires iteration because the values at $(i + 1, k + 1)$ initially are unknown. The three-point average values of c_k and Q are used as starting values in the four-point iteration method. Both methods, however, display some loss of volume in the routed outflow hydrograph, whereas the constant parameter method conserves volume. Ponce and Chaganti (1994) report slight improvements in volume conservation if c_k is computed from a three-point or four-point average value of Q rather than being itself averaged. On the other hand, the variable parameter methods reproduce the expected nonlinear steepening of the flood wave. In a comparison between an analytical solution and the constant parameter method of routing for a sinusoidal inflow hydrograph, Ponce, Lohani, and Scheyhing (1996) show that the peak outflow and the peak travel time vary between 1 and 2 percent from the analytical values.

Tang, Knight, and Samuels (1999) investigated the volume loss in the variable-parameter Muskingum-Cunge method in more detail and confirmed that the use of an average Q rather than an average c_k slightly improved the volume loss (by about 0.1 percent). In general, greater volume loss was reported for the three-point methods than the four-point methods, and routing on very mild slopes ($S = 0.0001$) produced the greatest volume loss, with values up to 8 percent. An attempt was made to incorporate the correction suggested by Cappaleare (1997) for including the effects of the pressure gradient terms in the estimation of routing parameters but only in an approximate way. Some improvement in volume loss was shown but it depended on an empirical adjustment factor in the pressure-gradient correction formula for Q.

If we return to the questions of stability and accuracy with respect to the Muskingum-Cunge method, it must be true that $X \leq 0.5$ for stability as shown by Cunge (1969), but the criterion for the routing coefficient C_0 to be greater than or equal to zero to avoid negative outflows (or a dip in the outflow hydrograph) can be expressed in a different way. Ponce (1989) referred to λ, defined by Equation 9.72, as a cell Reynolds number. In terms of λ, the accuracy criterion of $C_0 \geq 0$, which is equivalent to the criterion given by (9.15), becomes $(C_n + \lambda) \geq 1$ from Equations 9.64 and 9.67. Based on routing a large number of inflow hydrographs with a realistic shape given by the gamma function, Ponce and Theurer (1982) suggested a stronger condition of $C_0 \geq 0.33$ to ensure accuracy as well as consistency (in the sense of removing the sensitivity of the outflow hydrograph to grid size). As a practical matter, this criterion becomes $(C_n + \lambda) \geq 2$, which defines the maximum length of the routing reach, Δx, for given values of the time step and the wave celerity as well as the flow rate, channel top width, and slope, as follows:

$$\Delta x \leq \frac{1}{2} \left(c_k \Delta t + \frac{Q}{BS_0 c_k} \right) \tag{9.73}$$

Note from (9.64) that the simple criterion of maintaining C_0 greater than some positive constant based on the empirical studies of Ponce and Theurer (1982) does not preclude the value of X from becoming negative. Dooge (1973) as well as Strupczewski and Kundzewicz (1980) justified mathematically the possibility of $X < 0$. In the conventional Muskingum method, an additional criterion of $C_2 \geq 0$ often is specified to ensure the avoidance of negative outflows, which generally is satisfied for $C_n < 1$ and $X \leq 0.5$, but Hjelmfeldt (1985) demonstrated that this criterion can be relaxed for most realistic inflow sequences in agreement with Ponce and Theurer (1982). The additional criterion does guarantee, however, positive outflows for any possible positive inflow sequence.

Although the Muskingum-Cunge method gives the exact analytical solution of the linear kinematic wave routing problem for $C_n = 1$ and $X = Y = 0.5$, the more usual case is for $X < 0.5$. From Equation 9.62, we see that, for $Y = 0.5$, $X < 0.5$, and $C_n = 1.0$, the dispersion term is zero and the numerical diffusion coefficient exists, making the numerical method first order; that is, the approximation error is $O(\Delta x)$. However, under the same conditions but for the Courant number not equal to 1, numerical dispersion occurs as well as numerical diffusion. For this reason, Ponce (1989) recommends that the Courant number be kept as close to unity as possible, not for stability reasons but to limit numerical dispersion. In fact, stability conditions require no specific limits on the Courant number, as are dictated in explicit finite difference approximations of the hyperbolic dynamic equations.

The applicability of the Muskingum-Cunge method is limited to flood waves of the diffusion type with no significant dynamic effects due to backwater, such as looped stage-discharge rating curves. Ponce (1989) suggests the following criterion for applicability:

$$T_r S_0 \left[\frac{g}{y_0} \right]^{0.5} \geq 15 \tag{9.74}$$

in which T_r is the rise time of the hydrograph; S_0 is the bottom slope; y_0 is the average flow depth; and g is gravitational acceleration. Overall, the Muskingum-Cunge method is a significant improvement over the Muskingum method because hydrograph data are not required for calibration, so that it can be used on ungauged streams with known geometry and slope. The variable-parameter method may be useful where slopes are moderate to large, so that volume loss is acceptable, but large improvements in accuracy should not be expected over the constant-parameter approach. Corrections for the pressure-gradient term have the potential to improve diffusion routing so long as the numerical methods remain simpler than full dynamic routing; otherwise, dynamic routing should be used in the first place.

Example 9.6

An inflow hydrograph for a river reach has a peak discharge of 4500 cfs (128 m³/s) at a time to peak of 2 hr with a base time of 6 hr. Assume that the inflow hydrograph is triangular in shape with a base flow of 500 cfs (14 m³/s). The river reach has a length of 18,000 ft (5490 m) and a slope of 0.0005 ft/ft. The channel cross section is trapezoidal with a bottom width of 100 ft (30.5 m) and side slopes of 2:1. The Manning's n for the channel is 0.025. Find the outflow peak discharge and time of occurrence for the river reach using the Muskingum-Cunge method and compare them to the dynamic routing method using the method of characteristics.

Solution.   First the kinematic wave speed is calculated based on Manning's equation and a reference discharge of 2500 cfs (70.8 m³/s), which is midway between the base flow and the peak discharge. For the given conditions, the resulting normal depth is 5.71 ft (1.74 m) and the velocity, V_0, is 3.93 ft/s (1.20 m/s). If we consider the channel to be very wide as a first approximation, then $c_k = (5/3)V_0 = 6.55$ ft/s (2.00 m/s). The value of Δt is tentatively chosen to be 0.5 hr based on discretization of the time to peak, which is equal to 2 hr. Then Δx can be estimated from the inequality of (9.73):

$$\Delta x \leq \frac{1}{2}\left(c_k \Delta t + \frac{Q}{B S_0 c_k}\right) = \frac{1}{2}\left(6.55 \times 0.50 \times 3600 + \frac{2500}{122.8 \times 0.0005 \times 6.55}\right)$$

$$= 9003 \text{ ft } (2744 \text{ m})$$

Therefore, two routing reaches, each with a length of 9000 ft (2743 m), can be used. This gives a Courant number of $c_k \Delta t / \Delta x = 6.55 \times 1800/9000 = 1.31$, which is slightly greater than unity, and the value of X from (9.67) is

$$X = 0.5\left(1 - \frac{Q}{B S_0 c_k \Delta x}\right) = 0.5\left(1 - \frac{2500}{122.8 \times 0.0005 \times 6.55 \times 9000}\right) = 0.155$$

If the value of either C_n or X seems unsatisfactory, further slight adjustments of Δt and Δx are possible, since the criterion given by (9.73) is conservative and guarantees $C_0 \geq 0.33$ rather than $C_0 \geq 0$, which is all that really is required to avoid negative outflows

TABLE 9.7   Muskingum-Cunge routing ($C_0 = 0.333$; $C_1 = 0.540$; $C_2 = 0.127$) of Example 9.6

Time, hr	Inflow, cfs	$C_0 \times I_2$	$C_1 \times I_1$	$C_2 \times O_1$	$x = 9000$ ft; O_2, cfs	$x = 18000$ ft; O_2, cfs
0.0	500				500	500
0.5	1500	500	270	64	833	611
1.0	2500	833	810	106	1748	1110
1.5	3500	1166	1350	222	2738	1997
2.0	4500	1499	1890	348	3736	2976
2.5	4000	1332	2430	474	4236	3806
3.0	3500	1166	2160	538	3864	4058
3.5	3000	999	1890	491	3380	3727
4.0	2500	833	1620	429	2882	3258
4.5	2000	666	1350	366	2382	2763
5.0	1500	500	1080	303	1882	2264
5.5	1000	333	810	239	1382	1764
6.0	500	167	540	176	882	1264
6.5	500	167	270	112	549	819
7.0	500	167	270	70	506	569
7.5	500	167	270	64	501	512

in most cases. The values of the routing coefficients are computed from (9.64) through (9.66) to yield

$$C_0 = 0.333; \qquad C_1 = 0.540; \qquad C_2 = 0.127;$$

which sum to unity as required. The routing equation is solved step by step in Table 9.7 for the first subreach of 9000 ft (2743 m). Then the outflow becomes the inflow for the next subreach, for which only the final results are shown in the table without the intermediate computations. The Muskingum-Cunge results are compared with dynamic routing results from the method of characteristics (MOC) in Figure 9.12. It is apparent that the assumption of a constant wave celerity in the Muskingum-Cunge method fails to capture the nonlinear steepening and flattening on the rising and falling sides, respectively, of the outflow hydrograph. However, the peak outflow rates agree very well. The occurrence of the outflow peak is at $t = 3.0$ hr, but this is limited by the time step of the approximate routing method. The method of characteristics gives a peak time of 2.8 hr. Note that the criterion of Equation 9.74 for diffusion routing is not met, but reasonable agreement still is obtained between the two methods in this example.

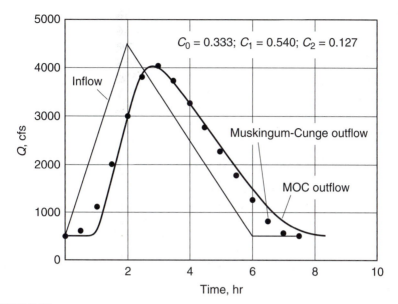

FIGURE 9.12
Inflow and outflow hydrographs for Muskingum-Cunge and MOC routing.

REFERENCES

Agsorn, S., and James C. I. Dooge. "Numerical Experiments on the Monoclinal Rising Wave." *J. Hydrology* 124 (1991), pp. 293–306.

Aldama, A. A. "Least-Squares Parameter Estimation for Muskingum Flood Routing." *J. Hydr. Engrg., ASCE* 116, no. 4 (1990), pp. 580–86.

Bras, R. L. *Hydrology.* Reading, MA: Addison-Wesley, 1990.

Cappelaere, B. "Accurate Diffusive Wave Routing." *J. Hydr. Engrg.,* ASCE 123, no. 3 (1997), pp. 174–81.

Carslaw, H. S., and J. C. Jaeger. *Conduction of Heat in Solids,* 2nd ed. New York: Oxford University Press, 1959.

Chow, Ven Te. *Open Channel Hydraulics.* New York: McGraw-Hill, 1959.

Chow, V. T., D. R. Maidment, and L. W. Mays. *Applied Hydrology.* New York: McGraw-Hill, 1988.

Cunge, J. A. "On the Subject of a Flood Propagation Computation Method." *J. Hyd. Res.* 7, no. 2 (1969), pp. 205–30.

Dooge, J. C. I. *Linear Theory of Hydrologic Systems.* U.S. Dept. Agriculture, ARS, Tech. Bull. No. 1468. Washington, DC: Government Printing Office, 1973.

Ferrick, M. G., and N. J. Goodman. "Analysis of Linear and Monoclinal River Wave Solutions." *J. Hydr. Engrg.,* ASCE 124, no. 7 (1998), pp. 728–41.

Gill, M. "Routing of Floods in River Channels." *Nordic Hydr.* 8 (1977), pp. 163–70.

Hayami, S. "On the Propagation of Flood Waves." *Bulletin No. 1,* Disaster Prevention Research Institute, Kyota University, Japan, 1951.

Henderson, F. M. *Open Channel Flow.* New York: Macmillan, 1966.

Hjelmfelt, A. T., Jr., and J. J. Cassidy. *Hydrology for Engineers and Planners.* Ames, Iowa: The Iowa State University Press, 1975.

Hjelmfelt, A. T., Jr. "Negative Outflows from Muskingum Routing." *J. Hydr. Engrg.*, ASCE 111, no. 6 (1985), pp. 1010–14.

Koussis, A. D. "Comparison of Muskingum Method Difference Schemes." *J. Hyd. Div.*, ASCE 106, no. HY5 (1980), pp. 925–29.

Koussis, A. D. "Theoretical Estimates of Flood Routing Parameters." *J. Hyd. Div.*, ASCE 104, no. HY1 (1978), pp. 109–15.

Lighthill, M. J., and G. B. Whitham. "On Kinematic Waves: Flood Movements in Long Rivers." *Proc. Roy. Soc. (A).* 229, no. 1178 (1955), pp. 281–316.

Miller, W. A., and J. A. Cunge. "Simplified Equations of Unsteady Flow." In *Unsteady Flow in Open Channels,* ed. K. Mahmood and V. Yevjevich, vol. 1, pp. 89–182. Fort Collins, CO: Water Resources Publications, 1975.

Nash, J. E. "A Note on the Muskingum Flood-Routing Method." *J. Geophys. Res.* 64, no. 8 (1959), pp. 1053–56.

Perumal, M. "Unification of Muskingum Difference Schemes." *J. Hydr. Engrg.*, ASCE 115, no. 4 (1989), pp. 536–43.

Ponce, V. M. *Engineering Hydrology.* Englewood Cliffs, NJ: Prentice-Hall, 1989.

Ponce, V. M., and P. V. Chaganti. "Variable-Parameter Muskingum Method Revisited." *J. Hydrology* 162 (1994), pp. 433–39.

Ponce, V. M., Y. H. Chen, and D. B. Simons. "Unconditional Stability in Convection Computations." *J. Hyd. Div.*, ASCE 105, no. HY9 (1979), pp. 1079–86.

Ponce, V. M., R-M. Li, and D. B. Simons. "Applicability of Kinematic and Diffusion Models." *J. Hyd. Div.*, ASCE 104, no. HY3 (1978), pp. 353–60.

Ponce, V. M., A. K. Lohani, and C. Scheyhing. "Analytical Verification of Muskingum-Cunge Routing." *J. Hydrology* 174 (1996), pp. 235–41.

Ponce, V. M., and F. D. Theurer. "Accuracy Criteria in Diffusion Routing." *J. Hyd. Div.*, ASCE 108, no. HY6 (1982), pp. 747–57.

Ponce, V. M., and V. Yevjevich. "Muskingum-Cunge Method with Variable Parameters." *J. Hyd. Div.*, ASCE 104, no. HY12 (1978), pp. 1663–67.

Singh, V. P., and R. McCann. "Some Notes on Muskingum Method of Flood Routing." *J. Hydrology* 48 (1980), pp. 343–61.

Strupczewski, W., and Z. Kundezewicz. "Muskingum Method Revisited." *J. Hydrology* 48 (1980), pp. 327–42.

Tang, X-N., D. W. Knight, and P. G. Samuels. "Volume Conservation in Variable Parameter Muskingum-Cunge Method." *J. Hydr. Engrg.*, ASCE 125, no. 6 (1999), pp. 610–20.

Weinmann, P. E., and E. M. Laurenson. "Approximate Flood Routing Methods: A Review." *J. Hyd. Div.*, ASCE 105, no. HY12 (1979), pp. 1521–36.

Woolhiser, D. A., and J. A. Liggett. "Unsteady, One-Dimensional Flow over a Plane—The Rising Hydrograph." *Water Resources Res.* 3, no. 3 (1967), pp. 753–71.

EXERCISES

9.1. A stormwater detention basin has bottom dimensions of 700 ft by 700 ft with interior 4:1 side slopes. The outflow structures consist of a 12 in. diameter pipe laid through the fill on a steep slope with an inlet invert elevation of 100 ft at the bottom of the basin and a broad-crested weir with a crest elevation of 107 ft and a crest length of 10 ft. The top of the berm is at elevation 109 ft. The design inflow hydrograph can be approximated as a triangular shape with a peak discharge of 125 cfs at a time of 8 hr and with a base time of 24 hr. If the detention basin initially is empty, route the inflow hydrograph through the basin and determine the peak outflow rate and stage. How could these be further reduced? Explain in detail.

9.2. Prove that the temporary storage in reservoir routing is given by the area between the inflow and outflow hydrographs. For triangular inflow and outflow hydrographs, derive a relationship for the detention storage as a function of inflow and outflow peak discharges and the base time of the inflow hydrograph.

9.3. The inflow hydrograph for a 25,000 ft reach of the Tallapoosa River follows. Route the hydrograph using Muskingum routing with $\Delta t = 2$ hr, $\Delta x = 25,000$ ft, $\theta = 3.6$ hr, and $X = 0.2$. Plot the inflow and outflow hydrographs and determine the percent reduction in the inflow peak as well as the travel time of the peak.

t, hr	Q, cfs	t, hr	Q, cfs
0	100	20	22000
2	500	22	17500
4	1500	24	14000
6	2500	26	10000
8	5000	28	7000
10	11000	30	4500
12	22000	32	2500
14	28000	34	1500
16	28500	36	1000
18	26000	38	500
		40	100

9.4. Route the inflow hydrograph of Example 9.6 through the same river, but for a total reach length of 27,000 ft, using the method of characteristics computer program CHAR (on the website). Then use this outflow hydrograph with the inflow hydrograph to derive θ, X, and the Muskingum routing coefficients using the least-squares fitting method. Finally do the Muskingum routing and compare the results with the outflow hydrograph from dynamic routing.

9.5. Derive the kinematic wave celerity for a trapezoidal channel. Plot the ratio of kinematic wave celerity to average channel flow velocity, c_k/V, as a function of the aspect ratio b/y for several values of side slope ratio m including $m = 0$ for a rectangular channel and discuss the results.

9.6. For uniform flow at a depth of 5 cm on a concrete parking lot having a slope of 0.01 and a flow length of 200 m, calculate the kinematic wave number, k. Would kinematic routing apply? Repeat for a very wide river channel with a slope of 0.001, $n = 0.035$, a depth of 5 m, and the same flow length of 200 m. Discuss the results.

9.7. For the same parking lot of Exercise 9.6, suppose the hydrograph rise time is 15 min. Would the kinematic wave routing method apply based on Equation 9.37?

9.8. Route an overland flow hydrograph over a distance of 300 m using the analytical kinematic wave solution with variable wave celerity. The overland flow occurs over a 100 m wide plane strip of constant slope 0.006 and $n = 0.015$. The inflow hydrograph is triangular with a peak of 1.25 m^3/s at a time to peak of 15 min and with a base time of 45 min. The initial base flow is 0.25 m^3/s. Plot the inflow and outflow hydrographs and discuss the shape of the outflow hydrograph.

9.9. Derive Equation 9.42 and then show that, as y_f approaches y_i, the monoclinal wave celerity approaches the initial kinematic wave celerity.

9.10. Compute and plot the monoclinal wave profile in a very wide river with an initial depth of 1.0 m and a final depth of 4.0 m. The slope of the river is 0.001 and the Manning's n is 0.040. Assume that the depth at $x = 0$ and $t = 0$ is 3.99 m. Also calculate the monoclinal wave celerity and compare this to the initial kinematic wave celerity.

9.11. Derive Equation 9.58, the linear diffusion routing equation.

9.12. Using the numerical approximation of Equation 9.60 for the kinematic wave equation with $X = 1$ and $Y = 0$, calculate the routing coefficients from (9.61) and route the hydrograph of Example 9.4 for $C_n = 1.5$. With reference to Equation 9.62, what happens to the outflow hydrograph if $C_n = 1.0$? Explain why the method becomes unstable for $C_n < 1.0$.

9.13. Using the numerical approximation of Equation 9.60 for the kinematic wave equation with $X = 0$ and $Y = 1$, calculate the routing coefficients from (9.61) and route the hydrograph of Example 9.4 for $C_n = 0.67$. With reference to Equation 9.62, what happens to the outflow hydrograph if $C_n = 1.0$? Explain why the method becomes unstable for $C_n > 1.0$.

9.14. Using the numerical approximation of Equation 9.60 for the kinematic wave equation with $X = 0$ and $Y = 0$, calculate the routing coefficients from (9.61) and route the hydrograph of Example 9.4 for $C_n = 1.0$. With reference to Equation 9.62, what happens to the outflow hydrograph if $C_n \neq 1.0$? Does the method become unstable for any value of C_n? Explain your answer.

9.15. Express the routing coefficients of the Muskingum-Cunge method in terms of λ and C_n.

9.16. Use the Muskingum-Cunge method to route the hydrograph of Example 9.6 for $S_0 = 0.001$ and compare with the results from the computer program CHAR.

9.17. Repeat the Muskingum-Cunge routing of Example 9.6 for a very wide channel using the three-point variable parameter method with c_k and λ calculated from three-point averages of Q at each time step. Compare the results with the constant parameter method.

9.18. Write a computer program for Muskingum-Cunge routing using the four-point variable parameter method and apply it to Example 9.6.

10

Flow in Alluvial Channels

10.1 INTRODUCTION

Rivers, which are natural open channels, often have movable alluvial sediment boundaries at the bed and banks that add another degree of complexity to the estimation of flow resistance. Because the sediment bed itself is subject to movement, the flow creates perturbations of that boundary, which amount to sand waves that propagate either downstream or upstream depending on the flow conditions and sediment properties. The amplitude of the perturbations affects the flow resistance and hence the stage at a given discharge, while at the same time the flow conditions control the amplitude and wavelength of the perturbations. For this reason alluvial rivers have been described as both *sculpture* and *sculptor* (Vanoni 1977).

Aside from the problem of additional flow resistance, the sediment regime of open channel flow in a river is responsible for bed and bank instability, scour around structures such as bridge piers and abutments, deposition and burial of fish habitat, loss of clarity in the water column and inhibition of photosynthesis, and transport of adsorbed contaminants. Such problems associated with sediment transport are intertwined with the pure hydraulic considerations of open channel flow and so deserve some attention in this text, especially with respect to flow-sediment interactions.

This chapter describes sediment properties and discusses methods for predicting bed and bank stability by identifying the threshold of sediment movement. Sediment in motion and the bed forms created are discussed next along with the coupled problem of flow resistance and stage-discharge prediction. A brief overview of bed-load and suspended load transport equations and the calculation of total sediment load are presented followed by a consideration of scour problems associated with bridges constructed across rivers.

10.2   SEDIMENT PROPERTIES

Some properties of individual sediment grains are important for cohesionless sediments (sands and gravels), such as grain size, shape, and specific gravity, as well as fall velocity, which is a function of all the previously mentioned properties. The behavior of sediment grains or particles in bulk may be of interest, too. The bulk specific weight of sediments deposited in a lake bed, for example, or the grain size distribution of sands and gravels in a well-graded streambed affect the behavior of the bed as a whole. In addition, for clay or cohesive sediments, identifying the interactions of plateletlike particles with variable surface charge is essential to an understanding of the stability of the bed with respect to erosion or resuspension (Dennett et al. 1998; Ravisangar, Sturm, and Amirtharajah 2005), but this chapter focuses on noncohesive sediments.

Particle Size

The grain size of a sediment particle is one of its most important properties. The American Geophysical Union (AGU) scale classifies size ranges as shown in Table 10.1 with each size class representing a geometric series in which the maximum and minimum sizes in the range differ by a factor of 2. The size of sand particles usually is measured as the sieve diameter, which is the length of the side of a square sieve opening through which the given particle will just pass. The size of silts and clays, on the other hand, often depends on sedimentation methods and the relationship between fall velocity and sedimentation diameter, which is defined as the diameter of a sphere of the same specific weight having the same terminal fall velocity as the given particle in the same sedimentation fluid. The relationship between sedimentation diameter and fall velocity is discussed in the section on fall velocity.

Particle Shape

Sand grains in particular have a shape that varies from angular to rounded depending on the fluvial environment in which they are found. River sands tend to be worn somewhat by fluvial action and deviate considerably from a spherical shape. One way of defining shape is the so-called shape factor (McKnown and Malaika 1950) given by

$$\text{S.F.} = \frac{a}{\sqrt{bc}} \qquad (10.1)$$

in which S.F. = shape factor, and the variables a, b, and c are the lengths of three mutually perpendicular axes such that a is the shortest axis. In other words, the shape factor is defined as the length of the shortest axis divided by the geometric mean length of the other two axes. A sphere obviously would have a shape factor of 1.0 with no preferential direction of axes. For an ellipsoid with axis lengths in the ratio of 1:1:3, the shape factor would be 0.577. A shape factor of 0.7 has been found to be about average for natural sands (U.S. Interagency Committee 1957). The shape factor can be determined using a microscope.

TABLE 10.1   Sediment grade scale (AGU)

Class name	Size range, mm
Very large boulders	4,096–2,048
Large boulders	2,048–1,024
Medium boulders	1,024–512
Small boulders	512–256
Large cobbles	256–128
Small cobbles	128–64
Very coarse gravel	64–32
Coarse gravel	32–16
Medium gravel	16–8
Fine gravel	8–4
Very fine gravel	4–2
Very coarse sand	2.0–1.0
Coarse sand	1.0–0.5
Medium sand	0.50–0.25
Fine sand	0.250–0.125
Very fine sand	0.125–0.062
Coarse silt	0.062–0.031
Medium silt	0.031–0.016
Fine silt	0.016–0.008
Very fine silt	0.008–0.004
Coarse clay	0.004–0.002
Medium clay	0.002–0.001
Fine clay	0.0010–0.0005
Very fine clay	0.0005–0.00024

Particle Specific Gravity

Because the predominant mineral in sand and gravel often is quartz, the specific gravity (SG) usually is taken to be 2.65. However, for less worn sediments that retain the mineralogy of the parent rock, several minerals such as feldspar, mica, barite, and magnetite, for example, still may be present in appreciable quantities, so that specific gravity may need to be measured at each investigation site. Clay sediments generally are hydrous aluminum silicates with a characteristic sheet structure having a specific gravity from 2.2 to 2.6 (Sowers 1979).

Once the specific gravity is known, the specific weight, γ_s, of the sediment solid is simply the specific gravity times the specific weight of water. Sand and gravel have a specific weight of approximately 165 lbs/ft^3 or 26.0 kN/m^3. The mass density, ρ_s, is the specific gravity times the mass density of water, so quartz sediments have a mass density of 5.14 slugs/ft^3 or 2650 kg/m^3.

Because the sediment grains of interest in sediment transport usually are submerged, another property of interest related to specific gravity is the submerged specific weight, which is given by $\gamma_s' = (\gamma_s - \gamma) = (\text{SG} - 1)\gamma$, in which γ_s = specific weight of the sediment solid and γ = specific weight of water. The submerged specific weight of sand grains, for example, is 103 lbs/ft³ or 16.2 kN/m³.

Bulk Specific Weight

As sediments are deposited in relatively quiescent environments, they occupy a volume that includes the pore space filled with water subject to consolidation over time. Estimates of sediment carried into a reservoir by weight can be translated into volume occupied only by use of the bulk specific weight. Such predictions of the volume of sediment deposited as a function of time are essential to estimates of the useful life of a reservoir, or the time between dredging events to maintain navigable waterways. The bulk specific weight of a sediment deposit is defined as the dry weight of sediment divided by the total volume occupied by both sediment and pore space. Lane and Koelzer (1953) proposed a relationship for the specific weight of deposits in reservoirs given by

$$\gamma_b = \gamma_{b1} + B \log t \qquad (10.2)$$

in which γ_b = bulk specific weight in lbs/ft³ of a deposit with an age of t years; γ_{b1} = initial bulk specific weight of the deposit in lbs/ft³ at the end of the first year; and B = constant (lbs/ft³). For a sediment that always is submerged or nearly submerged, γ_{b1} and B have values of 93 and 0 for sand, 65 and 5.7 for silt, and 30 and 16 for clay, respectively.

Fall Velocity

The fall velocity of sediment is defined as the terminal speed of a sediment grain in water at a specified temperature in an infinite expanse of quiescent water. It plays a very important role in distinguishing between suspended sediment load, in which the sediment grains are carried in the water column, and bedload, which consists of individual grains transported near the bed with intermittent or continuous contact with the bed itself. Fall velocity is closely related to the fluid mechanics problem of estimating drag around a submerged sphere due to a fluid flow of specified velocity. The only differences lie in the viewpoint of the observer (the sphere is moving and the fluid is at rest) and in which of the relevant quantities are unknown. In the case of flow around a fixed sphere, the unknown is the drag force; while for a sphere dropping at terminal speed in a fluid at rest, the unknown is the fall velocity. In the latter case, the drag force must be equal and opposite to the submerged weight of the sphere at terminal velocity to give

$$C_D \frac{\rho A_f w_f^2}{2} = (\gamma_s - \gamma)\frac{\pi d^3}{6} \qquad (10.3)$$

in which C_D = drag coefficient of the sphere; γ and ρ = specific weight and density of the fluid, respectively; γ_s = specific weight of the solid; A_f = frontal area of the sphere projected onto a plane perpendicular to the path of the falling sphere (= $\pi d^2/4$); d = diameter of the sphere; and w_f = fall velocity of the sphere. Solving for the fall velocity, we have

$$w_f = \sqrt{\frac{\frac{4}{3}(\gamma_s/\gamma - 1)gd}{C_D}} \qquad (10.4)$$

Unfortunately, Equation 10.4 cannot be solved explicitly for the fall velocity because the coefficient of drag, C_D, is a function of the Reynolds number ($\mathbf{Re} = w_f d/\nu$), where ν is the kinematic viscosity of the fluid. The Reynolds number obviously involves the unknown fall velocity. The C_D vs. $\mathbf{Re}$ diagram for a sphere is shown in Figure 10.1.

The dilemma of solving Equation 10.4 can be overcome in several ways. One approach is to assume a value of C_D, solve for the fall velocity from Equation 10.4 and compute the Reynolds number to use in Figure 10.1 to obtain the next value of C_D in an iterative process. To develop a numerical solution procedure involving a nonlinear algebraic equation solver, best-fit relationships are available for C_D, such as the one given in Figure 10.1 as suggested by White (2005):

$$C_D = \frac{24}{\mathbf{Re}} + \frac{6}{1 + \sqrt{\mathbf{Re}}} + 0.4 \qquad (10.5)$$

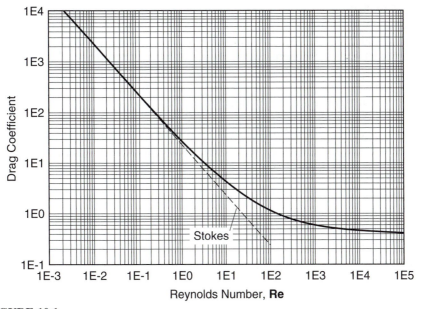

FIGURE 10.1
Coefficient of drag for spheres (best-fit equation from White 2005).

which is valid up to a Reynolds number of approximately 2×10^5 when the drag crisis occurs as the laminar boundary layer changes to a turbulent boundary layer and the separation point moves further downstream on the surface of the sphere. Iteration or a numerical solution of (10.4) is unnecessary, however, for the Stokes range ($\mathbf{Re} < 1$), for which there is an exact solution by Stokes for the drag force and coefficient of drag under the assumption of negligible inertia terms in the Navier-Stokes equations; that is, creeping motion. In this special case, $C_D = 24/\mathbf{Re}$ or the drag force $D = 3\pi\mu w_f d$. Substituting the Stokes solution for drag force on the left hand side of (10.3) and solving for the fall velocity gives Stokes' law for the fall velocity:

$$w_f = \frac{1}{18}\frac{(\gamma_s/\gamma - 1)gd^2}{\nu} \tag{10.6}$$

in which γ_s = specific weight of the sphere; γ = specific weight of the fluid; d = diameter of the sphere; and ν = kinematic viscosity of the fluid. Stokes' law is limited to $\mathbf{Re} < 1$. If the value of the fall velocity, w_f, is obtained from setting $\mathbf{Re} = 1.0$ and substituted into (10.6), the result is the maximum sphere size for which Stokes' law applies. The result for a quartz sphere falling in water at 20°C is $d_{max} = 0.1$ mm, which is a very fine sand.

For spherical particles outside the Stokes range, an alternative to the iterative solution involving Figure 10.1, or the numerical solution using Equation 10.5, is to rearrange the dimensional analysis of the problem. The difficulty with Figure 10.1 is that it was developed for predicting the drag force on a sphere, whereas the problem of interest here is the determination of fall velocity of the sphere, and the fall velocity appears in the definition of both C_D and $\mathbf{Re}$. However, according to the rules of dimensional analysis, any dimensionless group can be replaced by some combination of the other groups as discussed in Chapter 1. In this case, a good choice would be $C_D\mathbf{Re}^2$ because the fall velocity is eliminated in this group. The evaluation of a related dimensionless group can be obtained from

$$\frac{3}{4}C_D\mathbf{Re}^2 = \frac{(\gamma_s/\gamma - 1)gd^3}{\nu^2} \tag{10.7}$$

in which the constant of 4/3 on the right hand side has been moved to the left hand side. Now define a more convenient dimensionless number, d_*, given by

$$d_* = \left[\frac{(\gamma_s/\gamma - 1)gd^3}{\nu^2}\right]^{1/3} \tag{10.8}$$

Taking Equation 10.5 for the drag coefficient and plotting $\mathbf{Re}$ vs. d_* results in Figure 10.2, in which the abscissa is calculated from (10.8). The Reynolds number then can be read directly from the figure to determine the fall velocity outside the Stokes range.

It remains to apply the methods just developed for spheres to sediment particles that are not spherically shaped. One method for accomplishing this task is to define the sedimentation diameter as described in the section on sediment size, which relates the fall velocity to the diameter of a fictitious sphere having the same

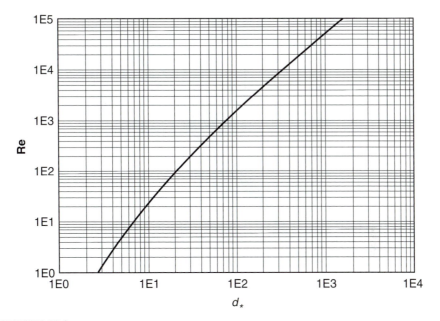

FIGURE 10.2
Fall velocity of a sphere as a function of dimensionless particle diameter d_*.

fall velocity as the given particle. Unfortunately, sedimentation diameter varies with Reynolds number, so it has been standardized for a fluid temperature of 24°C, and called the *standard fall diameter*. If the fall velocity of a sediment has been measured, its standard fall diameter can be determined from Figure 10.1 and Equation 10.4. However, for sand grains, the sieve diameter d_s usually is measured by taking the geometric mean of the sieve sizes just passing and retaining the given sand grain in a nest of sieves. What is needed then is a conversion from the sieve diameter of the actual sediment to the fall diameter, which depends on the shape factor, as shown in Figure 10.3. Once the fall diameter is known, any of the methods just discussed for spheres can be used to obtain the fall velocity. Fortunately, the fall diameter does not vary significantly from the standard fall diameter over a temperature range of 20° to 30°C.

As an alternative to using sedimentation diameter to find the fall velocity, the coefficient of drag of sand particles can be determined directly and given in a C_D vs. **Re** diagram like that of Figure 10.1. Engelund and Hansen (1967) have suggested the following best fit to the data for sand and gravel (**Re** $< 10^4$):

$$C_D = \frac{24}{\mathbf{Re}} + 1.5 \tag{10.9}$$

Equation 10.9 can be used in combination with Equation 10.4 for the fall velocity to obtain an exact solution for the fall velocity, which is given by (Julien 1995):

$$\mathbf{Re} = \frac{w_f d_s}{\nu} = 8\left[\sqrt{1 + 0.0139 d_*^3} - 1\right] \tag{10.10}$$

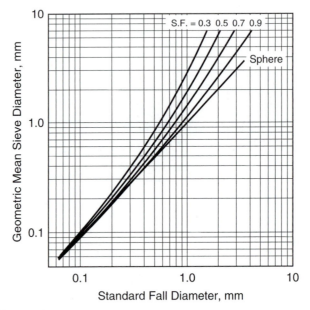

FIGURE 10.3
Relationship between fall diameter and sieve diameter for different shape factors of naturally worn sand particles (U.S. Interagency Committee 1957).

Example 10.1

Find the fall velocity of a medium sand with a sieve diameter of 0.50 mm (0.00164 ft) falling in water at 20°C by two methods: (1) using Figures 10.2 and 10.3 and (2) from Equation 10.10.

Solution.   From Figure 10.3, for a sieve diameter of 0.50 mm (0.00164 ft) and a shape factor of 0.7, the standard fall diameter is 0.47 mm (0.00154 ft). Then, we calculate d_* for the sphere with fall diameter, d_f, as

$$d_* = \left[\frac{1.65 \times 9.81 \times 0.00047^3}{(1 \times 10^{-6})^2} \right]^{1/3} = 11.9$$

From Figure 10.2, $\mathbf{Re} \approx 33$ so that $w_f = 33 \times (1 \times 10^{-6})/0.00047 = 7.0 \times 10^{-2}$ m/s (0.23 ft/s).

In the second method, which can be used only for sand grains, d_* is recalculated for the sieve diameter, d_s, of 0.5 mm to give a value of 12.6. Then, we substitute into (10.10) to obtain

$$\frac{w_f d_s}{\nu} = 8 \times \left[\sqrt{1 + 0.0139 \times 12.6^3} - 1 \right] = 35$$

from which $w_f = 35 \times (1 \times 10^{-6})/0.0005 = 7.0 \times 10^{-2}$ m/s (0.23 ft/s).

Grain Size Distribution

While some natural sorting occurs in rivers with the formation of a thin armor layer of coarser particles in the bed under conditions of degradation, generally a wide range of sizes can be found in transport and in the riverbed. Some measure of the degree of sorting of the grain sizes is required using statistical frequency distributions. The lognormal probability density function commonly is applied to river sands, with an estimate of its parameters (mean and standard deviation) being used to characterize the particle size distribution as obtained from sieve analysis. The lognormal probability density function simply is the normal probability density function applied to the logs of the sieve diameters, so it is given by

$$f(\zeta) = \frac{1}{\sqrt{2\pi}} e^{-\zeta^2/2} \tag{10.11}$$

in which $\zeta = (\log d_s - \mu)/\sigma$; d_s is sieve diameter; μ is the mean of the logs of the sieve diameters; and σ is the standard deviation of the logs of the sieve diameters. The geometric standard deviation, σ_g, is used more often to describe grain size distributions, and it is defined by $\log \sigma_g = \sigma$.

The cumulative distribution function, $F(\zeta)$, is used to relate the theoretical probability distribution of (10.11) to the results of a grain-size analysis. It represents the cumulative probability that a grain size is less than or equal to a given sieve diameter, and it is measured as the cumulative weight passing a given sieve size as a fraction of the total weight of the sediment sample. Mathematically, it is obtained from the area underneath the probability density function as

$$F(\zeta) = \frac{1}{\sqrt{2\pi}} \int_{-\infty}^{\zeta} e^{-t^2/2} \, dt \tag{10.12}$$

in which t is a dummy variable of integration, and $100 \times F(\zeta)$ = percent finer of the theoretical lognormal distribution. Shown in Figure 10.4 are the individual data points of a sieve analysis plotted on a lognormal grid. The abscissa values represent sieve sizes plotted on a log scale, while the ordinates are percent finer values plotted on a normal probability scale such that a theoretical lognormal cumulative distribution function plots as a straight line. The actual data show some curvature and deviation from the lognormal distribution, especially at the tails of the distribution. The data are fitted by drawing a straight line between the 84.1 percent finer size ($d_{84.1}$) and the 15.9 percent finer size ($d_{15.9}$), which represents the distance between plus or minus one standard deviation from the mean. Expressed in terms of σ_g, the distance is plus or minus one times log σ_g, as illustrated in Figure 10.4. The intersection of the straight line with the 50 percent finer ordinate is defined as the geometric mean sieve diameter, d_g, as shown in Figure 10.4, while the intersection of the curve connecting the data points and the 50 percent finer ordinate is the median size, d_{50}. These may or may not be the same, depending on the actual size distribution data.

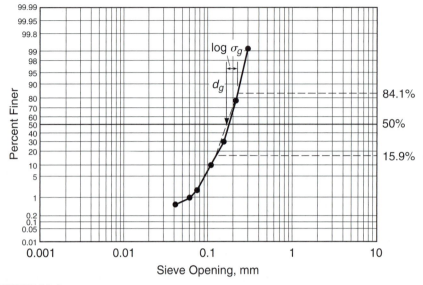

FIGURE 10.4

Size distribution of a sand sample on log-normal scale.

Both the geometric standard deviation and the geometric mean size can be expressed in terms of $d_{84.1}$ and $d_{15.9}$. By definition, $\log \sigma_g = (\log d_{84.1} - \log d_g) = (\log d_g - \log d_{15.9})$, which can be expressed as

$$\sigma_g = \frac{d_{84.1}}{d_g} = \frac{d_g}{d_{15.9}} \tag{10.13}$$

Then, by cross-multiplying, it is immediately apparent that $d_g = (d_{84.1}\, d_{15.9})^{1/2}$. Furthermore, by back substitution, the value of $\sigma_g = (d_{84.1}/d_{15.9})^{1/2}$.

10.3   INITIATION OF MOTION

Determining the stability of the bed and banks of a natural alluvial channel or of the rock riprap lining of a constructed channel as in Chapter 4 depends on the definition of the threshold of sediment movement. In a qualitative sense, sediment grains in a noncohesive sediment bed begin rolling and sliding at isolated, random locations on the bed as the threshold condition is just exceeded. The threshold condition can be described in terms of a critical shear stress or a critical velocity at which the forces or moments resisting motion of an individual grain are overcome. The forces resisting motion in a noncohesive sediment are due to the submerged weight of the grain, while in a cohesive sediment, physicochemical interparticle forces offer the primary resistance to sediment motion. This section focuses on the case of noncohesive sediments of relatively uniform size.

If the threshold of motion is defined in terms of a critical shear stress, τ_c, it can be given as a function of the following variables:

$$\tau_c = f_1(\gamma_s - \gamma, d, \rho, \mu) \tag{10.14}$$

in which $\gamma_s - \gamma$ = submerged specific weight of the sediment; d = sediment grain size; and ρ and μ = fluid density and dynamic viscosity, respectively. Dimensional analysis of (10.14) leads immediately to the result

$$\frac{\tau_c}{(\gamma_s - \gamma)d} = f_2\left(\frac{u_{*c}d}{\nu}\right) \tag{10.15}$$

in which $u_{*c} = (\tau_c/\rho)^{1/2}$ = critical value of the shear velocity; and $\nu = \mu/\rho$ = kinematic viscosity. This is the result that Shields (1936) obtained more indirectly. The dimensionless critical shear stress on the left of (10.15) is referred to as the *Shields parameter*, τ_{*c}, and the dimensionless parameter on the right of (10.15) has the form of a Reynolds number, which is called the *critical boundary* or *particle Reynolds number*, $\mathbf{Re}_{*c}$.

Shields was an American who, in Berlin in the 1930s, conducted flume experiments on initiation of motion and bedload transport of sediment as affected by the specific gravity of the sediment. He utilized sediments of barite, amber, lignite, and granite to obtain a wide range in the submerged specific weight of sediment from 590–32,000 N/m³ (4–200 lbs/ft³). The sediment grains were subangular to very angular, with median sizes ranging from 0.36 to 3.44 mm (0.0012 to 0.0113 ft). He combined his results with those of previous investigations at the same research institute that were conducted on river sands by Casey (1935) and Kramer (1935), as well as adding results of Gilbert (1914) and the U.S. Waterways Experiment Station (WES) for river sands. He presented the results according to the dimensionless groups given in Equation 10.15 as a shaded zone for the beginning of sediment motion in what has come to be called the *Shields diagram,* although it has undergone a number of revisions. Rouse (1939) first presented it in the English literature and replaced the shaded zone with a curve. The Shields diagram is given in Figure 10.5 with additional data and modifications proposed by Yalin and Karahan (1979). As given in Figure 10.5, the parameters have an instructive physical interpretation. The Shields parameter, τ_{*c}, can be interpreted as the ratio of the shear stress to the submerged weight of a grain per unit of surface area at critical conditions, while the boundary Reynolds number, $\mathbf{Re}_{*c}$, represents the ratio of the grain diameter to the viscous sublayer thickness (ignoring the constant in the expression for the viscous sublayer thickness $\delta = 11.6 \, \nu/u_*$). Accordingly, regions of smooth, transitional, and fully rough turbulent flow over a grain could be expected as shown in Figure 10.5 as the grain size becomes larger relative to the viscous sublayer thickness and the individual grains protrude from it, creating boundary-generated turbulence.

The data for $\mathbf{Re}_{*c} < 1$ in Figure 10.5 were obtained primarily for fine-grained silica solids that were cohesionless. For this range, in which the boundary layer is smooth-turbulent or laminar, Mantz (1977) proposed a relation given by

$$\tau_{*c} = 0.1(\mathbf{Re}_{*c})^{-0.3} \tag{10.16a}$$

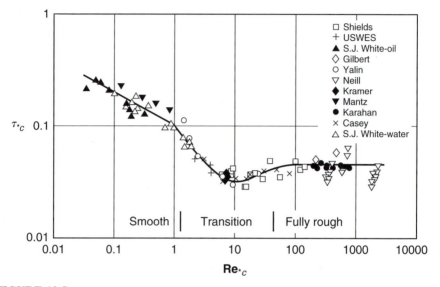

FIGURE 10.5

The Shields diagram as updated by Yalin and Karahan (1979). (*Source: M. S. Yalin and E. Karahan. "Inception of Sediment Transport," J. Hyd. Div., © 1979, ASCE. Reproduced by permission of ASCE.*)

Yalin and Karahan (1979) showed that a separate laminar flow curve, which is not shown in the figure, exists when the boundary Reynolds number exceeds unity and suggested that the laminar and smooth turbulent data coincide for Reynolds numbers less than unity because the grains are submerged in the viscous sublayer in both cases. For $\mathbf{Re}_{*c} > 1$, Yalin and Karahan (1979) added a considerable amount of additional data to the original Shields data, which includes data points labeled as Shields, Gilbert, Kramer, Casey, and USWES in Figure 10.5 as summarized by Buffington (1999). Based on the additional data, particularly in the fully rough region, the constant value of τ_{*c} in the fully rough turbulent region is 0.045, and the transition curve proposed by Yalin and Karahan (1979) can be fitted by

$$\tau_{*c} = \sum_{i=0}^{i=4} A_i (\log \mathbf{Re}_{*c})^i \tag{10.16b}$$

in which $A_0 = 0.100$, $A_1 = -0.1361$, $A_2 = 0.05977$, $A_3 = 0.01984$, and $A_4 = -0.01134$ for $1 < \mathbf{Re}_{*c} < 70$ with $\tau_{*c} = 0.045$ for $\mathbf{Re}_{*c} > 70$. However, the actual limits of the transition region are given by Yalin and Karahan (1979) as $1.5 < \mathbf{Re}_{*c} < 40$. Because $\mathbf{Re}_{*c}$ is defined usually in terms of d_{50}, and taking $k_s = 2d_{50}$, these limits correspond to $3 < u_*k_s/\nu < 80$, which is similar to the range of 5 to 70 for u_*k_s/ν given for the transition region in pipe flow by Schlichting (1979).

The manner in which Shields obtained the critical shear stress from both his experiments and those of others is a matter of some controversy (Kennedy 1995; Buffington 1999). Shields' original tabulated data were lost during World War II, and the descriptions of methodology in his doctoral thesis are vague and some-times contradictory. Because Shields continued his career in machine design in the United States after finishing his doctoral dissertation rather than in sediment transport, he was unaware of the impact of his work until near his death and so shed no light on the controversy. The critical shear stress can be obtained either from visual observation of the threshold of motion or from extrapolation of mea-sured sediment transport rates to zero. Kramer's work is based on the visual clas-sification of sediment motion as (1) weak movement, defined as the motion of a few or several sand particles at isolated points in the flume bed; (2) medium movement, described as motion of many sand grains too numerous to be counted but without appreciable sediment discharge; and (3) general movement, charac-terized as motion of grains of all sizes in all parts of the bed at all times. Kennedy (1995) suggested that Shields may have used the visual observation method developed by Kramer in previous experiments in the same flume, based on what appears to have been averaging by Shields of Kramer's widely varying data for critical shear stress. Based on analysis of the data of other investigators used by Shields, Buffington (1999) concluded that Shields probably did use the definition of "weak movement" as the criterion for threshold conditions for the data of Casey, Kramer, and WES, while it appears that he used "general movement" for Gilbert's data. On the other hand, Buffington surmises that Shields may have used the method of extrapolation of sediment discharge to zero for his own data because of the statement in his dissertation that this was the appropriate method for uniform sediments and references to his data elsewhere in the thesis as being representative of uniform sediments. Regardless of the method used for obtaining critical shear stress, additional uncertainties exist in Shields' original diagram as a result of the use of both mean and median grain sizes; the existence of bed forms in some of the data, which cause overestimation of critical shear stress; the lack of true uniformity of the sediment sizes; and the variability of sediment angularity of the sediments used (Buffington 1999). We can conclude that, although the Shields diagram is a valid representation of the physics of initiation of sediment motion, its users should recognize it as a band of data about a gen-eral relationship for incipient motion.

As presented in Figure 10.5, the Shields diagram is not very convenient for directly estimating the critical shear stress, because it appears in the definition of both the Shields parameter and the boundary Reynolds number. To use the Shields diagram to estimate critical shear stress, a third dimensionless parameter that elim-inates the critical shear stress can be introduced. Such a parameter is given, for example, by $[0.1\mathbf{Re}_{*c}^2/\tau_{*c}]^{1/2}$, so that an auxiliary set of curves can be constructed on the Shields diagram, the intersection of which with the Shields curve allows direct determination of the critical shear stress (see Vanoni 1977). On closer exam-ination, however, the auxiliary parameter can be recast as the dimensionless grain diameter $d_* = [\mathbf{Re}_{*c}^2/\tau_{*c}]^{1/3}$ that was encountered in the development of a relation-ship for fall velocity of sand grains. Accordingly, the Shields diagram is replotted in Figure 10.6 as τ_{*c} vs. d_*, as suggested by Julien (1995), so that the critical shear stress can be determined directly, since d_* is a function of only the grain diameter and specific weight, and the fluid specific weight and viscosity. The curve in

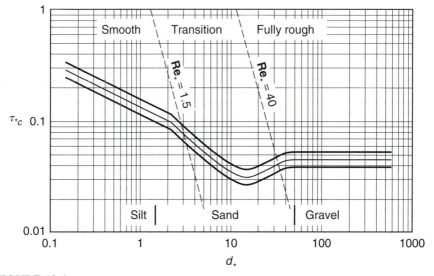

FIGURE 10.6

An alternate form of the Shields diagram for direct determination of critical shear stress (after Julien 1995). (*Source: P. Y. Julien, Erosion and Sedimentation, © 1995, Cambridge University Press. Reprinted with the permission of Cambridge University Press.*)

Figure 10.6 has been converted directly from the updated relationship proposed by Yalin and Karahan (1979) in Figure 10.5.

Of particular interest in Figure 10.5 or 10.6 for coarse sediments is the critical value of the Shields parameter in the region of fully rough turbulent flow, where it approaches a constant value. In this region, a constant value of the Shields parameter implies that the critical shear stress is linearly proportional to the grain diameter. Rouse (1939) initially indicated a constant value of $\tau_{*_c} = 0.060$ in his drawing of the Shields curve near the upper range of Shields data, although some extrapolation was involved. Laursen (1963), in his development of a prediction equation for bridge contraction scour, used a value of $\tau_{*_c} = 0.039$, while the value in Figure 10.5 from Yalin and Karahan is approximately 0.045. Julien suggested that the constant value of $\tau_{*_c} = 0.06 \tan \phi$, where ϕ = angle of internal friction to account for the size and angularity of the grains. In this formulation, τ_{*_c} varies from 0.039 for very fine gravel to 0.050 for very coarse gravel to 0.054 for boulders in the constant τ_{*_c} region in which d_* is greater than about 40.

The variability of the constant value of τ_{*_c} for large values of the boundary Reynolds number and the scatter of data in Figure 10.5 emphasize that a range of "critical conditions" should form the Shields diagram. Accordingly, two additional curves appear in Figure 10.6, which are defined by ± 1 times the standard error in log units between the curve in Figure 10.5 and the data given there.

Regardless of the value chosen for the Shields parameter, a corresponding value of critical velocity can be calculated from Keulegan's (1938) equation for fully rough turbulent flow as obtained from Equations 4.3 and 4.19. If the critical value of shear velocity, u_{*c}, is related to τ_{*c} with water as the fluid, Keulegan's equation becomes

$$V_c = 5.75 \sqrt{\tau_{*c}(SG - 1)gd_{50}} \log\left[\frac{12.2R}{k_s}\right] \tag{10.17}$$

in which SG = specific gravity of the sediment; R = hydraulic radius; and k_s = equivalent sand-grain roughness, which varies, as discussed in Chapter 4, from $1.4d_{84}$ to $3.5d_{84}$. It is of interest to note that the critical velocity, which is a mean cross-sectional velocity, varies with the hydraulic radius and therefore the flow depth for the same value of the Shields parameter. Hence, reports of critical velocity for sediments of varying grain size should correspond with a specific depth range over which they are applicable. If Manning's equation is used instead of Keulegan's equation with Manning's n expressed in terms of a Strickler-type expression $(n = c_n d_{50}^{1/6})$, then the critical water velocity for a very wide channel can be expressed as

$$V_c = \frac{K_n}{c_n}\sqrt{(SG - 1)\,\tau_{*c}}\;d_{50}^{1/3}y_0^{1/6} \tag{10.18}$$

in which K_n = 1.49 in English units and 1.0 in SI units; c_n = constant in Strickler-type relationship for Manning's n $(n = c_n d_{50}^{1/6})$, which is equal to 0.039 in English units and 0.0475 in SI units; SG = specific gravity of the sediment; τ_{*c} = critical value of the Shields parameter; d_{50} = median grain diameter; and y_0 = depth of uniform flow. (Note that a value of c_n = 0.034 in English units also is used for the Strickler constant, as discussed in Chapter 4.)

If the grain size is such that the flow is not fully rough turbulent, then the value of τ_{*c} is obtained from the Shields diagram and substituted into a Keulegan-type equation for velocity derived by Einstein (1950) and given by

$$V_c = 5.75u_{*c} \log\left[\frac{12.2R'x}{k_s}\right] \tag{10.19}$$

in which u_{*c} = critical value of the shear velocity = $[\tau_{*c}(SG - 1)gd_{50}]^{0.5}$; R' = hydraulic radius due to grain roughness, independent of form roughness caused by ripples and dunes (to be discussed in the next section); x = a correction factor for smooth and transitional turbulent flow, which is equal to unity for fully rough turbulent flow; and k_s = equivalent sand-grain roughness, which Einstein equated to d_{65}, the 65 percent finer grain size. The correction factor, x, is a function of k_s/δ, as shown in Figure 10.7, where δ = viscous sublayer thickness = $11.6\,\nu/u'_*$ and u'_* = shear velocity due only to grain or surface roughness = $(gR'S_0)^{0.5}$. Coarse sediments have no bed forms so the hydraulic radius $R = R'$, and furthermore $x = 1.0$ for fully rough turbulent flow, with the result that Equation 10.19 reduces to Equation 10.17 for sediments coarse enough to fall in the fully rough turbulent regime.

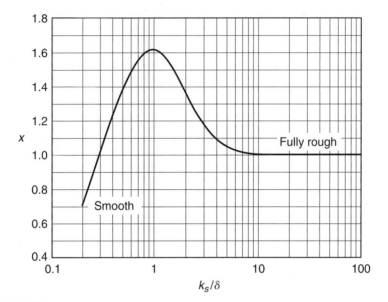

FIGURE 10.7
Einstein velocity correction factor, x, for calculating mean velocity in smooth and transition turbulent flow (Einstein 1950).

The relationships for critical velocity in Equations 10.17, 10.18, and 10.19 can be placed in dimensionless form in terms of a critical value of the sediment number, $\mathbf{N}_{sc}$, as defined by (Carstens 1966)

$$\mathbf{N}_{sc} = \frac{V_c}{\sqrt{(SG - 1)gd_{50}}} \tag{10.20}$$

Neill (1967) has done extensive experiments on "first displacement" of uniformly graded gravel and proposed a best fit relationship as shown in Figure 10.8 and given by

$$\mathbf{N}_{sc}^2 = 2.50 \left(\frac{d_g}{y_0} \right)^{-0.20} \tag{10.21}$$

in which d_g = geometric mean diameter and y_0 = depth of uniform flow. As reported by Pagán-Ortiz (1991), Parola obtained similar experimental results for uniform flow over a gravel bed when utilizing Neill's criterion of first displacement. Shown for comparison in Figure 10.8 are Equations 10.17 and 10.18 in terms of $\mathbf{N}_{sc}$ (with $\tau_{*c} = 0.045$; $k_s = 2d_{50}$; $d_{50} = d_g$; and the Strickler constant $c_n = 0.034$ in English units). For $d_{50}/y_0 > 0.1$, Manning's n begins to vary with depth as the roughness elements become large relative to the depth as discussed in Chapter 4. In this zone, Manning's equation tends to overestimate the critical velocity; while Keulegan's and Neill's equations underestimate it and so are on the conservative side. Manning's equation provides a more conservative estimate if $c_n = 0.039$ in English units. It is of interest to note that Neill's equation is very similar to that employed in the Corps of Engineers riprap design method as given by (4.58).

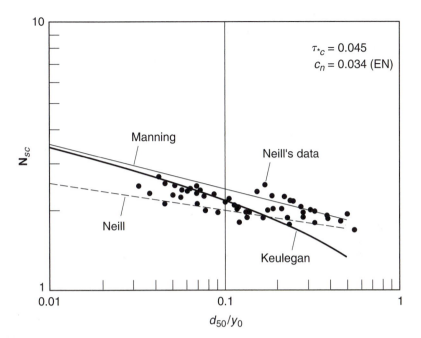

FIGURE 10.8
Critical sediment number for initiation of motion of coarse sediment (data from Neill 1967).

Example 10.2

Find the critical shear stress and critical velocity for a medium sand with $d_{50} = 0.3$ mm
$(9.8 \times 10^{-4}$ ft) and a medium gravel with $d_{50} = 10$ mm (0.0328 ft) for a uniform flow
depth of water (20°C) of 1.0 m (3.28 ft).

Solution.   First calculate the dimensionless sediment number, d_*, for both sediment
sizes. For sand with a specific gravity of 2.65 and water with a viscosity of 1×10^{-6} m²/s
$(1.08 \times 10^{-5}$ ft²/s), d_* is determined by

$$d_* = \left[\frac{(SG - 1)gd_{50}^3}{\nu^2}\right]^{1/3} = \left[\frac{1.65 \times 9.81 \times (0.0003)^3}{(1 \times 10^{-6})^2}\right]^{1/3} = 7.59$$

A similar calculation for the gravel yields $d_* = 253$. Then, from Figure 10.6,
$\tau_{*c} = 0.041$ for the sand and 0.045 for the gravel with the former in the transitional turbulent range and the latter in the fully rough turbulent range. The corresponding value
of critical shear stress for the sand is

$$\tau_c = (\gamma_s - \gamma)d_{50}\tau_{*c} = 1.65 \times 9810 \times 0.0003 \times 0.041$$

$$= 0.20 \text{ N/m}^2 \ (0.0042 \text{ lbs/ft}^2)$$

and for the gravel it is 7.28 N/m² or Pa (0.152 lbs/ft²).

To find the critical velocity for the sand, use Equation 10.19 with x determined
from Figure 10.7. Assume that no bed forms exist at initiation of motion, so that

$R' = R$. Take $k_s = 2d_{50} = 0.0006$ m (0.002 ft) and $\delta = 11.6 \, \nu/u_{*c} = 11.6 \times 10^{-6}/(0.20/1000)^{1/2} = 8.20 \times 10^{-4}$ m (2.69×10^{-3} ft). Then, $k_s/\delta = 0.73$, and from Figure 10.7, $x = 1.57$ so that the critical velocity is calculated from Equation 10.19 as

$$V_c = 5.75\sqrt{\frac{\tau_c}{\rho}} \log\left[\frac{12.2y_0x}{k_s}\right]$$

$$= 5.75 \times (0.20/1000)^{1/2} \times \log\left[\frac{(12.2 \times 1.0 \times 1.57)}{0.0006}\right] = 0.37 \text{ m/s } (1.2 \text{ ft/s})$$

For the gravel, use Equation 10.18 (Manning) with $c_n = 0.0414$ and $K_n = 1.0$ for SI units to obtain

$$V_c = \frac{K_n}{c_n}\sqrt{(SG - 1)\,\tau_{*c}}\; d_{50}^{1/3}y_0^{1/6}$$

$$= \frac{1.0}{0.0414} \times (1.65 \times 0.045)^{1/2} \times (0.01)^{1/3} \times (1.0)^{1/6} = 1.42 \text{ m/s}$$

or 4.66 ft/s. For comparison, the reader can confirm that the critical velocity for the gravel from Equation 10.17 (Keulegan) for the same value of τ_{*c} is 1.37 m/s (4.50 ft/s) and from Equation 10.21, it is 1.01 m/s (3.31 ft/s). The latter value from Neill's results is considerably more conservative than either Equation 10.17 or 10.18 for this value of $d_{50}/y_0 = 0.033$.

10.4   APPLICATION TO STABLE CHANNEL DESIGN

Once the critical shear stress for initiation of motion is evaluated, an obvious engineering application occurs in stable channel design, as discussed in Chapter 4. The design philosophy is to choose a rock-riprap lining of sufficient size that the maximum bed shear stress at the design flow does not exceed the critical shear stress. In the simplest problem, which is a very wide channel with stable banks, $\tau_c = \tau_0 = \gamma y_0 S_0$, in which y_0 is the normal depth and S_0 is the bed slope. If the channel lining material is coarse enough to be in the fully rough turbulent region of the Shields diagram, which is the usual case, then the critical value of the Shields parameter is a constant and the critical shear stress is directly proportional to the grain diameter. For $\tau_{*c} = 0.045$, it follows that $\tau_c = \tau_{*c}(\gamma_s - \gamma)d_{50}$. Therefore, setting $\tau_c = \tau_0$ results in

$$d_{50} = 13.5y_0S_0 \tag{10.22}$$

for quartz sediment in water. From (10.22) we can see that increasing the depth or the slope requires a proportionately larger rock-riprap size for stability of the channel bed. On the other hand, for a given native sediment size and channel slope, the flow depth must be limited to a specific maximum value, which requires the channel width to be larger to accommodate larger design flows while maintaining bed stability.

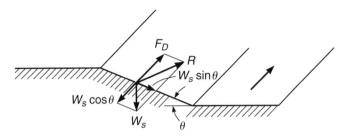

FIGURE 10.9
Stability of a particle on a channel bank.

If a canal in the existing native sediment becomes too wide to achieve bed stability at the design flow, then a larger sediment, used as a riprap lining of a narrower channel, is required. The channel shape usually is chosen to be trapezoidal for ease of construction, and stability of the rock-riprap lining on the sloping bank as well as on the bed becomes an issue. In general, the value of the critical shear stress on the banks of the channel is not the same as on the channel bed because of the additional force of gravity acting down the side slope of a trapezoidal channel. This gravity force component assists the hydrodynamic force in causing initiation of motion. In the simplest case, as shown in Figure 10.9, the drag force, F_D, coincides with the direction of motion and the submerged weight, W_s, has components down the side slope and perpendicular to the side slope. At the point of incipient motion on the channel bank, the ratio of the resultant force parallel to the bank, R, to the force normal to the bank, $W_s \cos \theta$, must equal the tangent of the angle of internal friction or angle of repose of the bank material based on soil mechanics principles. With reference to Figure 10.9, this is written as

$$\tan \phi = \frac{\sqrt{F_D^2 + (W_s \sin \theta)^2}}{W_s \cos \theta} \tag{10.23}$$

The drag force is the critical shear stress on the bank or wall, τ_c^w, times the surface area of a grain, A_s. Substituting into (10.23) and solving for the critical shear stress on the wall gives

$$\tau_c^w = \frac{W_s}{A_s} \cos \theta \tan \phi \sqrt{1 - \frac{\tan^2 \theta}{\tan^2 \phi}} \tag{10.24}$$

When applied to the bed, $\tan \theta = 0$ and $\cos \theta = 1$ corresponding to $\theta = 0$, so that Equation 10.24 implies τ_c on the bed is equal to $(W_s/A_s) \tan \phi$. If we then take the ratio of the shear stress on the wall to that on the bed, which is called the *tractive force ratio*, K_r, there results from dividing (10.24) by τ_c:

$$K_r = \frac{\tau_c^w}{\tau_c} = \cos \theta \sqrt{1 - \frac{\tan^2 \theta}{\tan^2 \phi}} = \sqrt{1 - \frac{\sin^2 \theta}{\sin^2 \phi}} \tag{10.25}$$

The last step on the right hand side of (10.25) is the result of a trigonometric identity. By definition, $\theta < \phi$ for gravitational stability of the side slope, so $K_r < 1$. For a given side slope angle, θ, and angle of repose, ϕ, which depends on the grain size and angularity, the critical shear stress on the side slope is obtained by multiplying the critical shear stress on the bed from Shields' diagram by the tractive force ratio, K_r. It remains only to compare the maximum shear stress on the side slope, which was given by Lane (1955a) as $0.75\gamma y_0 S_0$, with the critical shear stress on the side slope, $K_r \tau_c$, to determine stability. Setting them equal, as in the analysis that led to Equation 10.22, the result is

$$d_{50} = \frac{10.1}{K_r} y_0 S_0 \tag{10.26}$$

for quartz sediment in water and for $\tau_{*c} = 0.045$. From (10.26), we can see that the flatter is the side slope angle θ, the larger the value of K_r and the smaller the minimum sediment size, d_{50}, that will be stable on the side slope. The stability of the bed also must be checked; and for this purpose, Lane (1955a) gave the maximum shear stress on the bed as $0.97\gamma y_0 S_0$, which is compared with the critical bed shear stress from the Shields diagram.

The tractive force ratio and the expressions for maximum flow shear stress applied to the bed and sides of the channel as developed by Lane form the basis of the *Bureau of Reclamation procedure* for the design of rock riprap channel linings. Equations 10.22 and 10.26 show how the required rock riprap size is affected by channel slope and flow depth, and they demonstrate the application of the Shields diagram for estimation of the critical shear stress that cannot be exceeded for channel stability. However, the normal depth, which appears in both equations, is affected by the roughness coefficient of the chosen rock size and dependent on the selected channel bottom width and side slope. In addition, the tractive force ratio is also dependent on the rock size and angularity, which determine the angle of repose, and the channel side slope. Thus, rock riprap design for channel stability becomes a trial and error procedure as outlined in detail in Chapter 4. Several other design methods are also given in Chapter 4.

10.5   BED FORMS

Bed forms are irregularities in an alluvial channel bed with respect to a flat bed that are higher than the sediment size itself. The three main types of bed forms are ripples, dunes, and antidunes, each with a different physical origin. Ripples and dunes are called *lower regime bed forms,* because they occur generally in subcritical flow, while antidunes exist either near or in supercritical flow. As the discharge or Froude number increases, a transition zone forms between the lower regime ripples and dunes and the upper regime, which consists of a flat bed with sediment transport or antidunes. The transition zone can consist of several bed form types occurring in different parts of the bed simultaneously. Specifically, ripples or dunes and flat bed can occur together during transition.

Sketches of several bed forms are shown in Figure 10.10. Ripples are approximately triangular in shape with a long, flat upstream slope followed by an abrupt

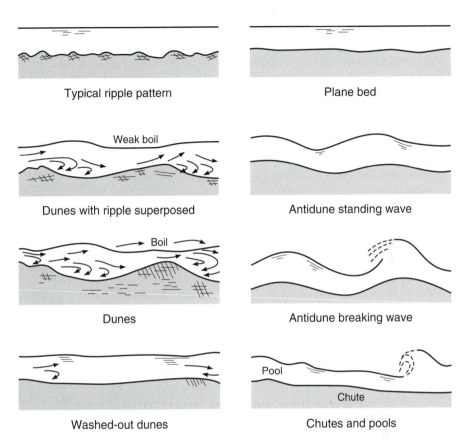

FIGURE 10.10

Forms of bed roughness in an alluvial channel (Simons and Richardson 1966).

steep slope approximately equal to the angle of repose of the sediment. However, ripples sometimes can be nearly sinusoidal in shape. Ripples and dunes move slowly downstream at a velocity much less than the flow velocity as erosion occurs on the upstream slope with deposition on the downstream slope. Ripples have amplitudes of approximately 3 cm (0.1 ft) and wavelengths on the order of 30 cm (1 ft); and they usually occur only in sands with grain sizes smaller than 0.6 mm (0.002 ft). Ripples can occur on the upstream slopes of dunes, which are much larger in amplitude and wavelength.

Whereas ripples are the result of the growth of any small discontinuity on the bed caused by deformation of the bed, dunes tend to be related to the largest scale turbulent eddies in the flow, with a height on the order of the flow depth. In both cases, alternating regions of scour and deposition are created in the flow direction that produce growth of the bed forms to some relatively stable shape. Yalin (1972) shows that the wavelength of dunes must be related to the depth, since the largest eddy sizes are depth dependent. Dunes occur at higher flow velocities than ripples, but they are similar in shape with a gentle, slightly convex upstream slope followed by an abrupt drop at the angle of repose. Dunes may be of sufficient height to cause surface waves, but these are of much smaller amplitude than the dunes and are out of phase with the dunes. Ripples can be two dimensional, with parallel crests transverse to the flow direction, or can exist as a three-dimensional array of individual ridges and valleys; while dunes tend to be three-dimensional except, possibly, in narrow laboratory flumes. As the flow velocity increases to cause ripples just beyond the threshold of motion and then dunes at higher velocities, sediment transport increases. With further increases in velocity or stream power, the dunes are washed out to form a plane bed with sediment transport.

In contrast to ripples and dunes, antidunes are not caused by either bed deformation or disturbance due to the largest-scale turbulent eddies, but rather by the standing surface waves that occur when the Froude number is near unity (Kennedy 1963; Yalin 1972). The alternating regions of scour and deposition in the flow direction due to the surface waves create antidunes in phase with the surface waves, which can become breaking waves. Antidunes can move upstream or downstream or remain stationary. Kennedy (1963) shows that the wavelength of antidunes depends on the Froude number of the flow, as given by

$$\frac{\lambda}{y_0} = 2\pi \frac{V^2}{g y_0} \tag{10.27}$$

in which λ = wavelength; y_0 = flow depth; and V = flow velocity. Sediment transport continues to increase as the bed passes through transition to flat bed and antidunes.

Several other bed form types have been classified (Vanoni 1977). Bars are bed forms having a triangular longitudinal profile, like dunes, but are of a scale comparable to the channel width and depth. Point bars occur on the inside of meander bends, while alternating bars occur in relatively straight river sections as the thalweg undulates from one bank to the other. Chutes and pools, as shown in Figure 10.10, consist of large deposits on which supercritical flow forms a chute that serves to connect deep pools.

Because bed forms depend on flow conditions in the river, they generate a variable form roughness due to flow separation in the lee of the bed form with attendant separation eddies and turbulent energy dissipation. This has led to the idea of separation of the total bed shear stress into a portion that can be attributed to form roughness (τ_0'') and the remainder due to surface or grain roughness (τ_0'). Assuming linear superposition of the shear stress components, this is written as

$$\tau_0 = \tau_0' + \tau_0'' \tag{10.28}$$

where τ_0 is the average boundary shear stress in uniform flow given by $\gamma R_0 S_0$, in which R_0 is hydraulic radius and S_0 is the channel slope. Such a separation has been found to be necessary to correctly predict stage-discharge relationships for alluvial channels, as described in the next section. The existence of a changing flow resistance due to variable bed forms as river discharge increases gives rise to discontinuous stage-discharge rating curves, as shown in Figure 10.11 for the Rio Grande, for example. In the lower regime, the flow resistance is high, but as the discharge or velocity continues to increase, the ripples and dunes are washed out in a transition

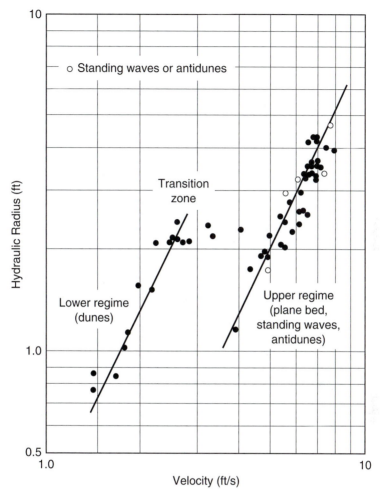

FIGURE 10.11
Relationship of hydraulic radius to velocity for Rio Grande near Bernalillo, New Mexico
(Nordin 1964).

zone to produce a plane bed with lower resistance, albeit with a larger value of sediment transport. Thus, for a given slope, more than one depth-velocity combination
is possible as the roughness and corresponding sediment transport rate change
(Vanoni and Brooks 1957; Kennedy 1963).

Numerous attempts have been made to identify bed forms as a function of flow
and sediment properties. The Simons-Richardson diagram (1966) shown in
Figure 10.12 plots stream power, defined as the product of mean bed shear stress
and velocity, against sediment size to identify regions of occurrence of various bed
forms. Figure 10.12 is based on extensive flume data collected by Simons and
Richardson, and reported by Guy, Simons, and Richardson (1966), on river data

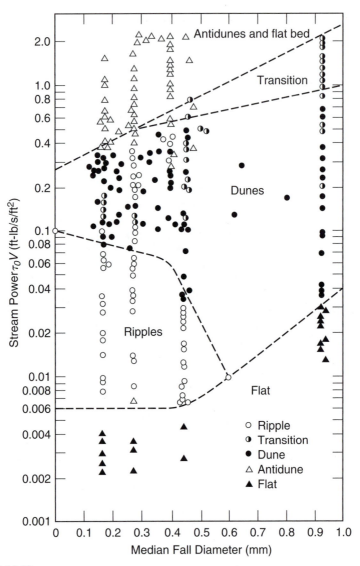

FIGURE 10.12
Prediction of bed form type from sediment fall diameter and stream power (Simons and Richardson 1966).

from the Rio Grande at several stations and the Middle Loup River at Dunning, Nebraska, and on irrigation canal data from India and Pakistan. Ripples occur at low values of stream power for fine sands, while lower regime dunes transition into antidunes and flat bed as the stream power increases. Simons and Senturk (1977) report that the diagram performs well on small natural streams, but on the Mississippi River, it predicts flat beds for cases that have been observed to be dunes.

Van Rijn (1984c) proposed a bed form classification system based on the dimensionless grain diameter, d_*, and a shear-stress parameter related to sediment transport and given by $T = (\tau_*'/\tau_{*_c} - 1)$, in which τ_*' = value of Shields' parameter for the grain shear stress and τ_{*_c} = critical value of Shields' parameter. As shown in Figure 10.13, Van Rijn suggested that ripples predominate when $d_* < 10$ and $T < 3$. Dunes fall in all other parts of the region $T < 15$. Transition is defined by $15 < T < 25$, and upper regime bed forms occur for $T > 25$. From laboratory and field data, including some of the same data used by Simons and Richardson, Van Rijn also developed a direct predictor for the height of dunes, Δ, given by

$$\frac{\Delta}{y_0} = 0.11\left(\frac{d_{50}}{y_0}\right)^{0.3}(1 - e^{-0.5T})(25 - T) \qquad (10.29)$$

in which y_0 is flow depth; d_{50} is median grain size; and T is the sediment transport variable defined previously. Clearly, if the value of T exceeds 25 in (10.29), then a flat bed is predicted with $\Delta = 0$. The wavelength of the dunes λ was given by Van Rijn to be $7.3y_0$, which is in good agreement with Yalin's (1972) theoretical value of $\lambda = 2\pi y_0$. Julien (1995) suggested that the van Rijn classification also suffers from poor prediction of the upper regime for very large rivers, such as the Mississippi, for which dune-covered beds have been observed for T values well in excess

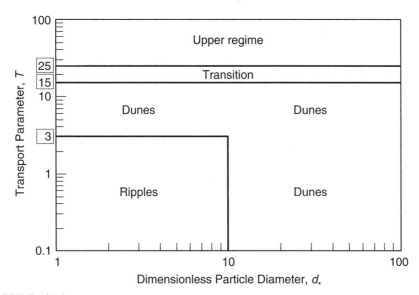

FIGURE 10.13

Prediction of type of bed form from transport parameter, T, and dimensionless particle diameter, d_* (Van Rijn 1984c). (*Source: L. C. van Rijn. "Sediment Transport III: Bed Forms and Alluvial Roughness," J. Hydr. Engrg.,* © *1984, ASCE. Reproduced by permission of ASCE.*)

of 25. This appears to be due to the fact that the Froude number approaches unity for T approximately equal to 25 in laboratory experiments, whereas in large sand-bed rivers, the Froude number is considerably less than 1 at $T = 25$. As a result, Julien and Klaassen (1995) proposed dropping the dependence of bed-form height on T and determined, from field data for several large rivers, that bed-form height can be given by

$$\frac{\Delta}{y_0} = 2.5 \left(\frac{d_{50}}{y_0} \right)^{0.3} \tag{10.30}$$

while bed-form length is approximately $\lambda = 6.25 y_0$.

Another difficulty in bed form prediction is caused by water temperature effects. As discussed by Shen, Mellema, and Harrison (1978), the Missouri River experiences a change from dune-covered bed to flat bed from summer to winter as the temperature decreases at the same values of discharge and slope. Obviously, the stream power in the Simons-Richardson diagram would remain the same even though the bed form changes with temperature.

Brownlie (1983) studied the transition regime and suggested that it can be delineated by the value of the grain Froude number or sediment number, $N_s = V/[(SG - 1)g d_{50}]^{0.50}$, and the ratio of grain diameter to viscous sublayer thickness, d_{50}/δ, where $\delta = 11.6\,\nu/u'_*$. For slopes greater than 0.006, Brownlie found that all the bed forms were in the upper regime, while for slopes less than 0.006 he suggested the following relationships for the lower limit of the upper regime based on both flume and river data:

$$\log \frac{N_s}{N_s^*} = -0.02469 + 0.1517 \log \frac{d_{50}}{\delta} + 0.8381 \left(\log \frac{d_{50}}{\delta} \right)^2 \quad \text{for} \quad \frac{d_{50}}{\delta} < 2 \tag{10.31a}$$

$$\log \frac{N_s}{N_s^*} = \log 1.25 \quad \text{for} \quad \frac{d_{50}}{\delta} \ge 2 \tag{10.31b}$$

where $N_s^* = 1.74\, S^{-1/3}$ and $S = $ slope. For the upper limit of the lower regime, Brownlie proposed the best-fit equations

$$\log \frac{N_s}{N_s^*} = -0.2026 + 0.07026 \log \frac{d_{50}}{\delta} + 0.9330 \left(\log \frac{d_{50}}{\delta} \right)^2 \quad \text{for} \quad \frac{d_{50}}{\delta} < 2 \tag{10.32a}$$

$$\log \frac{N_s}{N_s^*} = \log 0.8 \quad \text{for} \quad \frac{d_{50}}{\delta} \ge 2 \tag{10.32b}$$

These relationships for the transition zone are shown in Figure 10.14, and we can see that the variable d_{50}/δ, the ratio of grain size to viscous sublayer thickness, reflects the viscous influence near the bed and thus indicates a temperature dependence for the bed forms.

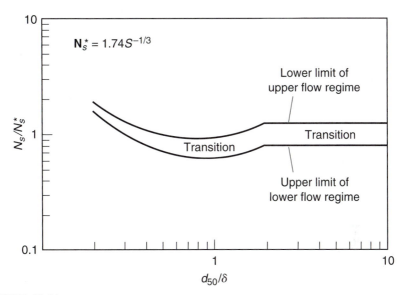

FIGURE 10.14

Delineation of bed form transition zone from lower regime to upper regime (Brownlie 1983). (*Source: W. R. Brownlie. "Flow Depth in Sand-Bed Channels," J. Hydr. Engrg., © 1983, ASCE. Reproduced by permission of ASCE.*)

10.6   STAGE-DISCHARGE RELATIONSHIPS

Perhaps the most fundamental difference between alluvial channel flows with movable beds and rigid-bed channels is the effect of variable bed forms on flow resistance and therefore on the stage-discharge relationship. It cannot be emphasized enough that the Manning's n values of Chapter 4 no longer are applicable in alluvial channels experiencing active sediment transport, because of the accompanying bed forms. During the rising side of large flood hydrographs, dunes formed at low discharges may be washed out to produce a flat bed at the flood peak, which buffers large variations in stage with discharge.

Many methods have been proposed to predict stage-discharge relationships for alluvial streams but only a limited number are presented here. The reader is referred to Vanoni (1977) and Brownlie (1983) for a more complete treatment. Einstein and Barbarossa (1952) were the first to separate flow resistance into form and surface (grain) resistance in alluvial channels, as indicated by Equation 10.28. The actual separation of shear stress into form and surface components was achieved through the definition of two additive components of the hydraulic radius, R' due to surface resistance and R'' due to form resistance, such that the total hydraulic radius $R = R' + R''$. The value of R' was determined from formulas for flow resistance in rigid-bed channels, while R'' came from the "bar resistance curve" relating V/u_*'', in which V = mean flow velocity and u_*'' = shear velocity due to bed forms $=(gR''S)^{1/2}$, to the Einstein sediment transport parameter $\psi' = (\gamma_s - \gamma)d_{35}/(\gamma R'S)$, which essentially is the inverse of the Shields parameter using d_{35} as the representative grain size. The physical reasoning behind the bar resistance curve was based on the inferred relationship between the rate of sediment transport and the bed form topography and, thus, the form resistance of ripples and dunes. Following Einstein and Barbarossa, others presented methods for separating the friction factor or the slope of the energy grade line into form and grain resistance components.

Engelund's Method

The method proposed by Engelund (1966, 1967) divided the slope of the energy grade line into two components as $S = S' + S''$, in which S' is the grain roughness slope and S'' is the additional slope due to form drag on the bed forms. The value of S'' is expressed in terms of an expansion loss due to the separation zone downstream of ripples and dunes. Engelund applied two similarity hypotheses given as follows: (1) In two dynamically similar streams, the Shields parameter τ_*' (due to grain resistance) has equal values and (2), in two dynamically similar streams, the expansion loss is the same fraction of the total energy loss. The latter hypothesis can be shown to imply that, for the two dynamically similar streams, $f_1'/f_1 = f_2'/f_2$, in which f_1 and f_2 are the total friction factors for streams 1 and 2, and f_1' and f_2' are the grain resistance friction factors. From the definition of the friction factor, this is equivalent to

$$\frac{\tau_{*1}}{\tau_{*2}} = \frac{\tau_{*1}'}{\tau_{*2}'} \tag{10.33}$$

However, according to the first similarity principle, the values of τ_*' on the right hand side of (10.33) are equal; therefore, so must the values of τ_* be equal. This can be true in general only if τ_* is a function of τ_*' alone. Engelund plotted flume results summarized by Guy, Simons, and Richardson (1966) according to this conclusion, as shown in Figure 10.15. It is evident from the data that separate curves for lower regime and upper regime bed forms exist with a transition between them, and apparent discontinuous stage-discharge relationships can occur in alluvial streams. In fact, Engelund (1967) showed that the use of Figure 10.15 produced close agreement with the measured stage-discharge relation for the Rio Grande given in Figure 10.11.

The lower regime curve in Figure 10.15 has an empirically fitted curve given by Engelund (1967) as

$$\tau_*' = 0.06 + 0.4\,\tau_*^2 \tag{10.34a}$$

while for the upper regime curve,

$$\tau_* = \tau_*' \qquad \text{for} \quad \tau_*' < 1 \tag{10.34b}$$

$$\tau_* = \left(1.425\,\tau_*'^{-1.8} - 0.425\right)^{-1/1.8} \qquad \text{for} \quad \tau_*' > 1 \tag{10.34c}$$

Equation 10.34b was given by Engelund (1967), while Equation 10.34c was proposed by Brownlie (1983) as an empirical fit of the data.

The application of Engelund's method requires the calculation of the velocity from a Keulegan-type relationship for fully rough turbulent flow given by Engelund as

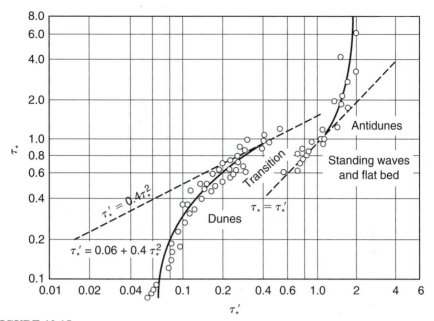

FIGURE 10.15

Engelund's τ'_* vs. τ_* diagram for stage-discharge prediction (Engelund and Hansen 1967; Engelund 1967, using data from Guy, Simons, and Richardson 1966). (*Source: F. Engelund. Closure to "Hydraulic Resistance of Alluvial Streams," J. Hyd. Div., © 1967, ASCE. Reproduced by permission of ASCE.*)

$$\frac{V}{u'_*} = 6 + 5.75 \log \frac{R'}{2d_{65}} \tag{10.35}$$

in which V = mean flow velocity; R' = hydraulic radius due to grain roughness; and u'_* = grain shear velocity = $(gR'S)^{1/2}$. Note that the equivalent sand grain roughness, k_s, is taken to be $2d_{65}$ in Equation 10.35. Implicit in the application of the Engelund method is a switch from dividing the slope of the energy grade line to dividing the hydraulic radius into form and grain resistance components. In order to determine the depth and velocity of flow for a given discharge, geometric mean grain size, bed slope and geometry, the following procedure is suggested:

1. Assume a value of R', or y'_0 for a wide channel, and calculate the velocity V from Equation 10.35 with $u'_* = (gR'S)^{1/2}$.
2. Calculate $\tau'_* = R'S/[(SG-1)d_g]$ and then obtain the value of τ_* from Figure 10.15 or Equation 10.34.
3. Solve for the total hydraulic radius, R, from the definition of τ_*:

$$R = \frac{\tau_*(SG-1)d_g}{S} \tag{10.36}$$

4. From the channel geometry, determine the depth, y_0, and cross-sectional area, A, for the value of R from (10.36), and calculate $Q = AV$, or for a wide channel, $R = y_0$ and $q = Vy_0$.

5. If the calculated discharge from step 4 does not agree with the given discharge, repeat steps 1 through 4 with a new value of R', or y_0' for a wide channel, until they do agree.

For the transition region of a discontinuous rating curve, such as the one shown in Figure 10.11, Brownlie (1983) suggested extending a horizontal line across from the depth at the upper limit of the lower regime to the upper regime curve for gradually increasing discharges. For gradually decreasing discharges, a horizontal line would extend from the lower limit of the upper regime to the lower regime curve. Alternatively, an average could be taken of upper and lower regime depths in the transition region, but ultimately more needs to be known about the dynamics of the transition itself recognizing the stochastic nature and three-dimensionality of the bed form formation.

Van Rijn's Method

An alternative approach for obtaining depth-discharge relationships was presented by van Rijn (1984c). He used the predicted height of the bed form to infer a form-resistance component of the equivalent sand-grain roughness such that $k_s = k_s' + k_s''$. The value of $k_s' = 3d_{90}$, which is an average value taken from a wide variation in laboratory and field data between 1 and $10d_{90}$ (van Rijn 1982), is substituted into the Keulegan equation to define u_*', the grain shear velocity:

$$u_*' = \frac{V}{5.75 \log \dfrac{12R}{3d_{90}}} \tag{10.37}$$

in which R = total hydraulic radius. This is a somewhat different definition of u_*' than in Engelund's method, but the final velocity for a given depth is computed from Keulegan's equation utilizing the total value of k_s and u_*:

$$V = 5.75u_* \log \frac{12R}{3d_{90} + k_s''} \tag{10.38}$$

in which $u_* = (gRS)^{0.5}$ and k_s'' for the form roughness is calculated from the bed form height Δ and steepness Δ/λ as

$$k_s'' = 1.1\Delta(1 - e^{-25\Delta/\lambda}) \tag{10.39}$$

The application of the method when discharge per unit of width q is given and both depth and velocity are unknown is as follows:

1. Estimate a value of the hydraulic radius, $R = y_0$.
2. Calculate $V = q/y_0$.
3. Solve for u_*' from Equation 10.37.

4. Calculate $T = u_*'^2/u_{*c}^2 - 1$ and the dimensionless grain diameter, d_*.
5. Calculate Δ from Equation 10.29 and $\lambda = 7.3\ y_0$.
6. Determine k_s'' from Equation 10.39 and velocity, V, from Equation 10.38, with $k_s = k_s' + k_s''$.
7. Calculate a new depth, $y_0 = q/V$, and repeat, starting from step 3.

If the value of $T > 25$, then a plane bed results and $k_s'' = 0$ in the original van Rijn method, but as discussed previously, the value of k_s'' may continue to increase beyond $T = 25$ in very large rivers, according to Julien.

Karim-Kennedy Method

A third approach to the stage-discharge problem in alluvial channels was presented by Karim and Kennedy (1990). Nonlinear regression analysis was applied to a database consisting of 339 river flows and 608 flume flows to determine the most significant dimensionless variables affecting depth-discharge as well as sediment transport relationships. The database included the laboratory data reported by Guy, Simons, and Richardson (1966) as well as field data for the Missouri River; Middle Loup, Niobrara, and Elkhorn Rivers in Nebraska; Rio Grande; Mississippi River; and canal data from Pakistan. Depths varied from 0.03 to 16 m (0.1 to 52 ft); velocities covered the range from 0.3 to 2.9 m/s (1.0 to 9.5 ft/s); and sediment sizes from 0.08 to 28.6 mm (2.6×10^{-4} to 9.4×10^{-2} ft) were included. The flow resistance was formulated in terms of the ratio of friction factors f/f_0 in which f is the friction factor for flow over a moving sediment bed, and f_0 is a reference friction factor for flow over a fixed sediment bed given by a Nikuradse-Keulegan type of relationship as

$$f_0 = \frac{8}{\left[5.75 \log \dfrac{12y_0}{2.5d_{50}}\right]^2} \tag{10.40}$$

in which $k_s = 2.5\ d_{50}$. It was assumed, based on Engelund's (1966) analysis of flow over lower regime beds, that f/f_0 varies linearly with the ratio of ripple or dune height to flow depth:

$$\frac{f}{f_0} = 1.20 + 8.92\frac{\Delta}{y_0} \tag{10.41}$$

with the coefficients determined from the river and flume data. It remains to obtain a relationship for Δ/y_0, which was developed in the original Karim-Kennedy method from data by Allen (1978) in terms of the Shields parameter. The best-fit relationship was given by

$$\frac{\Delta}{y_0} = 0.08 + 2.24\left(\frac{\tau_*}{3}\right) - 18.13\left(\frac{\tau_*}{3}\right)^2 + 70.9\left(\frac{\tau_*}{3}\right)^3 - 88.33\left(\frac{\tau_*}{3}\right)^4 \tag{10.42}$$

for $\tau_* \leq 1.5$ and $\Delta/y_0 = 0$ for $\tau_* > 1.5$. Karim and Kennedy then applied regression analysis to their data set to obtain a relationship for dimensionless velocity as a function of relative roughness, slope, and f/f_0, which is given by

$$\frac{V}{\sqrt{(SG-1)gd_{50}}} = 6.683\left(\frac{y_0}{d_{50}}\right)^{0.626} S^{0.503}\left(\frac{f}{f_0}\right)^{-0.465} \tag{10.43}$$

in which SG = the specific gravity of the sediment; d_{50} = the median sediment size; S = bed slope; y_0 = depth; and f/f_0 is obtained from Equations 10.41 and 10.42. For a given depth, the velocity can be calculated directly from Equation 10.43. The bed forms are identified as being in lower regime for $\tau_* < 1.2$, transition for $1.2 < \tau_* < 1.5$, and upper regime for $\tau_* > 1.5$.

The procedure for determining depth and velocity for a given value of discharge per unit width, q, for a wide channel can be summarized as follows:

1. Assume a value of depth, y_0, and calculate $\tau_* = y_0 S/[(SG-1)d_{50}]$.
2. Substitute τ_* into Equation 10.42 to obtain Δ/y_0 if $\tau_* \leq 1.5$; otherwise set $\Delta/y_0 = 0$ for $\tau_* > 1.5$.
3. Calculate f/f_0 from Equation 10.41, which will give a value of 1.20 if $\Delta/y_0 = 0$.
4. Substitute into Equation 10.43 to calculate the velocity, V.
5. Calculate Vy_0 and compare with the given value of q.
6. Repeat steps 1 through 5 with a new value of y_0 until the given and calculated values of q agree.

It is interesting to compare Equation 10.43 with Manning's equation for a wide channel by rearranging it for SG = 2.65 and $g = 9.81$ m/s² to yield an expression for Manning's n given by

$$n = 0.037 d_{50}^{0.126}\left(\frac{f}{f_0}\right)^{0.465} \tag{10.44}$$

in which very small exponents on S and y_0 have been neglected. Equation 10.44 is in SI units and similar to the Strickler equation with an exponent on d_{50} of 0.126, which is close to the Strickler value of $\frac{1}{6}$, but with the very important addition of the f/f_0 term, which reflects the resistance of the bed forms. This equation emphasizes the significant role played by bed forms in alluvial channel resistance and underscores the mistakes that can be made by applying estimates of Manning's n for fixed-bed channels from Chapter 4 to alluvial channels with movable beds.

Karim and Kennedy also developed a simplified procedure for computing the transition portion of discontinuous depth-velocity curves. The upper part of the lower regime is assumed to occur at about $\tau_* = 1.3$, while the lower part of the upper regime is defined at $\tau_* = 0.9$. The corresponding depths for these transition points then are calculated from the definition of τ_*. The lower regime relationship is constructed as a straight line on log-log scales from the computed depth-velocity point for the minimum depth to the lower-regime transition depth-velocity point with $f/f_0 = 4.5$, the maximum value. The upper regime relationship is developed in the same way from the maximum depth to the upper-regime transition depth with $f/f_0 = 1.2$. Horizontal lines are drawn from the lower to the upper regime relationships at both the transition points at the upper limit of the lower regime and the lower limit of the upper regime to represent the falling and rising portions, respectively, of the depth-velocity rating curves for rising and falling hydrographs.

Subsequent research by Karim (1995) revised the relationship for Δ/y_0 in terms of the ratio u_*/w_f, the ratio of shear velocity to sediment fall velocity, which is an indicator of the relative contribution of bedload and suspended sediment load to the total sediment load. One stated advantage of this change is to include the temperature effect on the bed form height, since fall velocity depends on the fluid temperature. The resulting relationship for Δ/y_0 is given by

$$\frac{\Delta}{y_0} = -0.04 + 0.294\left(\frac{u_*}{w_f}\right) + 0.00316\left(\frac{u_*}{w_f}\right)^2 - 0.0319\left(\frac{u_*}{w_f}\right)^3 + 0.00272\left(\frac{u_*}{w_f}\right)^4$$

(10.45)

for $0.15 < u_*/w_f < 3.64$, and $\Delta/y_0 = 0$ for $u_*/w_f < 0.15$ or $u_*/w_f > 3.64$. Equation 10.45 is based on only the laboratory flume data reported by Guy, Simons, and Richardson (1966) and some Missouri River data. Equation 10.45 in combination with Equations 10.40 through 10.42 is applied to the full data set of the Karim-Kennedy method as well as to 13 flows in the Ganges River, Rio Grande, and Mississippi River to predict depth-velocity rating curves. Mean normalized errors in both depth and velocity for all data sets are approximately 10 percent.

More recently, Karim (1999) developed another relationship for Δ/y_0 that provides a better fit than previous methods for a data set consisting of field data from the Missouri River, Jamuna River, Parana River, Zaire River, Bergshe Mass River, and the Rhine River as well as Pakistan canal data. The relationship of Julien and Klaassen (1995) given as Equation 10.30 also performed well for this data set.

Example 10.3

The Middle Loup River in Nebraska has a slope of 0.001 and a median grain size $d_{50} = 0.26$ mm (0.000852 ft). The values of $d_{65} = 0.32$ mm (0.00105 ft) and $d_{90} = 0.48$ mm (0.00157 ft). For a discharge per unit width of 3.0 ft²/s (0.28 m²/s), find the depth and velocity of flow using the Engelund method, van Rijn method, and Karim-Kennedy method.

Solution.   Assume that the channel is very wide so that $R = y_0$ in all the methods.

1. *Engelund Method.* Assume a value of $y_0' = 0.3$ ft (0.09 m). Then calculate τ_*' as

$$\tau_*' = \frac{\gamma y_0' S}{(\gamma_s - \gamma)d_{50}} = \frac{0.3 \times 0.001}{1.65 \times 0.000852} = 0.21$$

The velocity is given by

$$V = \sqrt{gy_0' S}\left[6 + 5.75 \log\frac{y_0'}{2d_{65}}\right]$$

$$= \sqrt{32.2 \times 0.3 \times 0.001} \times \left[6 + 5.75 \log\frac{0.3}{2 \times 0.00105}\right] = 1.81 \text{ ft/s}$$

or 0.55 m/s. From Figure 10.15, find τ_* or use Equation 10.34a assuming lower regime bed forms, from which

$$\tau_* = \sqrt{2.5(\tau_*' - 0.06)} = \sqrt{2.5 \times (0.21 - 0.06)} = 0.61$$

Now calculate y_0 from the definition of τ_* to give

$$y_0 = \frac{\tau_*(SG - 1)d_{50}}{S} = \frac{0.61 \times 1.65 \times 0.000852}{0.001} = 0.86 \text{ ft (0.26 m)}$$

Finally, calculate $q = Vy_0 = 1.81 \times 0.86 = 1.56$ ft²/s (0.145 m²/s). Because this is smaller than the given value of 3.0 ft²/s (0.28 m³/s), repeat for a larger value of y_0'. For $y_0' = 0.5$ ft (0.15 m), $\tau_*' = 0.36$ and $V = 2.50$ ft/s (0.76 m/s). Then $\tau_* = 0.86$ and $y_0 = 1.21$ ft (0.37 m) so that $q = 3.02$ ft²/s (0.281 m³/s). This is close enough, but check for lower regime bed forms. Calculate $\tau_0 = \gamma y_0 S = 62.4 \times 1.21 \times 0.001 = 0.076$ lbs/ft² (3.6 Pa) and stream power $= \tau_0 V = 0.076 \times 2.5 = 0.19$ lbs/(ft-s) (2.8 N/(m-s)). Then, for a fall diameter of 0.25 mm (see Figure 10.3), the Simons-Richardson diagram (Figure 10.12) indicates dunes, so this is a satisfactory solution: $y_0 = 1.21$ ft (0.37 m) and $V = 2.50$ ft/s (0.76 m/s).

2. *Van Rijn Method.* Assume a depth of 1.0 ft (0.30 m) and from continuity, $V = q/y_0 = 3.0/1.0 = 3.0$ ft/s (0.91 m/s). Then calculate u_*' from

$$u_*' = \frac{V}{5.75 \log \dfrac{12y_0}{3d_{90}}} = \frac{3.0}{5.75 \log \dfrac{12 \times 1.0}{3 \times 0.00157}} = 0.153 \text{ ft/s (0.0466 m/s)}$$

By definition, $\tau_*' = u_*'^2/[(SG - 1)gd_{50}] = 0.153^2/(1.65 \times 32.2 \times 0.000852) = 0.52$. Obtain τ_{*c} by first calculating d_* as

$$d_* = \left[\frac{(SG - 1)gd_{50}^3}{\nu^2}\right]^{1/3} = \left[\frac{1.65 \times 32.2}{(1.2 \times 10^{-5})^2}\right]^{1/3} \times 0.000852 = 6.1$$

so that $\tau_{*c} = 0.047$ from Figure 10.6 and $T = \tau_*'/\tau_{*c} - 1 = 0.52/0.047 - 1 = 10.1$. The height of the dunes is obtained from Equation 10.29 as

$$\frac{\Delta}{y_0} = 0.11\left(\frac{d_{50}}{y_0}\right)^{0.3}(1 - e^{-0.5T})(25 - T)$$

$$= 0.11 \times \left(\frac{0.000852}{1.0}\right)^{0.3}(1 - e^{-0.5 \times 10.1})(25 - 10.1) = 0.20$$

so that $\Delta = 0.20 \times 1.0 = 0.20$ ft (0.061 m). Having the dune height and with the wave length, $\lambda = 7.3y_0$, the equivalent sand-grain roughness height due to the bed forms can be estimated from Equation 10.39 as

$$k_s'' = 1.1\Delta(1 - e^{-25\Delta/\lambda}) = 1.1 \times 0.20 \times (1 - e^{(-25 \times 0.20/7.3)})$$

$$= 0.109 \text{ ft (0.033 m)}$$

Finally, the velocity can be obtained from Equation 10.38 based on the total shear velocity:

$$V = 5.75 u_* \log \frac{12R}{3d_{90} + k''_s}$$

$$= 5.75 \sqrt{32.2 \times 1.0 \times 0.001} \log \left[\frac{12 \times 1.0}{3 \times 0.00157 + 0.109} \right] = 2.09 \text{ ft/s}$$

or 0.64 m/s. The result for discharge per unit width is $q = V y_0 = 2.09$ ft²/s (0.194 m²/s), which requires a second iteration with a larger value of depth. For $y_0 = 1.3$ ft (0.40 m), the trial value of velocity is 2.31 ft/s (0.704 m/s) and $u'_* = 0.114$ ft/s (0.0347 m/s). Then $\tau'_* = 0.287$ and $T = 5.11$. This gives a dune height $\Delta = 0.291$ ft and $k''_s = 0.171$ ft (0.0521 m). Finally, the velocity is 2.29 ft/s (0.70 m/s), which is very close to the initial value, so the solution by the van Rijn method is $y_0 = 1.30$ ft (0.40 m) and $V = 2.30$ ft/s (0.70 m/s).

3. *Karim-Kennedy Method.* First calculate the value of the Shields parameter for an assumed depth of 1.3 ft (0.40 m) to give

$$\tau_* = \frac{y_0 S}{(SG - 1)d_{50}} = \frac{1.3 \times 0.001}{1.65 \times 0.000852} = 0.925$$

which is less than 1.2 and therefore in the lower regime. The relative dune height follows from Equation 10.42 into which the value of τ_* has been substituted:

$$\frac{\Delta}{y_0} = 0.08 + 2.24 \left(\frac{0.925}{3} \right) - 18.13 \left(\frac{0.925}{3} \right)^2$$

$$+ 70.9 \left(\frac{0.925}{3} \right)^3 - 88.33 \left(\frac{0.925}{3} \right)^4 = 0.327$$

Therefore, the relative value of the friction factor is obtained from Equation 10.41 as

$$\frac{f}{f_0} = 1.20 + 8.92 \frac{\Delta}{y_0} = 1.20 + 8.92 \times 0.327 = 4.12$$

Finally, the velocity comes from substituting into Equation 10.43 to give

$$\frac{V}{\sqrt{(SG - 1)g d_{50}}} = 6.683 \times \left(\frac{1.3}{0.000852} \right)^{0.626} \times 0.001^{0.503} \times 4.12^{-0.465} = 10.5$$

so that $V = 10.5 \times (1.65 \times 32.2 \times 0.000852)^{0.5} = 2.23$ ft/s (0.68 m/s). The discharge per unit of width then is 2.90 ft²/s (0.269 m²/s), which is close, but an adddi- tional iteration yields $y_0 = 1.33$ ft (0.41 m) and $V = 2.26$ ft/s (0.69 m/s).

The results of the van Rijn method and the Karim-Kennedy method are virtually identical, while the Engelund method gives a depth and velocity both of which are within about 8 percent of the values from the other two methods.

10.7   SEDIMENT DISCHARGE

The prediction of total sediment discharge in an alluvial stream is an important aspect of river engineering with applications from the assessment of changes in stream sediment regime due to urbanization to the evaluation of long-term bridge scour. This section focuses on the bed-material discharge; that is, the portion of the sediment discharge consisting of grain sizes found in the streambed as opposed to wash load, which is defined as the fine sediment resulting from erosion of the watershed.

Two distinct approaches are taken to the problem of determining total bed-material discharge. The first was pioneered by Einstein (1950), in which total bed-material discharge is divided into bed-load discharge and suspended-load discharge and summed to estimate total sediment discharge. The bed load is that portion of the sediment carried near the bed by the physical processes of intermittent rolling, sliding, and saltation (hopping) of individual grains at various random locations in the bed, so that the sediment remains in contact with the bed a large percentage of the time. Suspended load, on the other hand, is composed of sediment particles that are lifted into the body of the flow by turbulence, where they remain and are transported downstream. An equilibrium distribution of suspended sediment concentration develops as a result of the balance between turbulent diffusion of the grains upward and gravitational settling of the grains downward. The sediment concentration near the bed as determined by the bed-load discharge is the essential link to estimation of suspended-load discharge because it provides the boundary condition for the vertical distribution of suspended sediment concentration.

In general, the opposing forces of turbulent suspension and gravity are reflected by the dimensionless ratio u_*/w_f, in which u_* is shear velocity and w_f is the sediment fall velocity. Bed load is the dominant transport mechanism for $u_*/w_f < 0.4$, and suspended load is the primary contributor to sediment load for $u_*/w_f > 2.5$ (Julien 1995). In between these two limits, mixed load occurs, with components of both bed load and suspended load.

The second approach to determination of total sediment discharge is to directly relate the total rate of transport to hydraulic variables such as depth, velocity, and slope and to sediment properties. This method depends on large databases of flume and field data to be applicable to a wide variety of situations, and the best-fit relationship often is presented in terms of dimensionless variables for the same reason. In either approach, issues of water temperature, the effect of fine sediment, bed roughness, armoring, and the inherent difficulties of measuring total sediment discharge can cause significant deviations between estimates and measurements of total sediment discharge as demonstrated by Nakato (1990). Nevertheless, such estimates of sediment discharge must be made for engineering purposes. This often involves the use of several different formulas determined to be applicable to the situation of interest and reliance on engineering judgment to make the final estimate.

This section presents a few selected formulas for estimating sediment discharge and limited comparisons with field measurements. For a more complete treatment, refer to the references at the end of this chapter. The transport formulas are presented in terms of the volumetric transport rate of sediment per unit of stream width, q, with a subscript of b for bed load, s for suspended load, and t for total load. The sediment transport rate also can be expressed in terms of dry weight of sediment transported per unit of width and time as the symbol g with the same subscripts, so that $g_b = \gamma_s q_b$ for bed-load discharge, for example. Thus q_b has dimensions of L^2/T (ft²/s or m²/s), while g_b has dimensions of $F/T/L$ (lbs/s/ft or N/s/m). In the English system, the weight rate of transport will be used, but in the SI system a mass transport rate traditionally is used. The mass transport rate per unit of channel width can be obtained by dividing the corresponding weight rate by gravitational acceleration to obtain dimensions of $M/T/L$ (slugs/s/ft or kg/s/m). The sediment transport rate for the full stream width is obtained as the product of transport rate per unit of width and stream width, and the symbols Q and G are utilized for this purpose for volumetric and weight rates of transport, respectively, with the appropriate subscript to indicate bed load (b), suspended load (s), or total load (t).

Bed-Load Discharge

Bed-load formulas tend to be empirical by necessity, because of the complexity of the physics of movement of individual grains by rolling, sliding, and saltation. Early views of bed-load transport, such as that of DuBoys (1879), assumed that the bed load moved in sliding bed layers having a linear velocity distribution, which although incorrect in theory, nevertheless led to a useful transport formula dependent on shear stress with coefficients determined by experiments by Straub (Brown 1950). Graf (1971) shows that the same formula can be deduced from a power series expansion with respect to the bed shear stress and imposition of the boundary conditions of no transport rate for $\tau_0 = \tau_c$ and $\tau_0 = 0$, where τ_0 is the bed shear stress. The resulting bed-load transport formula is given by

$$q_b = \frac{C_{DB}}{(d_{50})^{0.75}} \tau_0(\tau_0 - \tau_c) \tag{10.46}$$

in which d_{50} = the median grain size in mm; τ_0 = the bed shear stress in lbs/ft²; τ_c = the critical shear stress also in lbs/ft²; q_b = the volumetric sediment transport rate per unit of width in ft²/s; and $C_{DB} = 0.17$ for this set of units. If the shear stresses are expressed in N/m² and q_b is in m²/s with d_{50} in mm, then $C_{DB} = 6.9 \times 10^{-6}$. The most important contribution of the DuBoys formula is the relation between sediment transport rate and an excess bed shear stress with respect to the critical value.

Several other bed-load transport formulas are of the DuBoys type. These include the formulas of Shields (1936) and Shoklitsch (Graf 1971), although the latter is expressed in terms of a difference between the actual water discharge per unit of width and the water discharge per unit width at incipient motion. The Meyer-Peter and Müller formula (1948) is the most well known formula of this type. It is based on laboratory experiments at ETH (Eidgenössische Technische Hochschule) in Zurich, Switzerland, for sediment sizes from 5 to 28.6 mm. It can be placed in the dimensionless form

$$\phi_b = \frac{q_b}{\sqrt{(\text{SG} - 1)g d_s^3}} = 8.0(\tau_* - 0.047)^{3/2} \tag{10.47}$$

in which d_s = the sediment size; SG = the specific gravity of the sediment; and τ_* = the Shields parameter. The value of 0.047 can be interpreted as the critical value of Shields' parameter, τ_{*_c}. Equation 10.47 applies to the case of no bed forms in coarse sediment.

Einstein (1942) introduced the concept of probability to the bed-load transport phenomenon, which resulted in a bed-load formula dependent on shear stress rather than excess shear stress. In other words, very small transport rates were predicted at shear stresses less than the critical value. Einstein assumed that individual sediment grains move in finite steps of length L_1 proportional to the grain size. Then, for a bed area of unit width and length L_1, the bed-load transport rate is taken as the product of the number of grains in this bed area, the volume of a grain, and the probability that a grain will be moved per unit time, p_s. The number of grains in the bed area is $L_1/C_1 d_s^2$, where C_1 is a constant of proportionality and d_s is the grain size, so the volumetric transport rate is

$$q_b = \frac{L_1}{C_1 d_s^2} C_2 d_s^3 p_s \tag{10.48}$$

where $C_2 d_s^3$ represents the volume of an individual grain and $L_1 = C_0 d_s$, in which C_0 is a constant of proportionality for the step length. The probability, p_s, is converted to a true probability by multiplying by a characteristic time scale, which Einstein expressed as d_s/w_f, or the time required for a grain to fall through a height equal to its own diameter, d_s, since w_f is the fall velocity. Substituting for L_1 and p_s results in

$$q_b = \frac{C_2}{C_1} C_0 d_s w_f p \tag{10.49}$$

in which the probability $p = p_s d_s/w_f$. Finally, Einstein assumed that the probability, p, was a function of the ratio of the lift force on a grain, taken to be proportional to $\tau_0 d_s^2$, to the submerged weight of the grain, which was assumed proportional to $(\gamma_s - \gamma)d_s^3$, so that

$$p = f\left[\frac{\tau_0}{(\gamma_s - \gamma)d_s}\right] \tag{10.50}$$

The resulting dimensionless parameter is seen to be the Shields parameter τ_*, which Einstein referred to as $1/\psi$. The fall velocity, w_f, was expressed by the Rubey equation (1933) as

$$w_f = F_*\sqrt{\left(\frac{\gamma_s}{\gamma} - 1\right)g d_s} \tag{10.51}$$

in which

$$F_* = \sqrt{\frac{2}{3} + \frac{36}{d_*^3}} - \sqrt{\frac{36}{d_*^3}} \tag{10.52}$$

and d_* is the dimensionless sediment diameter defined previously. Combining Equations 10.49 through 10.51 results in a bed-load discharge formula expressed in terms of dimensionless variables as

$$\phi_{EB} = \frac{q_b}{F_*\sqrt{\left(\dfrac{\gamma_s}{\gamma} - 1\right)gd_s^3}} = f(\psi) \tag{10.53}$$

in which

$$\psi = \frac{(\gamma_s - \gamma)d_s}{\tau_0} = \frac{1}{\tau_*} \tag{10.54}$$

and ϕ_{EB} is defined here (for reasons that will become apparent in the following discussion) as the Einstein-Brown dimensionless sediment transport rate such that $\phi_{EB} = \phi_b/F_*$. Einstein plotted the laboratory data of Gilbert (1914) and the ETH Zurich data (see Meyer-Peter and Müller 1948) for sand, gravel, and coal ($0.3 < d_{50} < 28.6$ mm) in terms of these dimensionless parameters and derived an exponential relation between them. It is given by

$$0.465\phi_{EB} = e^{-0.391\psi} \tag{10.55a}$$

However, in the chapter in *Engineering Hydraulics* written by Brown (1950) and edited by Rouse, the same data were fitted by the equation

$$\phi_{EB} = 40\left(\frac{1}{\psi}\right)^3 = 40(\tau_*)^3 \tag{10.55b}$$

Equations 10.55a and 10.55b, taken together, have come to be known as the Einstein-Brown bed-load transport formula, with Equation 10.55a applicable for $\psi > 5.5$ and Equation 10.55b applicable for $\psi < 5.5$, as shown in Figure 10.16. The Einstein-Brown formula was derived for uniform sediments with no appreciable bed forms but often is applied to other field conditions. If bed forms exist, then it is more appropriate to express ψ in terms of the grain shear stress, defined as ψ', because the grain shear stress is primarily responsible for sediment transport (Graf 1971).

Einstein (1950) further developed his bed-load relation into what now is called the *Einstein bed-load function*. In the expression for probability, the lift force was evaluated in terms of the shear velocity using the grain contribution to the hydraulic radius, R', in the definition of ψ'. In addition, a hiding factor and a pressure correction factor were introduced to obtain better agreement with data for sediment mixtures in which larger grains tend to shelter smaller grains in the bed from transport. The Meyer-Peter and Müller (1948) formula also has a more general form, in which only the contribution of grain resistance to the slope of the energy grade line is included for the case of significant bed forms. Chien (1956) shows that the more general form of the Meyer-Peter and Müller formula can be modified and expressed in terms of the Einstein variables as

$$\phi_b = \frac{q_b}{\sqrt{(SG - 1)gd_{50}^3}} = \left[\frac{4}{\psi'} - 0.188\right]^{3/2} \tag{10.56}$$

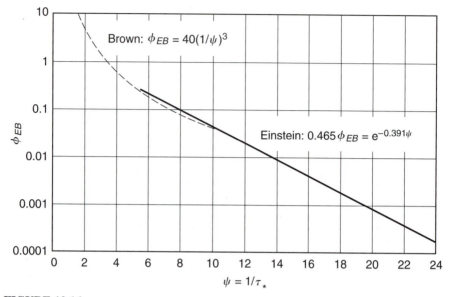

FIGURE 10.16
Einstein-Brown bedload transport formula (adapted from Brown 1950). (*Source: Figure used courtesy of Iowa Institute of Hydraulic Research.*)

in which $\psi' = (SG - 1)d_{50}/R'S$. Chien shows close agreement between this expression of the Meyer-Peter and Müller formula and the 1950 Einstein bed-load function, which uses d_{35} as the representative sediment size. For both formulas, only the grain shear stress is used in the definition of $\psi' = 1/\tau'_*$.

Van Rijn (1984a) took a different approach to the estimation of bed-load discharge by modeling the trajectories of saltating bed particles, assumed to be spherical in shape. He defined the bed-load transport rate as $q_b = u_b\delta_b c_b$, in which u_b is the particle velocity; δ_b is the saltation height; and c_b is the bedload concentration. The equations of motion for an individual saltating sphere were solved with a turbulent lift coefficient of 20, and an equivalent sand-grain roughness height of two or three grain diameters, which was used in the logarithmic velocity distribution for the fluid. These two parameters were obtained by calibration of the model with measured trajectories. Published values of the drag coefficient were used, and the initial velocities of the particles were assumed to be $2u_*$. The maximum thickness of the bed layer corresponding to the maximum saltation height was approximately 10 grain diameters, but the saltation height varied with the transport parameter, T, and the dimensionless grain diameter, d_*, introduced previously in van Rijn's method for depth-discharge prediction. From a set of computed trajectory data for particles from 0.2 to 2.0 mm in diameter and u_* values varying from 0.02 to 0.14 m/s (0.066 to 0.46 ft/s), the saltation height and particle velocity were correlated with the transport parameter, T, and the dimensionless grain diameter, d_*. The values for bed load concentration c_b were obtained from measurements of bed-load discharge in flumes and the corresponding computed values of u_b and δ_b; that is, $c_b = q_b/(u_b\delta_b)$. The results for c_b also were correlated with T and d_* and combined with regression relations for u_b and δ_b to obtain a bed-load transport formula given by

$$\phi_b = \frac{q_b}{\sqrt{(SG-1)gd_{50}^3}} = 0.053\,\frac{T^{2.1}}{d_*^{0.3}} \qquad (10.57)$$

The value of T in Equation 10.57 depends on u_*', which is calculated from Equation 10.37. In a verification data set including flume data and limited field data, the proposed bed-load discharge formula was found to predict the measured bed-load discharge within a factor of 2 for 77 percent of the data points, which was comparable to the variability in the data itself.

The bed-load formulas presented here are limited to predictions of bed-load discharge when bed load is the dominant mode of transport or to the prediction of the bed-load contribution to the total sediment discharge. They are not intended for predictions of total sediment discharge in cases where both bed load and suspended load are significant components of the total.

Suspended Sediment Discharge

In steady, uniform turbulent flow in a stream, turbulent velocity fluctuations in the vertical direction transport sediment particles upward. If the vertical velocity fluctuation is w', and the turbulent fluctuation in sediment concentration is c', then a positive correlation between c' and w' leads to a mean turbulent flux of sediment per unit of area given by $\overline{w'c'}$ as shown in Figure 10.17. The positive correlation results from positive (upward) values of w' bringing parcels of fluid with higher sediment concentration $(+c')$ with it since the suspended sediment concentration decreases in the upward direction. For an equilibrium sediment concentration profile in the vertical having no changes in the flow direction, the only other sediment

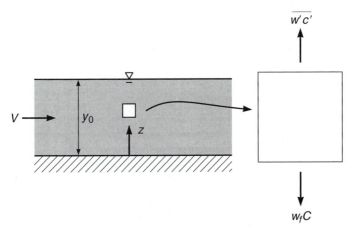

FIGURE 10.17
Suspended sediment flux balance.

flux, as shown in Figure 10.17, is due to the gravitational settling of sediment particles given by $w_f C$ for a unit area, in which w_f is the fall velocity of the sediment and C is the time-averaged concentration at a point in the vertical. Now the turbulent flux is assumed to behave like a Fickian diffusion process with a turbulent sediment diffusion coefficient of ε_s so that

$$\overline{w'c'} = -\varepsilon_s \frac{dC}{dz} \tag{10.58}$$

Equating the turbulent flux to the gravitational settling flux results in the following differential equation that governs the concentration distribution $C(z)$:

$$\varepsilon_s \frac{dC}{dz} + w_f C = 0 \tag{10.59}$$

This equation immediately is integrable by separation of variables if ε_s is a constant with respect to depth. The result is an exponential distribution given by

$$\frac{C}{C_a} = \exp\left[-\frac{w_f}{\varepsilon_s}(z-a)\right] \tag{10.60}$$

in which C_a = suspended sediment concentration at $z = a$. Unfortunately, ε_s is not a constant in alluvial channel flows, particularly near the bed, where the turbulence characteristics are changing with distance above the bed. The distribution of ε_s with the vertical coordinate z is deduced based on the vertical distribution of turbulent eddy viscosity, ε, defined by

$$\tau = \rho\varepsilon \frac{du}{dz} \tag{10.61}$$

in which τ and u represent the point shear stress and time-averaged velocity, respectively, at any distance z above the bed, as shown in Figure 10.18. First, we assume that $\varepsilon_s = \beta\varepsilon$, where β is a coefficient of proportionality. Second, we can show, from the Navier-Stokes equations, that the vertical shear stress distribution in a steady, uniform flow in an open channel is linear, as shown in Figure 10.18 and given by

$$\tau = \tau_0 \frac{(y_0 - z)}{y_0} \tag{10.62}$$

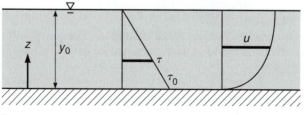

FIGURE 10.18
Shear stress and velocity distributions in steady, uniform turbulent flow.

in which τ_0 = the shear stress at the bed and y_0 = the depth of uniform flow. From the Prandtl-von Karman velocity defect law, given previously as Equation 4.14b but dropping the overbar on u, we can show that

$$\frac{du}{dz} = \frac{u_*}{\kappa z}$$

(10.63)

in which κ = von Karman's constant, having a value of 0.4 for clear fluids. Substituting $\varepsilon_s = \beta \varepsilon$ into Equation 10.61, along with Equations 10.62 and 10.63, and solving for ε_s gives

$$\varepsilon_s = \beta \kappa u_* \frac{z}{y_0} (y_0 - z)$$

(10.64)

which is a parabolic distribution with a maximum value of the sediment diffusion coefficient at mid-depth. Equation 10.64 can be substituted into the differential equation (10.59) so that it can be integrated to produce

$$\frac{C}{C_a} = \left[\frac{(y_0 - z)}{z} \frac{a}{(y_0 - a)} \right]^{\mathbf{R}_0}$$

(10.65)

in which C_a = the reference concentration at the distance $z = a$ above the bed and $\mathbf{R}_0 = w_f/(\beta \kappa u_*)$. Equation 10.65 was derived by Rouse (1937), and $\mathbf{R}_0$ is referred to as the *Rouse number* (Vanoni 1977; Julien 1995). Equation 10.65 is plotted in Figure 10.19 for different values of $\mathbf{R}_0$. The vertical coordinate is defined

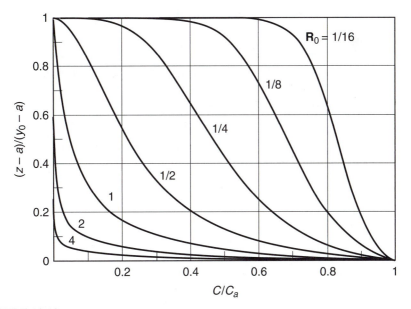

FIGURE 10.19
Rouse solution for vertical distribution of suspended sediment concentration (Vanoni 1977). (*Source: V. A. Vanoni, ed. Sedimentation Engineering, © 1977, ASCE. Reproduced by permission of ASCE.*)

as $(z - a)/(y_0 - a)$ in which, arbitrarily, $a = 0.05y_0$. As the value of $\mathbf{R}_0$ decreases, which would correspond to a finer sediment for the same flow conditions in the stream (u_*), the concentration distribution becomes more uniform. On the other hand, coarser sediment particles corresponding to larger values of $\mathbf{R}_0$ result in a suspended sediment concentration distribution carried in the lower portion of the flow.

The Rouse solution given by Equation 10.65 has been compared favorably with measured suspended sediment concentration distributions from rivers and flumes (Vanoni, 1977). Its application to measured suspended sand concentrations for the Rio Grande at Bernalillo, New Mexico (Nordin and Beverage 1965), is illustrated in Figure 10.20 at a single vertical location in the stream cross section. The concentration is plotted on the horizontal scale vs. the variable $(y_0 - z)/z$ on the vertical scale using log-log axes. From Equation 10.65, we see that the Rouse solution should plot as a straight line on log-log axes with an inverse slope of $\mathbf{R}_0$ as shown in Figure 10.20. Where there is a wide variation in the size distribution, Equation 10.65 can be applied to separate size fractions.

Values of $\mathbf{R}_0$ can be obtained from measured concentration profiles as shown in Figure 10.20, but the results do not always agree with predicted values of $\mathbf{R}_0 = w_f /(\beta \kappa u_*)$ with $\beta = 1.0$ and $\kappa = 0.4$. The von Karman constant can vary from its clear-water value of 0.4 to a value of 0.2 at high concentrations of suspended sediment as shown by Vanoni (1953) and Vanoni and Brooks (1957) from laboratory experiments. Einstein and Chien (1954) presented a method for predicting the von Karman constant in terms of the ratio of the power required to suspend sediment to the rate of doing work by the boundary resistance force. In addition, the value of β ordinarily is taken to be unity, but flume and river data (Chien 1956)

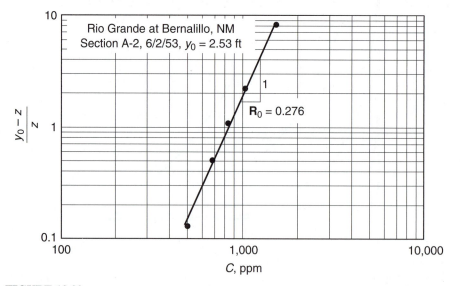

FIGURE 10.20

Measured suspended sediment concentration in the Rio Grande and determination of Rouse number, $\mathbf{R}_0$.

show that it can vary from 1 to 1.5 as $\mathbf{R}_0$ becomes larger (>2), which corresponds to coarser sediments. Finally, the estimation of the shear velocity, u_*, from a uniform flow formula as $(gy_0S)^{0.5}$ introduces errors because river flows seldom are uniform. The slope often is estimated as the water surface slope, but it is very difficult to measure accurately. The estimation of u_* affects the value of the von Karman constant if it is determined from measured velocity profiles as well as the value of $\mathbf{R}_0$ directly. These difficulties suggest that measured values of $\mathbf{R}_0$ may be more reliable than predicted ones.

The suspended sediment transport rate is computed from an integration of the product of the point velocity and concentration from the reference bed level at $z = a$ to the free surface where $z = y_0$:

$$g_s = \int_a^{y_0} uC_s\,dz \qquad (10.66)$$

in which C_s is the suspended sediment concentration usually given in mg/L or g/L so that g_s in this case often is expressed as kg/s/m in the SI system or converted to the English system as lbs/s/ft. The concentration also can be expressed as ppm (parts per million) by weight, $C_{s,\text{ppm}}$, which is related to the concentration in mg/L, $C_{s,\text{mg/L}}$, by

$$C_{s,\text{ppm}} = \frac{10^6(\text{SG})C_v}{1 + C_v(\text{SG} - 1)} = \frac{C_{s,\text{mg/L}}}{1 + C_v(\text{SG} - 1)} \qquad (10.67)$$

in which SG = the specific gravity of sediment; and C_v = the concentration by volume defined by the volume of sediment divided by the total volume of sediment and water. From Equation 10.67, it can be demonstrated that sediment concentration expressed in ppm is equivalent to the units of mg/L (within 5 percent) as long as $C_v < 0.032$ or $C_{s,\text{ppm}} < 80{,}500$ ppm.

Einstein (1950) substituted the Rouse solution for suspended sediment concentration (Equation 10.65) and the semi-logarithmic velocity distribution (Equation 4.16) into Equation 10.66 to obtain the suspended sediment transport rate. He assumed a value of $\kappa = 0.4$ and used u_*' in the calculation of $\mathbf{R}_0$. The reference concentration, C_a, was calculated for a bed layer with a thickness of two grain diameters having a bed-load transport rate determined from the Einstein bed-load function. The grain velocity in the bed layer was taken as the velocity at the edge of the viscous sublayer $(11.6u_*')$ so that

$$C_a = \frac{g_b}{2d_s(11.6\,u_*')} \qquad (10.68)$$

The integration of Equation 10.66 was done numerically and presented in graphical form. Furthermore, Einstein suggested that the grain size distribution be divided into size fractions each with a representative grain size, d_{si}, and that the suspended sediment discharge be computed for each size fraction. The total suspended sediment discharge then is $\Sigma\, p_i g_{si}$, in which p_i is the fraction by weight of the bed sediment with mean size d_{si}. The bed-load discharge for each size fraction also is weighted by p_i and added to the suspended sediment discharge to obtain the total bed-material discharge.

The principal criticism of the Einstein methodology is the use of u'_* and $\kappa = 0.4$ in the Rouse exponent $\mathbf{R}_0$. The grain shear velocity clearly is the appropriate choice for depth-discharge predictors and bed-load transport formulas when bed forms are present. However, the full contribution of turbulence to suspended load, as reflected by the value of u_*, should be used in the definition of $\mathbf{R}_0$. The decrease of the von Karman constant from 0.4 to values as small as 0.2 for heavy sediment concentrations also is not reflected in the Einstein methodology. Vanoni (1946) suggested that the decrease in κ results from damping of the turbulence by sediment, especially near the bed. Regardless of the criticism of the Einstein methodology, it is an important historical contribution because of its comprehensive approach and the introduction of the concept of probability applied to sediment discharge estimation.

Another comprehensive approach to the estimation of suspended sediment discharge and the corresponding total sediment discharge has been proposed by van Rijn (1984b). He employed the parabolic distribution of the sediment diffusion coefficient in the lower half of the flow (Equation 10.64) and a constant distribution in the upper half of the flow (equal to the maximum of the parabolic distribution), purportedly to obtain better agreement between measured and predicted distributions of suspended sediment. The resulting predicted concentration distribution is a combination of Equations 10.60 and 10.65 for the upper and lower halves of the flow depth, respectively. Van Rijn separated the effects of β and κ on the Rouse exponent $\mathbf{R}_0$. Based on the results of Coleman (1970) for the sediment diffusion coefficient in the upper half of the flow, van Rijn suggested a relationship for β given by

$$\beta = 1 + 2\left[\frac{w_f}{u_*}\right]^2 \tag{10.69}$$

for $0.1 < w_f/u_* < 1$. The effect of turbulence damping on reduction in mixing near the bed and the change in the velocity profile was treated by van Rijn by increasing the value of $\mathbf{R}_0$ instead of decreasing κ, so that $\mathbf{R}'_0 = \mathbf{R}_0 + \Delta\mathbf{R}_0$, in which $\mathbf{R}_0$ is defined with a value of $\kappa = 0.4$, while $\Delta\mathbf{R}_0$ represents a mixing correction factor. Ultimately, the value of $\Delta\mathbf{R}_0$ was obtained as a result of fitting velocity and concentration profiles from the laboratory data of Einstein and Chien (1955), Barton and Lin (1955), and Vanoni and Brooks (1957) for heavy sediment-laden flows and simplifying the results to obtain

$$\Delta\mathbf{R}_0 = 2.5\left[\frac{w_f}{u_*}\right]^{0.8}\left[\frac{C_a}{C_0}\right]^{0.4} \tag{10.70}$$

in which C_a = the reference concentration (volumetric) and C_0 = the maximum volumetric concentration taken to be 0.65. Equation 10.70 is valid for $0.01 < w_f/u_* < 1$. The reference concentration, C_a, is modified somewhat from the value used to develop van Rijn's bed-load transport formula. The reference level a for determining C_a is assumed to be half the bed-form height, Δ, or the equivalent sand-grain roughness height, k_s, if the former is unavailable. Based on only 20 flume and river data points, the expression for C_a was determined to be

$$C_a = 0.015\,\frac{d_{50}}{a}\,\frac{T^{1.5}}{d_*^{0.3}} \tag{10.71}$$

Finally, instead of using the Einstein approach of weighting the size fractions to determine the suspended sediment discharge for a sediment mixture, van Rijn developed an expression for an effective grain size of the suspended sediment, d_e, given by

$$\frac{d_e}{d_{50}} = 1 + 0.011(\sigma_g - 1)(T - 25) \tag{10.72}$$

This result was obtained by making several computations using the size-fractions method and then determining the effective grain size that would give the same value of suspended sediment discharge. Using the effective grain size (Equation 10.72) to obtain the fall velocity, the two-part solution for suspended sediment concentration with correction of $\mathbf{R}_0$ (Equation 10.70), the value of β given by (10.69), and the Nikuradse fully rough turbulent velocity profile, Equation 10.66 for suspended sediment discharge is integrated with a reference concentration given by (10.71). The numerical integration is simplified to obtain an approximate relationship for the suspended sediment discharge given by

$$q_s = I_f V y_0 C_a \tag{10.73a}$$

in which V = mean velocity; y_0 = depth; C_a = reference concentration; and the integration factor I_f is calculated from

$$I_f = \frac{\left[\dfrac{a}{y_0}\right]^{\mathbf{R}_0'} - \left[\dfrac{a}{y_0}\right]^{1.2}}{\left[1 - \dfrac{a}{y_0}\right]^{\mathbf{R}_0'}[1.2 - \mathbf{R}_0']} \tag{10.73b}$$

The van Rijn bed-load formula is used to calculate bed-load discharge, which is added to the suspended-sediment discharge to obtain the total bed-material discharge.

While the foregoing methodology contains several simplified expressions based on limited data to describe the very complicated interaction between turbulence and sediment particles, van Rijn obtained reasonable agreement between predicted and measured sediment discharges for several laboratory and field data sets. The agreement was shown to be comparable to the results from several other total sediment discharge formulas. Using a discrepancy ratio defined as the ratio of computed sediment discharge to measured sediment discharge, 76 percent of the computed values were in the range of discrepancy ratios from 0.5 to 2.0. For comparison, the Engelund-Hansen and Yang total sediment discharge formulas, described in the following section, had performance scores of 68 percent and 58 percent, respectively, of the computed values falling in the range of 0.5 to 2.0 times the measured values for the same data set.

Total Sediment Discharge

In contrast to the methodologies just described for separate calculations of bed-load discharge and suspended-sediment discharge, total sediment discharge formulas correlate total sediment transport rates directly with hydraulic variables without distinguishing between bed load and suspended load. This avoids the difficult problem of defining the difference between the two types of load and of determining the bed-load concentration at some reference level. If such formulas perform at least as well as the bed-load/suspended-load formulations, then there is much to commend their use, not the least of which is a greater degree of simplicity. However, for total load formulas to be successful, they must rely on as large a database of field and laboratory measurements as possible and be formulated in terms of physically meaningful dimensionless parameters.

The Engelund-Hansen formula (1967) for total sediment discharge, q_t, was derived from energy considerations and the similarity principles discussed previously in connection with the Engelund method for depth-discharge prediction. It is given by

$$c_f \phi_t = 0.1 \tau_*^{5/2} \tag{10.74}$$

in which $c_f = 2\tau_0/\rho V^2$; $\phi_t = q_t/[(SG - 1)gd_{50}^3]^{1/2}$; and $\tau_* = $ Shields' parameter, defined with the total bed shear stress $= \tau_0/[(\gamma_s - \gamma)d_{50}]$. The coefficient and exponent in Equation 10.74 were obtained from correlation of sediment transport data from the laboratory experiments reported by Guy, Simons, and Richardson (1966), and reasonably good correlation was found for dune bed forms as well as transition and upper regime bed forms (Engelund 1967).

Yang (1972, 1973) developed the concept of unit stream power as an important independent variable that determines total sediment discharge. The unit stream power is defined as the power available per unit weight of fluid to transport sediment and is equal to the product of velocity and energy slope, VS. A dimensional analysis that includes unit stream power, $VS;$ fall velocity, w_f; shear velocity, u_*; median grain size, d_{50}; and viscosity, ν, suggests that the independent dimensionless variables affecting total sediment discharge or concentration C_t are VS/w_f, $w_f d_{50}/\nu$, and u_*/w_f. Yang (1973) modified the dimensionless unit stream power, VS/w_f, by subtracting its critical value at the initiation of motion, $V_c S/w_f$, in which V_c is the critical velocity. A multiple regression analysis of 463 sets of laboratory data for sand transport in terms of these dimensionless variables gave the following relationship for total sediment discharge:

$$\log C_t = 5.435 - 0.286 \log \frac{w_f d_{50}}{\nu} - 0.457 \log \frac{u_*}{w_f} \tag{10.75}$$

$$+ \left(1.799 - 0.409 \log \frac{w_f d_{50}}{\nu} - 0.314 \log \frac{u_*}{w_f} \right) \log \left(\frac{VS}{w_f} - \frac{V_c S}{w_f} \right)$$

in which $C_t = $ total sand concentration by weight in ppm $= 10^6 \times \gamma_s q_t/\gamma q$. The dimensionless critical velocity is defined by

$$\frac{V_c}{w_f} = \frac{2.5}{\log(u_* d_{50}/\nu) - 0.06} + 0.66 \quad \text{for} \quad 1.2 < \frac{u_* d_{50}}{\nu} < 70 \tag{10.76}$$

and $V_c/w_f = 2.05$ for $u_* d_{50}/\nu \geq 70$. Yang's (1973) laboratory data set on which Equation 10.75 was based includes the data of Guy, Simons, and Richardson (1966), Williams (1967), Vanoni and Brooks (1957), and Kennedy (1961) as well as others for which flow depths were on the order of 0.03 to 0.30 m (0.1 to 1.0 ft). The coefficient of determination r^2 for the regression equation was 0.94. Equation 10.75 was verified with Gilbert's (1914) laboratory data and field data from the Niobrara River (Colby and Hembree 1955), Middle Loup River (Hubbell and Matejka 1959), and the Mississippi River (Jordan 1965), although the comparisons with the Middle Loup and Mississippi rivers were not quite as good as for the laboratory data sets.

The Karim-Kennedy (1990) methodology for depth-discharge predictors described previously also includes a total sediment discharge formula obtained from nonlinear regression using a database of 339 river flows and 608 flume flows. Several physically reasonable dimensionless ratios are used with a calibration data set (615 laboratory and field flows), and nonlinear regression analysis is carried out for the dimensionless sediment discharge and velocity. The resulting values of sediment discharge and velocity then are compared with measured values for a control data set and the least significant independent dimensionless variables removed from the analysis. This process is repeated several times until the final relationship is obtained as

$$
\begin{aligned}
\log \phi_t = \log \frac{q_t}{\sqrt{(SG-1)gd_{50}^3}} &= -2.279 + 2.972 \log\left[\frac{V}{\sqrt{(SG-1)gd_{50}}}\right] \\
&+ 1.060 \log\left[\frac{V}{\sqrt{(SG-1)gd_{50}}}\right] \log\left[\frac{u_* - u_{*c}}{\sqrt{(SG-1)gd_{50}}}\right] \\
&+ 0.299 \log\left(\frac{y_0}{d_{50}}\right) \log\left[\frac{u_* - u_{*c}}{\sqrt{(SG-1)gd_{50}}}\right]
\end{aligned}
\tag{10.77}
$$

in which q_t = total volumetric sediment discharge per unit width; V = flow velocity; y_0 = flow depth; SG = sediment specific gravity; d_{50} = median sediment grain size; u_* = shear velocity; and u_{*c} = critical shear velocity. The mean normalized error of Equation 10.77, defined as the mean of the ratios formed by the absolute values of the differences between predicted and measured sediment discharges over the measured values, is found to be approximately 43 percent for the control data set and 40 percent for the combined data set. The combined data set includes flow depths from 0.03 to 5.9 m (0.1 to 19 ft), velocities from 0.3 to 2.7 m/s (1.0 to 8.9 ft/s), d_{50} values from 0.08 to 28.6 mm (2.6 × 10^{-4} ft to 9.4 × 10^{-2} ft), and total sediment discharge concentrations from 9 to 49,300 ppm by weight.

Karim (1998) proposed a simpler power relationship for the same data sets as employed in the Karim-Kennedy analysis, with the result given by

$$
\frac{q_t}{\sqrt{(SG-1)gd_{50}^3}} = 0.00139\left[\frac{V}{\sqrt{(SG-1)gd_{50}}}\right]^{2.97}\left[\frac{u_*}{w_f}\right]^{1.47}
\tag{10.78}
$$

The mean normalized error for Equation 10.78 is 45 percent for the control data set, which is not significantly different from the performance of Equation 10.77. The mean normalized errors for the Yang formula and the Engelund-Hansen formula for the same control data set are 63 percent and 49 percent, respectively.

Karim applied Equation 10.78 to laboratory and field data having nonuniform sediments by dividing the sediment into size fractions. The sediment discharge is computed in each size fraction by Equation 10.78 multiplied by a partial bed armoring factor and a hiding factor. The partial armoring factor is intended to account for portions of the bed that are armored and unavailable for transport, while the hiding factor takes into account the sheltering effect of larger grains on smaller grains. The sediment discharges in each size fraction then are summed, and the total sediment discharge values found to be comparable to those computed from Equation 10.78 using only the median grain size, d_{50}.

Several other total sediment discharge formulas can be found in the literature, including those of Bagnold (1966), Laursen (1958b), Ackers and White (1973), and Brownlie (1981). A more complete review and ranking of various formulas for computation of total sediment discharge can be found in Alonso (1980), ASCE Task Committee (1982), Yang (1996), and Bechteler and Vetter (1989). In the last reference, the Karim-Kennedy formula was "recommended best for common use" while the formulas of Yang and Bagnold, "within the range of validity," were found to "yield the most reliable results."

Sediment transport formulas should be chosen that have a database within which the flow and sediment conditions of interest fit, and several formulas should be used and compared whenever possible. For example, the Engelund-Hansen formula is most appropriate for sand transport in the lower regime, while the Meyer-Peter and Müller formula should be chosen when there is coarse bed material in bed-load transport. On the other hand, the Einstein-Brown formula is not a good choice when appreciable bed material is carried in suspension. Where they exist, gauging stations are useful for developing sediment rating curves between measured sediment discharge and either water discharge or velocity. However, the wash load has to be subtracted from the measured suspended sediment discharge, and the bed load and unmeasured suspended sediment discharge usually have to be calculated and added to the measured suspended sediment discharge to obtain the total bed-material discharge (see Colby and Hembree 1955).

Example 10.4

The Niobrara River has a measured flow depth of 1.60 ft (0.49 m) and measured velocity of 3.57 ft/s (1.09 m/s) to give $q = 5.71$ ft²/s (0.53 m²/s) with an energy slope of 0.0017. The median sediment size $d_{50} = 0.27$ mm (0.000885 ft), $d_{90} = 0.48$ mm (0.00157 ft), and $\sigma_g = 1.58$. The temperature is 68° F. The mean total sediment concentration for these conditions was measured to be 1890 ppm by weight. Calculate the total sediment discharge using the van Rijn method, Yang method, and Karim-Kennedy method.

Solution.   First, calculate some quantities common to all three methods. For the given temperature, $\nu = 1.08 \times 10^{-5}$ ft²/s (1.0×10^{-6} m²/s) and d_* is obtained from

$$d_* = d_{50}[(SG - 1)g/\nu^2]^{1/3} = 0.000885 \times [1.65 \times 32.2/(1.08 \times 10^{-5})^2]^{1/3} = 6.81$$

The fall velocity then is

$$w_f = \frac{8v}{d_{50}}[(1 + 0.0139d_*^3)^{0.5} - 1]$$

$$= \frac{8 \times 1.08 \times 10^{-5}}{0.000885}[(1 + 0.0139 \times 6.81^3)^{0.5} - 1] = 0.129 \text{ ft/s}$$

or 0.0393 m/s, and the critical value of Shields' parameter is $\tau_{*c} = 0.045$ from Figure 10.6. The corresponding value of $u_{*c} = [\tau_{*c}(SG - 1)gd_{50}]^{0.5} = [0.045 \times 1.65 \times 32.2 \times 0.000885]^{0.5} = 0.046$ ft/s (0.014 m/s). The shear velocity is

$$u_* = \sqrt{gy_0 S} = \sqrt{32.2 \times 1.60 \times 0.0017} = 0.296 \text{ ft/s } (0.0902 \text{ m/s})$$

Note that $u_*/w_f = 2.3$ so that the sediment discharge is mostly suspended load.

1. *Van Rijn's Method.* The value of T is needed, and it depends on u_*'. As in Example 10.3, u_*' is obtained from Keulegan's equation using the measured velocity and $k_s' = 3d_{90}$:

$$u_*' = \frac{V}{5.75 \log \dfrac{12y_0}{3d_{90}}} = \frac{3.57}{5.75 \log \dfrac{12 \times 1.6}{3 \times 0.00157}} = 0.172 \text{ ft/s } (0.0524 \text{ m/s})$$

Then, by definition, $\tau_*' = u_*'^2/[(SG - 1)gd_{50}] = 0.172^2/(1.65 \times 32.2 \times 0.000885) = 0.63$ The resulting value of $T = \tau_*'/\tau_{*c} - 1 = 0.63/0.045 - 1 = 13.0$, and the bed-load discharge from (10.57) becomes

$$q_b = 0.053\sqrt{(SG - 1)gd_{50}^3}\frac{T^{2.1}}{d_*^{0.3}}$$

$$= 0.053\sqrt{1.65 \times 32.2 \times 0.000885^3}\frac{13.0^{2.1}}{6.8^{0.3}} = 0.00125 \text{ ft}^2/\text{s}$$

or 1.16×10^{-4} m²/s. For the suspended sediment discharge, values of β, $\mathbf{R}_0$, $\Delta\mathbf{R}_0$, a, and C_a are needed. For this example, Equation 10.73 gives a relatively small correction to d_{50} for the effective grain size, so the value of d_{50} is used. The value of β comes from Equation 10.69:

$$\beta = 1 + 2\left[\frac{w_f}{u_*}\right]^2 = 1 + 2\left[\frac{0.129}{0.296}\right]^2 = 1.38$$

and then from the definition of $\mathbf{R}_0$, we have

$$\mathbf{R}_0 = \frac{w_f}{\beta\kappa u_*} = \frac{0.129}{1.38 \times 0.4 \times 0.296} = 0.790$$

The reference concentration, C_a, is calculated from Equation 10.71, in which the reference level is taken as half the dune height from Equation 10.29 to give $a = 0.11$ ft (0.034 m). The value of C_a as a volumetric concentration from (10.71) is

$$C_a = 0.015\frac{d_{50}}{a}\frac{T^{1.5}}{d_*^{0.3}} = 0.015 \times \frac{0.000885}{0.11} \times \frac{13.0^{1.5}}{6.8^{0.3}} = 0.0032$$

Now the correction to $\mathbf{R}_0$ follows from Equation 10.70:

$$\Delta\mathbf{R}_0 = 2.5\left[\frac{w_f}{u_*}\right]^{0.8}\left[\frac{C_a}{C_0}\right]^{0.4} = 2.5\left[\frac{0.129}{0.296}\right]^{0.8}\left[\frac{0.0032}{0.65}\right]^{0.4} = 0.15$$

so that $R'_0 = R_0 + \Delta R_0 = 0.79 + 0.15 = 0.94$. The integration factor, I_f, to calculate the suspended sediment discharge comes from Equation 10.73b:

$$I_f = \frac{\left[\dfrac{a}{y_0}\right]^{R'_0} - \left[\dfrac{a}{y_0}\right]^{1.2}}{\left[1 - \dfrac{a}{y_0}\right]^{R'_0}[1.2 - R'_0]} = \frac{\left[\dfrac{0.11}{1.6}\right]^{0.94} - \left[\dfrac{0.11}{1.6}\right]^{1.2}}{\left[1 - \dfrac{0.11}{1.6}\right]^{0.94}[1.2 - 0.94]} = 0.166$$

Finally, the suspended sediment discharge is given by

$$q_s = I_f V y_0 C_a = 0.166 \times 3.57 \times 1.6 \times 0.0032$$
$$= 0.00303 \text{ ft}^2/\text{s} \; (2.82 \times 10^{-4} \text{ m}^2/\text{s})$$

The total sediment discharge, q_t, is the sum of the bed-load and suspended-load discharges and equal to $(0.00125 + 0.00303) = 0.00428$ ft²/s $(3.98 \times 10^{-4}$ m²/s). Converted to tons/day, $g_t = \gamma_s q_t = 2.65 \times 62.4 \times 0.00428 \times 86,400/2000 = 30.6$ tons/day $(28,000$ kg/day) and $C_t = 10^6 \, (\gamma_s/\gamma)(q_t/q) = 10^6 \times 2.65 \times 0.00428/5.71 = 1990$ ppm.

2. *Yang's Method.* First, the critical velocity from Equation 10.76 is needed, since $u_* d_{50}/\nu = 0.296 \times 0.000885/1.08 \times 10^{-5} = 24.3 < 70$, so

$$V_c = w_f \left[\frac{2.5}{\log(u_* d_{50}/\nu) - 0.06} + 0.66\right]$$

$$= 0.129 \times \left[\frac{2.5}{\log(24.3) - 0.06} + 0.66\right] = 0.328 \text{ ft/s} \; (0.10 \text{ m/s})$$

Then, $VS/w_f = 3.57 \times 0.0017/0.129 = 0.0470$ and $V_c S/w = 0.328 \times 0.0017/0.129 = 0.00432$. The other two independent variables required are $u_*/w_f = 0.296/0.129 = 2.30$ and $w_f d_{50}/\nu = 0.129 \times 0.000885/1.08 \times 10^{-5} = 10.6$. Substituting directly into Equation 10.75, we have

$$\log C_t = 5.435 - 0.286 \log(10.6) - 0.457 \log(2.30)$$
$$+ [1.799 - 0.409 \log(10.6) - 0.314 \log(2.30)][\log(0.0470 - 0.00432)]$$
$$= 3.24$$

and then $C_t = 1,740$ ppm.

3. *Karim-Kennedy's Method.* Three dimensionless variables are required for the total sediment discharge computation:

$$\frac{V}{\sqrt{(SG - 1)gd_{50}}} = \frac{3.57}{\sqrt{1.65 \times 32.2 \times 0.000885}} = 16.5$$

$$\frac{u_* - u_{*c}}{\sqrt{(SG - 1)gd_{50}}} = \frac{0.296 - 0.046}{\sqrt{1.65 \times 32.2 \times 0.000885}} = 1.15$$

$$\frac{y_0}{d_{50}} = \frac{1.6}{0.000885} = 1808$$

Substituting directly into Equation 10.77, we have

$$\log \frac{q_t}{\sqrt{(SG - 1)gd_{50}^3}} = -2.279 + 2.972 \log(16.5) + 1.060 \log(16.5) \log(1.15)$$

$$+ \ 0.299 \log(1808) \log(1.15) = 1.477$$

Taking the antilog and solving, we have $q_t = 0.00575$ ft²/s (5.34×10^{-4} m²/s). Converting to concentration, $C_t = 10^6 \times 2.65 \times 0.00575/5.71 = 2{,}670$ ppm. On the other hand, if we use the Karim power formula (Equation 10.78), we have

$$\frac{q_t}{\sqrt{(SG - 1)gd_{50}^3}} = 0.00139 \times (16.5)^{2.97} \times (2.30)^{1.47} = 19.5$$

with the result that $q_t = 0.00374$ ft²/s (3.47×10^{-4} m²/s) and $C_t = 1{,}740$ ppm.

No conclusions can be drawn about the accuracy of the methods in Example 10.4 based on a single data point for one river. The Niobrara River data are included in the control data set of 341 data points used by Karim (1998) to test his method as well as Yang's method, for which the mean normalized errors are 45 percent and 63 percent, respectively. Furthermore, note that the measured velocity and depth values are used in the sediment discharge predictions, but are predicted rather than measured in the general case.

10.8   STREAMBED ADJUSTMENTS AND SCOUR

The sediment transport relationships developed in previous sections of this chapter assumed equilibrium sediment transport conditions, for which the sediment transport rate into a river reach was considered identical to the sediment transport rate out of the reach with no net aggradation, degradation, or scour of the bed within the reach. The bed itself was considered movable with bed forms, but on average, the bed was assumed not to be undergoing significant changes in elevation on an engineering time scale, which may be on the order of several years. In the short term, however, sediment storage (plus or minus) compensates for imbalance in the inflow and outflow sediment discharges for a river reach. Under these circumstances, the independent variables are the stream slope and water discharge, in addition to the sediment properties; and the dependent variables are the depth, velocity, and sediment discharge, which are interrelated. The bed forms adjust themselves to provide a roughness consistent with the depth and velocity necessary to carry the equilibrium sediment discharge. On the other hand, there may be no depth-velocity combination for the given water discharge and slope to carry the equilibrium sediment discharge, so that in the short-term, local scour and deposition may occur, albeit without altering the stream slope over a long reach (Kennedy and Brooks 1965).

On a much longer time scale, on the order of hundreds of years, the water discharge and sediment discharge become the independent variables; and the stream width, slope, and stream planform adjust themselves so as just to be able to transport the water and sediment discharge delivered to the upstream end of the stream reach. This is Mackin's (1948) concept of the "graded stream." If, for example, the sediment discharge to a stream reach over many years is too large for the stream to transport, some sediment will deposit, steepening the reach, or the meander length or stream width will change, so that the stream equilibrium is restored.

In this section, applications of these concepts are considered for the important engineering problem of bridge scour. Both long-term and short-term channel bed adjustments as well as the scour caused by bridge obstructions can undermine bridge foundations, with possible failure and loss of life. First, long-term channel aggradation and degradation are discussed, then contraction scour caused by the restricted bridge opening is analyzed. Finally, local scour caused by bridge piers and abutments is considered.

Aggradation and Degradation

Long-term aggradation and degradation of an alluvial stream can occur at a proposed or existing bridge site. In addition to changes in bed elevation that can be in the form of either scour or fill, the stream planform can shift laterally away from the designed bridge opening and cause local scour around the abutments and embankments. Some brief discussion of different types of alluvial streams with respect to planform is needed to understand the various geomorphic changes that can occur in response to human activities such as building dams and bridges to cross the stream.

Alluvial streams can be classified as straight, meandering, or braided, with transitional forms between each type. The sinuosity of a stream, defined as the stream length divided by the valley length, is used to distinguish between straight and meandering streams. In general, a stream is considered to be meandering if the sinuosity exceeds a value of 1.5. Even straight streams can have an oscillating thalweg at low stages as the flow moves from one bank to the other around sandbars. In meandering streams, the oscillating thalweg initiates streambank erosion and the formation of a continuous series of bends connected by crossings, as shown in Figure 10.21. Erosion of the outside of a bend carries sediment to the inside of the next bend downstream where it is deposited as a point bar. In addition, because of the centripetal acceleration associated with the turning of the flow through the bend, a transverse pressure gradient manifested by a sloping water surface toward the center of the radius of curvature develops. The result is a secondary current with a transverse velocity component toward the inside of the bend at the stream bottom and a return circulation at the free surface toward the outside of the bend leading to a helical flow through the bend. Deeper pools develop at the outside of the bends, and they are connected by shallow crossings from one pool to the next. Meander loops can migrate downstream as well as laterally and form cutoffs across the neck leaving behind oxbow lakes. The rate of longitudinal and lateral migration of a meander depends on the erodibility of the sediments encountered, which can be quite heterogeneous. In a study of 50 different meandering rivers, Leopold, Wolman, and Miller (1964) found that the ratio of meander radius to stream width varied from 1.5 to 4.3 with a median value of 2.7. The actual cause of meandering has

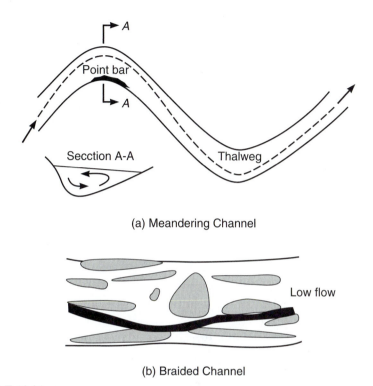

(a) Meandering Channel

(b) Braided Channel

FIGURE 10.21
Schematic of meandering and braided channels.

been attributed to various factors, including heterogeneity of bank sediments (Petersen 1986), secondary currents (Tanner 1960), and the need for a stream slope to be flatter than the valley slope to carry a lower sediment load than was available during the development of the valley slope (Chang 1988). Schumm (1971), on the other hand, argued that a change in the type of sediment load carried by a stream accounts for meandering behavior and the associated increase in sinuosity. A change from sand carried predominantly as bed load to wash load transported only in suspension would result in an excess slope of the energy grade line which might only be dissipated by the decrease in slope associated with an increase in sinuosity.

For larger stream slopes and increased bed load, the alluvial stream takes on a braided form that consists of multiple channels around numerous sandbars, as shown in Figure 10.21. The channels are connected in a network to form a wide shallow belt that is unstable with unpredictable rates of lateral migration. As the sandbars grow and form islands that are large relative to stream width, the braided stream becomes an anabranched stream with somewhat more permanent channels that can carry a substantial portion of the total flow.

Qualitatively, stream sinuosity first increases with slope and levels off before decreasing with further increases in slope for a characteristic discharge, which usually is taken to be the bank-full discharge with a return period of 1–2 years. Such a relationship has been confirmed for several natural streams, as reported by

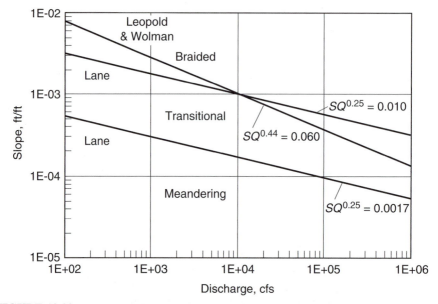

FIGURE 10.22
Changes in planform of streams with stream slope at a given characteristic discharge (Richardson and Davis 2001).

Schumm, Mosley, and Weaver (1987). For increasing values of slope, the straight channel transitions into a meandering thalweg channel with large values of sinuosity and then into a combination of meandering and braided forms until the channel becomes completely braided with low sinuosity. Changes in the planform of the stream can be analyzed with the help of Figure 10.22, which shows dividing lines or thresholds for distinguishing meandering and braided streams as given by Leopold and Wolman (1957, 1960) and Lane (1957). The significance of the relationship in Figure 10.22 is that any engineering changes that result in a change in slope can cause major changes in planform of the stream.

Characteristic widths and depths of alluvial streams have been related to the mean annual discharge to the power 0.5 for width and 0.4 for depth by Leopold and Maddock (1953). However, in a study of 36 stable alluvial rivers in the Great Plains of the United States and in the Riverine Plain of New South Wales, Australia, Schumm (1969) showed that the stream width and depth also were functions of the percent silt-clay, %SC, in the sediment forming the channel bed and banks. The channel width-depth ratio and sinuosity were found to be influenced primarily by %SC with little effect of the mean annual discharge. The larger is the percentage of silt-clay in the sediment, the smaller the width-depth ratio and the larger the sinuosity.

Analysis of long-term changes in the stream morphology can be achieved qualitatively with the aid of the approximate relationship proposed by Lane (1955b):

$$QS \sim Q_t d_{50} \qquad (10.79)$$

in which Q = water discharge; S = energy slope; Q_t = total sediment discharge; and d_{50} = median sediment size. In the channel downstream of a dam or sand and gravel mining operation, for example, the sediment supply is cut off so that there is a decrease in sediment discharge, which is balanced by a decrease in slope for the same water discharge and sediment size. The decrease in slope is accomplished by degradation of the channel bottom beginning from the dam and moving in the downstream direction with the largest scour and drop occurring in the channel bottom just downstream of the dam. As shown in Figure 10.22, a decrease in slope can cause a change in stream planform from braided to meandering, for example. On the other hand, a more realistic analysis would indicate that decreases in Q also occur due to flow regulation by the dam, and the sediment size may increase due to streambed armoring, in which the larger sizes of the size distribution are left behind in the degradation process. In this case, the slope may increase or decrease, depending on the relative magnitudes of the other changes, but degradation limited by both armoring and reductions in Q is a common result (Lagasse et al. 2001). Schumm's analysis (1969) further showed that long-term river metamorphosis can result from changes in water discharge and type of sediment load. Again, using the construction of a dam as an example, decreases in both water discharge and bed-material load (primarily sand) can result in decreases in width and width-depth ratio while sinuosity increases.

Quantitative analysis of long and short-term changes in stream morphology can be accomplished with a numerical solution of the sediment continuity equation (Exner equation) given by

$$B(1 - p_o) \frac{\partial z_b}{\partial t} + \frac{\partial Q_t}{\partial x} = 0 \tag{10.80}$$

in which B = stream width; p_o = porosity of the sediment bed; z_b = bed elevation; x = longitudinal distance along the stream; and Q_t = total volumetric sediment discharge. The equation can be solved simultaneously with the one-dimensional unsteady flow equations as described in Chapter 8, or if the changes in bed elevation are slow compared to the time scale of the changes in water surface elevations, a quasi-steady approach can be employed. In this approach, Equation 10.80 is solved and the sediment bed elevations are updated for the current quasi-steady, gradually varied flow profile. The change in bed elevation is assumed to be the same at all cross-sectional points within the specified movable-bed width. Then the water surface profile is recomputed with the new bed elevations, using the standard step method for the current quasi-steady water discharge. The sediment and flow equations are solved alternately in this uncoupled fashion at each time step to determine the development of bed elevation changes. A sediment transport relationship is required for the solution of Equation 10.80, and the roughness coefficient has to be specified. This is the basic approach used by the U.S. Corps of Engineers (1995) program HEC-6, which also accounts for bed armoring using the method proposed by Gessler (1970). Also see the implementation in HEC-RAS (U.S. Army Corps of Engineers 2008).

Chang (1982, 1984) proposed a similar water and sediment routing procedure, except that stream width changes are accounted for by minimizing the stream power per unit of length, γQS. This is equivalent to adjusting the width of adjacent cross sections until QS approaches a constant value along the stream. If Q is relatively constant along the stream, the result is to minimize the variation in the energy gradient, S, in the streamwise direction. In general, increasing the width at a cross section corresponds with larger values of S and vice versa. A weighted average energy gradient of adjacent cross sections is computed; and if the actual energy gradient is higher (lower), channel width at this cross section is decreased (increased) to decrease (increase) the energy gradient. Once the width adjustment has been made, the remaining change in sediment cross-sectional area is applied to the bed. For deposition, the bed is allowed to build up in horizontal layers, while scour is applied according to the distribution of the excess shear stress with respect to critical shear stress across the section. Chang (1985, 1986) applied his water and sediment routing model (FLUVIAL-12) with width adjustment and simplified modeling of bank erosion due to stream curvature to define thresholds for different planforms of rivers from meandering to braided.

The water and sediment routing model IALLUVIAL (Holly, Yang, and Karim 1984) is a one-dimensional model developed to predict long-term degradation of the Missouri River. Rather than specifying the value of Manning's n, the sediment discharge relationship and the friction-factor relationship are coupled and solved at each time step to model bed form changes and their interaction with the flow and sediment transport (Karim and Kennedy 1981, 1990). In the first stage of the time step, the water surface profile is obtained from a quasi-steady, simultaneous solution of the energy and continuity equations as well as the sediment discharge and friction-factor relationships. In the second stage, the sediment continuity equation is solved by an implicit finite-difference approximation to update the bed elevations uniformly. Bed armoring procedures and the option of specifying a known bank erosion rate are included in the model.

Several other numerical models of aggradation-degradation have been developed, but all are limited to varying degrees by an incomplete knowledge of the mechanics of bank erosion and width adjustment. Kovacs and Parker (1994) provided some insight by developing a vectorial bed-load formulation that takes into account the particle movement on steep, noncohesive banks, as influenced by gravity as well as fluid shear. They applied their bed-load formulation along with the sediment continuity equation and the momentum equation utilizing a simple algebraic turbulence closure model for steady, uniform flow. The initial trapezoidal channel evolved into an equilibrium cross-section shape consisting of a flat bed near the central part of the channel that connected smoothly to a curving, concave bank having a slope that approached the angle of repose. Comparisons with experimental measurements showed good agreement.

Several other models of width adjustment have been reviewed by the ASCE Task Committee on River Width Adjustment (ASCE 1998). Problems of a variety of different bank failure mechanisms, unknown shear stress distributions in the near-bank zone, limited understanding of the erosion behavior of cohesive sediment banks, lack of data on the longitudinal extent of mass failures of the bank, and the significance of overbank flows indicate that much remains to be learned about the mechanics of bank erosion and width adjustment. The computational tools presently available for predicting width adjustments are approximate at best. In spite of this, evaluation of a bridge-crossing site should include as much qualitative and quantitative information as possible on the current state of equilibrium of the stream or lack thereof, and possible consequences of the construction of a bridge crossing.

Bridge Contraction Scour

The acceleration of the flow caused by a bridge contraction can lead to scour in the bridge opening that extends across the entire contracted channel. The contraction can arise from a narrowing of the main channel as well as blockage of flow on the floodplain, if the abutments are at the banks of the main channel and overbank flow is occurring. If the abutments are set back from the edge of the main channel, contraction scour can occur on the floodplain in the setback area as well as in the main channel. Relief bridges on the floodplain or over a secondary stream in the overbank area also can cause contraction scour.

The type of contraction scour can be either clear water or live bed. In clearwater scour, the velocities and shear stresses in the approach cross-section upstream of the bridge are insufficient to initiate sediment motion, so no sediment transport is coming into the contracted area. In this case, scour continues in the contracted section until the enlargement of the cross-section is such that the velocity approaches the critical velocity and no additional sediment can be transported out of the contraction. This is the equilibrium condition that is approached asymptotically in time. Live-bed scour, on the other hand, occurs when sediment is being transported into the contraction from upstream. Scour continues until the sediment discharge out of the contracted section is equal to the sediment discharge into the section from upstream, at which time equilibrium conditions have been reached.

Laursen (1958a, 1960) developed expressions for both live-bed and clear-water contraction scour, assuming that the contraction is long so that the approach flow and the contracted flow both can be considered uniform. The live-bed case is considered first with reference to Figure 10.23, which shows the limiting case of the contracted section formed by the abutments set at the banks of the main channel in an idealized sketch. The approach main channel width is B_1, and the main channel in the contracted section has a width of B_2. The approach channel discharge is Q_c, and the overbank discharge is Q_0. From the continuity equation, it must be true that the total discharge in the contracted section $Q_T = Q_c + Q_0$. In addition, equilibrium of the live-bed scour process is reached when sediment continuity is satisfied; that is, assuming sediment transport occurs only in the main channel, we have

$$C_{t1}Q_c = C_{t2}Q_T \qquad (10.81)$$

in which C_{t1} = the mean sediment concentration in the approach section and C_{t2} = the mean sediment concentration in the contracted section. Laursen applied his total sediment discharge formula (Laursen 1958b) given by

$$C_{t,\,ppm} = (1 \times 10^4)\left[\frac{d_{50}}{y_0}\right]^{7/6}\left(\frac{\tau_0'}{\tau_c} - 1\right)f\left(\frac{u_*}{w_f}\right) \qquad (10.82)$$

in which C_t is the total sediment concentration in parts per million (ppm) by weight; d_{50} = median grain size; y_0 = depth of uniform flow; τ_0' = grain shear stress;

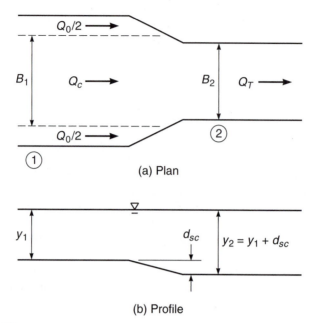

(a) Plan

(b) Profile

FIGURE 10.23
Scour in an idealized long contraction (Laursen 1958a).

τ_c = critical shear stress; and $f(u_*/w_f)$ is a specified graphical function of the ratio of shear velocity, u_*, to fall velocity, w_f, which Laursen determined from laboratory data. The ratio of grain shear stress to critical shear stress is evaluated from the Manning and Strickler equations and from the critical value of the Shields parameter, τ_{*c}, to yield

$$\frac{\tau_0'}{\tau_c} = \frac{c_n^2 g}{K_n^2 \tau_{*c}} \left[\frac{V^2}{(SG - 1)gy_0^{1/3}d_{50}^{2/3}} \right] \tag{10.83}$$

in which c_n = the constant in the Strickler equation; g = gravitational acceleration; K_n = the Manning equation constant = 1.49 in English units and 1.0 in SI units; V = mean velocity; and SG = specific gravity of the sediments, with all other variables defined as in the previous equation. Laursen applied English units and used τ_{*c} = 0.039, SG = 2.65, and c_n = 0.034 in English units (0.041 in SI units) to give

$$\frac{\tau_0'}{\tau_c} = \frac{V^2}{120y_0^{1/3}d_{50}^{2/3}} = \frac{Q^2}{120B^2 y_0^{7/3}d_{50}^{2/3}} \tag{10.84}$$

which is specific to English units. Furthermore, the shear velocity also is evaluated from Manning's equation to give

$$u_* = \sqrt{gy_0 S} = \frac{n\sqrt{gQ}}{K_n By_0^{7/6}} \tag{10.85}$$

Then, assuming that $\tau'_0/\tau_c \gg 1$, and that $f(u_*/w_f)$ is a power function $= k_p(u_*/w_f)^a$, the sediment transport formula for C_t (Equation 10.82) is substituted into Equation 10.81 along with Equations 10.84 and 10.85 to produce

$$\frac{y_2}{y_1} = \left(\frac{Q_T}{Q_c}\right)^{6/7} \left(\frac{B_1}{B_2}\right)^{\frac{6}{7}\frac{2+a}{3+a}} \left(\frac{n_2}{n_1}\right)^{\frac{6}{7}\frac{a}{3+a}} \tag{10.86}$$

The values of a are the exponent in the power fit to the graphical function of u_*/w_f and have the values $a = 0.25$ for $u_*/w_f < 0.5$; $a = 1$ for $0.5 < u_*/w_f < 2$; and $a = 2.25$ for $u_*/w_f > 2$. These ranges in u_*/w_f correspond to transport modes of mostly bed load, mixed load, and mostly suspended load, respectively.

The ratio of n values is assumed to be close to unity and so is neglected. Then special cases of Equation 10.86 can be identified. For an overbank contraction in which $B_1 = B_2$, the result for live-bed contraction scour is

$$\frac{y_2}{y_1} = \left(\frac{Q_T}{Q_c}\right)^{6/7} \tag{10.87}$$

while for a main channel contraction in which $Q_T = Q_c$, the equation for contraction scour becomes

$$\frac{y_2}{y_1} = \left(\frac{B_1}{B_2}\right)^{p_1} \tag{10.88}$$

in which p_1 has values of 0.59 (bed load), 0.64 (mixed load), or 0.69 (suspended load). Finally, Laursen assumed that, at the end of scour, both the change in velocity head and the friction loss from section 1 to section 2 were small, so that the energy equation reduces to $y_2/y_1 = d_{sc}/y_1 + 1$, in which $d_{sc} = $ depth of contraction scour as shown in Figure 10.23. It is interesting to observe that live-bed contraction scour for the overbank contraction, as given by (10.87), is independent of the mode of sediment tranport, while for main-channel contraction only, the mode of sediment transport makes some difference in the exponent p_1.

The clear-water contraction scour formula also can be derived from the long contraction theory as described by Laursen (1963) for relief bridge scour. Following a simplification of the derivation as presented in HEC-18 (Richardson and Davis 2001), the value of τ_0 is set equal to τ_c at the contracted section (2) at equilibrium when the sediment transport rate out of the contracted section approaches zero. Then Equation 10.83 is solved for depth y_2 and divided by depth y_1 to yield

$$\frac{y_2}{y_1} = \left(\frac{d_{sc}}{y_1} + 1\right) = \left(\frac{c_n^2 g}{K_n^2}\right)^{3/7} \left[\frac{q_2^2}{\tau_{*c}(SG - 1)g y_1^{7/3} d_{50}^{2/3}}\right]^{3/7} \tag{10.89}$$

in which $y_2 = $ depth after scour in the contracted section; $y_1 = $ depth before scour in the contraction; $d_{sc} = $ contraction scour depth; $q_2 = Q/B_2$; $B_2 = $ contracted width; $g = $ gravitational acceleration; $d_{50} = $ median sediment grain size; and $SG = $ specific gravity of the sediment. The coefficient in front of the square brackets has the same value in SI or English units, depending on the choice of the Strickler constant, c_n; and so Equation 10.89 is expressed in nondimensional form. For

$c_n = n/d_{50}^{1/6} = 0.0340$ in English units or 0.0414 in SI units, for example, $(c_n^2 g/K_n^2)^{3/7} = 0.174$. The value of τ_{*c} was taken equal to 0.039 by Laursen, but other values can be substituted into Equation 10.89.

Guidance is provided in HEC-18 (Richardson and Davis 2001) for the application of the contraction scour equations. The first step is to determine if live-bed or clear-water scour is occurring by comparing approach velocities with the critical velocity, which can be determined as described in a previous section. If there is an overbank contraction, heavy vegetation on the floodplain may prevent sediment transport and so the case may be one of clear-water scour even though the sediment itself has a critical velocity less than the floodplain velocity, based on sediment size alone. This often is the case for relief bridges on the floodplain. Significant backwater caused by the bridge can reduce velocities upstream so that what otherwise may have been live-bed conditions can be changed to clear-water scour in the contraction. Furthermore, if the value of u_*/w_f is very large, the incoming sediment discharge is likely to be washed through the contraction as suspended load only, and so in reality, this is a case of clear-water scour because there is no interaction between the sediment being scoured out of the contraction and the incoming sediment load.

For overbank contractions or main channel contractions with no setback of the abutments from the banks of the main channel, the application of either the live-bed or clear-water scour equations is relatively straightforward. For significant setback distances, separate contraction scour computations should be made for the main channel and the setback overbank areas, with the flow distribution between main channel and overbank area in the bridge contraction estimated by WSPRO, for example. If the setbank distance is less than three to five flow depths, it is likely that contraction scour and local abutment scour occur simultaneously and are not independent (Richardson and Davis 2001). This case will be considered further in the discussion of abutment scour.

Local Scour

Local scour around bridge piers and abutments is caused by obstruction and separation of the flow with attendant generation of a system of vortices. As the flow approaches a bridge pier, for example, the boundary layer encounters an adverse pressure gradient that causes flow separation at the bed and rolling up of the flow into a spiral horseshoe vortex with two legs that wrap around the base of the pier as shown in Figure 10.24. There is a stagnation line on the front of the pier with decreasing pressure downward due to the lower velocities near the bed. This causes a downflow directed toward the bed that feeds flow from the outer part of the boundary layer into the horseshoe vortex system. Above the downward flow is a surface roller of opposite rotation that can interact with the horseshoe vortex for shallow depths of flow. In reality, this picture of events is more complicated by the fact that it is not steady but rather oscillates in time as multiple vortices combine and then degenerate to form again as a horseshoe vortex system in front of the pier to create a broad range of turbulence scales that control the sediment transport rate and consequently lead to the growth of a scour hole in front of and around the sides of the pier (Dargahi 1989). In addition, there are wake vortices with vertical axes

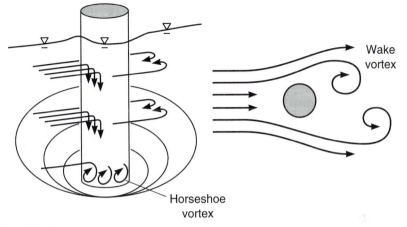

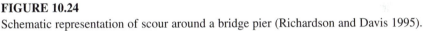

FIGURE 10.24
Schematic representation of scour around a bridge pier (Richardson and Davis 1995).

downstream of the pier, which carry sediment out of the scour zone. Both the primary horseshoe vortex and the wake vortices are illustrated in Figure 10.24 albeit as a steady-state representation of a very complex flow process that exhibits repeating patterns of large-scale unsteadiness, or *coherent turbulent structure*.

Either clear-water or live-bed local scour can occur just as for contraction scour. The main difference is in the time scale required to reach equilibrium scour. Because clear-water scour tends to occur in coarser bed material, it takes longer to reach equilibrium, as shown in Figure 10.25 for a bridge pier. The approach to equilibrium is asymptotic, so maximum clear-water scour depth is considered to occur when further changes in the bed elevations are negligibly small. Live-bed scour occurs more rapidly as shown in Figure 10.25 and tends to oscillate around the equilibrium depth due to passage of bed forms through the scour hole. Scour depth is shown as a function of the ratio of approach flow velocity to the critical velocity for initiation of sediment motion, V_1/V_c, in Figure 10.26. For $V_1/V_c < 1$, no sediment transport is occurring upstream of the pier for clear-water scour, while for $V_1/V_c > 1$ sediment is transported through the scour hole for the case of live-bed scour. A peak in scour depth occurs at $V_1/V_c = 1$, which is called the threshold peak. As V_1/V_c increases beyond the threshold peak, the scour depth begins to decrease as sediment from upstream is carried through the scour hole and creates bed forms. Continuing increases in V_1/V_c lead to an increase in scour depth to the live-bed peak, which corresponds to washing out of the bed forms and a plane bed (Sheppard and Miller 2006; Melville 2008).

Pier scour

Scour depth at a pier is a function of pier geometry, flow variables, fluid properties, and sediment properties:

$$d_s = f_1(K_s, K_\theta, b, V_1, y_1, g, \rho, \mu, (\rho_s - \rho), d_{50}, \sigma_g) \tag{10.90}$$

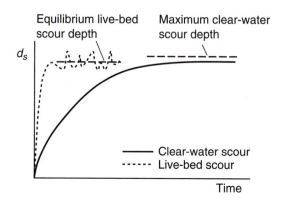

FIGURE 10.25
Illustration of the development of pier scour depth, d_s, with respect to time *(adapted from Sheppard and Miller 2006; Melville and Chiew 1999).*

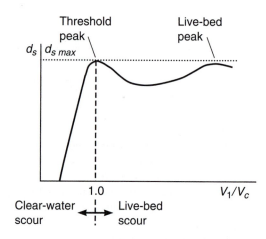

FIGURE 10.26
Illustration of clear-water and live-bed scour depth, d_s, at a pier with respect to the ratio of approach flow velocity V_1 to critical velocity V_c for sediment of uniform size *(adapted from Melville 1997, 2008).*

in which K_s = pier shape factor = 1.0 for cylindrical piers; K_θ = pier alignment factor; b = pier width; V_1 = approach velocity; y_1 = approach depth; g = gravitational acceleration; ρ = fluid density; ρ_s = sediment density; μ = fluid viscosity; d_{50} = median sediment size; and σ_g = geometric standard deviation of sediment size distribution. Choosing ρ, V_1, and b as repeating variables and carrying out the dimensional analysis results in

$$\frac{d_s}{b} = f_2\left[K_s, K_\theta, \frac{y_1}{b}, \frac{V_1}{\sqrt{gy_1}}, \frac{\rho V_1 b}{\mu}, \frac{\rho_s - \rho}{\rho}, \frac{y_1}{d_{50}}, \sigma_g\right] \qquad (10.91)$$

Combining the Froude number, $V_1/(gy_1)^{0.5}$, with $(\rho_s - \rho)/\rho$ and y_1/d_{50} results in the sediment number $\mathbf{N}_s = V_1/[(\mathrm{SG} - 1)gd_{50}]^{0.5}$, which can be replaced by V_1/u_{*c} from the Shields diagram in the absence of viscous effects ($\tau_{*c} =$ constant). Furthermore, it is apparent from the Keulegan equation written for critical velocity that y_1/d_{50} can be expressed in terms of V_c/u_{*c}, in which V_c is the critical velocity. Finally, the ratio of V_1/u_{*c} to V_c/u_{*c} just gives a sediment mobility parameter V_1/V_c. Thus, an alternative choice for the relative submerged density is the sediment mobility parameter. In addition, y_1/d_{50} can be replaced by b/d_{50}. With these substitutions and neglecting viscous effects, the result is

$$\frac{d_s}{b} = f_3\left[K_s, K_\theta, \frac{y_1}{b}, \frac{V_1}{\sqrt{gy_1}}, \frac{V_1}{V_c}, \frac{b}{d_{50}}, \sigma_g \right] \tag{10.92}$$

Most pier scour equations can be placed in this form, but some of the independent variables are neglected in all of them (Ettema, Melville, and Barkdoll 1998).

The pier scour equation recommended by the Federal Highway Administration in HEC-18 (Richardson and Davis 2001) is the Colorado State University (CSU) formula (Richardson, Simons, and Julien 1990) given by

$$\frac{d_s}{b} = 2.0K_sK_\theta K_bK_a\left(\frac{y_1}{b}\right)^{0.35} \mathbf{F}_1^{0.43} \tag{10.93}$$

in which $K_s =$ pier shape factor; $K_\theta =$ pier skewness factor; $K_b =$ correction factor for bed condition; $K_a =$ bed armoring factor; $y_1 =$ approach depth directly upstream of the pier; $b =$ pier width; and $\mathbf{F}_1 =$ approach flow Froude number. Equation 10.93 is based on laboratory data and recommended for both live-bed and clear-water scour. The value of $K_s = 1.0$ for round nose, cylindrical, and groups of cylindrical piers, while it has the value of 1.1 for square-nose piers and 0.9 for sharp-nose piers. The skewness factor is expressed as a function of the angle of attack, θ, of the flow direction relative to the longitudinal axis of the pier:

$$K_\theta = \left(\cos\theta + \frac{L_p}{b}\sin\theta \right)^{0.65} \tag{10.94}$$

in which $L_p =$ length of the pier and $b =$ width of pier. The maximum value of L_p/b is taken to be 12, even if the actual value exceeds 12. The value of $K_\theta = 1.0$ for $\theta = 0.0$, but it can be significantly different from unity. For $L_p/b = 4$, for example, and $\theta = 30°$, $K_\theta = 2.0$. Therefore, piers should be aligned with the flow direction during flood conditions. For attack angles greater than 5°, K_θ dominates K_s, which is taken to be 1.0 for this case. The value of K_b reflects the presence or absence of bed forms and so is related to maximum clear-water vs. live-bed scour. The value of $K_b = 1.1$ for clear-water scour and for live-bed scour with plane bed, antidunes, and small dunes ($0.6 < \Delta < 3.0$ m). For dune heights Δ from 3.0 to 9.0 m, $K_b = 1.1$ to 1.2, while for Δ greater than 9.0 m, $K_b = 1.3$. Finally, the armoring correction factor is defined by

$$K_a = [1.0 - 0.89(1.0 - V_R)^2]^{0.5} \tag{10.95}$$

in which $V_R = (V_1 - V_i)/(V_{c90} - V_i)$; V_1 = approach velocity in meters per second (m/s); V_{c90} = critical velocity for d_{90} bed material size in m/s; and V_i = approach velocity in m/s when sediment grains begin to move at the pier. The value of V_i in m/s is calculated from

$$V_i = 0.645 \left[\frac{d_{50}}{b} \right]^{0.053} V_{c50} \tag{10.96}$$

where V_{c50} = critical velocity for d_{50} bed material size in m/s. The factor K_a applies only for $d_{50} \geq 60$ mm. It has a minimum value of 0.7 and a maximum value of 1.0 when $V_R > 1.0$.

Laursen and Toch (1956) measured pier scour in the laboratory for conditions of live-bed scour around cylindrical piers with a subcritical approach flow and bed-load transport of sediment. They argued that neither the approach velocity nor the sediment size affected their results for depth of scour because a change in either one simply caused a proportional change in sediment transport rate both into and out of the scour hole to set up a new equilibrium in transport rate with essentially the same scour depth. The resulting pier-scour formula as given by Jain (1981) is

$$\frac{d_s}{b} = 1.35 \left[\frac{y_1}{b} \right]^{0.3} \tag{10.97}$$

The experiments covered the range of $1 \leq y_1/b \leq 4.5$, and d_{50} from 0.44 to 2.25 mm (medium to very coarse sand).

Jain (1981) proposed a formula for maximum clear-water scour around cylindrical piers that includes an effect of sediment size. In the dimensional analysis of Equation 10.92, $V_1/V_c = 1$ at maximum clear-water scour so that the Froude number $\mathbf{F}_1 = \mathbf{F}_c = V_c/(gy_1)^{0.5}$, which is the critical value of the Froude number calculated from the critical velocity evaluated from Keulegan's equation and the Shields diagram, as described previously. The resulting formula is based on the experimental data of Shen, Schneider, and Karaki (1969) and Chabert and Engeldinger (1956). It is given by

$$\frac{d_s}{b} = 1.84 \left[\frac{y_1}{b} \right]^{0.3} \mathbf{F}_c^{0.25} \tag{10.98}$$

The exponent on y_1/b is the same as for the Laursen and Toch formula. The range in y_1/b of the data varied from 0.7 to 7.0, while the mean sediment sizes of the data were between 0.24 and 3.0 mm. Equation 10.98 provides an upper envelope for the data.

Jain and Fischer (1980) investigated live-bed pier scour around cylindrical piers at high velocities. They measured the scour depths around piers in a flume using threads placed vertically in the sediment bed prior to scour. At the end of scour, the threads were excavated and the scour depth was measured at the elevation at which the threads were bent over. This procedure was intended to avoid the bias caused by partial infilling of the scour hole when the experimental flow was stopped. The resulting scour formula is similar to Equation 10.98, except that the scour depth is related to the excess Froude number $(\mathbf{F}_1 - \mathbf{F}_c)$, because the formula applies only to the live-bed case. The results showed a slight decrease in scour depth after maximum clear-water scour followed by increases in scour depth with increases in $(\mathbf{F}_1 - \mathbf{F}_c)$. The live-bed scour formula is

$$\frac{d_s}{b} = 2.0\left[\frac{y_1}{b}\right]^{0.5}[\mathbf{F}_1 - \mathbf{F}_c]^{0.25} \tag{10.99}$$

which provides an envelope of the data. Most of the data had y_1/b values of either 1 or 2 with three data points in the range of 4 to 5. Sediment sizes varied from 0.25 to 2.5 mm.

Melville and Sutherland (1988) and Melville (1997) developed an empirical pier scour equation based on a large number of laboratory experiments at the University of Auckland, New Zealand. It has the form

$$\frac{d_s}{b} = K_s K_\theta K_I K_y K_d K_\sigma \tag{10.100}$$

in which K_s and K_θ are the shape and skewness correction factors as before; K_I = expression for effect of flow intensity; K_y = expression for effect of flow depth; K_d = expression for effect of sediment size; and K_σ = expression for effect of sediment gradation. Raudkivi and Ettema (1983) showed that, for clear-water scour, sediment gradation caused a large reduction in scour depth due to armoring for $\sigma_g > 1.3$. However, Melville and Sutherland (1988) presented a method for accounting for sediment gradation effects by defining an armor velocity $V_a > V_c$ at which live-bed scour begins. The value of V_a is calculated as 0.8 V_{ca} in which V_{ca} is the critical velocity of the coarsest armor size given by $d_{max}/1.8$, where d_{max} is some representative maximum grain size in the sediment mixture. Then, the flow intensity expression, K_I, is evaluated from

$$K_I = 2.4\left[\frac{V_1 - (V_a - V_c)}{V_c}\right] \quad \text{if} \quad \frac{V_1 - (V_a - V_c)}{V_c} < 1 \tag{10.101a}$$

$$K_I = 2.4 \quad \text{if} \quad \frac{V_1 - (V_a - V_c)}{V_c} \geq 1 \tag{10.101b}$$

These expressions for K_I have the effect of collapsing the scour data for both uniform and nonuniform sediments in both clear-water and live-bed scour. For uniform sediments, $V_a = V_c$, so that the determining sediment mobility factor is $V_1/V_c < 1$ for clear-water scour. For nonuniform sediments, it must be true that $V_a > V_c$; otherwise, V_a is set equal to V_c. The depth effect, which is due to interaction of the surface roller and the downflow on the upstream face of the pier (Raudkivi and Ettema 1983), is accounted for by

$$K_y = 0.78\left(\frac{y_1}{b}\right)^{0.255} \quad \text{if} \quad \frac{y_1}{b} < 2.6 \tag{10.102a}$$

$$K_y = 1.0 \quad \text{if} \quad \frac{y_1}{b} \geq 2.6 \tag{10.102b}$$

The sediment size effect depends on the value of b/d_{50}, as given by

$$K_d = 0.57 \log\left[\frac{2.24b}{d_{50}}\right] \qquad \text{if} \quad \frac{b}{d_{50}} < 25 \qquad (10.103a)$$

$$K_d = 1.0 \qquad \text{if} \quad \frac{b}{d_{50}} \geq 25 \qquad (10.103b)$$

For nonuniform sediments, d_{50} is replaced by the armor sediment size, $d_{max}/1.8$. The maximum possible value of d_s/b is 2.4, and this formulation provides an upper envelope to the scour data. Experiments reported by Sheppard and Miller (2006) show decreasing values of scour depth at very large values of b/d_{50} that are more typical of the field. The data range for the Melville and Sutherland method includes sediment sizes from 0.24 to 5.24 mm, y_1/b values from 0.7 to 12, and V_1/V_c values between 0.4 and 5.2 (Melville 1997). Slight changes in the depth expression K_y were made by Melville (1997) to include wide piers ($y_1/b < 0.2$) as well as inter-mediate width and narrow piers.

Froehlich (1988) completed a regression analysis of live-bed scour at bridge piers at some 23 field sites. He presented a best-fit relationship given by

$$\frac{d_s}{b} = 0.32 K_s K_\theta \left[\frac{y_1}{b}\right]^{0.46} \mathbf{F}_1^{0.20} \left[\frac{b}{d_{50}}\right]^{0.08} \qquad (10.104)$$

in which K_s = pier shape factor; K_θ = skewness factor = $(b'/b)^{0.62}$; $b' = b \cos\theta + L_p \sin\theta$; b = pier width; L_p = pier length; y_1 = depth of approach flow; $\mathbf{F}_1$ = Froude number of approach flow; and d_{50} = median grain size. The skewness factor essentially is the same as Equation 10.94 used in the CSU formula. The power on b/d_{50} is very small, indicating a relatively minor influence. The coefficient of determination of Equation 10.104 is 0.75. Froehlich recommended an envelope curve obtained by adding a factor of safety of 1.0 to the right hand side of Equation 10.104.

Comparisons between several pier scour formulas and laboratory and field data have been made by Jones (1983), Johnson (1995), and Landers and Mueller (1996). Jones concluded that the CSU formula enveloped all of the laboratory and field data tested, but it gives smaller estimates of scour depth than the Laursen and Toch, Jain and Fischer, and Melville and Sutherland formulas at low values of the Froude number. Johnson found that all four of these scour formulas have high values of bias (ratio of predicted to measured scour depth) for $y_1/b < 1.5$, with high values of the coefficient of variation (COV) as well. For $y_1/b > 1.5$, the CSU formula performed well with a low value of COV and a bias from 1.5 to 1.8, providing a reasonable factor of safety. In general, the Melville and Sutherland formula overpredicted more than any of the formulas tested with bias values varying from 2.2 to 2.9 for $y_1/b > 1.5$, for example. Landers and Mueller (1996) evaluated pier-scour formulas on the basis of a much more extensive data set of 139 field pier-scour measurements from 90 piers at 44 bridges obtained during high-flow conditions. Data were separated into live-bed scour and clear-water scour measurements. Although the data showed considerable scatter, it was concluded that the influence of flow depth on scour depth did not become insignificant at large values of the ratio of flow depth to pier width. In addition, no influence of the Froude number and only a very weak influence of sediment size were found. Both the HEC-18 and Froehlich scour formulas performed as conservative design equations in general, but overpredicted the scour by large amounts for some cases.

Abutment scour

Melville (1992, 1997) proposed an abutment scour formula that is similar in form to the Melville and Sutherland pier scour formula, arguing that short abutments behave like piers. The abutment scour formula is given by

$$d_s = K_{yL} K_I K_d K_s K_\theta K_G \tag{10.105}$$

in which K represents expressions accounting for various influences on scour depth: K_{yL} = depth-size effect; K_I = flow intensity effect; K_d = sediment size effect; K_s = abutment shape factor; K_θ = skewness or alignment factor; and K_G = channel geometry factor. The depth-size factor is defined by the following expressions:

$$K_{yL} = 2L_a; \qquad L_a/y_1 \leq 1 \tag{10.106a}$$

$$K_{yL} = 2\sqrt{y_1 L_a}; \qquad 1 < \frac{L_a}{y_1} < 25 \tag{10.106b}$$

$$K_{yL} = 10y_1; \qquad \frac{L_a}{y_1} \geq 25 \tag{10.106c}$$

in which y_1 = approach flow depth and L_a = embankment or abutment length. These expressions indicate that scour depth is independent of depth for short abutments ($L_a/y_1 < 1$) and independent of abutment length for long abutments ($L_a/y_1 > 25$). The flow intensity factor essentially is the same as for piers, except that the maximum value of $d_s/b = 2.4$ for piers has been removed to give

$$K_I = \frac{V_1 - (V_a - V_c)}{V_c} \qquad \text{for} \qquad \frac{V_1 - (V_a - V_c)}{V_c} < 1 \tag{10.107a}$$

$$K_I = 1 \qquad \text{for} \qquad \frac{V_1 - (V_a - V_c)}{V_c} \geq 1 \tag{10.107b}$$

in which V_a = armor velocity defined in the same way as for piers; V_c = critical velocity; and V_1 = velocity in the bridge approach section. The sediment size factor, K_d, is the same as for piers, as expressed by Equations 10.103, except that the pier width, b, is replaced by the abutment length, L_a. The abutment shape factor is assumed to be 1.0 for vertical-wall abutments and 0.75 for wing-wall abutments. Spill-through abutments are assigned values of K_s = 0.6, 0.5, and 0.45 for 0.5:1 ($H:V$), 1:1, and 1.5:1 side slopes, respectively. These values of shape factor apply only to shorter abutments, for which $L_a/y_1 \leq 10$. Shape effects were found to be unimportant for longer abutments, so that $K_s = 1.0$ for $L_a/y_1 \geq 25$. For abutment lengths between these two extremes, a linear interpolation was suggested:

$$K_s^* = K_s + 0.667(1 - K_s)\left(0.1\frac{L_a}{y_1} - 1\right) \qquad \text{for} \quad 10 < \frac{L_a}{y_1} < 25$$

(10.108)

in which K_s represents the shape factor for short abutments, and K_s^* is the interpolated value for intermediate length abutments. Values of K_θ for flow alignment and K_G for abutments that protrude into the main channel from the floodplain are given by Melville (1997).

Froehlich (1989) applied regression analysis to a laboratory data set for live-bed abutment scour from several investigators to produce the relationship

$$\frac{d_s}{y_1} = 2.27 K_s K_\theta \left[\frac{L_a}{y_1}\right]^{0.43} \mathbf{F}_1^{0.61} + 1$$

(10.109)

in which d_s = local abutment scour depth; y_1 = approach flow depth; K_s = abutment shape factor (1.0 for vertical wall, 0.82 for wing-wall, and 0.55 for spill-through); K_θ = skewness factor; L_a = abutment length; and $\mathbf{F}_1$ = approach flow Froude number. A factor of safety of 1.0 is added to the right-hand side of Equation 10.109. Froehlich calculated the approach Froude number based on an average velocity and depth in the area obstructed by the embankment and abutment in the approach flow cross section. All the experimental results in the regression analysis came from experiments in rectangular flumes.

Richardson and Richardson (1998) argued that experimental results for rectangular flumes that depend on abutment length as an independent variable do not accurately reflect the abutment scour process for compound channels, which have a nonuniform velocity and discharge distribution across the channel. Sturm and Janjua (1994) demonstrated that a discharge contraction ratio, M, represents the redistribution of flow between main channel and floodplain as the flow passes through the bridge contraction. As shown in Figure 10.27, the discharge contraction ratio, M, is defined by

$$M = \frac{Q - Q_{\text{obst1}}}{Q}$$

(10.110)

in which Q_{obst1} = obstructed floodplain discharge over a length equal to the abutment length projected onto the approach cross section; and Q = total discharge through the bridge opening for an abutment on one side only, as in Figure 10.27, or Q = total discharge from the outer edge of the floodplain to the centerline of the main channel for abutments on both sides of the main channel. The variable $(1-M)$ was proposed by Kindsvater and Carter (1954) to characterize the backwater effect of a bridge obstruction, and it is used in the FHWA/USGS program WSPRO to calculate bridge backwater height (see Chapter 6). Sturm and Janjua (1994) showed

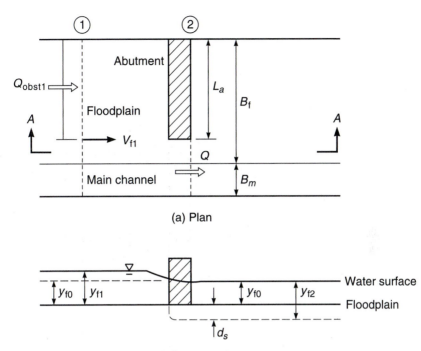

FIGURE 10.27
Definition sketch for idealized abutment scour in a compound channel (Sturm 1999b).

that M is approximately equal to the ratio of discharges per unit of width in the approach and contracted floodplain areas, q_{f1}/q_{f2}, for an abutment that terminates on the floodplain.

With reference to Figure 10.27, the idealized long contraction scour is formulated first, followed by equating the local abutment scour to some multiplier of the contraction scour as originally proposed by Laursen (1963). In two different compound channel geometries, Sturm and Sadiq (1996) and Sturm (1999a, 1999b, 2006) have shown that this approach to the problem results in a clear-water abutment scour equation given by

$$\frac{d_s}{y_{f0}} = 8.14K_s\left[\frac{q_{f1}}{MV_{0c}y_{f0}} - 0.4\right] + 1 \tag{10.111}$$

in which d_s = local clear-water abutment scour; y_{f0} = floodplain depth for uncontracted flow; K_s = abutment shape factor; q_{f1} = approach discharge per unit width in the floodplain = V_f1y_f1; M = discharge contraction ratio; V_{0c} = critical velocity in the floodplain at the unconstricted depth y_{f0} for setback abutments and critical velocity in the main channel at the unconstricted depth in the main channel for bankline abutments. The factor of 1 on the right hand side of Equation 10.111 is a factor of safety. If the approach floodplain velocity V_{f1} exceeds the critical value V_{1c}, then V_{f1} is set equal to V_{1c} for maximum clear-water scour. The shape factor $K_s = 1$ for vertical-wall abutments, while for spill-through abutments, it is given by

$$K_s = 1.52 \frac{\xi - 0.67}{\xi - 0.4} \qquad \text{for} \quad 0.67 \le \xi \le 1.2 \qquad (10.112)$$

where $\xi = q_{f1}/(MV_{0c}y_{f0})$, and $K_s = 1.0$ for $\xi > 1.2$ as the contraction effect becomes more important than the abutment shape. Equation 10.111 is compared with the experimental data for an asymmetric compound channel having a flood-plain width of 3.66 m and a main channel width of 0.55 m in Figure 10.28, which shows that d_s/y_{f0} has a maximum value of 10. The r^2 value for the best-fit equation without a factor of safety is 0.86 with a standard error of 0.74 in d_s/y_{f0}.

The success of clear-water abutment scour prediction formulas such as (10.105) or (10.111) depends upon an accurate characterization of both the hydraulic conditions causing the scour and the critical velocity of the sediment that is subject to scour. These variables may be difficult to measure in the field. Furthermore, geotechnical failure of an erodible embankment may modify the maximum abutment scour depth so that it is less than predicted by formulas based on experiments with rigid abutments having sheet pile foundations (Ettema et al. 2008).

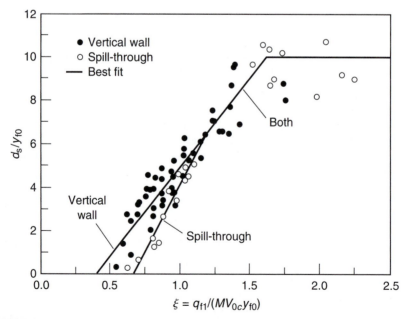

FIGURE 10.28
Abutment scour relationship for compound channels (Sturm 1999b, Sturm 2006).

Example 10.5

A bridge with a 228.6 m (750 ft) opening length spans Burdell Creek, which has a drainage area of 971 km^2 (375 mi^2). The exit and bridge cross sections are shown in Figure 10.29 with three subsections and their corresponding values of Manning's n. The slope of the stream reach at the bridge site is constant and equal to 0.001 m/m. The bridge has a deck elevation of 6.71 m (22.0 ft) and a bottom chord elevation of 5.49 m (18.0 ft). It is a Type 3 bridge (see Chapter 6) with 2:1 abutment and embankment slopes, and it is perpendicular to the flow direction (no skew). The tops of the left and right spill-through abutments are at X stations of 281.9 m (925 ft) and 510.5 m (1675 ft), and the abutments are set back from the banks of the main channel. There are six cylindrical bridge piers, each with a width of 1.52 m (5.00 ft). The sediment has a median grain diameter, d_{50}, of 2.0 mm (6.56 × 10^{-3} ft). Estimate the clear-water abutment scour and pier scour for the 100 yr design flood, which has a peak discharge of 397 m^3/s (14,000 cfs).

Solution.   The WSPRO option in HEC-RAS, described in Chapter 6, is run to obtain the hydraulic variables needed in the scour prediction formulas. The program actually is run twice, first to obtain the water surface elevations for the unconstricted flow at the approach cross section and, second, with the bridge in place for the constricted flow. The velocity distributions in the approach section for the unconstricted (undisturbed) water surface elevation of 4.038 m (13.25 ft) and the constricted water surface elevation of 4.157 m (13.64 ft) are computed by HEC-RAS by taking the ratio of conveyance in specified subsection slices to the total conveyance.

The scour parameters are determined from the WSPRO results. Calculations are made for the left abutment, which has a length, L_a, of 233 m (764 ft). From the computed velocity

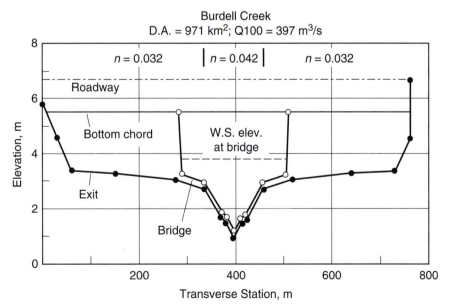

FIGURE 10.29
Bridge cross sections for Example 10.5.

distribution for the constricted flow, the blocked discharge in the approach section for the left abutment is 39.1 m³/s (1380 cfs) with a blocked cross-sectional area of 106.8 m² (1150 ft²). The discharge from the left edge of water in the approach cross section to the centerline of the main channel is 210 m³/s (7413 cfs). Then, the value of $M =$ (210 − 39.1)/210 = 0.81. Now we can calculate $V_{f1} = Q_{f1}/A_{f1}$ = 39.1/106.8 = 0.366 m/s (1.20 ft/s) and $y_{f1} = A_{f1}/L_a$ = 106.8/233 = 0.458 m (1.50 ft). In a similar way, the value of y_{f0} is found for the unconstricted cross section to be 0.357 m (1.17 ft).

The critical velocities for coarse sediments are determined by substituting into Equation 10.17 (Keulegan's equation). For the constricted approach section and for a floodplain depth of 0.458 m, we have

$$V_{f1c} = 5.75 \times \sqrt{(0.045)(2.65 - 1)(9.81)(0.002)} \times \log \frac{(12.2)(0.458)}{2 \times 0.002}$$

$$= 0.69 \text{ m/s } (2.3 \text{ ft/s})$$

in which the Shields parameter has been taken to be 0.045 for this sediment size and the equivalent sand-grain roughness $k_s = 2d_{50}$. Because $V_{f1} < V_{f1c}$, it is apparent that we have clear-water scour. In a similar manner, the value of V_{f0c} for an unconstricted floodplain depth of 0.357 m (1.17 ft) is 0.67 m/s (2.2 ft/s).

To compute the scour depth for the setback abutments, substitute into Equation 10.111 to obtain

$$\frac{d_s}{y_{f0}} = 8.14 \times 0.63 \times \left[\frac{(0.366)(0.458)}{(0.81)(0.67)(0.357)} - 0.4 \right] + 1.0 = 3.4$$

in which the shape factor $K_s = 0.63$ from Equation 10.112 and the safety factor of 1.0 has been included. Finally, the left abutment scour depth is 3.4 × 0.357 = 1.2 m (3.9 ft). In general, this calculation would be repeated for the right abutment, but this example has an essentially symmetric cross section.

Next, consider the scour around the bridge piers and use the largest flow depth in the cross section, assuming that the thalweg might migrate laterally. The WSPRO results give a water surface elevation of 3.80 m (12.5 ft) in the bridge section, which corresponds to a maximum depth of 2.63 m (8.63 ft). The maximum velocity in the bridge section is 1.68 m/s (5.51 ft/s). The resulting value of the pier approach Froude number is $V_1/(gy_1)^{0.5}$ = 1.68/(9.81 × 2.63)^{0.5} = 0.33. Substituting into the CSU pier scour formula and recalling that the pier width b = 1.52 m (5.0 ft), we have

$$d_s = b \times 2.0 K_s K_\theta K_b K_a \left[\frac{y_1}{b} \right]^{0.35} F_1^{0.43}$$

$$= 1.52 \times 2 \times (1.0)(1.0)(1.1)(1.0) \left[\frac{2.63}{1.52} \right]^{0.35} [0.33]^{0.43} = 2.5 \text{ m}$$

(or 8.2 ft), in which all the correction factors have the value of 1 except the bed correction, which is taken to be 1.1 for clear-water scour. The Laursen-Toch equation gives a pier scour depth of

$$d_s = b \times 1.35 \left[\frac{y_1}{b} \right]^{0.3} = 1.52 \times 1.35 \times \left[\frac{2.63}{1.52} \right]^{0.3} = 2.4 \text{ m } (7.9 \text{ ft})$$

Total Scour

It is recommended in HEC-18 (Richardson and Davis 2001) that degradation, contraction scour, and abutment or pier scour be added to produce a conservative total scour estimate. For setback abutments, contraction scour has to be calculated separately for the setback area and the main channel in the bridge section. Another conservative design suggestion is to use the calculated maximum scour depth at a pier in the main channel for a pier in the setback area as well, assuming lateral migration of the main channel into the setback area. For bankline abutments, contraction scour and abutment scour occur simultaneously rather than independently, so that adding abutment scour and contraction scour for this case may be overly conservative.

Pier and Abutment Scour Countermeasures

Scour countermeasures are methods designed either to reduce the risk of failure of a bridge foundation or to make a bridge safe from scour (Richardson and Davis 2001). In the latter case, structural modifications such as underpinning of pier and abutment foundations or altering the superstructure design to provide more bridge open area are likely to be site specific and can be quite expensive. This section focuses instead on countermeasures that are of the armoring type in that their purpose is to reduce the bed erodibility; in particular, only rock riprap protection is considered here. River training methods are a second type of countermeasure in which the flow is realigned and erosion/deposition patterns are altered using structures such as spurs and bendway weirs in order to prevent lateral migration of the river in the bridge opening. A related countermeasure is a guide bank constructed perpendicular to the highway embankment to move the area of scour away from a bridge abutment. Detailed design guidelines for several types of scour countermeasures can be found in the FHWA publication HEC-23 (Lagasse et al. 2001).

Because a method such as rock riprap protection cannot be considered a permanent countermeasure, its installation should be accompanied by inspection after each major flood event and/or the use of fixed instrumentation that can continuously monitor the integrity of the riprap protection. Sonar devices that continuously monitor scour can produce real-time data on bed elevations in the vicinity of a bridge pier or abutment during passage of a flood (Lee et al. 2004).

Design of rock riprap for mitigation or prevention of pier scour involves more than selecting the size of rock riprap because of different causes of riprap failures. Failure mechanisms under clear-water scour conditions include shear instability, winnowing, and edge failure (Chiew 1995). Shear instability refers to the case of the complex three-dimensional structure of the horseshoe vortex overcoming the resisting force afforded by the size and weight of individual stones and plucking them out of the riprap blanket. Winnowing, on the other hand, is associated with migration of finer sediment below the riprap blanket out through the interstices of the rock causing it to settle and possibly fail over a longer time period than for shear failure. Winnowing can be prevented by a combination of specifying sufficient thickness of the riprap layer possibly in combination with a geotextile filter below the riprap blanket. A thick, well-graded riprap layer can also prevent winnowing.

In edge failure, the shear at the edge of the riprap blanket moves some rocks and erodes the underlying sediment to form a trench along the sides of the riprap blanket. Some rocks roll into the trench and rearmor it, but the process can continue and lead to total failure if the riprap layer thickness and the spatial extent of the riprap protection are insufficient. In live-bed scour, a fourth failure mechanism occurs due to passage of bed forms, which cause undermining of the edges of the riprap blanket to the level of the trough of the bed forms (Chiew and Lim 2000).

Lagasse et al. (2006) tested several existing equations developed for the sizing of riprap for protection of bridge piers most of which are summarized by Melville and Coleman (2000). Based on comparisons of predicted and observed riprap failures using three sets of laboratory data, they recommended the HEC-23 equation. Its origins can be traced to the Isbash equation, which was first developed for sizing rock to be dumped in flowing water for closure of dams (Parola 1993). The equation is given in dimensionless form by

$$\frac{d_{50}}{y} = \frac{0.346(K_1 K_2 V)^2}{(SG-1)gy} = \frac{0.346(K_1 K_2)^2}{(SG-1)} \mathbf{F}^2 \tag{10.113}$$

in which d_{50} = the 50% finer size of the rock riprap; y = flow depth; V = mean cross-sectional flow velocity in the bridge section; K_1 = pier shape factor = 1.5 for round nose and 1.7 for rectangular; K_2 = velocity factor ranging from 0.9 for a pier near the bank in a straight reach to 1.7 for a pier in the main current of a river bend; SG = specific gravity of the rock; and $\mathbf{F}$ = Froude number. If the local depth-averaged velocity immediately upstream of the pier in the contracted bridge section is known, then it is used in the Froude number and K_2 = 1.0. For a cylindrical pier (K_1 = 1.5) near the channel thalweg such that K_2 = 1.2, Equation 10.113 can be rearranged into a limiting value of the critical sediment number given by

$$\mathbf{N}_{sc} = \frac{V}{\sqrt{(SG-1)gd_{50}}} \cong 1.0 \tag{10.114}$$

For comparison, see Figure 10.8 for the critical value of the sediment number for stability of rock riprap channel revetment. Equation 10.114 plots as a horizontal line in Figure 10.8 well below the stability limit for channel revetment and without any dependence on relative roughness.

An alternative equation for sizing riprap for piers is given by Lauchlan and Melville (2001) as

$$\frac{d_{50}}{y} = 0.3\left(1 - \frac{Y}{y}\right)^{2.75} \mathbf{F}^{1.2} \tag{10.115}$$

in which Y = depth of burial to the top of the riprap blanket; y = flow depth in the bridge section; and $\mathbf{F}$ = local flow Froude number in the bridge section. The equation is based on live-bed scour with bed forms in the approach flow and is limited to cylindrical piers with $Y/y < 0.6$. Equation 10.115 gives more conservative estimates of riprap size at low values of the Froude number in comparison with Equation 10.113.

Several construction issues are important in the placement of the riprap blanket. It should be placed in an excavated hole with the top of the riprap blanket coincident with the level of the river bed, and with a minimum blanket thickness of $3d_{50}$ and a lateral extent of at least two times the pier width in all directions around the pier. The riprap blanket should extend below the estimated depth of contraction scour. For live-bed scour conditions with the formation of dunes, a geotextile filter is recommended to be installed under the riprap, but it should extend a distance of only two-thirds the lateral width of the riprap blanket to avoid the possibility of dune troughs undermining the filter. More detailed guidelines on filters are available in Lagasse et al. (2006). Other types of pier scour countermeasures such as partially grouted riprap, articulated concrete block systems, gabion mattresses, and grout-filled mattresses are described in Lagasse et al. (2007).

Protection of abutments from scour using rock riprap requires a much more extensive coverage of scour-vulnerable areas than for bridge piers. Not only must the side slopes of the embankment and abutment be protected upstream and downstream, but because of the tendency for scour to occur at the toe of the slope and cause slope failure, a riprap apron around the toe of the abutment and embankment also is necessary. Recommended sizing of the rock in HEC-23 (Lagasse et al. 2001) is based on two equations developed from laboratory experiments on vertical wall and spill-through abutments (Pagán-Ortiz 1991; Atayee 1993). For Froude numbers less than or equal to 0.8, the following form of the Isbash relationship is suggested for the median rock diameter:

$$\frac{d_{50}}{y} = \frac{K}{(SG-1)} \mathbf{F}^2 \tag{10.116}$$

in which the Froude number is based on a characteristic average velocity in the contracted section; and $K = 0.89$ for a spill-through abutment and 1.02 for a vertical wall abutment. For Froude numbers greater than 0.8, the recommended rock sizing equation is

$$\frac{d_{50}}{y} = \frac{K}{(SG-1)} \mathbf{F}^{0.28} \tag{10.117}$$

in which $K = 0.61$ for spill-through abutments and 0.69 for vertical wall abutments. The characteristic contraction velocity is best determined from a two-dimensional or three-dimensional model. An approximate method is to use the average velocity through the bridge opening from HEC-RAS if the setback distance of the toe of the abutment from the top of the bank of the main channel is less than five times the main channel depth. Otherwise, an average velocity of the flow that goes through the floodplain in the contracted section is estimated from HEC-RAS.

Placement of the riprap is required all the way around the toe of the abutment up to the point of tangency with the embankment on the upstream side and the same on the downstream side plus an additional 7.5 m (25 ft) or two flow depths, whichever is larger (see Figure 10.30). Spill-through abutment slopes should be

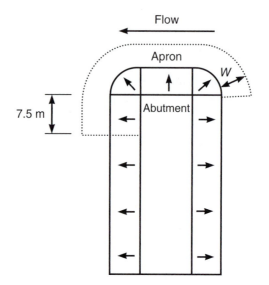

FIGURE 10.30

Riprap apron placement around spill-through abutments on the floodplain *(adapted from Lagasse et al. 2001)*.

protected to an elevation at least 0.6 m (2 ft) higher than the maximum water surface elevation for the design flood. The thickness of the rock riprap layer should be at least 1.5 times d_{50} or d_{100}, whichever is larger. Perhaps most important of all is the width of the riprap apron necessary to prevent exposure of the toe of the slope to scour. Laboratory experiments by Melville et al. (2006) on spill-through abutments in a compound channel at various setback distances from the main channel resulted in a suggested minimum width, W, of the riprap apron expressed in terms of the predicted abutment scour depth as

$$\frac{W}{y_f} = 0.5\left(\frac{d_{sf}}{y_f}\right)^{1.35} \tag{10.118}$$

in which y_f = flow depth on the floodplain; and d_{sf} = scour depth relative to the bed of the floodplain predicted from Equations 10.105 and 10.106b and adjusted by an empirical function if the maximum local scour depth is in the main channel. Larger scale laboratory experiments by Morales, Ettema, and Barkdoll (2008) on spill-through abutments indicate that for shorter aprons ($W/y_f < 1.0$), the apron width can be approximated as being equal to the estimated scour depth. For larger apron widths, the large-scale tests support the design recommendation of Lagasse et al. (2001) given by

$$W = 2y_f \tag{10.119}$$

Detailed recommendations for other types of abutment scour countermeasures can be found in Barkdoll, Ettema, and Melville (2007).

REFERENCES

Ackers, P., and W. R. White. "Sediment Transport: New Approach and Analysis." *J. Hyd. Div.,* ASCE 99, no. HY11 (1973), pp. 2041–60.

Allen, J. R. L. "Computational Methods for Dune Time Lag: Calculations Using Stein's Rule for Dune Height." *Sedimentary Geol.* 20, no. 3 (1978), pp. 165–216.

Alonso, C. V. "Selecting a Formula to Estimate Sediment Transport Capacity in Nonvegetated Channels." In *CREAMS,* ed. W. G. Knisel, Chapter 5, pp. 426–39. Conservation Research Report No. 26, U.S. Department of Agriculture, 1980.

ASCE Task Committee. "Relationships Between Morphology of Small Streams and Sediment Yields." *J. Hyd. Div.,* ASCE 108, no. HY11 (1982), pp. 1328–65.

ASCE Task Force on River Width Adjustment. "River Width Adjustment. II: Modeling." *J. Hydr. Engrg.* ASCE 124, no. 9 (1998), pp. 903–17.

Atayee, A. T. *Study of Riprap as Scour Protection for Spill-Through Abutments,* pp. 40–48. Transportation Research Record 1420. Washington, DC: Transportation Research Board, National Research Council, 1993.

Bagnold, R. A. *An Approach to the Sediment Transport Problem from General Physics.* Prof. Paper 422-I, Washington, DC: U. S. Geological Survey, 1966.

Barkdoll, B. D., R. Ettema, and B. W. Melville. *Countermeasures to Protect Bridge Abutments from Scour.* NCHRP Report 587. Washington, DC: Transportation Research Board, National Research Council, 2007.

Barton, J. K., and P. Lin. *A Study of the Sediment Transport in Alluvial Channels.* Report No. 55JRB2. Fort Collins, CO: Colorado Agricultural and Mechanical College, Dept. of Civil Engineering, 1955.

Bechteler, W., and M. Vetter. "The Computation of Total Sediment Transport in View of Changed Input Parameters." In *Proceedings of the ASCE International Symposium on Sediment Transport Modeling,* New Orleans, LA, 1989, pp. 548–53.

Biglari, B., and T. W. Sturm. "Numerical Modeling of Flow Around Bridge Abutments in Compound Channel." *J. Hydr. Engrg.,* ASCE 124, no. 2 (1998), pp. 156–64.

Brown, C. B. "Sediment Transportation." In *Engineering Hydraulics,* ed. H. Rouse, pp. 769–857. Iowa Institute of Hydraulic Research, Iowa City, Iowa, 1950.

Brownlie, W. R. "Flow Depth in Sand-Bed Channels." *J. Hydr. Engrg.,* ASCE 109, no. 7 (1983), pp. 959–90.

Brownlie, W. R. *Prediction of Flow Depth and Sediment Discharge in Open Channels.* Report No. KH-R-43A. Pasadena: W. M. Keck Laboratory, California Institute of Technology, 1981.

Buffington, J. M. "The Legend of A. F. Shields." *J. Hydr. Engrg.,* ASCE 125, no. 4 (1999), pp. 376–87.

Carstens, M. R. "Similarity Laws of Localized Scour." *J. Hydr. Div.,* ASCE 92, no. 3 (1966), pp. 13–36.

Casey, H. J. "Über Geschiebebewegung." *Mitteilungen der Preussischen Versuchsanstalt für Wasserbau und Schiffbau* [in German], Heft 19 (1935).

Chabert, J., and P. Engeldinger. *Etude des Affouillements Autour des Piles de Ponts.* Chatou, France: Laboratorie National d'Hydraulique, 1956.

Chang, H. H. *Fluvial Processes in River Engineering,* New York: John Wiley & Sons, 1988.

Chang, H. H. "Mathematical Model for Erodible Channels." *J. Hyd. Div.,* ASCE 108, no. HY5 (1982), pp. 678–89.

Chang, H. H. "Modeling of River Channel Changes." *J. Hydr. Engrg.,* ASCE 110, no. 2 (1984), pp. 157–72.

Chang, H. H. "River Channel Changes: Adjustments of Equilibrium." *J. Hydr. Engrg.,* ASCE 112, no. 1 (1986), pp. 43–55.

Chang, H. H. "River Morphology and Thresholds." *J. Hydr. Engrg.,* ASCE 111, no. 3 (1985), pp. 503–19.

Chien, N. "The Present Status of Research on Sediment Transport." *Transactions* [ASCE] 121, no. 2824 (1956), pp. 833–68.

Chiew, Y. M. "Mechanics of Riprap Failure at Bridge Piers." *J. Hydr. Engrg.,* ASCE 121, No. 9 (1995), pp. 635–43.

Chiew, Y. M., and F. H. Lim. "Failure Behavior of Riprap Layer at Bridge Piers under Live-Bed Conditions." *J. Hydr. Engrg.,* ASCE 126, No. 1 (2000), pp. 43–55.

Colby, B. R., and C. H. Hembree. *Computation of Total Sediment Discharge, Niobrara River near Cody, Nebraska.* Water Supply Paper 1357. Washington, DC: U.S. Geological Survey, 1955.

Coleman, N. L. "Flume Studies of the Sediment Transfer Coefficient." *Water Resources Research* 6, no. 3 (1970).

Dargahi, B. "Turbulent Flow Field Around a Circular Cylinder." *Exp. in Fluids,* Vol. 8, No. 1-2 (1989), pp. 1–12.

Dawdy, D. R. *Depth-Discharge Relations of Alluvial Streams—Discontinuous Rating Curves.* Water Supply Paper 1498-C. Washington, DC: U.S. Geological Survey, 1961.

Dennett, K. E., T. W. Sturm, A. Amirtharajah, and T. Mahmood. "Effect of Adsorbed Natural Organic Matter on the Erosion of Kaolinite Sediments." *Water Environment Research,* 70, no. 3 (1998), pp. 268–75.

DuBoys, P. "Le Rohne et les Rivieres a Lit Affouillable." *Annales des Pont et Chaussees,* Series 5, no. 18 (1879), pp. 141–95.

Einstein, H. A. "Formulas for the Transportation of Bed Load." *Transactions* [ASCE] 107, no. 2140 (1942), pp. 561–73.

Einstein, H. A. "The Bed Load Function for Sediment Transportation in Open Channels." Technical Bulletin 1026. Washington, DC: USDA, Soil Conservation Service, 1950.

Einstein, H. A., and N. Barbarossa. "River-Channel Roughness." *Transactions* [ASCE] 117, no. 2528 (1952), pp. 1121–46.

Einstein, H. A., and N. Chien. *Effects of Heavy Sediment Concentration near the Bed on Velocity and Sediment Distribution.* MRD Series Report No. 8. University of California at Berkeley and Missouri River Division, U.S. Army Corps of Engineers, Omaha, Neb., 1955.

Einstein, H. A., and N. Chien. *Second Approximation to the Solution of the Suspended Load Theory.* MRD Series Report No. 2. University of California at Berkeley and Missouri River Division, U.S. Army Corps of Engineers, Omaha, Neb., 1954.

Engelund, F. "Hydraulic Resistance of Alluvial Streams." *J. Hyd. Div.,* ASCE 92, no. HY2 (1966), pp. 315–26.

Engelund, F. Closure to "Hydraulic Resistance of Alluvial Streams." *J. Hyd. Div.,* ASCE 93, no. HY7 (1967), pp. 287–96.

Engelund, F., and E. Hansen. *A Monograph on Sediment Transport to Alluvial Streams.* Copenhagen: Teknik Vorlag, 1967.

Ettema, R., B. W. Melville, and B. Barkdoll. "Scale Effect in Pier-Scour Experiments." *J. Hydr. Engrg.,* ASCE 124, no. 6 (1998), pp. 639–42.

Ettema, R., T. Nakato, A. Yorozuya, and M. Muste. "Three Abutment Scour Conditions Investigated with Laboratory Flumes." *Proc. 4th International Conference on Scour and Erosion,* Japanese Technical Society, Tokyo 2008, pp. 208–13.

Froehlich, D. C. "Analysis of On-Site Measurements of Scour at Piers." In *Proceedings of the 1988 National Conference on Hydraulic Engineering,* ed. S. R. Abt and J. Gessler, pp. 534–39, Colorado Springs, ASCE, 1988.

Froehlich, D. C. "Local Scour at Bridge Abutments." In *Proceedings of National Conference on Hydraulic Engineering,* ed. M. A. Ports, pp. 13–18, New Orleans, ASCE, 1989.

Gessler, J. "Beginning and Ceasing of Sediment Motion." in *River Mechanics,* ed. H .W. Shen, pp. 7-1 to 7-22. Fort Collins, CO: Colorado State University, 1970.

Gilbert, G. K. *Transportation of Debris by Running Water.* Prof. Paper No. 86. Washington, DC: U.S. Geological Survey, 1914.

Graf, W. H. *Hydraulics of Sediment Transport.* New York: McGraw-Hill, 1971.

Guy, H. P., D. B. Simons, and E. V. Richardson. *Summary of Alluvial Channel Data from Flume Experiments, 1956–61.* Prof. Paper 462-I. U. S. Washington, DC: U.S. Geological Survey, 1966.

Holly, F. M. Jr., J. C. Yang, and M. F. Karim. *Computer-Based Prognosis of Missouri River Bed Degradation-Refinement of Computational Procedures.* IIHR Report No. 281. Iowa City: Iowa Institute of Hydraulic Research, 1984.

Hubbell, D. W., and D. Q. Matejka. *Investigations of Sediment Transportation, Middle Loup River at Dunning, Nebraska.* Water Supply Paper 1476. Washington, DC: U.S. Geological Survey, 1959.

Jain, S. C. "Maximum Clear-Water Scour Around Circular Piers." *J. Hyd. Div.,* ASCE 107, no. HY5 (1981), pp. 611–26.

Jain, S. C., and E. E. Fischer. "Scour Around Bridge Piers at High Flow Velocities." *J. Hyd. Div.,* ASCE 106, no. HY11 (1980), pp. 1827–42.

Johnson, P. A. "Comparison of Pier-Scour Equations Using Field Data." *J. Hydr. Engrg.,* ASCE 121, no. 8 (1995), pp. 626–29.

Jones, J. S. "Comparison of Prediction Equations for Bridge Pier and Abutment Scour." In *Transportation Research Record 950,* Second Bridge Engineering Conference, vol. 2. Washington, DC: Transportation Research Board, 1983.

Julien, P. Y. *Erosion and Sedimentation.* New York: Cambridge University Press, 1995.

Julien, P. Y., and G. J. Klaassen. "Sand-Dune Geometry of Large Rivers During Floods." *J. Hydr. Engrg.,* ASCE 121, no. 9 (1995), pp. 657–63.

Karim, F. "Bed Configuration and Hydraulic Resistance in Alluvial-Channel Flows." *J. Hydr. Engrg.,* ASCE 121, no. 1 (1995), pp. 15–25.

Karim, F. "Bed-Form Geometry in Sand-Bed Flows." *J. Hydr. Engrg.,* ASCE 125, no. 12 (1999), pp. 1253–61.

Karim, F. "Bed Material Discharge Prediction for Nonuniform Bed Sediments." *J. Hydr. Engrg.,* ASCE 124, no. 6 (1988), pp. 597–604.

Karim, F., and J. F. Kennedy. *Computer-Based Predictors for Sediment Discharges and Friction Factors of Alluvial Streams.* IIHR Report No. 242. Iowa City: Iowa Institute of Hydraulic Research, University of Iowa, 1981.

Karim, M. F., and J. F. Kennedy. "Menu of Coupled Velocity and Sediment Discharge Relations for Rivers." *J. Hydr. Engrg.,* ASCE 116, no. 8 (1990), pp. 978–96.

Kennedy, J. F. "Mechanics of Dunes and Antidunes in Erodible-Bed Channels." *J. Fl. Mech.,* 16, no. 4 (1963), pp. 521–44.

Kennedy, J. F. "Stationary Waves and Antidunes in Alluvial Channels." Report No. KH-R-2. Pasadena: W. M. Keck Laboratory, California Institute of Technology, 1961.

Kennedy, J. F. "The Albert Shields Story." *J. Hydr. Engrg.,* ASCE 121, no. 11 (1995), pp. 766–72.

Kennedy, J. F., and N. H. Brooks. "Laboratory Study of an Alluvial Stream at Constant Discharge." In *Proceedings of the Federal Interagency Sedimentation Conference,* Misc. Publ. No. 970, pp. 320–30. Washington, DC: Agricultural Research Service, U.S. Department of Agriculture, 1965.

Keulegan, G. H. "Laws of Turbulent Flow in Open Channels." Research Paper RP1151. *J. of Research of National Bureau of Standards* (1938).

Kindsvater, C. E., and R. W. Carter. "Tranquil Flow Through Open-Channel Constrictions." *Transactions* [ASCE] (1954), pp. 955–1005.

Kovacs, A., and G. Parker. "A New Vectorial Bedload Formulation and Its Application to the Time Evolution of Straight River Channels." *J. Fl. Mech.* 267 (1994), pp. 153–83.

Kramer, H. "Sand Mixtures and Sand Movement in Fluvial Models." *Transactions* [ASCE] 100, no. 1909 (1935), pp. 798–838.

Lagasse, P. F., L. W. Zevenbergen, J. D. Schall, and P. E. Clopper. *Bridge Scour and Stream Stability Countermeasures.* HEC-23, 2nd ed. Washington, DC: Federal Highway Administration, 2001.

Lagasse, P. F., J. D. Schall, and E. V. Richardson. *Stream Stability at Highway Structures.* Report No. HEC-20. Washington, DC: Federal Highway Administration, U.S. Department of Transportation, 2001.

Lagasse, P. F., P. E. Clopper, L. W. Zevenbergen, J. F. Ruff. *Riprap Design Criteria, Recommended Specifications, and Quality Control.* NCHRP Report 568. Washington, DC: Transportation Research Board, National Research Council, 2006.

Lagasse, P. F., P. E. Clopper, L. W. Zevenbergen, L. G. Girard. *Countermeasures to Protect Bridge Piers from Scour.* NCHRP Report 593. Washington, DC: Transportation Research Board, National Research Council, 2007.

Landers, M. N., and D. S. Mueller. Evaluation of Selected Pier-Scour Equations Using Field Data, pp. 186–95. *Transportation Research Record 1523.* Washington, DC: National Research Council, 1996.

Lane, E. W. *A Study of the Shape of Channels Formed by Natural Streams Flowing in Erodible Material.* MRD Series Report No. 9, Missouri River Division, U.S. Army Corps of Engineers, Omaha, Neb., 1957.

Lane, E. W. "Design of Stable Channels." *Transactions* [ASCE] 120, no. 2776 (1955a), pp. 1234–79.

Lane, E. W. "The Importance of Fluvial Geomorphology in Hydraulic Engineering." *Proc. ASCE* 81, no. 795 (1955b), pp. 1–17.

Lane, E. W., and V. A. Koelzer. *Density of Sediments Deposited in Reservoirs.* Report No. 9 of a Study of Methods Used in Measurement and Analysis of Sediment Loads in Streams. St. Paul, MN: U.S. Army Corps of Engineers, 1953.

Lauchlan, C. S., and B. W. Melville. "Riprap Protection at Bridge Piers." *J. Hydr. Engrg., ASCE* 127, No. 5 (2001), pp. 412–18.

Laursen, E. M. "An Analysis of Relief Bridge Scour." *J. Hyd. Div., ASCE* 89, no. HY3 (1963), pp. 93–117.

Laursen, E. M. "Scour at Bridge Crossings." Iowa Highway Research Board Bulletin No. 8. Iowa City: Iowa Institute of Hydraulic Research, University of Iowa, 1958a.

Laursen, E. M. "Scour at Bridge Crossings." *J. Hyd. Div., ASCE* 86, no. HY2 (1960), pp. 39–54.

Laursen, E. M. "The Total Sediment Load of Streams." *J. Hyd. Div., ASCE* 54, no. HY1 (1958b), pp. 1–36.

Laursen, E. M., and A. Toch. *Scour Around Bridge Piers and Abutments.* Iowa Highway Research Board, Bulletin No. 4. Iowa City: Iowa Institute of Hydraulic Research, University of Iowa, 1956.

Lee, Seung Oh, T. W. Sturm, A. Gotvald, and M. Landers. "Comparison of Laboratory and Field Measurements of Bridge Pier Scour," *Proceedings. 2nd Int. Conf. on Scour and Erosion,* ed. by Y. M. Chiew, S. Y. Lim, and N. S. Cheng, Singapore, 2004, pp. 231–39.

Leopold, L. B., and T. Maddock Jr. "The Hydraulic Geometry of Stream Channels and Some Physiographic Implications." USGS Professional Paper 252. Washington, DC: U.S. Government Printing Office, 1953.

Leopold, L. B., and M. G. Wolman. *River Channel Patterns: Braided, Meandering, and Straight.* USGS Professional Paper 282-B. Washington, DC: U.S. Government Printing Office, 1957.

Leopold, L. B., and M. G. Wolman. "River Meanders." *Geological Society of America Bulletin,* 71 (1960).

Leopold, L. G., M. G. Wolman, and J. P. Miller. (1964). *Fluvial Processes in Geomorphology.* San Francisco: W. H. Freeman, 1964.

Mackin, H. H. "Concept of the Graded River." *Geological Society of America Bulletin* 59 (1948) pp. 463–512.

Mantz, P. A. "Incipient Transport of Fine Grains and Flakes by Fluids—Extended Shields Diagram." *J. Hyd. Div.,* ASCE, 103, no. HY6 (1977), pp. 601–15.

McKnown, J. S., and J. Malaika. "Effect of Particle Shape on Settling Velocity at Low Reynolds Numbers." *Transactions* [American Geophysical Union] 31 (1950), pp. 74–82.

Melville, B. W. "Local Scour at Bridge Abutments." *J. Hydr. Engrg.,* ASCE 118, no. 4 (1992), pp. 615–31.

Melville, B. W. "Pier and Abutment Scour: Integrated Approach." *J. Hydr. Engrg.,* ASCE 123, no. 2 (1997), pp. 125–35.

Melville, B. W. "The Physics of Local Scour at Bridge Piers." *Proc. 4th International Conference on Scour and Erosion,* Japanese Geotechnical Society, Tokyo 2008, pp. 28–40.

Melville, B. W., and A. J. Sutherland. "Design Method for Local Scour at Bridge Piers." *J. Hydr. Engrg.,* ASCE 114, no. 10 (1988), pp. 1210–26.

Melville, B. W., and S. Coleman. *Bridge Scour.* Water Resources Publications LLC, Highlands Ranch, Colorado, 2000.

Melville, B. W., and Y. M. Chiew. "Time Scale for Local Scour." *J. Hydr. Engrg.,* ASCE 125, no. 1 (1999), pp. 59–65.

Melville, B. W., S. van Ballegooy, S. Coleman, and B. Barkdoll. "Countermeasure Toe Protection at Spill-Through Abutments." *J. Hydr. Engrg.,* ASCE 132, No. 3 (2006), pp. 235–45.

Meyer-Peter, E., and R. Muller. "Formulas for Bed-Load Transport." Paper at the Second Meeting of International Association for Hydraulic Research, Stockholm, 1948.

Morales, R., R. Ettema, and B. Barkdoll. "Large-Scale Flume Tests of Riprap-Apron Performance at a Bridge Abutment on a Floodplain." *J. Hydr. Engrg.,* ASCE 134, No. 6 (2008), pp. 800–809.

Nakato, T. "Tests of Selected Sediment-Transport Formulas." *J. Hydr. Engrg.,* ASCE 116, no. 3 (1990), pp. 362–79.

Neill, C. R. "Mean-Velocity Criterion for Scour of Coarse Uniform Bed-Material." In *Proceedings of the Twelfth Congress of the International Assoc. for Hydr. Research,* Fort Collins, CO, 1967.

Nordin, C. F., Jr. *Aspects of Flow Resistance and Sediment Transport, Rio Grande Near Bernalillo, New Mexico.* Water Supply Paper 1498-H. Washington, DC: U.S. Geological Survey, 1964.

Nordin, C. F., Jr., and J. P. Beverage. *Sediment Transport in the Rio Grande New Mexico.* Prof. Paper 462-F. Washington, DC: U.S. Geological Survey, 1965.

Pagán-Ortiz, J. E. *Stability of Rock Riprap for Protection at the Toe of Abutments Located at the Floodplain.* Publication No. FHWA-RD-91-057. Washington, DC: Federal Highway Administration, U.S. Department of Transportation, 1991.

Parola, A. C. "Stability of Riprap at Bridge Piers." *J. Hydr. Engrg.,* ASCE 119, No. 10 (1993), pp. 1080–93.

Petersen, M. S. *River Engineering.* Englewood Cliffs, NJ: Prentice-Hall, 1986.

Raudkivi, A. J. "Functional Trends of Scour at Bridge Piers." *J. Hydr. Engrg.,* ASCE 112, no. 1 (1986), pp. 1–13.

Raudkivi, A. J., and R. Ettema. "Clear-Water Scour at Cylindrical Piers." *J. Hydr. Engrg.,* ASCE 109, no. 3 (1983), pp. 339–50.

Ravisangar, V., T. W. Sturm, and A. Amirtharajah. "Influence of Sediment Structure on Erosional Strength and Density of Kaolinite Sediment Beds," *J. Hydr. Engrg.,* ASCE 131, No. 5 (2005), pp. 356–65.

Richardson, E. V., and S. R. Davis. *Evaluating Scour at Bridges.* Report No. HEC-18. Washington, DC: Federal Highway Administration, U.S. Department of Transportation, 2001.

Richardson, E. V., and J. R. Richardson. "Discussion of 'Pier and Abutment Scour: Integrated Approach.' by B.W. Melville." *J. Hydr. Engrg.,* ASCE 124, no. 7 (1998), pp. 771–72.

Richardson, E. V., D. B. Simons, and P. Lagasse. *Highways in the River Environment.* Report No. HDS 6. Washington, DC: Federal Highway Administration, U.S. Department of Transportation, 2001.

Rouse, H. "Modern Conceptions of the Mechanics of Fluid Turbulence." *Transactions* [ASCE] 102, no. 1965 (1937), pp. 463–543.

Rouse, H. *An Analysis of Sediment Transportation in Light of Fluid Turbulence.* SCS-TP-25. Washington, DC: Soil Conservation Service, U.S. Department of Agriculture, 1939.

Rubey, W. W. "Settling Velocities of Gravel, Sand, and Silt Particles." *American Journal of Science,* 5th Series, 25, no. 148 (1933), pp. 325–38.

Schlichting, H. *Boundary-Layer Theory.* New York: McGraw-Hill, 1979.

Schumm, S. A. "Fluvial Geomorphology: The Historical Perspective." In *River Mechanics,* Vol. 1, ed. H. W. Shen, pp. 4.1–4.30. Fort Collins, CO: H. W. Shen, 1971.

Schumm, S. A. "River Metamorphosis." *J. Hyd. Div.,* ASCE 95, no. HY1 (1969), pp. 255–73.

Schumm, S. A., M. P. Mosley, and W. E. Weaver. *Experimental Fluvial Geomorphology.* New York: John Wiley & Sons, 1987.

Shen, H. W., V. R. Schneider, and S. Karaki. "Local Scour Around Bridge Piers." *J. Hyd. Div.,* ASCE 95, no. HY6 (1969), pp. 1919–40.

Shen, H. W., W. J. Mellema, and A. S. Harrison. "Temperature and Missouri River Stages Near Omaha." *J. Hyd. Div.,* ASCE 104, no. HY1 (1978), pp. 1–20.

Sheppard, D. Max, and W. Miller Jr. "Live-Bed Local Pier Scour Experiments." *J. Hydr. Engrg.,* ASCE 132, No. 7 (2006), pp. 635–42.

Sheppard, D. Max, M. Odeh, and T. Glasser. "Large-Scale Clear-Water Local Pier Scour Experiments." *J. Hydr. Engrg.,* ASCE 130, No. 10 (2004), pp. 957–63.

Shields, A. *Application of Similarity Principles and Turbulence Research to Bed-Load Movement.* trans. W. P. Ott and J. C. van Uchelen. Hydrodynamics Laboratory Publ. No. 167. Pasadena: USDA, Soil Conservation Service Cooperative Laboratory, California Institute of Technology, 1936.

Simons, D. B., and E. V. Richardson. *Resistance to Flow in Alluvial Channels.* Prof. Paper 422-J. Washington, DC: U.S. Geological Survey, 1966.

Simons, D. B., and F. Senturk. *Sediment Transport Technology.* Fort Collins, CO: Water Resources Publications, 1977.

Sowers, G. F. *Introductory Soil Mechanics and Foundations: Geotechnical Engineering,* 4th ed. New York: MacMillan, 1979.

Sturm, T. W. "Abutment Scour in Compound Channels." In *Stream Stability and Scour at Highway Bridges,* ed. E. V. Richardson and P. F. Lagasse, pp. 443–56. ASCE, 1999a.

Sturm, T. W. *Abutment Scour Studies for Compound Channels.* Report No. FHWA-RD-99-156. Washington, DC: Federal Highway Administration, U.S. Department of Transportation, 1999b.

Sturm, T. W. "Scour Around Bankline and Setback Abutments in Compound Channels." *J. Hydr. Engrg.,* ASCE 132, No. 1 (2006), pp. 21–32.

Sturm, T. W., and A. Chrisochoides. *One-Dimensional and Two-Dimensional Estimates of Abutment Scour Prediction Variables,* pp. 18–26. Transportation Research Record 1647. Washington, DC: Transportation Research Board, National Research Council, 1998.

Sturm, T. W., and N. S. Janjua. "Clear-Water Scour Around Abutments in Floodplains." *J. Hydr. Engrg.,* ASCE 120, no. 8 (1994), pp. 956–72.

Sturm, T. W., and Aftab Sadiq. *Clear-Water Scour Around Bridge Abutments Under Backwater Conditions,* pp. 196–202. Transportation Research Record 1523. Washington, DC: Transportation Research Board, National Research Council, 1996.

Tanner, W. F. "Helical Flow, a Possible Cause of Meandering." *J. Geophys. Res.* 65 (1960), pp. 993–95.

U.S. Army Corps of Engineers. *HEC-RAS River Analysis System.* Hydraulic Reference Manual. Davis, CA: Hydrologic Engineering Center, U.S. Army Corps of Engineers. (2008).

U.S. Army Corps of Engineers. *Scour and Deposition in Rivers and Reservoirs (HEC-6).* Davis, CA: Hydrologic Engineering Center, U.S. Army Corps of Engineers (1993).

U.S. Interagency Committee on Water Resources. *Some Fundamentals of Particle Size Analysis, A Study of Methods Used in Measurement and Analysis of Sediment Loads in Streams.* Report No. 12. Subcommittee on Sedimentation, Water Resources Council. Washington, DC: Government Printing Office, 1957.

U.S. Waterways Experiment Station. *Study of Riverbed Material and their Use with Special Reference to the Lower Mississippi River.* Paper 17. Vicksburg, MS: U.S. Waterways Experiment Station, 1935.

Vanoni, V. A. "Some Effects of Suspended Sediment on Flow Characteristics." In *Proceedings, Fifth Hydraulic Conference,* pp. 137–58. Bulletin 34, Iowa City: University of Iowa, 1953.

Vanoni, V. A. "Transportation of Suspended Sediment by Water." *Transactions* [ASCE] 111, no. 2267 (1946), pp. 67–133.

Vanoni, V. A., and N. H. Brooks. *Laboratory Studies of Roughness and Suspended Load of Alluvial Streams.* Sedimentation Laboratory Report No. E68. Pasadena: California Institute of Technology, 1957.

Vanoni, Vito A., ed. *Sedimentation Engineering.* New York: ASCE Task Committee for the Preparation of the Manual on Sedimentation, ASCE, 1977.

van Rijn, L. C. "Equivalent Roughness of Alluvial Bed." *J. Hydr. Engrg.,* ASCE 108, no. HY10 (1982), pp. 1215–18.

van Rijn, L. C. "Sediment Transport, Part I: Bed Load Transport." *J. Hydr. Engrg.,* ASCE 110, no. 10 (1984a), pp. 1431–56.

van Rijn, L. C. "Sediment Transport, Part II: Suspended Load Transport." *J. Hydr. Engrg.,* ASCE 110, no. 11 (1984b), pp. 1613–41.

van Rijn, L. C. "Sediment Transport III: Bed Forms and Alluvial Roughness." *J. Hydr. Engrg.,* ASCE 110, no. 12 (1984c), pp. 1733–54.

White, F. M. *Viscous Fluid Flow.* New York: McGraw-Hill, 2005.

Williams, G. P. *Flume Experiments on the Transport of a Coarse Sand.* Prof. Paper 562-B. Washington, DC: U.S. Geological Survey, 1967.

Yalin, M. S. *Mechanics of Sediment Transport.* New York: Pergamon Press, 1972.

Yalin, M. S., and E. Karahan. "Inception of Sediment Transport." *J. Hyd. Div.,* ASCE 105, no. HY11 (1979), pp. 1433–43.

Yang, C. T. "Incipient Motion and Sediment Transport." *J. Hyd. Div.,* ASCE 99, no. HY10 (1973), pp. 1679–1704.

Yang, C. T. *Sediment Transport Theory and Practice.* New York: McGraw-Hill, 1996.

Yang, C. T. "Unit Stream Power and Sediment Transport." *J. Hyd. Div.,* ASCE 98, no. HY10 (1972), pp. 1805–26.

EXERCISES

10.1. Derive the Engelund-Hansen equation (10–10) for fall velocity of natural sands and gravels from their proposed relationship for C_D ($= 24/$ **Re** $+ 1.5$). Then plot the Engelund-Hansen equation for fall velocity (cm/s) vs. particle diameter (mm) for sand to gravel size ranges in water at 20°C and compare with the plot of calculated fall velocity of *quartz spheres* in the same sedimentation fluid (on the same graph).

10.2. Derive Rubey's equation from his proposed relationship for C_D ($= 24/$ **Re** $+ 2$). Plot the results for fall velocity for sand to gravel size ranges and compare with the results from the Engelund-Hansen equation on the same graph.

10.3. A river sand has a sieve diameter of 0.3 mm. Find the fall velocity at 20°C using two methods: (1) use Figures 10.2 and 10.3, (2) use the Engelund-Hansen equation plotted in Exercise 10.1.

10.4. A river sand has a measured fall velocity in water of 10 cm/s at 20°C. Find the sedimentation diameter.

10.5. Plot the results of the sieve analysis given below on lognormal paper. Find $d_{84.1}$, $d_{15.9}$, d_{50}, d_g, and σ_g. Derive a formula for d_{65} and d_{90} in terms of σ_g and d_g using the theoretical lognormal distribution:

Sieve diameter, mm	Grams retained
0.495	0.36
0.417	1.52
0.351	4.32
0.295	7.40
0.246	8.80
0.208	7.04
0.175	4.56
0.147	2.88
0.124	1.52
0.104	0.80
0.088	0.44
0.074	0.20
Pan	0.16

10.6. For a slope of 0.002 in a wide stream, calculate the depth at which sediment motion begins for a fine gravel with $d_{50} = 3.3$ mm.

10.7. Bank-full depth for a stream in well-rounded alluvium with $d_{50} = 12$ mm is 1.5 m. The stream slope is 0.0005. The channel shape is approximately trapezoidal with 2:1 side slopes and a bottom width of 10 m. Calculate the bank-full discharge and determine whether the bed and banks are stable at bank-full conditions.

10.8. A roadside drainage ditch has to carry a discharge of 100 cfs. It has a bottom width of 5 ft and side slopes of 2:1. The ditch is to be lined with locally available gravel, which has $d_{50} = 25$ mm. At what maximum slope can the channel be constructed for stability of the bed and banks?

10.9. Determine the critical velocity of a uniform flow of water (20°C) with a depth of 3 ft over a bed of very coarse sand with a median size of 2.0 mm using Keulegan's equation. Repeat for a laboratory uniform flow depth of 0.3 ft. Let $k_s = 2.5d_{50}$.

10.10. Calculate the critical velocities in Exercise 10.9 using Manning's equation with the Strickler equation for Manning's n. For coarse sediments and fully rough turbulent flow, derive a general relationship for the critical sediment number as a function of d_{50}/y_0, where y_0 is flow depth, and the Shields' parameter using Manning's equation with Strickler's equation for n. The critical sediment number is defined as

$$\mathbf{N}_{sc} = \frac{V_c}{\sqrt{(SG - 1)gd_{50}}}$$

in which V_c = critical velocity and SG = specific gravity of the sediment.

10.11. The Republican River is flowing at a depth of 3.0 ft with a velocity of 7 ft/s for a slope of 0.0017. The sediment sizes are $d_{50} = 0.32$ mm, $d_{65} = 0.39$ mm, and $d_{90} = 0.59$ mm. Determine the bed form type using the Simons-Richardson diagram and the van Rijn diagram.

10.12. The Missouri River at Omaha flows at a depth of 10.1 ft with a velocity of 4.49 ft/s and a slope of 0.000155 in summer ($T = 71°F$). At approximately the same discharge of 32,000 cfs, it flows at a depth of 9.1 ft with a velocity of 5.49 ft/s and a slope of 0.00016 in winter ($T = 41°F$). The measured bed sediment sizes are $d_{50} = 0.199$ mm and $d_{90} = 0.286$ mm in summer, while in winter, $d_{50} = 0.224$ mm and $d_{90} = 0.306$ mm. Predict the bed form type for both cases using Figure 10.14 and explain your results. (Use the van Rijn method to estimate u'_*.)

10.13. A 7 mi reach of the Chattahoochee River downstream of the Buford dam is subject to a maximum discharge of 8,000 cfs due to hydropower releases and a minimum discharge of 550 cfs. The average width of the river is 180 ft, and it can be considered very wide. The bed sediment is coarse sand with $d_{50} = 1.0$ mm and $d_{65} = 1.2$ mm. The average slope is 0.00031 ft/ft.
    (a) Calculate the velocity and depth at low and high flow using the Karim-Kennedy method and the Engelund method.
    (b) Predict the dominant bed form at low and high flow using the Simons-Richardson diagram and the van Rijn diagram. What are the bed form dimensions at low and high flow?

10.14. Calculate the velocity and flow depth for the Republican River using the measured discharge per unit width of 21 ft²/s as given in Exercise 10.11. Use the van Rijn method and the Karim-Kennedy method.

10.15. The Rio Grande at Bernalillo, New Mexico has the following characteristics:

Depth range:	$y_0 = 0.6$–5.0 ft
Slope:	$S = 0.00086$ ft/ft
Temperature:	$T = 60°$F
Bed sediment:	$d_{35} = 0.24$ mm
	$d_{50} = 0.29$ mm
	$d_{65} = 0.35$ mm
	$d_{90} = 0.53$ mm

Use a spreadsheet to prepare a depth-velocity curve using three methods: (1) Engelund, (2) van Rijn, and (3) Karim-Kennedy. Plot the results on log-log scales along with the measured data from the table given below. Compare the results and discuss.

Rio Grande near Bernalillo, New Mexico, Sec. A2

Velocity, ft/s	Depth, ft	Velocity, ft/s	Depth, ft
4.06	2.46	2.09	1.94
6.57	3.63	1.62	0.93
5.96	3.79	1.17	0.52
5.09	3.49	1.30	0.58
3.71	2.76	3.73	1.54
2.84	2.69	5.00	3.06
2.65	2.15	5.17	2.19
1.66	1.25	4.31	2.44
3.58	2.66	5.86	3.25
3.11	2.56	5.56	3.03
3.12	2.48	6.91	3.68
2.34	2.14	6.88	4.46
2.23	1.44	7.82	4.11
1.98	1.18	7.71	4.80
1.91	1.23	6.92	4.34
2.05	1.29	6.27	3.43
1.94	1.32	6.10	2.67
2.37	1.42	5.06	2.96
2.51	1.50	6.50	3.40
2.90	1.29	2.55	1.84
1.44	1.12	3.04	2.56
1.70	1.48	2.71	3.44
1.40	0.79	2.89	2.93
2.01	1.07	3.16	2.64
1.97	1.70	3.99	3.12
1.85	1.03	3.62	2.33
1.76	1.31	6.01	3.22
1.89	1.10	2.88	1.54
1.76	1.70	2.00	1.00

Source: Nordin and Beverage (1965).

10.16. Calculate the bed load discharge per unit width in a wide gravel-bed stream with a slope of 0.005 and a flow depth of 1.0 m. The median bed sediment size is 20 mm. Use the Meyer-Peter and Müller formula and the Einstein-Brown formula. Give the results in m^2/s and in metric tons/day. (1 metric ton = 1,000 kg).

10.17. Calculate the bed load discharge per unit width in a canal constructed in coarse sand (d_{50} = 1.0 mm; d_{90} = 1.8 mm) at a slope of 0.001, which is flowing at a depth of 2.0 m and a velocity of 1.5 m/s. Use the van Rijn bed load formula and the Einstein-Brown formula. Discuss the results.

10.18. The following table gives the velocity distribution and the suspended sand concentration distribution for the size fraction between the 0.074 mm and the 0.104 mm sieve sizes for vertical C-3 on the Missouri River at Omaha on November 4, 1952. On this day, the slope of the stream was 0.00012, the depth was 7.8 ft, the width was 800 ft, the water temperature was 45°F, and the flow was approximately uniform.

(*a*) Plot the velocity on semi-log scales and the concentration profile on log-log scales. Do a regression analysis to obtain the best-fit straight lines.

(*b*) Compute from the data and the graphs the following quantities:

u_* = shear velocity,
V = mean velocity,
κ = von Karman constant,
f = Darcy-Weisbach friction factor,
$\mathbf{R}_0$ = Rouse number,
g_s = suspended sediment discharge per unit width in lbs/ft/s.

z, distance above bottom (ft)	u, velocity (ft/s)	C, concentration (mg/L)
0.7	4.30	411
0.9	4.50	380
1.2	4.64	305
1.4	4.77	299
1.7	4.83	277
2.2	5.12	238
2.7	5.30	217
2.9	5.40	—
3.2	5.42	196
3.4	5.42	—
3.7	5.50	184
4.2	5.60	—
4.8	5.60	148
5.8	5.70	130
6.8	5.95	—
7.8	—	—

10.19. The Mississippi River at Arkansas City is very wide and has a bed sediment with d_{50} = 0.30 mm and d_{90} = 0.60 mm. For a flow depth of 45 ft, the water surface slope is measured to be 0.0001, and the water temperature is 80°F. The mean velocity is

3.0 ft/s, and the sediment concentration at a location of 5 ft above the channel bottom is 1,000 mg/L.

(a) What is the dominant mode of sediment transport?

(b) Calculate the middepth suspended sediment concentration.

(c) If the water temperature were 40°F while all other factors remained the same, would the suspended sediment discharge change? Explain.

10.20. A reach of Peachtree Creek has a slope of 0.0005, and a sand bed with $d_{50} = 0.5$ mm, $d_{65} = 0.6$ mm, and $d_{90} = 1.0$ mm. The measured water discharge per unit of width $q = 11.5$ ft²/s.

(a) Estimate the depth and velocity using the Engelund method. What is the equivalent value of Manning's n and how does it compare with the Strickler value? Explain.

(b) Determine the bed form type, and its height and wavelength from van Rijn's method. Use the value of τ'_* determined in part (a).

(c) What is the mode of sediment transport; that is, is it primarily bed load, mixed load, or suspended load?

(d) Calculate the suspended sediment concentration in ppm at a distance of 1 ft above the bed using a reference concentration, C_a, and the methodology of van Rijn.

(e) Calculate the total load from Karim's method, Yang's method, Engelund's method, and Karim-Kennedy's method in ft²/s and ppm by weight.

(f) Calculate the bed load transport rate in ft²/s using the Meyer-Peter and Müller formula, first using τ_* and then τ'_*. Which result is the correct one and why?

10.21. Calculate the total sediment discharge in the Niobrara River using the Engelund-Hansen formula for the conditions given in Example 10.4.

10.22. For the given water discharge in the Niobrara River in Example 10.4, calculate the depth and velocity using the Engelund method and the Karim-Kennedy method. Then determine the sensitivity of the calculated sediment discharge to the depth and velocity using one of the sediment discharge formulas in the example.

10.23. The Niobrara River in Nebraska has the following sediment sizes:

Bed sediment: $d_{50} = 0.27$ mm
                $d_{65} = 0.34$ mm
                $d_{90} = 0.48$ mm

For each of the measured data points in the table given below, apply the van Rijn method, Yang formula, Karim-Kennedy formula, and Karim formula to calculate the total sediment discharge per unit of width g_t in lbs/sec/ft for comparison with measured sediment discharges per unit width (convert from C_t in ppm). Use a spreadsheet or write a computer program. For each method plot a graph of measured vs. calculated sediment discharges and discuss the results.

Niobrara River

Depth, ft	Velocity, ft/s	Slope	Conc., ppm	Viscosity, ft²/s
1.44	3.20	0.001439	1610	1.61E-5
1.53	2.47	0.001344	970	0.99E-5
1.44	2.26	0.001250	1140	1.08E-5
1.60	3.57	0.001704	1890	1.64E-5
1.62	3.61	0.001704	1770	1.55E-5
1.89	4.17	0.001799	1780	1.34E-5
1.57	2.45	0.001269	780	1.13E-5
1.43	2.33	0.001288	790	1.00E-5
1.51	2.41	0.001288	910	1.03E-5
1.44	2.52	0.001174	1000	1.16E-5
1.53	3.21	0.001420	1780	1.21E-5
1.56	3.07	0.001401	1490	1.19E-5
1.73	3.70	0.001685	1900	1.25E-5
1.38	3.34	0.001553	1710	1.82E-5
1.67	2.95	0.001250	893	1.76E-5
1.70	3.42	0.001477	1820	1.76E-5
1.56	2.15	0.001250	754	1.08E-5
1.61	2.45	0.001287	934	0.95E-5
1.41	2.35	0.001136	503	1.05E-5
1.38	2.19	0.001250	392	0.88E-5
1.44	2.05	0.001212	429	1.01E-5
1.54	2.20	0.001136	736	1.19E-5
1.42	3.33	0.001609	1520	1.79E-5
1.36	3.11	0.001610	1660	1.72E-5
1.62	3.61	0.001667	2200	1.55E-5

Source: Colby and Hembree (1955); Karim and Kennedy (1981).

10.24. Discuss the expected channel adjustments in a sand-bed stream that is in equilibrium as a result of the following changes:

(*a*) Sand-dredging in a localized reach upstream of a bridge,

(*b*) Rapid residential development adjacent to the stream in an urban area,

(*c*) Excavation of a wider channel in a bridge cross section to reduce backwater,

(*d*) Cutoff of a meander bend,

(*e*) Channel improvement by removal of brush on the banks and deadfalls in the channel.

10.25. Develop the complete derivation of the equation for live-bed contraction scour given by Equation 10.86.

10.26. Consider the case of a very long contraction with overbank flow in the approach channel and a constant main channel width. The bridge abutments extend to the edge of the main channel. Assume that there is no change in the channel roughness from the approach through the contraction and that the sediment is coarse without bed forms with bed-load transport only in the main channel and no sediment transport in the floodplain. For the live-bed case, derive an expression for the contraction scour using the Meyer-Peter and Müller bed-load formula.

10.27. Calculate the live-bed contraction scour depths for bankline abutments in overbank flow as a function of the percent of total flow in the main channel, which varies from 50 to 90 percent. Assume a constant main-channel width and an approach flow depth of 10 ft in the main channel. Plot the results.

10.28. Estimate the maximum scour depth around cylindrical piers in a riverbed if the bed sediment has a median size of 1.0 mm, and the bridge piers are 1.2 m in diameter. The approach flow just upstream of the piers has a depth of 3.0 m and a velocity of 1.0 m/s. Use the following formulas: (1) CSU, (2) Laursen and Toch, (3) Jain, (4) Melville and Sutherland, and (5) Froehlich. What value of scour depth would you report to the geotechnical engineer designing the pier foundations?

10.29. Find the abutment scour depth and pier scour depth for Example 10.5 using Melville's formula and Froehlich's formula.

10.30. The following field data are given for pier diameter, b; approach velocity, V_1; approach depth, y_1; and median sediment diameter, d_{50}. Calculate the maximum pier scour depth for a cylindrical pier having no skew using the equations of Froehlich; Melville and Sutherland; Jain, Laursen and Toch; and CSU. Compare with the measured scour depths, d_s.

River	b, ft	y_1, ft	V_1, ft/s	d_{50}, mm	Measured d_s, ft
Red	26.9	14.1	2.0	0.060	14.1
Oahu	4.9	10.2	7.8	20.0	4.3
Knik	5.0	3.9	1.6	0.5	1.0

10.31. The main bridge pier for a bridge over the Peace River is cylindrical with a diameter of 3.14 m. The local flow velocity just upstream of the pier in the bridge section is 2.2 m/s, and the flow depth there is 13.4 m. Determine the rock size, and the extent and thickness of the riprap layer necessary to protect the pier from scour. The available rock has a specific gravity of 2.65. Use both the Melville and Lauchlan equations and the HEC-23 equation.

10.32. The left bridge abutment (spill-through) for the Towaliga River is set back 300 ft from the main channel. For the design flood, the left floodplain upstream of the bridge carries a discharge of 10,000 cfs. The flow depth in the left floodplain at the bridge abutment is 15 ft, and the depth of flow in the main channel is 26 ft. The flow area in the left floodplain in the bridge cross-section is 1900 ft². Determine the rock size, and the extent and thickness of a rock riprap apron to protect the abutment from scour.

10.33. Write a computer program with interactive input that computes the depth and velocity in an alluvial river for a given discharge per unit width. Implement all methods given in this chapter.

11

Three-Dimensional CFD Modeling for Open Channel Flows

11.1 INTRODUCTION

The use of computational fluid dynamics (CFD) to predict internal and external flows has risen dramatically in the past three decades. Roache first used CFD in the title of his 1972 book, *Computational Fluid Dynamics,* which revolutionized the scientific community by adding the third category "computational" to the traditional categories of "theoretical" and "experimental" in fluid flow research and engineering (Roache 1998). The solution of fluid flow problems by means of CFD is possible because of the recent dramatic increases in computer power and storage capabilities.

Nowadays, the widespread availability of engineering workstations together with efficient solution algorithms and sophisticated pre- and postprocessing software enable the use of CFD in both research and engineering applications. The area of CFD includes aerodynamics, aeroacoustics, surface-water flows and transport, groundwater flows and transport, electrodynamics, plasma dynamics, combustion modeling, free convection heat transfer, neutron transport, magnetohydrodynamics, chemistry, meteorology, oil recovery processes, and many more. In recent years a number of commercial CFD packages have been brought to the market for application to open channel flows by researchers and engineers. Enormous progress has been made in making these programs user friendly, robust, and as accurate as possible for a wide range of geophysical flows. Improving the accuracy of CFD codes, verification and validation of new algorithms and mathematical methods, as well as implementation of new physics into these models, however, is still an ongoing process and will continue for many years to come.

What makes predictions of open channel flows so difficult is the fact that they are three-dimensional and turbulent. Engineers and scientists need to understand turbulent flows in order to control them, design for their adverse effects, or utilize them to best effect (as in mixing processes). In fact, all flows over and through hydraulic structures, such as spillways, weirs, conveyance channels, and irrigation networks, are three dimensional and turbulent. In all cases, prediction is the key element in engineering design and it is essential to investigate, understand, and predict the structures of open channel turbulence (Nezu and Nakagawa 1993). Turbulence structures associated with channel bed interaction, flow separation, shear layers, secondary currents, and large-scale unsteadiness all contribute to the complexity of modeling open channel flows.

Another important aspect besides three-dimensionality and turbulence is the mathematical treatment of channel roughness, which is exceedingly difficult to characterize as discussed in Chapter 4 for fixed-bed channels and in Chapter 10 for mobile alluvial bed channels. Since channel bed roughness can vary (e.g., from cohesive fine sediments in river deltas to vegetated banks or floodplains in rivers), the mathematical roughness closure schemes involve calibration (i.e., make use of empirical formulas or parameters such that the physics of the flow is simulated correctly) or need thorough validation. The disadvantage of calibrating empirical parameters (representing the sources of sediment grain roughness, bedform roughness, channel three-dimensionality, and other factors as well) is that no distinction between the diverse roughness contributions can be made. This can lead to predictions of calibrated models that are prone to large inaccuracies. Instead of calibrating roughness, it is essential that roughness closure schemes be validated, which requires experimental or field data collected by accurate and reliable techniques. For example, Manning's equation often lumps all the effects of channel roughness, three-dimensionality, and turbulence into one resistance coefficient, although some progress has been made in separating surface and form roughness of alluvial channel bedforms as described in Chapter 10.

In terms of turbulence, rather more progress has been made. At an early stage in the development of turbulence research, Osborne Reynolds averaged the exact equations of fluid flow over time such that the mean velocity and pressure are described and thus the quantities of interest to the practicing engineer are obtained. Since averaging the equations over time no longer yields a closed set of equations (see Section 11.2), an empirical approach is needed in the form of a mathematical model for the unknown turbulence transport terms (Rodi 1980). In the past 30 years, as computers have become sufficiently powerful, a number of turbulence models have been developed to model the effects of turbulence on the mean flow behavior.

The purpose of this chapter is to introduce the engineer to CFD methods as applied to open channel flows. The treatment is necessarily an overview, but references for further reading are given at the end of the chapter. The use of CFD methods in the solution of engineering problems associated with open channel flow is expected to become more and more prevalent in the future. A typical example of the limitations of one-dimensional methods becomes apparent when examining the complexities of the flow patterns and bed scour around obstructions such as bridge piers as described in Chapter 10. As CFD models become more sophisticated and available to practicing engineers, their contribution is expected to grow hand-in-hand with experimental techniques to solve open channel flow problems.

In the following sections of this chapter, the basic equations for the computation of fluid flow in practical situations will be introduced and will be extended with the theory of turbulence models. The concept of discretization using the popular finite volume method will be introduced and the treatment of boundary conditions will be discussed. The application of a 3-D CFD code to a practical flow problem will close this chapter.

11.2   GOVERNING EQUATIONS

Conservation of Mass and Momentum

Simulating fluid flow on a computer is, in simplest terms, the solution of basic equations that describe the conservation of mass and momentum of a flowing fluid. These equations were discovered independently more than a century and a half ago by the French engineer Claude Navier and the Irish mathematician George Stokes and are consequently named the Navier-Stokes equations (NS-equations). The *conservation of mass* equation can be stated as

The change of mass in a control volume is equal to the total mass which leaves through its faces minus the total mass entering through its faces in a specified time interval.

With the assumption of an incompressible fluid with a constant density ρ, the conservation of mass can be expressed by the continuity equation in the three-dimensional Cartesian coordinate system as

$$\frac{\partial u_i}{\partial x_i} = 0 \tag{11.1}$$

where x_i ($i = 1, 2, 3$) or x, y, z are the Cartesian coordinates and u_i ($i = 1, 2, 3$) or u, v, w are the Cartesian components of the velocity vector **V**.

In fluid mechanics the analysis of motion is performed in the same way as in solid mechanics—using Newton's second law of motion:

The rate of change of momentum of a body is equal to the resultant force acting on the body, and takes place in the direction of the force.

Forces that may act on the fluid are, for example, surface forces such as pressure, normal and shear forces, or body forces like gravity. Expressing the normal and shear stresses in terms of velocity gradients, or the rates of fluid strain, in a flowing Newtonian fluid, and relating the sum of forces acting on a fluid element in an incompressible fluid to its acceleration or rate of change of momentum, the momentum equation becomes (in Cartesian form):

$$\frac{\partial u_i}{\partial t} + \frac{\partial(u_i u_j)}{\partial x_j} = -\frac{1}{\rho}\frac{\partial p}{\partial x_i} + \nu\frac{\partial^2 u_i}{\partial^2 x_j} + \rho g_i \qquad (11.2)$$

where p is the pressure, ν is the kinematic viscosity, and g_i is the component of the gravitational acceleration in the direction of the Cartesian coordinate x_i. Usually gravity is assumed to act in the negative z-direction and the gravity term can be written as:

$$\rho g_i = -grad[\rho g(z - z_b)] \qquad (11.3)$$

where z_b is the reference level or datum. For a numerical solution it is more efficient to combine the pressure and gravity terms into the piezometric pressure $\tilde{p}$:

$$\tilde{p} = p + \rho g(z - z_b) \qquad (11.4)$$

and use it in place of the pressure (Ferziger and Peric 1996).

Equations (11.1) and (11.2) form the well-known Navier-Stokes equations for an incompressible fluid. Before the Navier-Stokes equations can be applied to an open channel simulation, the bathymetry (i.e., the surfaces, boundaries, and space around and between the boundaries) has to be represented in a form usable by the computer. This can be done as a series of regularly or irregularly spaced points known as the computational grid. In effect, the computational grid discretizes the computational problem in space; the calculations are carried out at regular intervals to simulate the passage of time, so the simulation is temporally discrete as well. The closer together and, therefore, the more numerous the points are in the computational grid, and the more often they are computed (the shorter the time interval), the more accurate and realistic the simulation is. In fact, for objects with complex shapes, even defining the surface and generating a computational grid can be a challenge.

Turbulence and Turbulence Modeling

For laminar flow of a viscous, incompressible, Newtonian fluid, the Navier-Stokes equations are valid as given by Equations (11.1) and (11.2) without additional input, and a discretization method can be applied to the governing equation set. In reality, the flow will remain laminar only up to a certain critical value of the Reynolds number $\mathbf{Re} = (V L)/\nu$, which is defined by the mean velocity V and a characteristic length scale L (typically the flow depth or hydraulic radius for open channel flow) for the flow system to be simulated. For practical problems, Reynolds numbers in open channel flows have values far above this critical value and the flow conditions are considered turbulent.

Turbulence is a property of the flow that can be visualized as three-dimensional, unsteady eddy-structures. These eddies remove kinetic energy from the flow and are split from larger structures into smaller and smaller structures, which eventually dissipate the kinetic energy into heat energy. In principle, an incompressible, viscous, turbulent flow can be described with the Navier-Stokes equations. The method of direct numerical simulation (DNS) computes the evolution of all significant scales of motion. Therefore, the whole spectrum of turbulence length and time scales has to be contained within a very accurate modeling of the geometry of the flow. The resolution requirements for a DNS of isotropic turbulence that captures the viscous dissipation can be estimated to be four times the Kolmogorov length scale $\ell_K = (\nu^3/\varepsilon)^{1/4}$, which is the scale at which turbulent energy is dissipated, where ν is the kinematic viscosity and ε is the rate of dissipation of turbulent kinetic energy per unit mass (Reynolds 1984). However, the required number of mesh points to resolve even the smallest length scale in a three-dimensional mesh is proportional to $\mathbf{Re}^{9/4}$ (Nezu and Nakagawa 1993). According to this estimation, a DNS of homogeneous flows with the Reynolds number of the large-scale turbulence of $\mathbf{Re}_\tau = u_* L/\nu = 500$, a mesh of 128^3 grid points is necessary to simulate the turbulent eddies down to the dissipation length scale (Rogallo et al. 1984). For an open channel flow, the resolution requirements are dependent on the channel Reynolds number $\mathbf{Re}$ (Kim, Moin, and Moser 1987). The number of mesh points required for an open channel flow computation can be estimated to be $N_{xyz} = 2 \times 10^6\ (\mathbf{Re}/3300)^{2/7}$ (Reynolds 1989). Thus, a simulation of a relatively low Reynolds number ($\mathbf{Re} = 42,000$) lab experiment of a compound channel flow would require 2×10^9 mesh points. Furthermore, for time-accurate resolution of the smallest eddies, the time step must not carry a fluid particle across more than a mesh length, so that the time step has to be decreased as the mesh width is decreased (Reynolds 1989). In general, a DNS simulation of the flow field is desired because it is the most accurate. Although DNS can be used for research purposes, it still is not feasible for the complex geometries of practical engineering. "A DNS can be used to study the effects of turbulence in simple geometries, providing new insights in basic turbulence physics and guidance in turbulence modeling. . . .DNS will remain limited to relatively low Reynolds numbers for the foreseeable future and perhaps forever" (Reynolds 1989).

Lower levels of approximation to compromise between accuracy and computational costs of a simulation have to be applied. Figure 11.1 gives an overview of the methodologies that are currently applied to account for turbulence in numerical models. As can be seen from Figure 11.1, one way to calculate turbulent flow is a time averaging of the Navier-Stokes equations, which leads to the so-called Reynolds-averaged-Navier-Stokes equations (RANS). The instantaneous flow vector u_i and pressure vector p are averaged over time T, so that the mean components $\bar{u}$ and $\bar{p}$ and the fluctuating components u' and p' arise (see also Figure 4.2 in Chapter 4). The Reynolds-averaged Navier-Stokes equations (RANS) are written in the following form:

$$\frac{\partial \bar{u}_i}{\partial x_i} = 0 \tag{11.5}$$

$$\bar{u}_j \frac{\partial \bar{u}_i}{\partial x_j} + \frac{\partial \bar{u}_i}{\partial t} = -\frac{1}{\rho}\frac{\partial \bar{p}}{\partial x_i} + \frac{\partial}{\partial x_j}\left(\nu \frac{\partial \bar{u}_i}{\partial x_j} - \overline{u'_i u'_j} \right) + \bar{f}_e \tag{11.6}$$

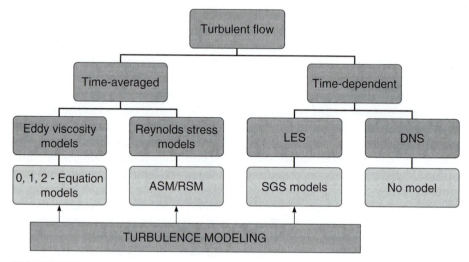

FIGURE 11.1

Treatment of turbulent flows (LES— large eddy simulation; DNS— direct numerical simulation; ASM—algebraic stress model; RSM—Reynolds stress model; SGS—sub grid scale).

The terms $\overline{u_i'^2}$ and $\overline{u_i'u_j'}$ in the second term on the right-hand side of Equation 11.6 are the correlations between the fluctuating velocity components arising from averaging the original Navier-Stokes equations. These components act on the fluid particles like stresses and upon being multiplied by the fluid density are thus called the Reynolds-stresses:

$$\tau_{ij}^R = -\rho\overline{u_i'u_j'} \tag{11.7}$$

The terms like $\rho\overline{u_i'^2}$ are normal stresses but also can be interpreted as components of the turbulent kinetic energy per unit volume, whereas the terms of $\rho\overline{u_i'u_j'}$ are the Reynolds shear stresses. The term $\overline{f_e}$ represents external forces like gravity or the Coriolis force. Note that the St. Venant equations introduced in Chapter 7 are a special 1-D case of the RANS in which the total shear stress is evaluated in terms of the friction slope estimated by Manning's equation and Manning's n. It can be seen from the RANS that there are only four equations with five unknown parameters. To overcome this closure problem, the Reynolds stresses must be estimated with the help of a turbulence model. A number of turbulence models for the Reynolds-averaged Navier-Stokes equations do exist. They will be introduced further below.

The mixed quality of the results obtained with RANS turbulence models has led to an increased interest in using Large Eddy Simulation (LES) as a tool to predict turbulent flows. A time-dependent filtering procedure has to be applied, which makes use of the fact that the large-scale motions in a flow are much more energetic than the small-scale ones (Leonard 1974). LES was first applied by Deardoff

(1970), who simulated a small straight channel flow at high Reynolds numbers. LES in its early days was limited to small scale simulations and performed on supercomputers only. With increasing computer speed it has become possible to apply LES to more complex geometries. However, according to Rodi et al. (1997), there is the need for further assessments of LES models in terms of numerical methods and geometrical treatments for complex flows, such that LES promises to be a successful alternative to DNS. The objective of LES is to explicitly simulate large scale eddies, which are responsible for the majority of momentum transport and turbulent diffusion (Ferziger and Peric 1996). The smaller scale eddies are treated by an approximate parameterization or model, usually called sub grid scale model (SGS model). It has been shown by a number of researchers that the results of LES are relatively insensitive to the contribution of the small-scale eddies, which can be determined with a model of minor quality. This is due to the fact that the statistics of small-scale turbulence are expected to be more universal than those of the large scales (Ferziger 1993). Hence, LES promises a wider generality and reduced modeling uncertainty and thus greater accuracy compared to a solution of the Reynolds-averaged Navier-Stokes equations. Even though computational resources are increasing at a rapid pace, a fully resolved LES of a practically relevant open-channel flow requires computing resources far from being available at the time this book chapter was written. To date, a few LES of open channel flows have been reported with Reynolds numbers close to 100,000. This is still below typical **Re** numbers occurring in nature where $\mathbf{Re} >> 1 \times 10^6$.

RANS Turbulence Closure with the Eddy Viscosity Concept

Most of the turbulence models used to date for calculating open channel flows employ the eddy-viscosity concept of Boussinesq (1877). It is based on the assumption that the turbulent Reynolds shear stresses are proportional to the mean velocity gradients in analogy to the viscous stresses in laminar flow. For the various stress components, the Boussinesq concept is written as:

$$-\overline{u_i' u_j'} = v_t \left(\frac{\partial \overline{u_i}}{\partial x_j} + \frac{\partial \overline{u_j}}{\partial x_i} \right) - \frac{2}{3} k \delta_{ij} \tag{11.8}$$

where k is the turbulent kinetic energy per unit mass and δ_{ij} is the Kronecker delta, which is defined as:

$$\delta_{ij} = 0 \; for \; i \neq j \quad and \quad \delta_{ij} = 1 \quad for \; i = j$$

Using the Kronecker delta, the expression involving the turbulent kinetic energy k can be applied to normal stresses (i.e., $i = j$). The definition of the turbulent kinetic energy reads:

$$k = \frac{1}{2} (\overline{u_i' u_i'}) \tag{11.9}$$

The factor v_t in Equation 11.8 is the turbulent viscosity, or so-called eddy viscosity, which is different from the molecular viscosity v. Since v_t is a scalar quantity and hence is the same for all stress components, the eddy viscosity is isotropic. The turbulent viscosity is not a fluid property but depends strongly on the nature of turbulence (Nezu and Nakagawa 1993). The introduction of Equation 11.8 alone does not constitute a turbulence model but must be extended to include the distribution of v_t in the flow field. The eddy viscosity v_t is considered to be proportional to a velocity characteristic of the fluctuating motion and to a typical length scale, which Prandtl called the mixing length. The Prandtl-Kolmogorov expression relates the characteristic velocity $\hat{v}$ and the length scale L to the eddy viscosity as:

$$v_t \propto L\hat{v} \tag{11.10}$$

Equation 11.8 also implies the assumption of an isotropic eddy viscosity, which is a simplification and is unrealistic in complex flows, since the velocities and length scales differ at every point in the flow field and hence the viscosity differs in different directions. However, the eddy viscosity model has proved successful in many practical calculations and is still the basis of most turbulence models in use today (Rodi 1980). The turbulent eddy viscosity v_t can be determined with many different turbulence closure strategies. Eddy viscosity models are generally characterized by their number of transport equations (see Figure 11.1).

A few calculation methods for large water bodies include a constant eddy viscosity for the entire flow field. The values of the eddy viscosity can be determined from experiments. This is not considered turbulence modeling, due to the assumption of constant values for the eddy viscosity and is not sufficient to describe the flow correctly when the turbulence terms are important to the behavior of the flow. In open channel flow, the turbulent viscosity is usually 100–1000 times greater than its laminar counterpart.

The first model to describe the distribution of the eddy viscosity, and thus the first proper turbulence model, is the mixing-length model proposed by Prandtl in 1925. Prandtl assumed that the turbulent viscosity v_t is proportional to the mixing length l_m and a velocity scale given by the product of the mean velocity gradient and the mixing length l_m:

$$v_t = l_m^2 \left| \frac{\partial \overline{u_i}}{\partial z} \right| \tag{11.11}$$

The mixing-length l_m is described as follows: If a fluid particle, flowing at its original mean velocity $\overline{u}$, is displaced due to the turbulent motion in the vertical direction from z_1 to z_2, the velocity at z_2 differs from the surrounding velocity by Δu. The distance $(z_2 - z_1)$ at which Δu is equal to the mean transverse fluctuation velocity (i.e., the velocity in the x-direction) is equal to the mixing-length l_m (Schlichting 1958). The mixing-length approach has been applied with great success for relatively simple flows because the distribution of l_m can be prescribed empirically from measurements in open channel flows or specified from simple empirical formulas for different flow situations (Nezu and Nakagawa 1993). The mixing-length model implies that the turbulence is in a state of local equilibrium at each point in the flow and the production and dissipation of turbulent energy are in balance, so that there is no transport of turbulent quantities. Thus, the mixing-length model neglects processes of convective or diffusive transport of turbulence. In open channel flows, the turbulence is produced mainly near the walls and is transported to the center. For this reason the mixing-length model is not suitable because it predicts zero turbulence in the center of the flow and hence zero turbulent viscosity v_t. Therefore, the model is of little use in complex flows because of the great difficulties in specifying the value of l_m. For many simple shear-layer flows where l_m can be specified empirically, the mixing-length model is very useful and therefore a popular tool (Rodi 1980).

Since an empirical length scale is difficult to prescribe for flows that are more complex than unidirectional channel flows, turbulence models were developed to determine the length scale distribution. Therefore, it is assumed that the length scale that characterizes the size of the large turbulent eddies is subject to transport processes in a similar manner to the energy k. Influences on the length scale of the eddies are the downstream convection and the dissipation of the eddies. The balance of these processes has to be expressed in a model transport equation for the length scale L, which can be used to calculate the distribution of L. In the following discussion, two such models that make use of the eddy-viscosity concept of Boussinesq and the Prandtl-Kolmogorov expression are introduced.

The k-ε Model

The k-ε model is the most commonly applied two-equation model. In addition to solving an equation for the kinetic energy k and, hence, for the turbulent velocity scale, the k-ε model employs for the length scale determination not the length L itself as dependent variable, but an equation for the turbulent energy dissipation $\varepsilon \sim k^{3/2}/L$. The semiempirical k-equation and the ε-equation, which are employed in the k-ε model, are written in the form:

$$\underbrace{\frac{\partial k}{\partial t}}_{\substack{\text{rate of} \\ \text{change}}} + \underbrace{\overline{u}_i \frac{\partial k}{\partial x_i}}_{\substack{\text{convective} \\ \text{transport}}} = \underbrace{\frac{\partial}{\partial x_i}\left(\frac{v_t}{\sigma_k}\frac{\partial k}{\partial x_i}\right)}_{\text{diffusive transport}} + \underbrace{v_t\left(\frac{\partial \overline{u}_i}{\partial x_j} + \frac{\partial \overline{u}_j}{\partial x_i}\right)\frac{\partial \overline{u}_i}{\partial x_j}}_{\text{production by shear}} - \underbrace{\frac{\varepsilon}{}}_{\text{dissipation}} \quad (11.12)$$

for the transport of the kinetic energy and:

$$\underbrace{\frac{\partial \varepsilon}{\partial t}}_{\substack{\text{rate of} \\ \text{change}}} + \underbrace{\overline{u}_i \frac{\partial \varepsilon}{\partial x_i}}_{\substack{\text{convective} \\ \text{transport}}} = \underbrace{\frac{\partial}{\partial x_i}\left(\frac{v_t}{\sigma_\varepsilon}\frac{\partial \varepsilon}{\partial x_i}\right)}_{\text{diffusive transport}} + \underbrace{c_{1\varepsilon}\frac{\varepsilon}{k}(P) - c_{2\varepsilon}\frac{\varepsilon^2}{k}}_{\text{generation} - \text{destruction}} \quad (11.13)$$

for the transport of the dissipation rate ε. The term $P = v_t\left(\dfrac{\partial \overline{u}_i}{\partial x_j} + \dfrac{\partial \overline{u}_j}{\partial x_i}\right)\dfrac{\partial \overline{u}_i}{\partial x_j}$ is the generation term of the turbulent kinetic energy and $c_{1\varepsilon}$, $c_{2\varepsilon}$, and σ_ε are empirical constants. Once the distributions of the quantities of k and ε have been determined,

TABLE 11.1   Constants of the k-ε model

c_μ	σ_k	σ_ε	$c_{1\varepsilon}$	$c_{2\varepsilon}$
0.09	1.0	1.3	1.44	1.92

the turbulent eddy viscosity is calculated with the Prandtl-Kolmogorov expression given as:

$$\nu_t = c_\mu \left(\frac{k^2}{\varepsilon} \right) \tag{11.14}$$

The values given in Table 11.1 have been determined for the various constants needed.

These values are based on extensive examination (Launder and Spalding 1974) of free turbulent flows, but can also be used for wall flows. They have been chosen to enable the compatibility of the model with the logarithmic velocity distribution found near the wall when using the von Karman constant of $\kappa = 0.43$. Using a von Karman constant of $\kappa = 0.41$ as measured in open channel flow, values of $\sigma_k = \sigma_\varepsilon = 1.2$ should be adopted (Nezu and Nakagawa 1993).

The k-ω Model

A second rather popular two-equation model that has been increasingly used in engineering applications is the k-ω model of Wilcox (2000). In this model the standard k-equation is solved, but as a length scale equation, the variable ω instead of ε is used. This quantity is called specific dissipation since $\omega \sim \varepsilon/k$. The omega equation reads:

$$\underbrace{\frac{\partial \omega}{\partial t}}_{\substack{\text{rate of} \\ \text{change}}} + \underbrace{\bar{u}_i \frac{\partial \omega}{\partial x_i}}_{\substack{\text{convective} \\ \text{transport}}} = \underbrace{\frac{\partial}{\partial x_i}\left[\left(\nu + \frac{\nu_t}{\sigma_\omega} \right) \frac{\partial \omega}{\partial x_i} \right]}_{\text{diffusive transport}} + \underbrace{\alpha \frac{\omega}{k}(P) - \beta \omega^2}_{\text{generation} - \text{destruction}} \tag{11.15}$$

The eddy-viscosity is computed from the following equation instead of Equation 11.14:

$$\nu_t = \frac{k}{\omega} \tag{11.16}$$

Details on the determination of the empirical parameters α, β, σ_ω, and P in Equation 11.15 are given by Wilcox (2000).

11.3   DISCRETIZATION OF THE GOVERNING EQUATIONS

The first step of a 3-D simulation of open channel flow is to generate a numerical mesh or grid that represents the bathymetry of the channel together with the water body inside. Once a mesh has been defined, the equations can be discretized by replacing the partial derivatives appearing in the governing equations

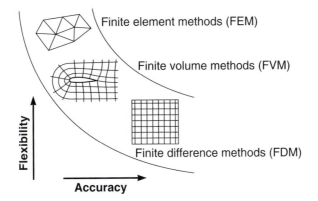

FIGURE 11.2
Classification of discretization methods (Oertel and Laurien 1995).

with algebraic difference quotients. This leads to a system of linear or nonlinear algebraic equations that can be solved for the flow field variables at specific, discrete grid points in the flow. The three most common methods of discretizing the differential operators are finite difference method (FDM), finite element method (FEM), and finite volume method (FVM).

The classification of these methods in terms of flexibility and accuracy can be seen in Figure 11.2. The finite element method is highly flexible, because the discretization in space is based on unstructured grids. Finite volume and finite difference techniques are based on structured grids; finite volumes are more flexible than finite differences because the volume of a cell can be chosen arbitrarily, curvilinear coordinates can be used, and thus boundary-fitted solution domains are possible. However, the finite difference method requires a Cartesian coordinate system for the computations and no coordinate transformations (as in FVM) or weighting functions (as in FEM) are needed. Hence the calculations are the most accurate since round-off errors are minimized. Recently, unstructured and/or block-structured body fitted curvilinear grids within the FV method have been successfully adapted (Olsen 2000; Lilek et al. 1997). These recent advancements in grid adaptability of the finite volume method provide the desired flexibility, and together with the method's accuracy and numerical efficiency, the FV method is employed in the majority of commercial CFD codes applied in research and engineering. In the following, the mathematical formulations and concepts of the FV method are introduced.

Finite Volume Method (FVM)

The finite volume method was first introduced by McDonald (1971) and MacCormack and Paullay (1972) for the solution of the 2-D Euler equations. The domain is divided into a grid with finite volumes represented as quadrangles for two-dimensional calculations or as hexahedral control volumes for the three-dimensional representation (Figure 11.3) lying around the grid points. This method can use both types of meshes, structured or unstructured grids, and several options are available for the definition of

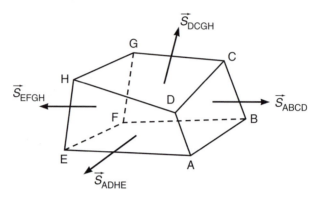

FIGURE 11.3
Finite volume cell for 3-D calculations.

the control volumes around which the conservation laws are expressed. The partial differential equations are integrated over finite control volumes by using the Gaussian theorem. With that, the volume integral of the diverse terms in the Navier-Stokes equations is converted into surface integrals over the faces of the finite volumes. The Gaussian integral theorem is written for a scalar quantity ϕ within the volume Ω as:

$$\int_\Omega (\nabla\phi)d\Omega = \int_S (\mathbf{n}\,\phi)dS = \oint_S \phi\,d\vec{S} \qquad (11.17)$$

where S is the surface of the control volume Ω; $\mathbf{n}$ is the unit outward normal vector; and $d\vec{S}$ is the surface vector pointing outward from the volume. The Gaussian theorem states that the gradient of a scalar quantity ϕ in a control volume Ω is equal to the fluxes going through the surfaces S of this control volume. The governing conservation equations integrated over the finite volume Ω reads:

$$\int_\Omega \frac{\partial}{\partial t}U\,d\Omega + \int_\Omega \frac{\partial}{\partial x_i}(C_i\,U + D_i)d\Omega = \int_\Omega S_i\,d\Omega \qquad (11.18)$$

where $C_i\,U$ are convective terms, D_i are diffusive terms, and S_i are source terms in the conservation laws. The conservation of mass (i.e., the continuity equation) can be obtained by setting the parameters as $C_i = 1$, $D_i = 0$, and $S_i = 0$.

Applying now the Gaussian theorem, the integral can be split into a sum over the six faces of a volume cell (Figure 11.3). The surface vector $\vec{S}_l$ is defined as:

$$\vec{S}_l = \begin{bmatrix} S_1 \\ S_2 \\ S_3 \end{bmatrix} = \mathbf{n}_l A_l \qquad l = 1, 2, \dots, 6 \qquad (11.19)$$

where $\mathbf{n}_l$ is the unit outward normal vector of face l with area A. For each volume cell, the governing equations can be approximated now as:

$$\frac{\partial}{\partial t}U\Omega_{ijk} + \sum_{m=1}^{3}\sum_{l=1}^{6}[(C_i\,U + D_i)_{lm}\,\vec{S}_{lm}]_{ijk} = 0 \qquad (11.20)$$

where Ω_{ijk} is the volume of cell i,j,k in the domain. The sum of the flux terms $[(C_i\,U + D_i)_{lm}\,\vec{S}_l]$ refers to all the external sides of the control cell Ω. The volumes and cell surface areas must be evaluated carefully in order to ensure that the sum of the computed volumes of adjacent cells is indeed equal to the total volume of the combined cells (i.e., the computational domain). For a general quadrilateral ABCD, the area A can be evaluated from the vector products of the diagonals as follows:

$$A_{ABCD} = \frac{1}{2}|\vec{x}_{AC} \times \vec{x}_{BD}|$$

$$= \frac{1}{2}[(x_C - x_A)(y_D - y_B) - (y_C - y_A)(x_D - x_B)] \tag{11.21}$$

The right-hand side of Equation 11.21 should be positive for a face ABCD, where A, B, C, D are located counterclockwise. The outward surface vector $\vec{S}_l$ can be calculated as:

$$\vec{S}_{ABCD} = \frac{1}{2}|\vec{x}_{AC} \times \vec{x}_{BD}| \tag{11.22}$$

The evaluation of the convective flux components over the finite volume sides depends on the selected scheme as well as on the location of the flow variable with respect to the mesh. Direct integration of the Navier-Stokes equations and the simplicity of the Gauss theorem are major advantages of FV methods.

11.4   BOUNDARY CONDITIONS

In a numerical solution of the NS-equations, each control volume provides one algebraic equation expressing fluxes through the CV faces and body forces for every unknown variable. However, control volumes that coincide with the domain boundaries require specification of boundary conditions. The boundary fluxes must either be known or be expressed as a combination of interior data and boundary data. The boundary conditions in the Navier-Stokes equations can be set either as Dirichlet or von Neuman types. The first condition is a direct specification of known values at a boundary, while the latter involves derivatives of the variables to be calculated wherever a direct specification is not possible. For the simulation of an open channel flow, four distinct boundary types have to be considered: (1) inflow boundary, (2) outflow boundary, (3) free surface boundary, and (4) wall boundary.

Inflow Boundary Conditions

When modeling an open channel flow, the discharge of the flow is known and convective fluxes are usually prescribed at the inlet. By use of the continuity equation, the cross-sectional average velocity over the inlet area can be set as a Dirichlet boundary condition (i.e., it is specified directly). If wall-bounded flows are modeled, the specification of a velocity profile (instead of an average value) at the inlet may be more appropriate, since for transportive flows (i.e., flows where convection dominates over diffusion), a certain number of grid points in the downstream direction are needed until the boundary layer has been fully developed. The following vertical velocity profile of a fully developed channel flow can be used (Zierep 1982):

$$\bar{u} = u_0 \left(\frac{z}{h}\right)^{1/n} \tag{11.23}$$

where h is the average water depth and u_0 is the maximum velocity. The range of the power $1/n$ can be chosen according to the Reynolds number and roughness condition.

This velocity profile formulation is only a rather coarse approximation of the boundary layer in the channel. However, its simplicity (with only the power $1/n$ being unknown a priori) has made it a popular choice as an inlet boundary condition because more complex formulations like the log-law involve unknown parameters such as the shear velocity and roughness length. It is recommended to position the inlet boundary as far upstream of the region of interest as possible so that the flow can become fully developed. From the velocity distribution at the inlet, convective fluxes can be calculated directly at the inlet plane. Diffusive fluxes can also be specified at the inlet. Usually, diffusive fluxes are calculated with the central differencing scheme where the flux between two cells is calculated at the center of the two adjacent nodes by linear interpolation. For a boundary node, where the center of the cell is the boundary node itself, diffusive fluxes are calculated as one-sided differences and no explicit modification is necessary. This is an approximation of first-order accuracy, which again requires the inlet boundary to be placed as far upstream from the region of interest as possible.

Outflow Boundary Conditions

At an outflow boundary, where the flow leaves the calculation domain, neither the value of the flow variable nor its flux is known. According to Patankar (1980), no boundary information is needed for convection-dominated flows (as is the case in open channel flow), since the values at the boundary do not influence the solution in the rest of the domain. However, the outflow boundary has to be placed at an appropriate location where the flow is everywhere directed outward so that any inaccuracy in estimating the outlet conditions will not propagate upstream (see Figure 11.4).

The easiest way to approximate the values at the outlet boundary is that of zero gradients along gridlines. For the convective terms this treatment gives:

$$\frac{\partial u_i}{\partial x_i} = 0 \tag{11.24}$$

Diffusive fluxes are treated similarly to the inlet (i.e., with one-sided differences). Another important feature of the outlet condition is the satisfaction of overall

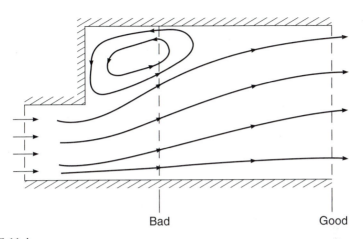

Bad   Good

FIGURE 11.4
Good and bad choice for the location of outlet boundary (Patankar 1980).

continuity (i.e., the sum of outlet mass fluxes $\dot{m}_{out}$ equals the sum of inlet mass fluxes $\dot{m}_{in}$). This can be done by correcting the outflow velocity as follows (Nilsson and Davidson 1998):

$$\left(u_{i,out}\right)^{l+1} = \left(u_{i,out}\right)^{l} + u_{i,incr} \tag{11.25}$$

where the velocity at the outlet of the previous iteration is corrected with an incremental velocity defined as:

$$u_{i,incr} = \frac{\dot{m}_{in} - \dot{m}_{out}}{\rho\, A_{out}} \tag{11.26}$$

where A_{out} is the cross-sectional area of the outlet.

Free Surface Boundary

The only free surface boundary in the domain of an open channel flow is the water surface. If wind-induced shear stresses on the surface are neglected, the water surface can be considered as a symmetry plane. Thus, the normal gradients as well as convective fluxes are zero for all quantities with symmetrical character such as a tracer, or the turbulence model variables k and ε. Velocity components normal to the water surface and shear stresses are also zero. When a shear layer is created by forces near the surface (e.g., wind-shear, ice-cover, or a moving lid), then the boundary conditions given above for wall boundaries can be applied and the free-surface is considered as a moving wall. In open channel flows, the free surface represents a boundary between water and air, where the following boundary conditions apply (Ferziger and Peric 1996): (1) *the kinematic condition* states that there is no convective mass transfer through the free surface, hence the fluid velocity component normal to the free surface is equal to the free surface velocity; and (2) *the dynamic condition* requires that forces acting on two fluids at a sharp interface are in equilibrium. In the absence of surface-tension effects, this condition reduces to the statement that the fluid stresses on both sides of the free surface in its immediate vicinity are equal. If a further assumption is made that viscous effects are negligible, as is often the case, the dynamic condition reduces to the statement that the pressure on either side of the interface is equal.

There are two principal possibilities for the calculation of the free surface (Ferziger and Peric 1996). *Interface tracking methods* are schemes that define the free surface as a sharp boundary whose motion is followed. Boundary-fitted, moving grids are required along with a solution algorithm to readjust the grid each time the surface has moved. *Interface capturing methods* are methods that do not define a sharp boundary between two surfaces. The grid of the calculation domain extends beyond the free surface and the free surface position is determined by a fraction *FA* ($0 \leq FA \leq 1$), which defines the percentage of the cells being filled with water. The volume of fluid method (VOF) is such an approach, where the fraction of the liquid phase is determined by the solution of a transport equation.

Wall Boundary Conditions

At an impermeable wall, the following no-slip condition applies:

$$u_i \bigg|_{\text{wall}} = 0 \tag{11.27}$$

Thus, the convective fluxes in normal and tangential direction of all quantities are zero at the wall. Diffusive fluxes in the momentum equations are nonzero. This follows since the Navier-Stokes equations include the viscosity of the fluid, so that along solid walls this no-slip condition must be set. At the wall the velocity has to be arranged into a component normal to the wall (v_n) and a component tangential to the wall (v_t). If the first grid point is placed in the viscous sublayer (i.e., $y^+ < 11$), the viscous stresses at a wall in normal and tangential directions, respectively, are:

$$\tau_{nn} = 2\mu \left(\frac{\partial v_n}{\partial n} \right)_{\text{wall}} = 0, \quad \tau_{nt} = \mu \left(\frac{\partial v_t}{\partial n} \right)_{\text{wall}} \tag{11.28}$$

Here it is assumed that the shear force that acts in the coordinate direction t is parallel to the projection of the velocity vector onto the wall. As discussed in Chapter 4, the vertical velocity profile near a wall is divided into the viscous sublayer, a buffer layer, and a logarithmic layer (see Figure 4.2). For high Reynolds number flows (as in open channels), resolving the viscous sublayer is undesirable since steep gradients prevail, so that for proper resolution, many grid points have to be placed in this layer and the computation becomes very expensive. Moreover, viscous effects are very important in this layer so that for high Reynolds numbers the layer is so thin that it is difficult to use enough grid points to resolve it.

Launder and Spalding (1974) suggest that integration through the sublayer is not necessary, since the universal law of the wall (Equation 4.13c), which is of sufficient generality, can be taken to connect wall shear stresses to the dependent variables outside the viscous sublayer.

Surface Roughness of Rough Open Channels

Grain and bedform roughness represent the boundary roughness of natural rivers and are responsible for the boundary layer. Many empirical formulas have been developed in the past to account for the roughness of channels (e.g., Chezy's [Equation 4.7] or Manning's [Equation 4.9]) in which the roughness coefficients such as the Manning's n account for the energy losses due to friction as well as due to bedforms. Many tables and guidelines do exist for the choice of the coefficients, but the empirical values can give only a rather coarse estimation of the roughness in a channel and values may vary for different water depths, discharges, and so on. As a result, the greatest difficulty lies in the determination of the roughness coefficients. Friction losses caused by grains covering a channel bed and the resulting bedforms have been investigated in great detail in the past. The range of grains covering a channel bed varies from very fine sand in lowland areas with dunes up to cobbles in mountainous gravel bed rivers with step-pool systems. An easy and probably the most detailed way to identify the roughness of natural rivers is by taking samples out of the river bed and running a sieve analysis so that the different characteristic diameters can be obtained. With that an estimation of the equivalent sandgrain roughness k_s can be gained using an appropriate conversion formula (see Table 11.2). With knowledge of k_s, Schlichting's (1958) suggestion of the log-law for rough walls can be employed (see Equation 4.13c):

$$\frac{\bar{u}}{u_*} = \frac{1}{\kappa}\ln\left(\frac{30z}{k_s}\right)$$
(11.29)

It has been shown that the inclusion of this formula in a CFD model gives a fairly accurate prediction of the vertical velocity profile over walls roughened by sand-grains with little calibration (Olsen 1991).

TABLE 11.2   Approaches to determine k_s depending on bed material

Author	Approach	Typical bed material
Einstein (1942)	$k_s = d_{65}$	Sand
Engelund / Hansen (1966)	$k_s = 2\ d_{65}$	Sand and gravel
Garbrecht (1961)	$k_s = d_{90}$	Sand
Hey (1979)	$k_s = 3.5\ d_{84}$	Coarse gravel
Mertens (1997)	$k_s = 2.5\ d_{50}$	Sand and gravel
Dittrich (1998)	$k_s = 3.5\ d_m$	Middle coarse gravel
Dittrich (1998)	$k_s = 3.5\ d_{84}$	Coarse gravel

Source: Dittrich 1998.

With Table 11.2, a fairly good assessment of equivalent roughness in natural channels can be given independent of stage or discharge. Moreover, a number of published guidelines and tables exist, similar to the tables for Manning's n (see Table 4.1), to estimate the roughness height in artificial channels (e.g., channels with a concrete bed) or in rivers where there is no information about the grain size.

Vegetative Roughness of Natural Rivers

The determination of vegetative roughness in natural rivers is probably the most difficult case of river roughness closure. In the past, vegetation in open channels was treated as additional flow resistance to be added to the bed roughness. Cowan (1956) and Chow (1959) first introduced this approach by producing recommended values for the Manning's coefficient to account for the additional form resistance due to grass, bushes, or trees. More recently, Masterman and Thorne (1982) introduced the conveyance method, where the roughness is varied across the channel width by subdividing into main channel and vegetated floodplains. As discussed in Section 4.12, many variations of this method exist to additionally account for the production of turbulence energy within the contact region between main channel and floodplains. However, these methods are considered to be one-dimensional, and the flow resistance values have to be calibrated. For multidimensional models, Shimizu and Tsujimoto (1994), Lopez and Garcia (1997), Fischer-Antze et al. (2001), and Stoesser et al. (2003) used an approach that introduces drag forces as sink terms into the Navier-Stokes equations to account for vegetational flow resistance. Drag on a vegetative element can be calculated as:

$$F_P = \rho \frac{u^2}{2} C_D \lambda_P \tag{11.30}$$

where C_D = drag coefficient and the vegetative coefficient λ_P is defined as:

$$\lambda_P = \frac{projected\ area\ of\ plant}{total\ volume} = \frac{A_P}{h \cdot a_x \cdot a_y} \tag{11.31}$$

in which A_p is the momentum absorbing area of a plant, a_x and a_y are characteristic lengths, and h is the local water depth. The characteristic lengths and the momentum absorbing area can be determined according to Figure 11.5 (DVWK 1991).

The drag force approach is used in analogy to the flow around a rigid cylinder, assuming uniform and undisturbed approach flow conditions. In such a case the C_D value is known to be unity at high **Re** numbers, and usually this has been used due to the lack of knowledge about C_D values under certain conditions. More sophisticated numerical model approaches to account for vegetative resistance, such as an explicit treatment of the plant stems and a modified porosity drag model to account for biomechanics of the momentum absorbing leaf area, are the subject of ongoing research. In the following practical example, the cylinder-based drag force model for vegetative resistance was used with rather good success.

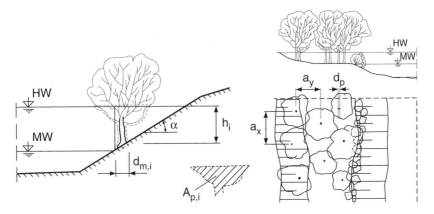

FIGURE 11.5
Definition of momentum absorbing area (left) and relative location of vegetative elements in the flow domain (right) (DVWK 1991).

11.5 CASE STUDY

This section describes the application of a three-dimensional computational fluid dynamics (CFD) model to a large-scale river restoration and flood protection scheme. A 100-year flood event on a considerable reach length (3500 m) of the lower River Rhine in southwest Germany was simulated in order to demonstrate that the desired retention volume could be met. The model was validated with mean floodplain velocity data that were measured using dilution gauging techniques during a 100-year stormwater event.

Background

River restoration is an area in its infancy and few recommendations are available to aid the practitioner to assess the impact of restoring river meanders, floodplains, and, in particular, vegetation on the local velocity distribution and the flood conveyance of the system. In the CFD study described here, the flow in a vegetated two-stage channel of a 3500 m reach of the lower River Rhine, Germany, was simulated in order to predict the response of a river restoration scheme to a stormwater event. The restoration of the Rhine over a 42 km reach had two main objectives: First, it aimed to provide flood protection for the River Rhine for a 200 year flood event. Second, and of equal importance, it aimed to reinstate a wetland and freshwater environment rich in species diversity. Previously, the upper reach of the Rhine had been regulated by dams, and this had not only reduced its flood conveyance capacity but also caused a considerable decline in both aquatic species and plant communities. The design storage volume can only be achieved by the establishment of a predesigned water level along the reach during a flood event to be attained by considerably higher roughness on the restored floodplains due to riparian forest vegetation. In addition to providing a flood retention volume, it is necessary to ensure that the floodplains are designed in such a manner that both the propagating floodwave front and its recession are damped.

Thus, an integrated hydraulic and ecological approach was chosen to establish both the magnitude of hydraulic resistance needed to reach the desired retention capacity and the corresponding type of vegetation necessary to fulfill this task. Furthermore, it was essential to predict the resulting floodplain flow properties and to assess whether these flow conditions are appropriate for the natural establishment, preservation, and succession of the riparian flora and fauna.

Model and Setup

A 3-D numerical model developed at Bristol University (Stoesser 2002) was applied to six relevant reaches of the Rhine. A 3-D approach was considered appropriate in order to capture the variation in streamwise and cross-streamwise velocity with depth that is induced by channel curvature, natural bed forms, and riparian vegetation. The model was validated in two stages. First, computed water levels were compared against measured water levels available for two high flow scenarios. Second, measurement of the floodplain flow velocity was conducted during a 100-year flood by dilution gauging to allow the computed floodplain velocities to be verified. This two-stage validation process ensured that the model accurately described the hydraulic resistance of a vegetated floodplain and correctly predicted the fully three-dimensional velocity distribution characteristic for vegetated floodplain flows.

The CFD model calculates hydrodynamics for a general three-dimensional geometry, discretized by the finite volume method on curvilinear coordinates. The Reynolds-averaged Navier-Stokes equations are solved, and turbulence is closed with a two-equation k-ε turbulence model. The following boundary conditions (b/c) were used. The inflow boundary was set as a Dirichlet b/c; since only the upstream discharge is known, it is prescribed at the inlet. At the outflow boundary, neither the value of the flow variable nor the flux is known. Therefore, the water level is kept constant and a zero variable gradient condition (von Neumann b/c) was used. The water surface was treated as a free surface boundary condition and was modeled with an interface tracking method as described in Stoesser (2002). The impermeable walls (i.e., the river bed and banks) were treated with the log-law wall b/c that connects wall shear stresses to the dependent variables. This in turn relies on bed roughness expressed through the equivalent grain roughness k_s (Stoesser 2002). To model the additional flow resistance of submerged or emergent vegetation, the drag force on a rigid obstacle was introduced as a sink term into the RANS equations. The corresponding vegetation geometric factors (i.e., plant diameter and spacing) were determined through fieldwork. An established groin field was used as a reference site in order to quantify the vegetation parameters of the existing plant communities.

Floodplain velocities were measured through tracer tests during a flood with a peak discharge of 3600 m^3/s that occurred on the River Rhine in 1999 as a result of heavy rainfalls and snowmelt in the Alps. This corresponded to a flood event of a 100-year return period. For the dilution gauging, the tracer used was Uranine, which is green-yellow in appearance. A gulp injection method was employed and the tracer was released from a boat. The average value of the mean floodplain flow velocity was found to be $V_{fp} = 1.10$ m/s with a standard deviation of 0.07 m/s.

The reach of the Rhine between km 190.0 and km 193.46 was discretized according to surveyed cross sections. For the CFD simulations, a mesh of 198,144 cells was constructed comprising $258 \times 64 \times 12$ cells in the stream-wise, cross-streamwise, and vertical directions, respectively (Figure 11.6). This gave an approximate cell size of 13 m by 3 m by 0.5 m. The grain roughness k_s was calculated from the average gradation curve of the bed material determined through particle sieve analysis. The equivalent grain roughness height k_s varied in the range 0.31–0.67 m depending on the formula employed (Table 11.2). The bed boundary roughness of the river channel was initially set to a roughness height $k_s = 0.45$ m evaluated following a formula suggested by Dittrich (1998) for a gravel bed river.

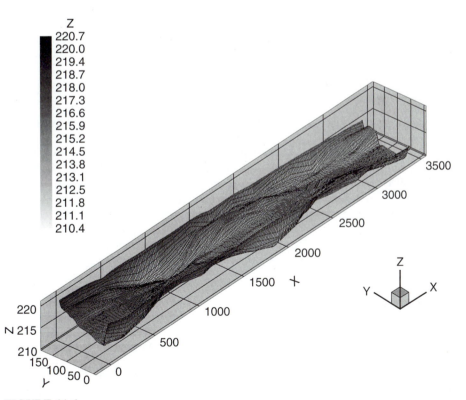

FIGURE 11.6

Computational grid used for the simulations, where z is the bed elevation (all dimensions are given in meters).

Results and Discussion

Water Level Predictions

While the aim of the model had been to utilize a physically based approach, the parameterization of grain roughness and vegetation geometry from fieldwork undoubtedly contained some error. For example, Table 11.2 demonstrates that even in this well-defined case where particle sieve analysis had been conducted, the choice of bed roughness value is dependent on the choice of formula. Hence, to provide increased confidence in the hydraulic resistance values, the bed roughness of the main channel and the form roughness of the vegetated floodplains were independently verified using two different flow conditions. These correspond to flow situations where both the water level observations and discharge measurements were available. The event in 1998 had a discharge of $Q = 700$ m³/s, and the flood event in 1994 had a discharge of $Q = 3040$ m³/s. During the 1998 event, most of the vegetated floodplains were not inundated, hence this flow condition was used to verify the value of the main channel bed roughness. For this flow the bed friction parameter was calibrated, and a value of $k_s = 0.35$ m was found, which is equivalent to the value suggested by Mertens' formula. Good correspondence could be obtained between the measured and computed water surface profiles for this event (see Figure 11.7). The vegetative roughness coefficient λ_p was verified using the larger flow event (3040 m³/s) where all floodplains were inundated. The parameter λ_p

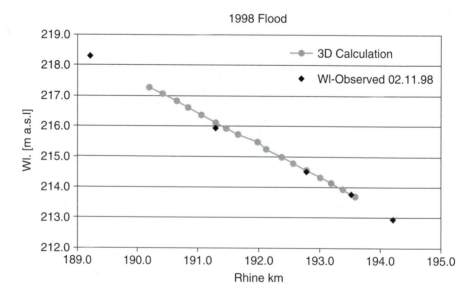

FIGURE 11.7

Comparison of observed and computed water surface elevation after the main channel bed roughness was calibrated for the 700 m³/s flood event in 1998. (The vertical axis denotes water level in meters above sea level.)

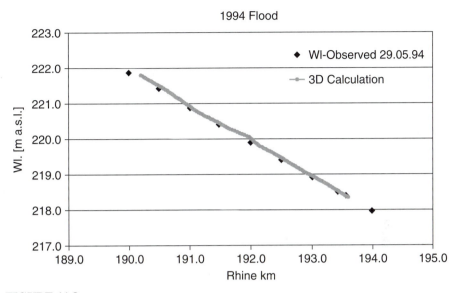

FIGURE 11.8

Comparison of observed and computed water surface elevation after the floodplain vegetative parameters were calibrated for the 3050 m³s⁻¹ flood event in 1994. (The vertical axis denotes water level in meters above sea level.)

was adjusted from 0.11 m⁻¹ to 0.082 m⁻¹ by the variation of plant diameter D in order to achieve a good match between the observed and computed water levels (see Figure 11.8). This need for recalibration was predominately due to errors in determining vegetation geometric and biomechanic properties and indicated the level of precision necessary to achieve these measurements. With the bed roughness calibrated from the 700 m³/s event and the vegetation roughness calibrated from the Q = 3040 m³/s event, the 100-year flood was simulated. As Figure 11.9 shows, the agreement between observed and calculated water levels was likewise satisfying. This illustrates that this method requires a minimal calibration effort, since the determination of roughness parameters is based on the physics of the flow resistance for both bed roughness and vegetation roughness.

Flow Field

Figure 11.10 presents streamwise velocities in the selected river reach. The presence of the vegetated groin fields on alternating sides along the reach cause a reduction in velocity within the floodplain relative to the main channel (Figures 11.9 and 11.10). The computed near surface velocities are up to 3.8 m/s. There is a marked reduction in velocity at the main channel/riparian floodplain interface, where floodplain flow velocities are approximately 1.0 m/s to 1.5 m/s. On the floodplain, the surface velocity distribution is fairly uniform and in the order of 1.0 to 1.2 m/s corresponding well with the measured value from the dilution gauging experiments. Figure 11.11 presents streamwise velocity distributions at selected

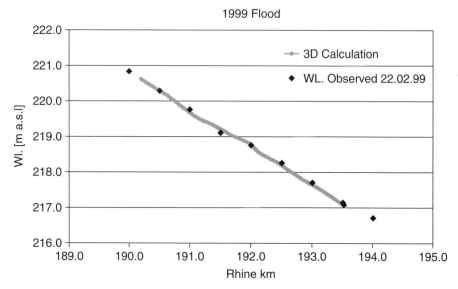

FIGURE 11.9

Independent verification of the computed water surface elevation against the observed water surface elevation for the 100-year flood using the calibrated parameters obtained from the 1994 and 1998 events. (The vertical axis denotes water level in meters above sea level.)

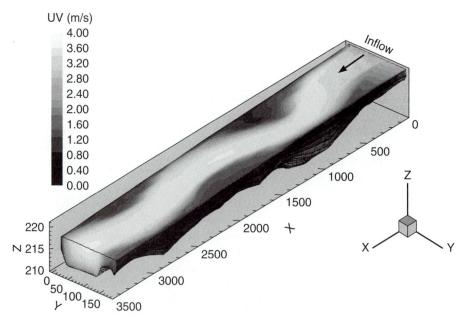

FIGURE 11.10

Flow velocities of the River Rhine reach (length scales given in meters).

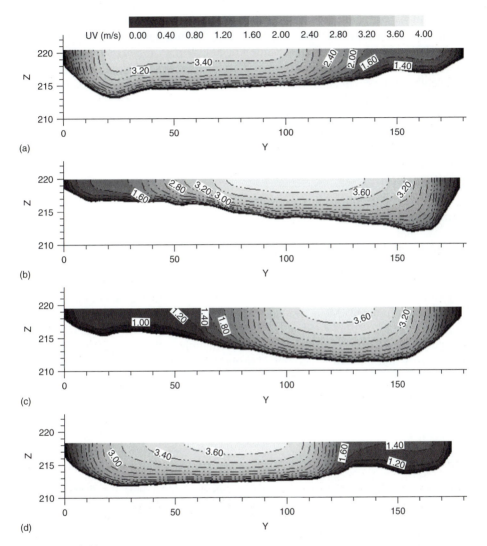

FIGURE 11.11
Cross-sectional distribution of the resultant velocity in (a) Rhine km: 190.42, (b) Rhine km 190.84, (c) Rhine km 191.29, (d) Rhine km 191.94 (all length scales in meters).

cross sections (looking in the upstream direction and exaggerated in the vertical by a factor of 3). In contrast to the main channel velocity distribution with depth, the floodplain velocity follows a relatively uniform vertical velocity profile. A submerged vegetation layer can be regarded analogous to a roughness sublayer, where the velocity distribution is rather linear than logarithmic. The reason for this is that the flow resistance of emergent vegetation acts over the whole water depth and retards the flow considerably. This is in agreement with findings from experimental investigations where the velocity profile within both rigid and flexible vegetation layers has been shown to no longer follow the logarithmic law profile (e.g., Tsujimoto et al. 1991). The cross section at Rhine km 191.29 (i.e., Fig. 11.11c) corresponds to the tracer test site. Here, a uniformly distributed velocity (over floodplain width and depth) of approximately 1.0 m/s was computed, which corresponds well with the observed mean velocity from the tracer field tests. Figures 11.10 and 11.11 also show the tendency of the relatively straight river reach to meander due to the alternating location of the floodplains and the vegetation on the river banks. There is a trend for the higher velocities to be predicted mainly in the deeper parts of the main channel.

Conclusions

It was shown that while the roughness closure approach presented here does not eliminate entirely the process of calibration, the numerical method significantly reduces the modeling uncertainties associated with traditional 1-D methods, which rely entirely on calibration. A 100-year flood event on a river reach of considerable length was simulated. Mean floodplain velocities were measured using dilution gauging techniques and good agreement was found between the measured and computed floodplain velocities. Given information such as vegetation distribution/density and plan view geometric properties of the vegetation, the proposed 3-D model was able to predict floodplain velocities, water elevations, and hydrodynamic features indicative of vegetated two-stage channel flow.

REFERENCES

Boussinesq, J., "Essai sur la théorie des eaux courantes." Mém. prés. par divers savants à l'Acad. des Sci. de l'Inst. Nat. de France, 23 (1) , 1–680, 1877.

Chow, Ven Te. *Open Channel Hydraulics*. New York: McGraw-Hill, 1959.

Cowan, W. L. "Estimating Hydraulic Roughness Coefficients." *Agricultural Engineering*, 37, no. 7 (1956), pp. 473–75.

Deardoff, J. W. "A Numerical Study of Three-Dimensional Turbulent Channel Flow at Large Reynold Numbers." *J. Fluid Mech.* 41, no. 2 (1970).

Dittrich, A. "Wechselwirkung Morphologie/Stroemung naturnaher Fliessgewaesser. Habilitationsschrift. " Mitteilungen des Institutes fuer Wasserwirtschaft und Kulturtechnik. No. 198. Universität Karlsruhe, 1998.

DVWK. "Hydraulische Berechnung von Fließgewaessern." Merkblaetter. 220/1991. Hamburg und Berlin: Verlag Paul Parey, 1991.

Ferziger, J. H. "Subgrid-Scale Modeling." In *Large Eddy Simulation of Complex Engineering and Geophysical Flows,* ed. B. Galperin, and S. A. Orszag. Cambridge University Press, 1993.

Ferziger J. H., and M. Peric. *Computational Methods for Fluid Dynamics*. Heidelberg: Springer Verlag, 1996.

Fischer-Antze, T., T. Stoesser, P. D. Bates, and N. R. B. Olsen. "3D Numerical Modelling of Submerged Vegetation." *J. Hydr. Res.* 39. no.3 (2001).

Kim, J., P. Moin, and R. Moser. (1987). "Turbulence Statistics In Fully Developed Channel Flow At Low Reynolds Number." *J. Fluid Mech.* 177, (1987), p. 133.

Launder, B. E., and D. B. Spalding. "Progress in the Development of a Reynolds Stress Turbulence Closure." *J. Fluid Mech.* 68, part 3 (1974), pp. 537–66.

Leonard, A. "On the Energy Cascade in Large-Eddy Simulations of Turbulent Fluid Flows." *Advanced Geophysics* 18A (1974), p. 237.

Lilek, Z., S. Mustaferija, M. Peric, and V. Seidl. "Computation of Unsteady Flows Using Non-Matching Blocks of Structured Grid." *Numerical Heat Transfer,* Part B, no. 27. (1997).

Lopez, F., and M. Garcia. "Open Channel Flow Through Simulated Vegetation: Turbulence Modelling and Sediment Transport." Hydrosystems Laboratory, Department of Civil Engineering, University of Illinois, 1997.

MacCormack, R. W., and A. J. Paullay. "Computational Efficiency Achieved by Time Splitting of Finite Difference Operators." American Institute of Aeronautics and Astronautics. Paper 72-154, 1972.

McDonald, P. W. "The Computation of Transonic Flow Through Two-Dimensional Gas Turbine Cascades." American Society of Mechanical Engineers, Paper 71-GT-89, 1971.

Masterman, R., and C. R. Thorne. "Predicting Influence of Bank Vegetation on Channel Capacity." *J. Hydr. Engrg.* ASCE 118, no.7 (1982), pp. 1052–58.

Nezu, I., and H. Nakagawa. *Turbulence in Open-Channel Flows.* IAHR Monograph Series. Rotterdam, Brookfield: A.A. Balkema, 1993.

Nilsson, H., and L. Davidson. *CALC-PVM: A Parallel SIMPLEC Multiblock Solver for Turbulent Flow in Complex Domains.* Internal Report Nr. 98/12. Chalmers University of Technology, Goeteborg, 1998.

Oertel, H., and E. Laurien. *Numerische Stroemungsmechanik.* Berlin, Heidelberg, New York, Tokyo: Springer Verlag, 1995.

Olsen N. R. B. "A Three Dimensional Numerical Model for Simulation of Sediment Movements in Water Intakes." Dissertation for the Dr. Ing. Degree. The Norwegian Institute of Technology, Division of Hydraulic Engineering, University of Trondheim, 1991.

Olsen, N. R. B. "Unstructured Hexahedral 3D Grids for CFD Modelling in Fluvial Geomorphology." *Hydroinformatics* 2000, Iowa, 2000.

Patankar, S. V. *Numerical Heat Transfer and Fluid Flow.* Washington, New York, London: Hemisphere Publishing Corporation, 1980.

Petryk, S., and G. Bosmajian. "Analysis of Flow Through Vegetation." *J. Hydr. Engrg.,* ASCE 101, no.7 (1975), pp. 871–84.

Prandtl, L. "Bericht über Untersuchungen zur ausgebildeten Turbulenz. Z. Angew." *Math. Mech.* 5 (1925).

Reynolds, W. C. "The Potential and Limitations of Direct and Large Eddy Simulations." *Whither Turbulence? Turbulence at the Crossroads.* Proceedings of a Workshop Held at Cornell University, Ithaca, NY, March 22–24, 1989. Ed. J. L. Lumley, *Lecture Notes in Physics,* vol. 357, pp. 313–43.

Roache, P. J. *Verification and Validation in Computational Science and Engineering.* Albuquerque: Hermosa Publishers, 1998.

Rodi W. *Turbulence Models and Their Applications in Hydraulics.* IAHR Monograph Series. Rotterdam, Brookfield: A.A. Balkema, 1980.

Rodi, W., et al. "Status of Large-Eddy Simulation: Results of a Workshop." *J. Fluids Engrg.* 119 (1997), pp. 248–62.

Schlichting, H. *Grenzschichttheorie.* Verlag G. Braun, Karlsruhe, 1958.

Stoesser, T. "Development and Validation of a CFD Code for Open-Channel Flows." PhD Thesis, Department of Civil Engineering, University of Bristol, 2002.

Stoesser, T., C.A.M.E. Wilson, P. D. Bates, and A. Dittrich. "Application of a 3D Numerical Model to a River with Vegetated Floodplains." *IAHR/IWA Journal of Hydroinformatics* 5 (2003), pp. 99–112.

Shimizu, Y., and T. Tsujimoto. "Numerical Analysis of Turbulent Open Channel Flow Over a Vegetation Layer Using a k-ε Turbulence Model." *Journal of Hydroscience and Hydraulic Engineering* 11, no.2 (1994), pp. 57–67.

Tsujimoto, T., T. Shimizu, and T. Okada. *Turbulent Structure of Flow over Rigid Vegetation-Covered Bed in Open Channels.* KHL Progressive Report 1. Hydraulic Laboratory, Kanazawa University, Japan, 1991.

Versteeg, H. K., and W. Malalasekera. *An Introduction to Computational Fluid Dynamics.* 2nd ed. Essex, England: Pearson Education Limited, 2007.

Wilcox, D. C. *Turbulence Modeling for CFD.* 2nd ed. DCW Industries, La Canada, USA, 2000.

Zierep, J. *Grundzuege der Stroemungslehre.* Karlsruhe: Verlag G. Braun, 1982.

APPENDIX A

Numerical Methods

A.1 INTRODUCTION

Numerical techniques are methods and algorithms for obtaining approximate solutions of algebraic and differential equations that can be programmed in computer language. Such algorithms must be efficient and accurate. The efficiency of an algorithm refers to it having as small a number of logical steps and execution time as possible. Efficiency also can entail minimizing computer memory requirements. The efficiency of the program that actualizes the algorithm also is important and has resulted in structured programming concepts. Programs developed in structured modules, which avoid indiscriminate branching, are easier for the user to read, understand, and modify, if necessary.

The accuracy of a numerical algorithm is essential. Roundoff error or truncation error can become so large as to "swamp" the numerical solution and make it numerically unstable. Error analysis can provide some help in identifying instability problems, but improving accuracy sometimes is a matter of experience obtained from numerical experiments with a particular algorithm.

Numerical methods that are iterative in nature must satisfy some expectation of convergence toward the true solution. Sometimes a trade-off must be made between efficiency and accuracy. The most efficient algorithm may diverge in some kinds of problems. There is no hope of accuracy in this situation, which is described more aptly in terms of the reliability of the algorithm.

This appendix provides only a few numerical techniques that are used throughout the text. For a more complete discussion of numerical techniques, refer to the list of references following this appendix. The numerical techniques presented here are given as procedures (subprograms) for use in standard modules of Visual BASIC. The procedures are easily translated to other languages.

A.2   NONLINEAR ALGEBRAIC EQUATIONS

Nonlinear algebraic equations are encountered often in open channel hydraulics. For example, the problems of critical and normal depth determination require either a trial-and-error or graphical solution without a numerical solver. Nonlinear algebraic equations can be placed in the form $F(y) = 0$, the root of which is the solution of the equation. Consider the following algebraic equation in the unknown y, for example, in which a and c are positive constants:

$$y + \frac{a}{y^2} = c \tag{A.1}$$

This equation can be rearranged in the form

$$F(y) = y + \frac{a}{y^2} - c = 0 \tag{A.2}$$

which can be solved graphically by finding the point at which the function $F(y)$ crosses the y axis or, in other words, by finding the zero of the function. This equation actually has two roots, which can complicate the root search. It represents the equation for alternate depths associated with a known specific energy in a rectangular channel. In this case, knowledge of the critical depth, which separates the two roots, aids the solution process. Generally, such knowledge of the physical basis of an equation, or at least of its graph, can be of great assistance in choosing an appropriate algorithm.

Three methods for solving such equations in the form $F(y) = 0$, in which y is the root, are given in this section. There are several additional methods, but the methods chosen illustrate the trade-off between efficiency and reliability.

Interval Halving Method

The first method is not very sophisticated but very reliable. Called the *interval halving* or *bisection method,* it easily is illustrated by the high-low number guessing game. For example, a professor thinks of a number between 1 and 100, asks a student to guess the number, and then provides feedback to the student as to whether the guess was high or low. This information provided to the student brackets the range within which the unknown number lies. The straight-A student then halves the interval after receiving the "high-low" feedback for each guess and converges rapidly toward the final number in this way.

The general strategy of interval halving then is to divide an interval, in which a root of the equation is known to occur, into equal parts. Which half of the interval the root actually is in can be determined by comparing the sign of the function, F, at the midpoint with the sign at the left boundary of the interval. The interval is

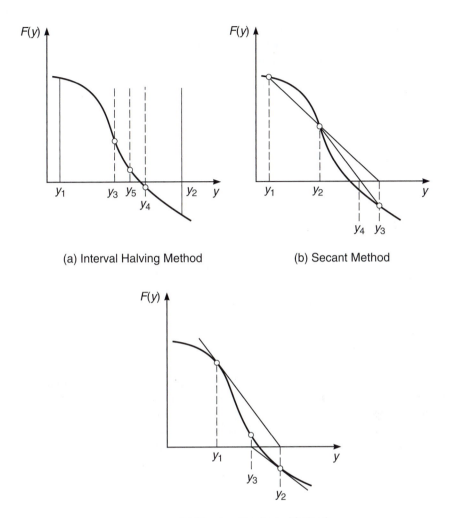

(a) Interval Halving Method

(b) Secant Method

(c) Newton-Raphson Method

FIGURE A.1
Graphical depiction of methods for solution of nonlinear algebraic equations.

halved again, and this process is continued until the root is approached as shown in Figure A.1a. The general procedure is summarized as follows:

1. Form a function, F, from the equation to be solved such that the root of $F(y) = 0$ is the solution.
2. Find the value of the function at the left boundary of the interval from y_1 to y_2 and store it in FY1.
3. Halve the interval between y_1 and y_2 by computing $y = (y_1 + y_2)/2$.
4. Evaluate the function $F(y)$ and store it in FY.

5. Compare the signs of FY1 and FY by multiplying. If they have the same sign, y is the new left boundary. If they are opposite in sign, y is the new right boundary.
6. Repeat steps 3–5 until some error criterion is met.

The absolute error ERABS in the interval halving method is reduced by half in each iteration so that it is given by

$$\text{ERABS} = \frac{y_2 - y_1}{2^n} \tag{A.3}$$

in which n = number of iterations and $(y_2 - y_1)$ = the original interval size. The error can be controlled by choosing the number of iterations, but an approximate relative error criterion may be more useful for stopping the computations. The difference between the current estimate of the root and the most recent estimate is divided by the latter to establish an approximate relative error. When this value falls below a specified value, the computations are stopped.

The advantage of the interval halving method is that it always brackets the root (provided a single root exists in the original interval). This avoids divergence and makes it very reliable. Its efficiency, however, is not as good as the other two methods presented here, because it usually requires more iterations to achieve the same accuracy.

The interval halving method is presented in Figure A.2a in Visual BASIC code as a procedure that can be placed in a standard code module. This procedure can be invoked from a form module, which also handles the chores of data input, printing output, and graphing the function, if desired. The procedure requires input values for the ends of the original interval Y1 and Y2. An additional procedure that evaluates the function $F(y)$ must also be supplied. Note that only one functional evaluation is required in each iteration and that its sign is compared with the sign of the functional value at the left boundary through the computation of FZ. First the value of FZ is checked to see if it is zero. A value of zero for FZ obviously means that the root has been found "exactly," which may seem unlikely, but this can happen if the error criterion is so restrictive that the significant figure limit of the computer has been reached in two successive computations. If FZ is negative, then the function has crossed the y axis and the midpoint becomes the new right hand endpoint (Y2 = Y3). A positive value of FZ, on the other hand, signifies that the midpoint should become the new left hand endpoint (Y1 = Y3). The algorithm works equally well if the function is increasing or decreasing with y. Some care must be exercised, however, in defining the function, so that it does not have a singularity at either endpoint.

An error message can be written in case the error criterion is not met in the specified number of iterations (I = 50). In this event, all that is necessary is to increase the maximum number of iterations. In fact, the maximum number of iterations can be made large enough using (A.3) that this will never happen for a specific problem of interest so that exit from the FOR-NEXT loop always occurs through the relative error check with the specified value of ER. If no root exists on the chosen interval, the algorithm converges toward the right hand endpoint and may even satisfy the error criterion. This eventuality is avoided by checking prior to entering the loop to determine if a root exists on the interval and sending an error message if it does not.

```
(a) *****BISECTION METHOD*****

Sub BISECTION (Y1, Y2, ER, Y3)
Dim FY1 As Single, FY2 As Single, FY3 As Single, FZ As Single
Dim I As Integer
        FY1 = F(Y1)
        FY2 = F(Y2)
        If FY1 * FY2 > 0 Then Exit Sub
For I = 1 to 50
        Y3 = (Y1 +Y2)/2
        FY3 = F(Y3)
        FZ = FY1 * FY3
        If FZ = 0 Then Exit Sub
        If FZ < 0 Then Y2 = Y3 Else Y1 = Y3
        If Abs ((Y2 - Y1) / Y3) < ER Then Exit Sub
Next I
End Sub


(b) *****SECANT METHOD*****

Sub SECANT (Y1, Y2, ER, Y3)
Dim FY1 As Single, FY2 As Single, FY3 As Single
Dim I As Integer
        FY1 = F(Y1)
        FY2 = F(Y2)
For I = 1 to 50
        Y3 = Y2 + FY2 * (Y2 - Y1)/(FY1 - FY2)
        FY3 = F(Y3)
        If Abs ((Y3 - Y2)/Y3) < ER Then Exit Sub
        Y1 = Y2
        FY1 = FY2
        Y2 = Y3
        FY2 = FY3
Next I
End Sub


(c) *****NEWTON-RAPHSON METHOD*****

Sub NEWT (Y1, ER, Y2)
Dim FY1 As Single, FY2 As Single, FPR1 As Single, FPR2 As Single
Dim I As Integer
        FY1 = F(Y1)
        FPR1 = FPR(Y1)
For I = 1 to 50
        Y2 = Y1 -FY1/FPR1
        FY2 = F(Y2)
        FPR2 = FPR(Y2)
        If Abs ((Y2 - Y1)/Y2) < ER Then Exit Sub
        Y1 = Y2
        FY1 = FY2
        FPR1 = FPR2
Next I
```

FIGURE A.2
Visual BASIC procedures for nonlinear algebraic equation solvers.

Secant Method

The second algorithm to be discussed is called the secant method. As shown in Figure A.1b, two starting values of the root, y_1 and y_2, are guessed. Then a straight line is drawn through the points $[y_1, F(y_1)]$ and $[y_2, F(y_2)]$. Its intersection with the y axis is the next guess for the root, y_3. Algebraically, y_3 is evaluated from the equation of the straight line:

$$y_3 = y_2 - \frac{F(y_2)(y_2 - y_1)}{F(y_2) - F(y_1)} \tag{A.4}$$

Then, y_2 and y_3 become the next two guesses for the root. This is continued until an error criterion is met.

The Visual BASIC code for the secant method is given as a procedure in Figure A.2b. It is similar in several respects to the interval halving program, but note that the two initial estimates for the root do not have to bracket the true value of the root. The evaluation of the next estimate of the root uses Equation A.4, and only one functional evaluation occurs for each iteration. Both the roots and their functional values are updated after the error check to prepare for the next iteration.

The secant method converges much more rapidly than the interval halving method. Its disadvantage is that it may diverge for some functions because it is open-ended in the sense that the root does not have to be bracketed. An example of the divergence of the secant method is shown in Figure A.3a.

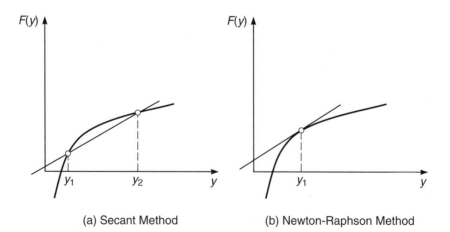

(a) Secant Method   (b) Newton-Raphson Method

FIGURE A.3
Divergence of numerical methods.

Newton-Raphson Method

The final method to be considered is the Newton-Raphson method, which really is a refinement of the secant method as shown in Figure A.1c. In this method, the tangent to the curve given by $F(y)$ is extended to the y axis to give the next guess for the root. This requires that the derivative of $F(y)$ be determined. From the Taylor series expansion for $F(y)$, we can see that, if higher-order terms are neglected, we can express the value of the function at y_2 in terms of the value and its derivative at y_1:

$$F(y_2) \approx F(y_1) + \frac{dF(y_1)}{dy_1}(y_2 - y_1) \tag{A.5}$$

in which terms beyond the first derivative term have been dropped. Now, we are seeking the intersection with the y axis at which $F(y_2) = 0$, so, for the next estimate of the root, we have

$$y_2 = y_1 - \frac{F(y_1)}{F'(y_1)} \tag{A.6}$$

in which $F'(y_1)$ is the first derivative evaluated at y_1. The Newton-Raphson technique is very powerful because of its fast convergence and capacity for extension to multiple-variable problems, but it requires a function for which the derivative can be evaluated. The procedure for the Newton-Raphson method is shown in Figure A.2c. It is nearly identical to that for the secant method except for the estimate of the next root by Equation A.6. Note that the SUB procedure refers not only to a functional evaluation F(Y) but also to an evaluation of the derivative of the function FPR(Y). The main disadvantage of the Newton-Raphson method is that it can be divergent under some circumstances as illustrated in Figure A.3b.

A.3   FINITE DIFFERENCE APPROXIMATIONS

In this textbook, the finite difference method is used to solve the equation of gradually varied flow, which is a first-order ordinary differential equation, and the Saint-Venant equations, which form a pair of nonlinear hyperbolic partial differential equations. In both cases, the derivatives are approximated over a small interval by finite differences. If we seek an approximation of the derivative dy/dx, for example, we begin by writing the Taylor's series expansion for the value of y at grid point $i + 1$ in terms of the value at point i, as shown in Figure A.4. The result is

$$y_{i+1} = y_i + f'(x_i)(x_{i+1} - x_i) + f''(x_i)\frac{(x_{i+1} - x_i)^2}{2!} + \ldots \tag{A.7}$$

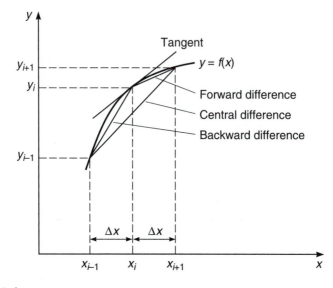

FIGURE A.4
Finite difference approximations of the First Derivative at x_i.

in which $y = f(x)$ and the primes refer to derivatives. All terms including and beyond the second derivative term can be referred to as higher order terms (HOT) that are $O(x_{i+1} - x_i)^2$, which means that they are "on the order of" Δx^2, where $\Delta x = (x_{i+1} - x_i)$. If we neglect all HOTs beyond the first derivative term, there will be a truncation error that is $O(\Delta x)^2$, so that halving the interval size Δx quarters the truncation error. Writing the truncation error in this way and solving for the first derivative results in

$$f'(x_i) = \frac{y_{i+1} - y_i}{x_{i+1} - x_i} + O(\Delta x) \tag{A.8}$$

If we drop the truncation error to approximate the derivative by

$$f'(x_i) \approx \frac{y_{i+1} - y_i}{x_{i+1} - x_i} \tag{A.9}$$

then this is referred to as a *first-order approximation* of the derivative, since the truncation error is proportional to Δx to the first power. This approximation of the derivative also is called a *forward difference,* because it utilizes data at points i and $i + 1$ in Figure A.4 to estimate the derivative.

In the same manner as for the forward Taylor's series expansion, we can write a backward expansion for y_{i-1} in terms of y_i as

$$y_{i-1} = y_i - f'(x_i)(x_i - x_{i-1}) + f''(x_i) \frac{(x_i - x_{i-1})^2}{2!} - \dots \tag{A.10}$$

Again, truncating the terms beyond the first derivative term and solving for the first derivative, we have

$$f'(x_i) \approx \frac{y_i - y_{i-1}}{x_i - x_{i-1}} \tag{A.11}$$

This is called the *backward difference,* and it too is first order.

Now if the backward Taylor's series (Equation A.10) is subtracted from the forward series (Equation A.7) and we solve for the first derivative, the result is

$$f'(x) = \frac{y_{i+1} - y_{i-1}}{2\Delta x} + O(\Delta x)^2 \tag{A.12}$$

in which $\Delta x = (x_{i+1} - x_i) = (x_i - x_{i-1})$. The second derivative terms cancel, and the third-order terms when divided by Δx result in higher-order terms that are of order $(\Delta x)^2$. Then, when the higher-order terms are dropped, the second-order approximation of the first derivative becomes

$$f'(x) \approx \frac{y_{i+1} - y_{i-1}}{2\Delta x} \tag{A.13}$$

Second-order approximations sometimes are used to obtain a more accurate representation of the first derivative. Equation A.13 is a *central difference* representation of the first derivative.

Forward, backward, and central difference approximations of the first derivative are shown graphically in Figure A.4. It is obvious that the central difference representation gives a slope or derivative closer to the true value.

Although illustrated for ordinary derivatives, the same derivative approximations can be made for the partial derivatives in the Saint-Venant equations (see Chapter 8). A first-order approximation of the time derivative, for example, becomes

$$\frac{\partial y}{\partial t} \approx \frac{y_i^{k+1} - y_i^k}{\Delta t} \tag{A.14}$$

in which Δt is the time interval, and the superscripts refer to values evaluated at times of $(k + 1)\Delta t$ and $k\Delta t$ at the spatial grid point located at x_i.

REFERENCES

Abbott, M. B., and D. R. Basco. *Computational Fluid Dynamics.* UK: Longman Scientific & Technical, and New York: John Wiley & Sons, 1989.

Ames, W. F. *Numerical Methods for Partial Differential Equations.* New York: Barnes and Noble, 1969.

Beckett, R., and J. Hurt. *Numerical Calculations and Algorithms.* New York: McGraw-Hill, 1967.

Chapra, S. C., and R. P. Canale. *Numerical Methods for Engineers,* 6th ed. New York: McGraw-Hill, 2009.

Index